From Chemical Topology
to
Three-Dimensional Geometry

TOPICS IN APPLIED CHEMISTRY

Series Editors: **Alan R. Katritzky, FRS**
Kenan Professor of Chemistry
University of Florida, Gainesville, Florida

Gebran J. Sabongi
Laboratory Manager, Encapsulation Technology Center
3-M, St. Paul, Minnesota

A Continuation Order Plan is available for this series. A continuation order will bring delivery of each new volume immediately upon publication. Volumes are billed only upon actual shipment. For further information please contact the publisher.

From Chemical Topology
to
Three-Dimensional Geometry

Edited by

Alexandru T. Balaban

Polytechnic University
Bucharest, Romania
and Academia Română
Bucharest, Romania

Plenum Press • New York and London

Library of Congress Cataloging in Publication Data

From chemical topology to three-dimensional geometry / edited by Alexandru T. Balaban.
    p.   cm.—(Topics in applied chemistry)
    Includes bibliographical references (p. –) and index.
    ISBN 0-306-45462-9
    1. Molecular structure—Mathematical models. I. Balaban, Alexandru T. II. Series.
QD461.F928   1997
541.2′2′01516—dc21
                                 96-51630
                                    CIP

ISBN 0-306-45462-9

© 1997 Plenum Press, New York
A Division of Plenum Publishing Corporation
233 Spring Street, New York, N. Y. 10013

http://www.plenum.com

10 9 8 7 6 5 4 3 2 1

Printed in the United States of America

Contributors

Alexandru T. Balaban, Polytechnic University, Organic Chemistry Department, Bucharest, Romania; and Academia Română, Calea Victoriei 125, 77207 Bucharest, Romania

Subhash C. Basak, Center for Water and the Environment, Natural Resources Research Institute, University of Minnesota at Duluth, Duluth, Minnesota 55811

Patrick W. Fowler, Department of Chemistry, University of Exeter, Exeter EX4 4QD, England

Gregory D. Grunwald, Center for Water and the Environment, Natural Resources Research Institute, University of Minnesota at Duluth, Duluth, Minnesota 55811

Davor Juretić, Department of Physics, Faculty of Science, HR-21001 Split, The Republic of Croatia

R. Bruce King, Department of Chemistry, The University of Georgia, Athens, Georgia 30602-2556

E. C. Kirby, Resource Use Institute, 14 Lower Oakfield, Pitlochry, Perthshire PH16 5DS, Scotland

Douglas J. Klein, Texas A&M University at Galveston, Galveston, Texas 77553-1675

Bono Lučić, The Rugjer Bošković Institute, P. O. Box 1016, HR-10001 Zagreb, The Republic of Croatia

Ovanes G. Mekenyan, Department of Physical Chemistry, Bourgas University "As Zlatarov" 8010 Bourgas, Bulgaria; and Lake Superior Research Institute, University of Wisconsin–Superior, Superior, Wisconsin 54880

Paul G. Mezey, Mathematical Chemistry Research Unit, Department of Chemistry and Department of Mathematics and Statistics, University of Saskatchewan, Saskatoon S7N 5C9, Canada

Gerald J. Niemi, Center for Water and the Environment, Natural Resources Research Institute, University of Minnesota at Duluth, Duluth, Minnesota 55811

Milan Randić, Department of Mathematics and Computer Science, Drake University, Des Moines, Iowa 50311

Marko Razinger,[†] National Institute of Chemistry, 61115 Ljubljana, P. O. Box 30, Republic of Slovenia

Nenad Trinajstić, The Rugjer Bošković Institute, P. O. Box 1016, HR-10001 Zagreb, The Republic of Croatia

Gilman D. Veith, National Health and Environmental Effects Research Laboratory, U. S. Environmental Protection Agency, Research Triangle Park, North Carolina 27711

Hongyao Zhu, Texas A&M University at Galveston, Galveston, Texas 77553-1675

[†]Deceased

Preface

Graph theory and topology, when applied to structural formulas in chemistry, have traditionally ignored geometrical angles and distances, and have paid attention only to vicinity (neighborhood) relationships, i.e., to adjacency between directly connected nodes (vertices or points, symbolizing atoms), and to incidence between edges (or lines, symbolizing covalent bonds) and their end points. The definition of a graph encountered in mathematical textbooks is: "a set of vertices, and a set (called edges) of unordered pairs of vertices"; for directed graphs (digraphs), edges are ordered pairs of vertices. However, chemists are more familiar with the representation of graphs as collections of lines and points: molecular (constitutional) hydrogen-depleted formulas of hydrocarbons are excellent examples for molecular graphs.

Constitutional formulas have been used to predict the number of constitutional isomers since chemical structure theory was created by Kekulé, Couper, Crum-Brown, and Butlerov, in 1850–1870. The mathematical background of graph theory was formulated around the same time by Cayley and Sylvester who had as incentive the problem of computing the number of constitutional isomers of alkanes. Thus, chemistry (and in particular organic chemistry) is entitled to be considered as one of the three independent sources for the birth of graph theory; the two other sources are the theory of electrical networks through Kirchhoff, and the theory of mathematical games or puzzles through Euler's problem of the seven Königsberg bridges (the earliest of the three sources).

Since the advent of stereochemical formulas introduced into organic chemistry by van't Hoff and Le Bel, and into inorganic chemistry by Werner, the numbers and structures of all chemical isomers (having the same molecular formula but differing in properties) can be reliably predicted. At present there exist computer programs using graph-theoretical and stereochemical approaches that can supply all of these numbers and structures for a given molecular formula.

In modern times, the search for new medicinal drugs is the primary incentive responsible for the synthesis of about 4 to 7×10^5 new, documented, and characterized compounds that are added each year to the 15×10^6 known compounds. Recently, with emerging combinatorial libraries, each containing thousands of congeners that are no longer fully characterized individually, a different outlook is beginning to take shape. Chemical documentation is faced with previously unforeseen challenges simply because of the sheer magnitude of the problem. For instance, although one is now able to fully optimize molecular geometries (energetically by quantum chemical methods, and sterically by molecular mechanics), the conversion of 2D data bases into 3D data bases cannot be based on such optimizations, which would be too time-consuming even with high-speed supercomputers; various noniterative approaches had to be invented for this purpose. With the advent of three-dimensional molecular design methods, medicinal chemists and pharmacologists can make considerable advances over traditional methods of drug design: one can devise structures that fit into receptor cavities so as to act either as enzyme promoters or as enzyme inhibitors. Electrostatic and hydrophilic/hydrophobic interactions are taken into account by various computational methods such as CoMFA.

Chemical graph theory was successful not only for counting and predicting correctly constitutional isomers, but also for enabling easily performed quantitative structure–activity and structure–property relationships (QSAR and QSPR, respectively) by means of topological indices associated with these isomers. Such topological indices are useful for QSAR studies, and are being used, along with other molecular descriptors, in various types of molecular design for reducing the cost of finding new drugs. It is particularly in this area that mathematical chemistry played a useful part. In the 1970s through the 1990s, topological indices (along with other molecular descriptors which can be computed even for unknown compounds, such as electronic, steric, and hydrophobic parameters) have helped to design new constitutional formulas for substances to be synthesized and screened for biological activity.

However, nowadays, drug design also needs more accurate descriptors, extending into the geometric description of relative atomic positions, of electrical charge distribution and dipole moments, and of hydrophilic and hydrophobic regions of molecules. Pathogens are known to develop resistance against medicines; therefore, such descriptors should allow the prediction of new molecules needed in the continuous fight against human or animal diseases, or against various pests, in order to produce new medicinal drugs or pesticides.

An area in which experimental and theoretical work has brought about an explosion of publications is that of cage molecules, among which fullerenes are the most prominent; this area also includes carbon tubules, cones, tori, and infinite arrays of carbon and/or other elements. The fascinating aspects of this third form of elemental carbon (in addition to the well-known forms of diamond and graphite) encompass superconductivity and biochemistry, and constitute a meeting ground for organic, inorganic, and theoretical chemists. The 1996 Nobel Prize for chemistry, awarded to the three discoverers of fullerenes, bears witness to the importance of this discovery.

The aim of this book, whose contributors are personally responsible for having made significant advances in their respective areas, is to provide the reader with an up-to-date account of how one can provide mathematically nonsophisticated molecular descriptors encompassing 3D aspects of molecules. The advantages of such descriptors are an easy intuitive grasp of their significance, the possibility to compute them for any imaginable structure, and their power to be used in QSAR studies and in molecular modeling for drug design.

What can mathematical chemistry offer in this field during the last decade of this century and the first decade of the next one? *This is the topic of the present book: mathematical methods, including topological ones, for taking into account the third dimension of the chemical world.*

A legitimate question raised by these challenges is the following: is there a future of topological methods and approaches for such types of problems? The great advantage of such methods and approaches is their simplicity and efficiency (in terms of computation time) in comparison with more sophisticated methods that are currently used. Thus, for searching large data bases only such fast approaches are acceptable.

Some readers may wonder whether there may be a contradiction between the two terms in the title of this book, namely "chemical topology," and "three-dimensional geometry." In a certain sense there is indeed a contradiction, because chemical topology (or chemical graph theory) traditionally encompasses only chemical constitution, ignoring 3D molecular features.

The present collection of chapters has been written by research scientists who are active in the areas where these challenges appeared; they try to provide an answer to this question. The reader will be able to see to what degree an affirmative reply is justified.

The first four chapters and Chapter 6 are connected with 3D molecular descriptors and their uses for QSAR and molecular similarity studies associated with molecular modeling of agonist–receptor interactions, with drug design, and with the discovery of new "lead compounds" for various types of biological activities.

Chapter 5 discusses problems associated with the poorly understood correlation between the primary structure of proteins, i.e., the sequence of amino acids, and the secondary and tertiary protein structures, which determine their biological activities.

Chapters 7, 8, and 9 discuss topological methods applied to the area of fullerenes and congeneric structures, some of which have not yet been obtained experimentally, but which are predicted to be as stable as fullerenes. It is the opinion of all authors contributing to this book that fullerene research is not a fleeting fashion, but is to stay and become even more developed when the full implications for the science of materials and for biological applications will be understood.

The final chapter is dedicated to applications of graph theory and topology to inorganic clusters and congeners.

This book is addressed to graduate students and research scientists of various profiles (chemists, biochemists, medicinal chemists, mathematicians, and statisticians) who are interested in molecular modeling, in fullerene research, in drug design, and in modern mathematical chemistry.

Alexandru T. Balaban

Contents

3. 3D Molecular Design: Searching for Active Conformers in QSAR

Ovanes G. Mekenyan and Gilman D. Veith

4. Use of Graph-Theoretic and Geometrical Molecular Descriptors in Structure–Activity Relationships

Subhash C. Basak, Gregory D. Grunwald, and Gerald J. Niemi

5. Recognition of Membrane Protein Structure from Amino Acid Sequence

Bono Lučić, Nenad Trinajstić, and Davor Juretić

6. On Characterization of 3D Molecular Structure

Milan Randić and Marko Razinger

7. Chemical Graph Theory of Fullerenes

Patrick W. Fowler

8. Recent Work on Toroidal and Other Exotic Fullerene Structures

E. C. Kirby

9. All-Conjugated Carbon Species

Douglas J. Klein and Hongyao Zhu

10. Applications of Topology and Graph Theory in Understanding Inorganic Molecules

R. Bruce King

1

From Chemical Graphs to 3D Molecular Modeling

ALEXANDRU T. BALABAN

1.1. CONSTITUTIONAL ISOMERS, STEREOISOMERS, AND CONFORMERS

Constitutional isomerism is defined as that type of isomerism (i.e., different structures corresponding to the same molecular formula) resulting from differences in vicinity relationships between atoms. Examples of pairs of constitutional isomers are *n*-butane and isobutane [CCCC versus CC(C)C in Smiles notation], ethanol and dimethyl ether (CCO versus COC), 1- and 2-methylbutene (C=CCC versus CC=CC), and 1- and 2-propanol [CCCO versus CC(O)C]. Constitutional isomerism is adequately accounted for in chemical graph theory by the adjacency or distance matrices, which consider only the vicinity relationships.[1,2]

Chemical structure also encompasses *stereoisomerism*,[3] in addition to constitutional isomerism. The two main types of stereoisomerism are *diastereomerism* (associated with differences in physicochemical properties which are as marked as for constitutional isomerism) and *enantiomerism* (or mirror-image stereoisomerism, associated with slight differences in physicochemical properties and manifested only in interactions with chiral forms of matter or energy, e.g., solubilites in chiral solvents, or rotation of plane-polarized light); on the other hand, since biological receptors are chiral, enantiomers show marked differences in their biological activity, e.g., smell or taste, or physiological activity. It is expected that in the future no medicinal drugs will

ALEXANDRU T. BALABAN • Polytechnic University, Organic Chemistry Department, Bucharest, Romania; and Academia Română, Calea Victoriei 125, 77207 Bucharest, Romania.

From Chemical Topology to Three-Dimensional Geometry, edited by Balaban. Plenum Press, New York, 1997

1

be accepted in racemic form, so that tragedies like that of malformations caused by thalidomide will no longer be possible. Examples for enantiomers are *R*-2-butanol and *S*-2-butanol, or D- and L-glyceraldehyde, or the two mirror images of helicenes. Diastereomers in turn are of two main types: polychiral stereoisomers (e.g., *meso*- and *R*-tartaric acids) and geometric (*Z/E*) stereoisomers, differing in interatomic distances between nonbonded atoms (e.g., *cis*- and *trans*-2-butene, *syn*- and *anti*-benzaldehyde oxime, or *cis*- and *trans*-dimethylcyclopropane).

The arrangement of bonds around a chiral carbon atom or a double bond is called *configuration*. When two structures differ by rotation around one or more single bonds, they are called *conformers* or *rotamers*, and usually they are more easily interconvertible than stereoisomers; rotamers can, however, be stable compounds when considerable steric hindrance is present.[4–6] In ethane viewed straight-on with superimposed carbon atoms, when the configurations of the two carbon atoms overlap, the confor-

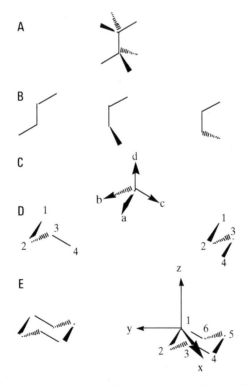

Figure 1. Top row: staggered conformation of ethane; row B: the three conformers of *n*-butane (only carbon atoms are shown in this and subsequent drawings); row C: the four orientations of vectors (grid coordinates *a–d*) around the tetrahedral central carbon atom of neopentane; row D: one of the chiral *gauche* rotamers, and the achiral *anti* rotamer of *n*-butane; row E: chair-cyclohexane, and its Cartesian coordinates x, y, z.

mation is called *eclipsed*; when bonds are intercalated, the conformation is called *staggered*. The top row of Figure 1 represents the staggered conformation of ethane, C_2H_6.

The enumeration of constitutional isomers by Cayley, then by Henze and Blair, and finally by means of Polya's theorem or double cosets was reviewed earlier.[1a,b,2b,7,8] Also, diastereomers of alkenes were enumerated by means of Polya's theorem.[2a] Alkanes can be chiral if they have at least seven carbon atoms (there are two isomers of heptane with a stereogenic center, i.e., two pairs of enantiomeric heptanes); if they have at least eight carbon atoms, alkanes can have polychiral diastereomers. The enumeration of stereoisomeric alkanes can be accomplished by means of Polya's theorem[2a]; achiral alkanes were enumerated for the first time by Robinson *et al.*[9] on applying Polya's theorem and Otter's dissimilarity characteristic equation (review in Ref. 7).

1.2. STAGGERED CONFORMERS OF ALKANES AND DIAMOND HYDROCARBONS

Saturated carbon chains, when they are free conformationally, adopt staggered conformations because these are energy minima. The enumeration of staggered conformers of linear alkanes was first reported by Funck in 1958,[10] but this contribution passed unnoticed. In 1976, Balaban[11] reraised this problem and, in addition to enumerating linear and branched staggered alkane rotamers, devised a system for their coding. Four years later, Milan Randić[12] published an alternative approach based on graphical enumeration. Recently, Cyvin[13] formulated the following open problem: deduce an algebraic solution for the number of nonisomorphic staggered alkane rotamers ("alkanoids") that can be embedded on a diamond lattice.

If there is no symmetry in the staggered conformation of a linear alkane, then the numbers of conformers increase by 3 for each additional carbon atom, because a terminal carbon has three possible staggered conformations. When symmetry elements are present, the numbers of conformers are lower.[14] Thus, *n*-butane has three conformers, namely, the achiral *trans*, and the two *gauche* enantiomers, as seen in Figure 1. Diamond hydrocarbons have dualist graphs that are hydrogen-depleted alkane rotamers as mentioned further below.[31]

1.3. 2D AND 3D TOPOLOGICAL INDICES

On the basis of the adjacency or distance matrices, many numerical constitutional descriptors associated with each molecule [called topological indices (TIs)] were described starting with Wiener's[15] (W) and Hosoya's indices (Z)[16] which are integer numbers used in quantitative structure–activity (or structure–property) relationships (QSARs or QSPRs, respectively). Since these TIs are derived from constitutional graphs which represent atoms by vertices, and covalent bonds by edges, on formulas that are embedded in a two-dimensional space, they can be called 2D TIs. No TI characterizes uniquely constitutional isomers because several nonisomorphic systems may correspond to one and the same TI, in which case one considers this TI to be degenerate. In order to reduce the degeneracy, recent efforts were devoted to developing real-number TIs such as Randić's molecular connectivity (χ, also denoted by X or ^1X),[17] the generalization[18] based on paths of length s (^sX) or on valence (X^v), Balaban's distance-based average connectivity (J),[19–21] and other TIs.

First-generation TIs were integer numbers based on integer local vertex invariants (LOVIs), such as the Wiener and Hosoya indices. They have a high degeneracy, and therefore are little used nowadays. Second-generation TIs are real numbers based on integer LOVIs, such as Randić, Kier–Hall, or Balaban indices; they are extensively used at present for QSAR and QSPR studies; Katritzky and Gordeeva[22] showed that they account for most of the variance in QSPR studies, and that for QSAR studies additional parameters such as hydrophobicity or quantum-chemical data are needed. Third-generation LOVIs are of recent date; they are real numbers based on real-number LOVIs, have very low degeneracy, and offer the possibility of a wide selection.[23,24] By combining two column vectors with the adjacency or distance matrix, one can obtain a system of linear equations whose solutions yield real-number LOVIs for chemical (constitutional) graphs; starting from these LOVIs one can obtain real-number TIs which are useful for QSAR/QSPR.[25] Another type of third-generation TI is based on information theory applied to summands of distance sums.[26]

The three-dimensional Wiener number (3-W) was introduced by Trinajstić and his co-workers[28,29] and is based on the topographic (geometric) distance matrix. Thus, one can encode information on geometric diastereomers and conformers. The original Wiener number (two-dimensional Wiener index, denoted by W or 2-W)[14] was based on the distance matrix whose entries are topological distances between nonhydrogen atoms (2-D_{ij}) and is the half-sum of all of these entries:

$$(2\text{-}W) = 1/2 \sum_i \sum_j (2\text{-}D_{ij})$$

To construct the topographic distance matrix of a staggered alkane, one can include or exclude the hydrogen atoms. The hydrogen-inclusive topographic matrix has as entries 3-D_{ij}) the geometric distances (in angstroms, with two decimals) for the conformer in the optimized geometry determined by molecular mechanics (MM2) computations. The half-sum of all entries, $1/2\Sigma\Sigma(3\text{-}D_{ij})$, is 3-$W$. Analogously, the hydrogen-depleted topographic matrix and the corresponding 3-W index result by deleting from the preceding matrix the entries corresponding to hydrogen atoms.

The 2-W indices for all heptanes present two pairs of degenerate values: 3-ethylheptane and 2,4-dimethylpentane have 2-W equal to 48; 2,2- and 2,3-dimethylpentane have 2-W equal to 46. Neither the hydrogen-inclusive nor the hydrogen-depleted 3-W values are degenerate for any conformer of heptanes. Actually, so far no degeneracy was found for 3-W in the case of staggered hydrogen-inclusive alkanes.

The 3-W numbers are useful for QSAR/QSPR: normal boiling points (bp) for the first 54 alkanes are correlated by the following equation with $r = 0.998$, $s = 5.8$ °C:

$$bp = 395 \, (3\text{-}W)^{0.0986} - 682$$

where 3-W is the hydrogen-inclusive 3-W index.[28]

Satisfactory correlations are also obtained[29] for the enthalpies of the first 21 alkanes (excluding methane) both in terms of 2-W and hydrogen-depleted 3-W numbers, and the best fit is obtained for the following type of equation:

$$\ln(H^0 - H^0_0)/T = A + B \ln(3\text{-}W)$$

$$3\text{-}W = 0.0849(2\text{-}W) - 13; \quad r = 0.997, \quad s = 3.66 \, °C$$

1.4. GRID COORDINATES AND MOLECULAR ID NUMBERS

As advocated by Randić et al.,[30] an undesirable feature of any type of distance matrix (be it topological for constitutional graphs, or topographical for geometrical isomers) is that the major contribution in W-type indices results from distant atoms which do not interact with each other. To counteract this drawback, a weighting procedure is introduced on multiplying entries by $(mn)^{-1/2}$ where m and n are the valencies of the atoms making up each bond. In addition, a single CC bond is considered to have a bond distance equal to the nondimensional number 1, instead of 1.54 Å. For hydrogen-depleted staggered alkane rotamers one obtains atomic path numbers and atomic ID numbers (which are LOVIs) as well as molecular ID numbers (MIDs, which are global indices). On a diamond lattice, one can assign grid coordinates to carbon atoms in staggered conformation of alkane rotamers, or in cyclohexane rings in chair conformation (which can be extended to alkylcyclohexanes, to cis- or trans-decalin, and so forth). The four tetrahedral directions in neopentane are described by the four unit vectors 1000, 0100, 0010, 0001, if the quaternary carbon is at the origin. With the tetrahedral angle $\alpha = 109.5$ we have $\cos \alpha = -1/3$. Figure 1, row C, illustrates the grid unit vectors.

Cartesian coordinates of the six carbon atoms in chair-cyclohexane are obtained from the grid coordinates $(a–d)$ along with, and oriented according to, the unit vectors by the geometrical relationships:

$$x = (2/3)^{1/2}a - (2/3)^{1/2}b$$

$$y = 2^{5/2}a - 2^{5/2}b - 8^{5/3}c$$

$$z = -a/3 - b/3 - c/3 - d$$

Adjacent grid points, corresponding to bonded carbon atoms, can differ only in one grid coordinate. We illustrate the above equations with the chair conformation of cyclohexane (Figure 1 and Table 1) and the origin at carbon atom numbered 1.

Table 1. Coordinates for Carbon Atoms in Chair-Cyclohexane

	Grid coordinates				Cartesian coordinates		
Atom No.	a	b	c	d	x	y	z
1	0	0	0	0	0	0	0
2	1	0	0	0	0.816	0.471	−0.333
3	0	−1	0	0	1.633	0	0
4	1	−1	1	0	1.633	−0.943	−0.333
5	0	−1	1	0	0.816	−1.414	0
6	0	0	1	0	0	−0.943	−0.333

Table 2. Grid Coordinates for Carbon Atoms in Staggered n-Butane Conformers

	gauche				anti			
Atom No.	a	b	c	d	a	b	c	d
1	0	0	0	0	0	0	0	0
2	1	0	0	0	1	0	0	0
3	1	−1	0	0	1	−1	0	0
4	1	−1	1	0	2	−1	0	0

Analogously, the grid coordinates for two staggered rotamers of n-butane (the *gauche* rotamer is chiral) are presented in Table 2; the origin, as before, is at carbon atom numbered 1.

The derived molecular numbers (MID) are 7.722 for the *gauche* rotamer and 7.680 for the *anti* rotamer. If a methyl group is attached to carbon atom 1 in chair-cyclohexane, the grid coordinates for the methyl carbon atom are 0,0,0,1 for the axial conformer (resulting in MID = 16.599) and 0,1,0,0 for the equatorial conformer (resulting in MID = 16.530), and thus differentiating these geometrical diastereomers.

The grid coordinates for the ten carbon atoms of adamantane, with the same conventions as those in Figure 1, are presented in Figure 2 (see also Ref. 31 for a graph-theoretical enumeration of isomers of diamond hydrocarbons in terms of their symmetries).

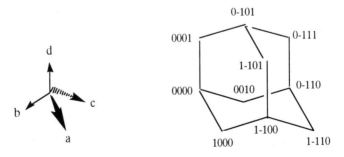

Figure 2. The four unit vectors around a tetrahedral carbon atom, and the grid coordinates of the carbon atoms in adamantane.

The MID values for hydrogen-depleted staggered rotamers of *n*-alkanes are about twice as large as the corresponding number of carbon atoms. Among staggered rotamers of the same linear alkane, the maximum MID corresponds to the most coiled rotamer, and the minimum MID to the most extended (all-*anti*) rotamer.[32]

1.5. *cis/trans* ISOMERS OF ALKENES, CYCLOALKANES, AND CONGENERS

Another approach for a stereochemical index Ω encoding information on *cis–trans* isomerism in alkenes was described by Estrada[32]; on including a corrected electron charge density $\delta^c(q_i)$, calculated with the MOPAC version 6.0, the following relationships were used for hydrogen-depleted graphs:

$$q_i = Z_i - \sum_{\mu \in i} (\mathbf{P\,S})$$

$$\delta_i(q) = q_i - h_i$$

$$\delta^c(q_i) = q_i - \sum_k q_{Hk}$$

$$\Omega(q) = \sum_{\text{edges } ij} [\delta_i(q)\, \delta_j(q)]^{-1/2}$$

The electron charge density on atom i is q_i; Z_i is the nuclear charge; $\mathbf{P}$ and $\mathbf{S}$ are the density and overlap matrices, respectively; the sum in the first equation is the Mulliken population, i.e., the number of electrons in each atomic orbital μ; $\delta_i(q)$ is the electron charge density connectivity; h_i is the number of hydrogen atoms attached to atom i; the electron charge density q_{Hk} is for the kth hydrogen atom bonded to atom i, and this differentiates *cis/trans* alkene diastereomers.

Geometry optimizations were carried out[32] by using the PM3 or AM1 approaches. Satisfactory linear correlations with normal boiling points for 53 alkenes were obtained ($r = 0.984$, $s = 5.6\ °C$) with the $\Omega(q)$ and $\Omega^c(q)$ indices [in the last case, $\delta(q)$ values are replaced by $\delta^c(q)$ values], as well as with the Kier–Hall 2D-index χ^v; a better linear correlation was obtained by including a second variable, namely, the number of methyl groups attached to the double bond of the alkene.

1.6. MOLECULAR TOPOGRAPHIC INDICES AND CONFIGURATION OF ANNULENES

Balaban,[33,34] following the method first used by Gordon,[35] used the three possible orientations of a planar honeycomb (hexagonal) lattice with some of the bonds in vertical position to define (1) the boundary code of polycyclic aromatic hydrocarbons and (2) the configurations of annulenes. Here we shall not review the former application (which was considerably developed later in a slightly different form by Trinajstić and co-workers,[8,36] by using digits 1 through 6 instead of 1, 2, 3, −1, −2, −3), but will concentrate instead on the latter application. Figure 3 presents the two notation systems for the orientation of bonds in a hexagonal lattice.

The configuration of any annulene that is superimposable on this lattice can be specified by listing sequentially the bond orientations; the initial bond (marked by an asterisk) is determined unambiguously by the convention to choose the lexicographic maximum. Using Trinajstić's notation,[36] [14]annulene in two different configurations is coded as shown in Figure 3.

A different code for configurations of annulenes, advocated by Oth *et al.*,[37,38] employs letters C and T (for *cis* and *trans*); on converting them into 0 and 1, respectively, and on adopting the convention for lexicographic minimum, one obtains for two limiting structures of the same [14]annulene the two different codes indicated in Figure 4, revealing the ambiguity of this code.

A nonambiguous code is obtained according to a convention introduced by Balaban[39] for indicating the direction of annelation using dualist graphs in the sequential "aufbau" of catafusenes: 1 and 2 denote *cis* and *trans* triplets of edges, respectively, together with the convention for lexicographic minimum; in all cases where the configuration of [n]annulenes is known, when $n = 4k + 2$ in [n]annulenes (satisfying Hückel's rule, with $k = 4, 5, 6, \ldots$), the stable configuration has the

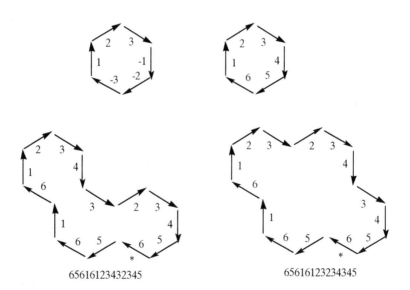

Figure 3. Top row: Balaban's and Trinajstić's notation of the six directions on a hexagonal net; bottom row: two different configurations of [14]annulene with codes in Trinajstić's notation.

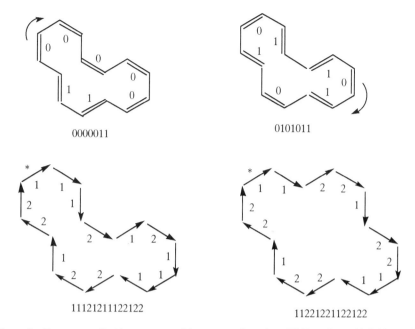

Figure 4. Top row: two limiting structures of the same configuration of [14]annulene with Oth's notation; bottom row: two different configurations of [14]annulene with corresponding codes after Balaban.

maximum area and is superimposable on the hexagonal lattice.[34,40] Thus, angle strain is reduced to a minimum because sp^2-hybridized carbon atoms have 120° bond angles. On the other hand, the [4k]annulenes have localized single and double bonds and manifest bond-shift isomerizations, in addition to ring inversion and (in some cases such as that of cyclo-octatetraene) valence isomerization.[41] A different characterization of molecular shapes was proposed by Randić and Razinger[42] with the canonical binary periphery code which consists in digit 0 for moving **O**utside of rings (left kink on adopting a clockwise direction) and digit 1 for moving toward the **I**nside, together with the convention for lexicographic minimum. This code is exemplified in Figure 5 for several configurations of [22]annulene; numbers d^2 on peripheries are sums of squared geometrical distances (which are integers for the hexagonal grid) to all other vertices; the inscribed d^2 index is the sum of all peripheral numbers.

For distinguishing *cis/trans* isomers of conjugated polyenes or of alkenes, Randić *et al.*[12,30,43,44] and then Pogliani[45] started from the observation that in an *s-cis* butadienic or a *cis*-2-butenic fragment the 1,4-carbon atoms can be connected by a virtual bond to form a virtual 4-membered ring, raising thereby the valence of the newly connected vertices. One can define the following *cis* $^n\chi_c$ and *cis–trans* X_{ct} connectivity indices, by analogy with Kier and Hall's $^n\chi_{ch}$ chain index[46]:

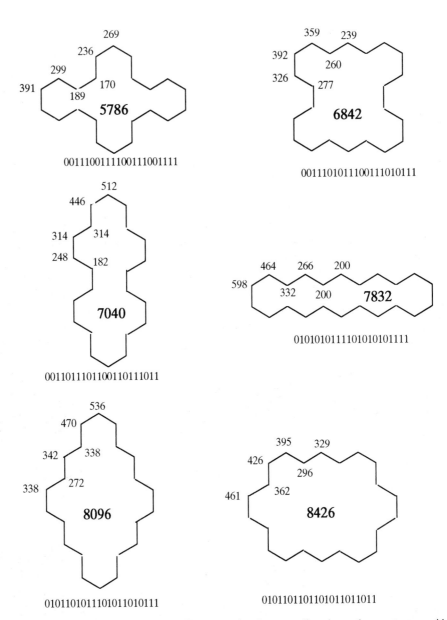

Figure 5. Six configurations of [22]annulene possessing three mutually orthogonal symmetry axes with inscribed D^2 index (boldface) which is the sum of all individual squared geometrical distance sums d^2, and with Randić and Razinger's periphery code under each structure.

$$^n\chi_c = \sum (\delta_1^r \delta_2^r \cdots \delta_v^r)^{-6/n}$$

$$X_{ct} = {}^1X - {}^n X_c$$

Here c and t stand for *cis* and *trans*, respectively, $n = 4$ for the virtual ring, and ^1X is Randić's first-order molecular connectivity:

$$^1X = \sum_{\text{edges } ij} (d_i\, d_j)^{-1/2}$$

defined in terms of the vertex degrees $d_i\, d_j$ for each edge ij of the graph.

For *trans* double bonds $^n X_c = 0$ and $X_{ct} = {}^1X$, but for *cis* double bonds the index X_c is smaller than ^1X, achieving thereby a differentiation between TIs of *cis/trans* diastereomers. Index X_{ct} is linearly related to Randić's *XX* index:

$$XX = -1.709X_{ct} + 7.932 \quad (r = 0.979, s = 0.003)$$

For a series of physical properties such as boiling points, refractive index, density, and molar refractivity, X_{ct} scores better than other indices such as the GAI index.[47]

Schultz *et al.*[48] devised a method based on weighted matrices for obtaining geometrically modified topological indices *S, PRS*, as well as indices using the determinant (det), permanent (per), and long hafnian (lhaf) of the matrices. The weighting employed geometric factors for alkanic or cycloalkanic *cis–trans* diastereoisomerism: factors +1 and −1 were assigned to *cis* and *trans* geometry, respectively. The new indices are integers (very large numbers, except for *S*). Schultz and co-workers extended their treatment to enantiomers by assigning chiral factors of +1 and −1 to each chiral stereocenter *R* and *S*, respectively (the remaining vertices of the graph had factors equal to zero). As in the previous case, the matrix **D** for the vertex and valence weighted graphs was constructed, this matrix **D** was transposed to yield $\mathbf{D}^T$, and then the **D** and $\mathbf{D}^T$ matrices were summed. A diagonal chiral factor matrix was constructed and premultiplied into the summed $\mathbf{D} + \mathbf{D}^T$ symmetrical matrix; the resultant product matrix row sum of the chiral vertex(es) was employed as a chiral modifier to adjust the original topological index for reflecting the chirality of the molecule. For illustration, one should consult the original paper, which includes tables with indices for *cis/trans* alkenes, for molecules with stereocenters, heteroatoms, double bonds, or rings, as well as for monosaccharides. Correlations with molar refractivities were also presented.

Mekenyan and co-workers[49] developed an algorithm (3DGEN) for generating all but high-energy conformers of molecules proceeding from the molecular topology. This algorithm performs the following main tasks: (1) generation of all possible stereoisomers (diastereomers and enantiomers); (2) generation of all possible conformers under some constraints; (3) design of 3D models of all generated stereoisomers and conformers; and (4) screening of the generated 3D molecular models according to chemical expertise. The approach is based on the hierarchically ordered extended connectivities (HOC) algorithm exployed earlier by Balaban *et al.*,[50] on the ring perception algorithm for the smallest ring incident to each pair of neighboring edges, and on optimized QSAR approach based on structural indices set (OASIS) algorithm. For modeling biologically active molecules in receptor cavities, it is essential to consider not only the conformation of minimal energy but also other conformations that may be imposed by the geometry of the receptor cavity. In QSAR studies using the 3DGEN + OASIS algorithm, or the CODESSA algorithm,[22] the quantum-chemical, topological, geometrical, or hydrophobicity parameters characterizing a molecular structure will have a range of values for the possible conformations. The biological activities, like any measured quantity, will also have a range of values corresponding to the measured value and corresponding standard error. Thus, in QSAR studies based on linear correlations, instead of drawing a straight line through series of points, the points are replaced by a series of ellipses. These ideas were validated by several QSAR studies for the toxicity of polychloro-biphenyl derivatives and other compounds that are described in detail in the chapter by Mekenyan and Veith in this book.

Novel matrices (distance–distance or **DD** matrices) for graphs embedded on 2D or 3D grids were introduced by Randić *et al.*[51] The two aims were: (1) to counteract the drawback of the graph-theoretical distance matrix and the geometrical distance matrix consisting in the large entries corresponding to the largest separations between atoms, and (2) to include both topological and geometrical information.

For a graph superimposable on the hexagonal (honeycomb) lattice, entries in **DD** = **D′T** are Hadamard products of the **D′** matrix (with diagonal entries equal to 1, and off-diagonal entries equal to reciprocals of entries in the distance matrix, as used earlier by Balaban *et al.*[52] and Trinajstić *et al.*[53]) with entries in the topographic matrix **T** corresponding to geometric interatomic distances relative to the single bond. For *s-cis* and *s-trans* 1,3-butadiene we present the **T** and **DD** matrices in Figure 6. It is seen that all entries in **DD** are numbers smaller than 1. One can reconstruct the graph from the **DD** matrix by obtaining the adjacency matrix (and hence the distance matrix) on setting all values that are different from 1 to be equal to zero. It is then easy to obtain matrix **T** and thus to find the embedding on the 2D or 3D grid.

The same authors discussed in detail various invariants of **DD** matrices. Sums of the entries over rows or columns define atomic ID numbers (LOVIs). The first eigenvalue normalized by dividing it by the number of vertices is a folding index for the configuration (the largest value of this index corresponds to the more "linear" configuration). This index allows a quantitative measure of similarity between chains of the same length but with different geometries.

Ugi *et al.*[54–56] developed a new computer-assisted treatment of chemical structure by means of bond- and electron-matrices (*be*-matrices) and chemical reactions by means of reaction matrices (*r*-matrices). Stereochemical *be*- and *r*-matrices provide information on stereochemistry; operations with stereochemical vectors and reaction vectors are carried out, and this algebra unifies the theory of the chemical group and

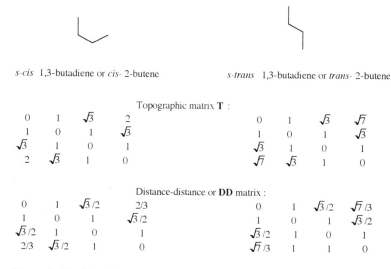

s-cis 1,3-butadiene or *cis*- 2-butene   *s-trans* 1,3-butadiene or *trans*- 2-butene

Topographic matrix **T** :

$$
\begin{array}{cccc}
0 & 1 & \sqrt{3} & 2 \\
1 & 0 & 1 & \sqrt{3} \\
\sqrt{3} & 1 & 0 & 1 \\
2 & \sqrt{3} & 1 & 0
\end{array}
\qquad
\begin{array}{cccc}
0 & 1 & \sqrt{3} & \sqrt{7} \\
1 & 0 & 1 & \sqrt{3} \\
\sqrt{3} & 1 & 0 & 1 \\
\sqrt{7} & \sqrt{3} & 1 & 0
\end{array}
$$

Distance-distance or **DD** matrix :

$$
\begin{array}{cccc}
0 & 1 & \sqrt{3}/2 & 2/3 \\
1 & 0 & 1 & \sqrt{3}/2 \\
\sqrt{3}/2 & 1 & 0 & 1 \\
2/3 & \sqrt{3}/2 & 1 & 0
\end{array}
\qquad
\begin{array}{cccc}
0 & 1 & \sqrt{3}/2 & \sqrt{7}/3 \\
1 & 0 & 1 & \sqrt{3}/2 \\
\sqrt{3}/2 & 1 & 0 & 1 \\
\sqrt{7}/3 & 1 & 1 & 0
\end{array}
$$

Figure 6. The **T** and **DD** matrices of *cisoid/transoid* 1,3-butadiene, or *cis/trans*- 2-butene.

the algebra of *be-* and *r*-matrices. The Dugundji–Ugi theory can be used to create a hierarchic classification of chemical reactions.[57]

An interesting novel approach for encoding stereochemical information in linear notation based on the "configurational paddle wheel" division of space around a chemical bond was published by Dietz.[58]

A new paradigm, called the electrotopological state for the electronic and topological characterizing of atoms, was recently introduced and reviewed by Kier and Hall.[46,59–61] Each atom in the molecular graph is represented by an electrotopological state variable, and this variable includes the perturbation of the given by all other atoms in the molecule. Thus, a third-generation LOVI is obtained, and it can be used for constructing TIs and then as a basis for molecular similarity judgment.

In recent papers, Balaban, Diudea *et al.*[62,63] constructed new nonsymmetrical matrices encoding the environment of matrices in layers at various topological distances from each given vertex. By grouping in a different fashion the topological distances according to layers at distance 1, 2, 3, . . . , around each vertex one obtains the nonsymmetrical layer matrix; sums over rows are the same as for distance sums. By reducing progressively the contribution of vertices at increasing distances from a given vertex, new LOVIs (regressive vertex degrees, regressive distance sums) and 2D TIs were obtained on the basis of such layer matrices. Diudea *et al.*,[64] proceeding in a similar fashion with geometric distances obtained by molecular mechanics calculations, were able to obtain 3D distance matrices; from these matrices two types of 3D TIs were constructed: centricity indices and centrocomplexity indices. They were validated by inter- and intramolecular comparisons, and by QSAR studies with toxicities (for series of ethers), van der Waals areas, and conformational energies.

Kunz[65] made the interesting observation that operations such as inversion and/or squaring of topological distance and incidence matrices may yield angles between graph edges. The same author also discussed operations with distance matrices of line graphs.[66]

1.7. DATA BASES FOR 3D CHEMICAL STRUCTURES

3D search queries, frequently representing pharmacophores derived from modeling studies, consist of 3D structural templates, topological substructures, geometric data (e.g., planes, lines, points), and geometric constraints (distances, angles, exclusion volumes). When molecules are flexible, they can be accommodated in the receptor cavity (often in an enzyme) in an alternative conformation of slightly higher energy than the ground-state conformation.[67] In addition to contributions described by Mekenyan and Veith in this book, one should mention papers by Murrall and Davies,[68] Güner et al.,[69,70] Clark et al.,[71,72] and Moock et al.[73] on representing conformational flexibility inquiries for searching in 3D data bases.

From the Chemical Abstracts compound registry with $> 10^7$ substances, in order to convert a 2D data base into a 3D one, one needs a program that iterates neither energy minimizations, nor molecular mechanics calculations. Instead, the program CONCORD of the University of Texas at Austin and TRIPOS Associates[74] avoids bad intramolecular contacts in terms of van der Waals radii. This program was also able to convert 2D into 3D structures in about 50% of the cases; among all atoms heavier than chlorine, only Br and I can be handled, and molecules with flexible bonds are sometimes obtained in conformations that differ considerably from real structures in crystals, as shown by comparison with the Cambridge Structural Database, which contains about 10^5 substances. Despite the fact that "crystal structures" will not necessarily coincide with molecular structures in solution, X-ray data furnish precious information.

During the past 40 years, the National Cancer Institute (NCI) has examined almost half a million compounds for anticancer activity, and since 1988 about 25,000 of the compounds have also been examined for anti-HIV activity; although the yield (the ratio between new "lead compounds" and all of the investigated compounds) was low, this program succeeded in putting more than a third of all 30 cytotoxic antitumor drugs on the market. The NCI data base (Drug Information Service) contains more than 450,000 structures (mostly organic ones); about half of it is open to public access via Telnet, and the remainder ("discreet structures") has restricted access. This data base was first converted into 3D searchable data base by using the program CONCORD; then another program was tested for 3D structure generation from connection tables, namely ChemModel[75] (within Chem-X, a product of Chemical Design Ltd.), and the results were somewhat better than those of CONCORD. The reason is that ChemModel is able to generate multiple structures covering the entire conformational space, as shown in a series of papers by Milne and co-workers[76–79]: Chem-X builds for flexible molecules multiple conformers by allowing torsion angles of 60, 180, and 300° for sp^3–sp^3 bonds (up to 15 rotor bonds per compound); for sp^2–sp^2 bonds, it allows two to six torsion angles. Large rings, however, have to be processed by the less performant CONCORD program.

1.8. SIMILARITY SEARCHES IN FILES OF 3D CHEMICAL STRUCTURES

The following hierarchy of molecular features is probable in ascertaining similarity in the context of biological activity:

1. Disposition of electrical charges
2. Disposition of polar (electronegative/electropositive) groups
3. Disposition of hydrogen-bond donor and acceptor groups
4. Disposition of hydrophobic (lipophilic) and hydrophilic regions
5. Disposition of unsaturated/aromatic groups
6. Disposition of bulky groups
7. Disposition of flexible chains
8. Shape similarity

Under *disposition* one looks for angles and distances between specified groups or atoms of given types.

Johnson and Maggiora[80] formulated the *similar property principle* according to which structurally similar molecules exhibit similar properties. Therefore, once a molecule is known to possess a certain biological activity, a 3D similarity search may identify other molecules in the search for better medicinal drugs, or pesticides. Vice versa, in looking for new lead compounds in screening large varieties of chemical structures, it is desirable to test substances that are as dissimilar as possible. Therefore, methods for quantitative assessment of 2D or 3D molecular similarity are of great importance.

Among many methods for 3D similarity relevant to the present chapter, Bath *et al.*[81] focused their attention on angle- and distance-based similarity measures. They defined a generalized torsion angle (well suited for 3D searches) by considering four atoms, which need not be all connected; if they are, we have a BBB torsion angle; if one or two bonds are missing, we have BNB or NBN torsion angles (B and N stand for the presence or absence of a bond).

If one wishes to obtain 3D similarity descriptors ignoring atom types, one starts from distorted tetrahedra (sets of four atoms) with BNB or NBN bonding patterns; these (together with distances between pairs of atoms) give the most effective similarity searches, validated by the Kendall test statistic, on using the Cambridge Structural Database (CSD) for carbohydrates.

An analysis of 25,322 organic molecules with just one molecule in the unit cell from the CSD showed that there is a low degree of correlation ($r^2 < 0.42$) between 2D and 3D shape measures, even for molecules with no conformational flexibility; the graph-theoretical (2D) and geometrical (3D) shape descriptors I were defined as

$$I = D/R - 1$$

where D and R are the graph diameter and radius, respectively (for 2D), or the corresponding geometric radius and diameter. The resulting 2D and 3D indices are denoted by I_2 and I_3, respectively. Even for 188 molecules containing one unsubstituted aromatic ring and no flexible bonds, the correlation between 2D and 3D descriptors is low ($r^2 = 0.53$). Petitjean and Dubois[82] examined the similarity of shapes of over 4 million compounds from the Chemical Abstracts Service data base in terms of the topological index I and concluded that: (1) many shapes are absent; (2) prior to 1978, organic chemistry had evolved only in a few specialized directions; and (3) one can conveniently characterize a large set of compounds by the pair of I_2 and I_3 indices. Both research groups[81,82] agree that 2D and 3D descriptors convey useful information, being complementary to each other.

The similarity between two structures is calculated by means of an associated coefficient, usually the Tanimoto coefficient. Valence angles are 2D, and interatomic distances are 1D. Most of the 3D searches of data bases have concentrated on interatomic distance information, e.g., hash codes for all distinct triplets of nonhydrogen atoms in pairs of isomeric structures. If one uses frequency distributions instead of hash codes, one can also compare nonisomeric molecules.

Nilakantan et al.[83] described a similarity measure based also on the interatomic distances between all sets of three nonhydrogen atoms, using a bit-string representation of molecular shape. It is also possible to consider doublets or quadruplets of nonhydrogen atoms (the doublets give rise to similarity measures which are related to the distance-distribution measure described by Pepperrell and Willett[84,85]; see also Ref. 86).

In 3D searching (pharmacophore searching) based on distance or angle constraints, the usual procedure is to use the Ullmann algorithm[87] or its variants. Several groups of authors, namely, Murrall and Davies,[68] Clark, Willett, and co-workers,[71,72] as well as Hurst,[88] addressed the problem of flexible 3D searching for structures that have rotatable bonds. Hurst's directed tweak technique allows 3D flexible searching on an interactive time scale, and proves to be one of the most effective approaches; it can also address van der Waals interactions and ring flexibility. The next best technique is the genetic algorithm used by Clark et al.[72] who compared these methods with the effectiveness of the distance geometry and systematic or random searches, concluding on the effectiveness and efficiency of each method; they used the POMONA 89 data base with 1538 structures and the 9886 CONCORD structures from the Chemical Abstracts Service data base.

Yvonne Martin and her associates[89–94] implemented a 3D similarity searching program ALADDIN using a computer algorithm *dis*tance *co*mparisons (DISCO) and two derivative programs, FAMILY and COMPAIR. The ALADDIN program[90,94] generates points Y at the center-of-mass of collections of atoms such as rings, and projections from atoms to hypothetical hydrogen-bond donating or accepting sites. Running DISCO requires a reference structure, which is compared to every conformation of each compound in the set under study. The first step is the calculation of distances between all pairs of points (nonhydrogen atoms or points Y mentioned above) being considered in each structure. Distances within a given tolerance specified by the user are noted and stored in a correspondence table.

The Bron-Kerbosh clique detection[95,96] is used iteratively to find sets of common distances, i.e., cliques, obtaining thereby DISCO solutions, e.g., pharmacophore maps; chirality is also considered by evaluating the sign of the torsion angle for quadruplets of non-coplanar points. In FAMILY only one pair of structures is composed at a time for grouping structures into families. The program COMPAIR makes similar pairwise comparisons employing a user-supplied reference structure each time.

Several of the most used similarity measures are the comparison of atom pairs,[97–99] of molecular shape and volume,[100–104] of electrostatic potentials,[105] and of molecular fields.[106]

In his book on molecular similarity,[107] Willett suggested that there are four types of molecular structure representation: (1) systematic nomenclature (e.g., the IUPAC, Chemical Abstracts, or Beilstein nomenclature systems); (2) line notation (e.g., the Wiswesser Line Notation); (3) connection tables (e.g., adjacency matrices for constitutional graphs, or more elaborate systems such as the Morgan-type connection tables used by Chemical Abstracts Service, or tables including various kinds of extra information such as those proposed by Wipke and Dyott,[108] or by Balaban *et al.*[50]); and (4) fragmentation codes. A fifth category, as advocated by Hall *et al.*[60] can be considered to be represented by topological indices (TIs). They do not characterize molecules up to isomorphism, but there is a continuous trend toward reducing the degeneracy of TIs. Willett's group, which is one of the most active in this field, has published several other reviews,[109–111] and numerous papers.[112–116]

There are four methods for 3D data base searching: distance distributions, individual distances, maximum common substructure, and atom mapping; in her book,[85] Catherine Pepperrell emphasizes the latter method, and also addresses dissimilarity (which matters when searching for new "lead compounds," and one wishes to screen as many different classes as possible).

The similarity test employed by Herndon and his co-workers[117,118] is based on the notation adopted for representing the molecular structure (both constitution and stereochemistry). For canonical numbering of atoms, an algorithm based on extended connectivities assigns priority to atoms of high connectivity and centricity in the molecule. The matrix incorporating graph-theoretical and chemical information can be converted into a linear notation (termed LN1). Metrics of molecular similarity

(similarity indices S_{ij} varying in the range from 0 to 1) are obtained by standard text comparison procedures from the linear notation

$$S_{ij} = 1 - (indels)/(N_i - N_j)$$

For a pair of molecules i and j, one counts the number of *indels* (insertions plus deletions) required to convert one linear notation into the other while preserving the alignment; N_i and N_j are the numbers of terms in each linear notation.

These ideas were applied for comparing aliphatic alcohols, carcinogenic polycyclic benzenoid aromatic hydrocarbons, and binding constants between human corticosteroid binding globulin and a set of 47 steroids. The last set of compounds was also investigated by *c*omparative *m*olecular *f*ield *a*nalysis (CoMFA) with comparable results.[106] A similar approach was used by Klopman and Raychaudhury[119,120] on the basis of the Wiswesser Line Notation system.

1.9. 3D PROTEIN STRUCTURE

The advent of nuclear magnetic resonance methods for elucidating the structure of proteins labeled with stable isotopes (nitrogen-15 and carbon-13) in solution will add to our understanding of enzymatic mechanisms; until recently, X-ray diffraction was the only method for obtaining information on the secondary and tertiary structure of proteins, but in the crystal the structure may differ from the active form in solution because of packing forces. Serine proteases have been investigated extensively; on the basis of the X-ray crystal structure for a peptidic inhibitor bound to human leukocyte elastase (a hydrolase for elastin), Greene *et al.*[121] explored a combined data base of over 200,000 compounds and obtained a few thousand hits; these authors proposed generalized function definitions for hydrogen-bond acceptors or donors, charge centers, and hydrophobes.

Willett, Artimiuk, and co-workers[122–127] used graph-theoretical techniques for searching 3D protein structures. Points (nodes) of the first type of graph describe secondary-structure elements, and edges denote the geometric relationships between pairs of these elements. For β-sheets, each β-strand is denoted by 0 or 1, depending on whether it is parallel or antiparallel to the first strand, which is always assigned the value of 1. Note that on reverting 1 and 0, or on starting from either end of the sheet, one selects a canonical description of the β-sheet.

In a second type of graph, nodes represent the amino acid residues in protein structures, and the edges indicate the geometric relationship between pairs of these residues. Ullmann's subgraph isomorphism algorithm[87] with the latter type of graphs allows the selection of ψ-loops in proteins similar to aspartic proteinase; with the former type of graphs, it was possible to show, by using the Protein Topographic Exploration Program, that only a small fraction (0.5%) of the possible β-motifs are actually encountered in the Protein Data Bank structures; some of Richardson's results[128] are confirmed. Thus, graph theoretical techniques are shown to be useful for the analysis of the secondary and tertiary structure of proteins.[129]

As a side remark, following a recent comment by Mislow,[130] the molecular graphs of proteins, like those of all other organic compounds, are usually composed of nodes (vertices) representing atoms and of lines (edges) representing *covalent* bonds. However, the secondary and tertiary structure of proteins is determined by hydrogen bonding, disulfide bridges, and cofactors; all of these contribute to the formation of interchain cross-links and are thermodynamically, but not kinetically, stable. Their inclusion into molecular graphs, advocated by some authors, would lead to novel types of chirality and topological linking (catenane and knot formation). Predictions for still unknown topological stereoisomers were made by Walba[131] and Balaban.[132]

Note added in proof. Randić and Razinger[133] used the periphery code for benzenoid contours in order to characterize molecular shape and two-dimensional chirality. Gasteiger and co-workers[134,135] continued to use their 3D-structure generator CORINA and developed a 3D-MoRSE code (**Mo**lecule **R**epresentation of **S**tructures based on **E**lectron diffraction) which can be applied in QSAR studies. Balasubramanian[136a] published a procedure based on 3D molecular structure for obtaining NMR equivalence classes for atoms; he also devised[136b] a program for computer perception of molecular symmetry. Toropov and co-workers[137] introduced 3D-weights (using interatomic distances in optimized molecular geometries) for all atoms, including hydrogens, and employed these 3D-weights instead of vertex degrees for 3D-molecular descriptors.

On using extensions of the atom pair method[138] and of topological torsion,[83,139] Kearsley *et al.*[140–142] developed fuzzy descriptors for chemical similarity based on the CONCORD program, and explored their applicability on the Derwent Standard Drug File. Willett and co-workers combined a genetic algorithm and aligned molecular electrostatic field potentials for similarity searching of 3D chemical structures,[143] complementing their earlier 3D substructure dissimilarity approach[144] using angle and torsional descriptors.

A *de novo* 3D database can be generated using the MOLMAKER program and a 2D–3D conversion program.[145,146]

Milne and co-workers[147] continued to explore the discovery of drug leads by characterizing the enzymatic active sites which complements the pharmacophore; 3D searches can be made by electronic screening with extra requirements for solubility and hydrophobicity. The presence of a pharmacophore is, however, a necessary but insufficient property because of conformational flexibility.

Interesting and surprising results were obtained[148] from a comparison of molecular similarity measurements based on topological indices,[149] or on the occurrence of common structural fragments. These two measures agree in some cases and disagree in other ones.

A clear discussion of the distinction between geometrical and topological chirality of molecules was presented by Mislow[150]: geometrical (or Euclidean) chirality is associated with chirality of the full symmetry group of the molecule. Geometrically achiral objects are necessarily also achiral, but the reverse is not true: geometrically chiral objects may be either topologically chiral or achiral. Graph-theoretical non-planarity is a necessary but insufficient condition for topological chirality in knots or catenanes. Experimental results for such molecular systems were reviewed by Sauvage and co-workers,[151,152] as well as by Stoddart and co-workers.[153]

Tratch and Zefirov[154] have recently presented algebraic chirality criteria, applying them to classify chirality and to develop a computer program for this purpose.

Skein polynomials[155-157] are invariants which can distinguish between enantiomers in most cases. However, it appears that chemically relevant and easy-to-use characterization and codification of chirality (for electronic inclusion in data bases and QSAR) remains a challenge to chemists and mathematicians for the future.

REFERENCES

1. A. T. Balaban, ed., *Chemical Applications of Graph Theory*, Academic Press, New York (1976): (a) Chapter 4, by R. C. Read; (b) Chapter 5, by A. T. Balaban; (c) Chapter 6, by J. Dugundji, P. Gillespie, D. Marquarding, I. Ugi, and F. Ramirez; (d) Chapter 7, by D. H. Rouvray.
2. N. Trinajstić, *Chemical Graph Theory,* 2nd ed., CRC Press, Boca Raton, Florida (1992): (a) Chapter 11, p. 275; (b) Chapter 10, p. 225.
3. E. L. Eliel and S. H. Willen, *Stereochemistry of Organic Compounds,* John Wiley & Sons, New York (1994).
4. M. Oki, *Angew. Chem. Int. Ed. Engl. 15,* 87 (1976).
5. H. Iwamura and K. Mislow, *Acc. Chem. Res. 21,* 175 (1988).
6. C. Uncuta, I. Paun, A. Ghitescu, C. Deleanu, T. S. Balaban, F. Chiraleu, M. D. Gheorghiu, and A. T. Balaban, *Tetrahedron Lett. 21,* 5645 (1990).
7. A. T. Balaban, in: *Chemical Graph Theory: Introduction and Fundamentals* (D. Bonchev and D. H. Rouvray, eds.), Chapter 5, p. 177, Abacus Press/Gordon & Breach, New York (1991).
8. J. V. Knop, W. R. Müller, K. Szymanski, and N. Trinajstić, *Computer Generation of Certain Classes of Molecules,* Assoc. Chem. & Technol. of Croatia, Zagreb (1985).
9. R. W. Robinson, F. Harary, and A. T. Balaban, *Tetrahedron 32,* 355 (1976).
10. E. Funck, Z. *Elektrochem. 62,* 901 (1958).
11. A. T. Balaban, *Rev. Roum. Chim. 21,* 1049 (1976).
12. M. Randić, *Int. J. Quantum Chem. Quantum Biol. Symp. 7,* 187 (1980).
13. S. J. Cyvin, *J. Math. Chem. 17,* 291 (1995).
14. A. T. Balaban, in: *Mathematical Chemistry* (D. Bonchev and D. H. Rouvray, eds.), Gordon & Breach, New York (1994).
15. H. Wiener, *J. Am. Chem. Soc. 69,* 17 (1947).
16. H. Hosoya, *Bull. Chem. Soc. Jpn. 44,* 2332 (1971).
17. M. Randić, *J. Amer. Chem. Soc. 97,* 6609 (1975).
18. L. B. Kier and L. H. Hall, *J. Pharm. Sci. 70,* 583 (1981); *Molecular Connectivity in Structure–Activity Analysis,* Research Studies Press–Wiley, New York (1986); *Molecular Connectivity in Chemistry and Drug Research,* Academic Press, New York (1976).
19. A. T. Balaban, *Chem. Phys. Lett. 80,* 399 (1982).
20. A. T. Balaban, *Pure Appl. Chem. 55,* 199 (1983).

21. A. T. Balaban, *Math. Chem. (MATCH) 21*, 115 (1986).
22. A. R. Katritzky and E. V. Gordeeva, *J. Chem. Inf. Comput. Sci. 33*, 835 (1993).
23. A. T. Balaban, *J. Chem. Inf. Comput. Sci. 32*, 23 (1992).
24. A. T. Balaban, *J. Chem. Inf. Comput. Sci. 34*, 398 (1994).
25. A. T. Balaban, *Croat. Chem. Acta 66*, 447 (1993).
26. P. Filip, T. S. Balaban, and A. T. Balaban, *J. Math. Chem. 1*, 61 (1987).
27. A. T. Balaban and T. S. Balaban, *J. Math. Chem. 8*, 383 (1991).
28. Z. Mihalić and N. Trinajstić, *J. Mol. Struct. (Theochem.) 232*, 65 (1991).
29. B. Bogdanov, S. Nikolić, and N. Trinajstić, *J. Math. Chem. 3*, 299 (1989); *5*, 305 (1990).
30. M. Randić, B. Jerman-Blazić, and N. Trinajstić, *Comput. Chem. 14*, 237 (1990).
31. A. T. Balaban and P. von Ragué Schleyer, *Tetrahedron 34*, 3597 (1978).
32. E. Estrada, *J. Chem. Inf. Comput. Sci. 35*, 708 (1995).
33. A. T. Balaban and F. Harary, *Tetrahedron 24*, 2505 (1968).
34. A. T. Balaban, *Tetrahedron 27*, 6115 (1971).
35. M. Gordon and W. H. T. Davison, *J. Chem. Phys. 20*, 428 (1952).
36. B. Dzonova-Jerman-Blazić and N. Trinajstić, *Comput. Chem. 6*, 121 (1982).
37. J. F. M. Oth and J.-M. Gilles, *Tetrahedron Lett. 1968*, 6259 (1968).
38. J. F. M. Oth, G. Anthoine, and J.-M. Gilles, *Tetrahedron Lett. 1968*, 6265 (1968).
39. A. T. Balaban, *Tetrahedron 25*, 2949 (1969).
40. A. T. Balaban, *Pure Appl. Chem. 54*, 1075 (1982).
41. A. T. Balaban, M. Banciu, and V. Ciorba, *Annulenes, Benzo-, Hetero-, Homo-Derivatives and Their Valence Isomers*, CRC Press, Boca Raton, Florida (1987).
42. M. Randić and M. Razinger, *J. Chem. Inf. Comput. Sci. 35*, 140 (1995).
43. M. Randić, in: *MATH/CHEM/COMP 1987* (R. C. Lacher, ed.), p. 101, Elsevier, Amsterdam (1988).
44. M. Randić, *Stud. Phys. Theor. Chem. 54*, 101 (1988).
45. L. Pogliani, *J. Chem. Inf. Comput. Sci. 34*, 801 (1994).
46. L. B. Kier and L. H. Hall, *Pharm. Res. 7*, 801 (1990).
47. L. Xu, H. Y. Wang, and Q. Su, *Comput. Chem. 16*, 187 (1992).
48. H. P. Schultz, E. S. Schultz, and T. P. Schultz, *J. Chem. Inf. Comput. Sci. 35*, 864 (1995).
49. J. Ivanov, S. Karabunarliev, and O. Mekenyan, *J. Chem. Inf. Comput. Sci. 34*, 234 (1994).
50. A. T. Balaban, O. Mekenyan, and D. Bonchev, *J. Comput. Chem. 6*, 538 (1985) and subsequent papers in the series.
51. M. Randić, A. F. Kleiner, and L. M. DeAlba, *J. Chem. Inf. Comput. Sci. 35*, 366 (1995).
52. O. Ivanciuc, T. S. Balaban, and A. T. Balaban, *J. Math. Chem. 12*, 309 (1993).
53. D. Plavsić, S. Nikolić, N. Trinajstić, and Z. Mihalić, *J. Math. Chem. 12*, 235 (1993).
54. I. Ugi, J. Bauer, C. Blomberger, J. Brandt, A. Dietz, E. Fontain, B. Gruber, A. v. Scholley-Pfab, A. Senff, and N. Stein, *J. Chem. Inf. Comput. Sci. 34*, 3 (1994).
55. I. Ugi, J. Bauer, K. Bley, A. Dengler, A. Dietz, E. Fontain, B. Gruber, R. Herges, M. Knauer, K. Reitsam, and N. Stein, *Angew. Chem. Int. Ed. Engl. 32*, 201 (1993).
56. I. Ugi, J. Brandt, J. Friedrich, J. Gasteiger, C. Jochum, P. Lemmen, and W. Schubert, *Pure Appl. Chem. 50*, 1303 (1978).
57. J. Dugundji and I. Ugi, *Top. Curr. Chem. 39*, 19 (1973).
58. A. Dietz, *J. Chem. Inf. Comput. Sci. 35*, 787 (1995).
59. L. H. Hall and L. B. Kier, *J. Chem. Inf. Comput. Sci. 35*, 1039 (1995).
60. L. H. Hall, L. B. Kier, and B. B. Brown, *J. Chem. Inf. Comput. Sci. 35*, 1074 (1995).
61. L. H. Hall, B. Mohney, and L. B. Kier, *Quant. Struct. Act. Relat. 10*, 43 (1991); *12*, 44 (1993); *J. Chem. Inf. Comput. Sci. 31*, 76 (1991).
62. A. T. Balaban and M. V. Diudea, *J. Chem. Inf. Comput. Sci. 33*, 421 (1993).
63. M. V. Diudea, O. Minailiuc, and A. T. Balaban, *J. Comput. Chem. 12*, 527 (1991).
64. M. V. Diudea, D. Horvath, and A. Graovac, *J. Chem. Inf. Comput. Sci. 35*, 129 (1995).
65. M. Kunz, *J. Math. Chem. 13*, 145 (1993).

66. M. Kunz, *J. Chem. Inf. Comput. Sci. 34*, 957 (1994).

67. W. L. Jorgenson, *Science 254*, 954 (1991).

68. N. W. Murrall and E. K. Davies, in: *Chemical Structures, 11*, Springer-Verlag, Berlin (1993); *J. Chem. Inf. Comput. Sci. 30*, 312 (1990).

69. O. F. Güner, D. R. Henry, T. E. Moock, and R. S. Pearlman, *Tetrahedron Comput. Methodol. 3*, 557 (1990).

70. O. F. Güner, D. R. Henry, and R. S. Pearlman, *J. Chem. Inf. Comput. Sci. 32*, 101 (1992).

71. D. Clark, P. Willett, and P. Kenny, *J. Mol. Graphics 10*, 194 (1992).

72. D. E. Clark, G. Jones, P. Willett, P. W. Kenny, and R. C. Glen, *J. Chem. Inf. Comput. Sci. 34*, 197 (1994).

73. T. E. Moock, D. R. Henry, A. G. Ozkabak, and M. Alamgir, *J. Chem. Inf. Comput. Sci. 34*, 184 (1994).

74. R. S. Pearlman and A. Rusinko III, *Chem. Des. Auto. News 2*, 1 (1987).

75. D. P. Dolata, A. R. Leach, and K. Proud, *J. Comput. Aided Mol. Des. 1*, 73 (1987).

76. G. W. A. Milne and J. A. Miller, *J. Chem. Inf. Comput. Sci. 26*, 154 (1986) and subsequent papers in the same issue.

77. M. A. Hendrickson, M. C. Nicklaus, G. W. A. Milne, and D. Zaharevitz, *J. Chem. Inf. Comput. Sci. 33*, 155 (1993).

78. M. C. Nicklaus, G. W. A. Milne, and D. Zaharevitz, *J. Chem. Inf. Comput. Sci. 33*, 639 (1993).

79. G. W. A. Milne, M. C. Nicklaus, J. S. Driscoll, S. Wang, and D. Zaharevitz, *J. Chem. Inf. Comput. Sci. 34*, 1219 (1994).

80. M. A. Johnson and G. M. Maggiora, eds., *Concepts and Applications of Molecular Similarity*, John Wiley & Sons, New York (1990).

81. P. A. Bath, A. R. Poirrette, P. Willett, and F. H. Allen, *J. Chem. Inf. Comput. Sci. 34*, 141 (1994); *35*, 714, 1081 (1995).

82. M. Petitjean and J.-E. Dubois, *J. Chem. Inf. Comput. Sci. 30*, 332 (1990); M. Petitjean, *J. Chem. Inf. Comput. Sci. 32*, 331 (1992).

83. R. Nilakantan, N. Bauman, and R. Venkataraghavagan, *J. Chem. Inf. Comput. Sci. 33*, 79 (1993); R. Sheridan, R. Nilakantan, A. Rusinko III, N. Bauman, K. Haraki, and R. Venkataraghavagan, *J. Chem. Inf. Comput. Sci. 29*, 255 (1989).

84. C. A. Pepperrell and P. Willett, *J. Comput. Aided Mol. Des. 5*, 455 (1991); C. A. Pepperrell, R. Taylor, and P. Willett, *Tetrahedron Comput. Methodol. 3*, 575 (1990).

85. C. A. Pepperrell, *Three-Dimensional Chemical Similarity Searching*, Research Studies Press–Wiley, New York (1994).

86. A. R. Poirrette, P. Willett, and F. H. Allen, *J. Mol. Graphics 9*, 203 (1991); *11*, 2 (1993).

87. J. R. Ullmann, *J. Assoc. Comput. Mach. 16*, 31 (1976).

88. T. Hurst, *J. Chem. Inf. Comput. Sci. 34*, 190 (1994).

89. Y. C. Martin, M. G. Burres, and P. Willett, in: *Reviews in Computational Chemistry* (K. B. Lipkowitz and D. B. Boyd, eds.), p. 213, VCH Publishers, New York (1990).

90. J. Van Drie, D. Weininger and Y. Martin, *J. Comp. Aided Mol. Des. 3*, 225 (1989).

91. Y. Martin, *J. Med. Chem. 35*, 2145 (1992); *Tetrahedron Comput. Methodol. 3*, 15 (1990).

92. M. G. Burres, E. Danaher, J. DeLazzer, Y. C. Martin, D. Clark, P. Willett, and P. Kenny, *J. Mol. Graphics 10*, 194 (1992).

93. Y. Martin, M. Bures, E. Danaher, J. DeLazzer, in: *Trends in QSAR and Molecular Modelling 92* (C. Wermuth, ed.), p. 20, ESCOM, Leiden (1993).

94. Y. C. Martin, M. G. Burres, E. A. Danaher, J. DeLazzer, I. Lico, and P. A. Pavlik, *J. Comput. Aided Mol. Des. 7*, 83 (1993).

95. A. T. Brint and P. Willett, *J. Chem. Inf. Comput. Sci. 27*, 152 (1987).

96. F. S. Kuhl, G. M. Crippen, and D. K. Friesen, *J. Comput. Chem. 5*, 24 (1984).

97. M. T. Bakarat and P. M. Dean, *J. Comput. Aided Mol. Des. 5*, 107 (1991).

98. S. Basak, S. Bertelsen, and G. D. Grunwald, *J. Chem. Inf. Comput. Sci. 34*, 270 (1994).

99. S. Basak and G. D. Grunwald, *J. Chem. Inf. Comput. Sci. 35*, 366 (1995).

100. A. M. Meyer and W. G. Richards, *J. Comput. Aided Mol. Des.* 5, 426 (1991).
101. P. G. Mezey, in: *Structure and Dynamics of Molecular Systems* (R. Daudel, J.-P. Korb, J.-P. Lemaistre, and J. Maruani, eds.), Reidel, Dordrecht (1985).
102. P. G. Mezey, *Potential Energy Hypersurfaces*, Elsevier, Amsterdam (1987).
103. P. G. Mezey, *Shape in Chemistry: An Introduction to Molecular Shape and Topology*, VCH Publishers, New York (1993).
104. P. D. Walker, G. M. Maggiora, M. A. Johnson, J. D. Petke, and P. G. Mezey, *J. Chem. Inf. Comput. Sci.* 35, 568 (1995).
105. A. M. Richard, *J. Comput. Chem.* 12, 959 (1991).
106. R. D. Cramer III, D. E. Patterson, and J. D. Bunce, *J. Am. Chem. Soc.* 110, 5959 (1988).
107. P. Willett, *Three-Dimensional Chemical Structure Handling*, Research Studies Press–Wiley, New York (1991).
108. W. T. Wipke and T. M. Dyott, *J. Am. Chem. Soc.* 96, 4825, 4834 (1974).
109. M. G. Bates, Y. C. Martin, and P. Willett, *Top. Stereochem.* 21, 467 (1994).
110. P. Willett, *Similarity and Clustering in Chemical Information Systems*, Research Studies Press–Wiley, New York (1987).
111. J. E. Ash, W. A. Warr, and P. Willett, eds., *Chemical Structure Systems*, Ellis Horwood Chichester (1991).
112. P. Willett, T. Wilson, and S. F. Reddaway, *J. Chem. Inf. Comput. Sci.* 31, 225 (1991).
113. S. E. Jakes, N. Watts, P. Willett, D. Bawden, and J. S. Fisher, *J. Mol. Graphics* 5, 41 (1987).
114. S. E. Jakes and P. Willett, *J. Mol. Graphics* 4, 12 (1986).
115. D. A. Thorner, D. J. Wild, P. Willett, and P. M. Wright, *J. Chem. Inf. Comput. Sci.* 36, 900 (1996).
116. D. J. Wild and P. Willett, *J. Chem. Inf. Comput. Sci.* 34, 224 (1994).
117. W. C. Herndon and S. H. Bertz, *J. Comput. Chem.* 8, 367 (1987).
118. G. Rum and W. C. Herndon, *J. Am. Chem. Soc.* 113, 9055 (1991).
119. G. Klopman and C. Raychaudhury, *J. Chem. Inf. Comput. Sci.* 30, 12 (1990).
120. G. Klopman and C. Raychaudhury, *J. Comput. Chem.* 9, 232 (1988).
121. J. Greene, S. Kahn, H. Savoj, P. Sprague, and S. Teig, *J. Chem. Inf. Comput. Sci.* 34, 224 (1994).
122. P. J. Artimiuk, H. M. Grindley, D. W. Rice, E. C. Ujah, and P. Willett, in: *Recent Advances in Chemical Information*, R. Soc. Chem. Spec. Publ. No. 120 (H. Collier, ed.), p. 91, Royal Society of Chemistry, Cambridge (1991).
123. P. J. Artimiuk, E. M. Mitchell, D. W. Rice, and P. Willett, *J. Inf. Sci.* 15, 287 (1989).
124. E. M. Mitchell, P. J. Artimiuk, D. W. Rice, and P. Willett, *Protein Eng.* 4, 39 (1990).
125. P. J. Artimiuk, P. A. Bath, H. M. Grindley, C. A. Pepperrell, A. R. Poirrette, D. W. Rice, D. A. Thorner, D. J. Wild, P. Willett, A. F. Allen, and R. Taylor, *J. Chem. Inf. Comput. Sci.* 32, 617 (1992).
126. H. M. Grindley, P. J. Artimiuk, D. W. Rice, and P. Willett, *J. Mol. Biol.* 33, 707 (1993).
127. P. J. Artimiuk, H. M. Grindley, A. R. Poirrette, D. W. Rice, E. C. Ujah, and P. Willett, *J. Chem. Inf. Comput. Sci.* 34, 54 (1994).
128. J. S. Richardson, *Nature* 268, 495 (1977).
129. I. Koch, F. Kaden, and J. Selbig, *Proteins Struct. Funct. Genet.* 12, 314 (1992).
130. C. Liang and K. Mislow, *J. Am. Chem. Soc.* 117, 4201 (1995).
131. D. M. Walba, in: *Chemical Applications of Topology and Graph Theory* (R. B. King, ed.), p. 17, Elsevier, Amsterdam (1983); *Tetrahedron 41*, 316 (1985).
132. A. T. Balaban, *Rev. Roum. Chim.* 33, 699 (1988).
133. M. Randić and M. Razinger, *J. Chem. Inf. Comput. Sci.* 36, 429 (1996).
134. J. K. Schur, P. Selzer, and J. Gasteiger, *J. Chem. Inf. Comput. Sci.* 36, 334 (1996).
135. J. Sadowski and J. Gasteiger, *Chem. Rev.* 93, 2567 (1993).
136. (a) K. Balasubramanian, *J. Chem. Inf. Comput. Sci.* 32 243 (1995); (b) *J. Chem. Inf. Comput. Sci. 32*, 761 (1995).
137. A. Toropov, A. Toropov, T. Ismailov, and D. Bonchev, *J. Chem. Inf. Comput. Sci.* (in press).
138. R. E. Carhart, D. H. Smith, and R. Ventkataraghavan, *J. Chem. Inf. Comput. Sci.* 25, 64 (1985).

139. R. Nilakantan, N. Bauman, J.S. Dixon, and R. Ventkataraghavan, *J. Chem. Inf. Comput. Sci. 27*, 82 (1987).

140. S. K. Kearsley, S. Sallamack, E. M. Fluder, J. D. Androse, R. T. Mosley, and R. P. Sheridan, *J. Chem. Inf. Comput. Sci. 36*, 118 (1996).

141. R. P. Sheridan, M. D. Miller, D. J. Underwood, and S. K. Kearsley, *J. Chem. Inf. Comput. Sci. 36*, 128 (1996).

142. S. K. Kearsley, D. J. Underwood, R. P. Sheidan, and M. D. Miller, *J. Comput. Aided Molec. Des. 8*, 565 (1994).

143. D. J. Wild and P. Willett, *J. Chem. Inf. Comput. Sci. 36*, 159 (1996).

144. F. H. Allen, P. A. Bath, and P. Willett, *J. Chem. Inf. Comput. Sci. 32*, 261 (1995).

145. D. E. Clark, M. A. Firth, and C. W. Murray, *J. Chem. Inf. Comput. Sci. 36*, 137 (1996).

146. L. H. Hall and J. B. Fisk, *J. Chem. Inf. Comput. Sci. 34*, 1184 (1994).

147. G. W. A. Milne, S. Wang, and M. C. Nicklaus, *J. Chem. Inf. Comput. Sci. 36*, 726 (1996).

148. C. Cheng, G. Maggiora, M. Lajiness, and M. Johnson, *J. Chem. Inf. Comput. Sci. 36*, 909 (1996).

149. S. C. Basak, V. R. Magnuson, G. J. Niemi, and R. R. Regal, *Discrete Appl. Math. 19*, 17 (1988).

150. K. Mislow, *Croat. Chem. Acta 69*, 485 (1996).

151. J. C. Chambron, C. Dietrich, and J. P. Sauvage, *Top. Curr. Chem. 165*, 131 (1993).

152. J. P. Sauvage, *Acc. Chem. Res. 23*, 319 (1990).

153. D. B. Amabilino and J. F. Stoddart, *Chem. Rev. 95*, 2725 (1995).

154. S. S. Tratch and N. S. Zefirov, *J. Chem. Inf. Comput. Sci. 36*, 448 (1996).

155. V. F. R. Jones, *Bull. Amer. Math. Soc. 12*, 103 (1985).

156. K. D. Millet, *Croat. Chem. Acta 59*, 669 (1986); *Bull. Amer. Math. Soc. 12*, 239 (1985); K. C. Millett, in *New Developments of Molecular Chirality* (P. Mezey, ed.), Kluwer, Dordrecht (1991); W. B. R. Lickorish and K. C. Millet, *Topology 26*, 107 (1987); *Math. Mag. 61*, 3 (1988).

157. L. H. Kauffman, *On Knots*, Princeton University Press, Princeton, New Jersey (1987); *Trans. Amer. Math. Soc. 318*, 417 (1990); *Amer. Math. Monthly 95*, 195 (1988).

2

Descriptors of Molecular Shape in 3D

PAUL G. MEZEY

2.1. INTRODUCTION

Molecular shape has a fundamental influence on both the static and dynamic properties of molecules; for example, the shapes of the nuclear and electronic distributions determine the molecular dipole moments as well as the likely sites of approach by a nucleophilic reagent. The evolution of concepts and models used by chemists and physicists for the description of molecular shape closely mirrors the advances made in our understanding of molecular behavior. Whereas most of the early models focused on the nuclear arrangements, the more advanced recent approaches have placed increasingly more emphasis on the electronic distribution. Molecules consist of interacting nuclear and electronic distributions, where the nuclear distribution is fully reflected in the electronic density. This fact allows one to obtain a complete description of molecular shapes in terms of the electronic density.[1]

There are, however, important advantages in focusing on the nuclear arrangements. Nuclei within a molecule are more "particlelike" than electrons. Whereas the Heisenberg uncertainty relation renders the concept of precise position of an electron within a molecule nearly meaningless, the concept of nuclear position within a molecule is a useful one, even if one must modify the classical idea of position with qualifiers such as zero-point vibration and tunneling.

PAUL G. MEZEY • Mathematical Chemistry Research Unit, Department of Chemistry and Department of Mathematics and Statistics, University of Saskatchewan, Saskatoon S7N 5C9, Canada.

From Chemical Topology to Three-Dimensional Geometry, edited by Balaban. Plenum Press, New York, 1997

The intimate relation between the nuclear distribution and the electronic density distribution is a natural bridge that connects the more conventional, essentially classical, ball-and-stick models, and the more accurate, quantum-chemical electronic density descriptors of molecular shape. It is somewhat surprising that relatively little effort has been devoted to the natural relation between the purely nuclear interactions and the electronic density. In this contribution this connection will be discussed from a specific viewpoint, leading to a 3D representation of molecular shape and to an interpretation of chemical bonding.

2.2. ELECTRONIC DENSITY, THE COMPOSITE NUCLEAR POTENTIAL, AND THE SOMOYAI FUNCTION

An early advance in the study of the connection between electronic density and nuclear interactions was made by Parr, Gadre, and Bartolotti,[2] followed by Parr and Berk,[3] who pointed out the striking similarities between quantum-chemical electronic densities of molecules and another, simple molecular function that can be easily obtained from an essentially classical model based solely on the nuclear arrangement: the composite nuclear potential. (In the original work of Parr and Berk[3] the term *bare nuclear potential* was used.) In what follows, the basic relations between electronic density and the composite nuclear potential will be reviewed.

Whereas the concepts and method described in this contribution are equally applicable to various approximate and more advanced quantum-chemical representations, the basic concepts will be discussed and illustrated within the framework of the conventional Hartree–Fock–Roothaan–Hall SCF LCAO *ab initio* representation of molecular wave functions and electronic densities,[4–7] as can be computed, for example, using the Gaussian family of computer programs of Pople and co-workers.[8] The essence of the shape analysis methods will be discussed with respect to some fixed nuclear arrangement K; note, however, that the generalizations will involve changes in the nuclear arrangement K.

The SCF LCAO *ab initio* representation of the molecular electronic density $\rho(\mathbf{r})$ is a function of the 3D position variable $\mathbf{r}$, and is defined in terms of a set of n atomic orbitals $\varphi_i(\mathbf{r})$, $i = 1, 2, \ldots, n$. The $n \times n$ dimensional density matrix $\mathbf{P}$ can be computed from the set of self-consistent coefficients of atomic orbitals in the occupied molecular orbitals. In terms of this density matrix $\mathbf{P}$, the electronic density $\rho(\mathbf{r})$ of the molecule can be written as

$$(1) \qquad \rho(\mathbf{r}) = \sum_{i=1}^{n} \sum_{j=1}^{n} P_{ij}\, \varphi_i(\mathbf{r})\, \varphi_j(\mathbf{r})$$

The *ab initio* molecular electronic density $\rho(\mathbf{r})$, representing the fuzzy, electronic charge cloud, provides a detailed representation of the shape of the actual, fuzzy "body" of the molecule.

A simple molecular property that can be calculated easily and can also be used for shape representation is the composite nuclear potential, denoted by $V_n(\mathbf{r})$. This 3D molecular function was first compared to 3D molecular electronic densities in a systematic study by Parr and Berk.[3] The $V_n(\mathbf{r})$ potential can be determined easily for a large number of nuclear arrangements of large molecules, providing a valid shape representation and a tool for shape comparisons of chemical species.

For a given collection of the nuclei of the molecule, the composite nuclear potential $V_n(\mathbf{r})$ (the "bare nuclear potential" in the terminology of Parr and Berk[3]) is defined as

$$(2) \qquad\qquad V_n(\mathbf{r}) = \sum_i Z_i \,/\, |\mathbf{r} - \mathbf{R}_i|$$

where the customary notations are used for nuclear charges Z_i at the formal nuclear positions $\mathbf{R}_i$. If we assume that the electronic density is removed without changing the nuclear arrangement, then the composite nuclear potential $V_n(\mathbf{r})$ is the potential experienced by a unit charge at location $\mathbf{r}$.

The computation of the composite nuclear potential $V_n(\mathbf{r})$ involves only a simple, repeated application of Coulomb's law, an important computational advantage if one requires a large number of shape comparisons on a nearly real-time scale.

An important fact has been pointed out by Parr and Berk[3]: the bare nuclear potential $V_n(\mathbf{r})$ shows many similarities with the electronic density function $\rho(\mathbf{r})$. The computed isopotential contours of the composite nuclear potential $V_n(\mathbf{r})$ were remarkably similar to some of the molecular isodensity contours (MIDCOs) of the electronic ground states in several simple molecules. One may regard the composite nuclear potential as the "harbinger" of electronic density, and isopotential contours of the composite nuclear potential $V_n(\mathbf{r})$ can serve as surprisingly good approximations of MIDCOs. The nuclear potential contours (NUPCOs) are suitable for an inexpensive, approximate shape representation of molecules.

These two functions also show important differences. Naturally, the essentially exponential behavior of electronic density at large distances from the nuclei cannot be described precisely by the essentially hyperbolic behavior of the composite nuclear potential. At short range, the shape of those parts of the molecular electronic density that involve strong contributions from formal π-bonds cannot be expected to have a faithful representation by NUPCOs; furthermore, at the nuclear locations the electronic density is finite but the composite nuclear potential must have singularities. In spite of these fundamental differences, the similarities between the composite nuclear potential $V_n(\mathbf{r})$ and the electronic density function $\rho(\mathbf{r})$ are striking.

Neither the composite nuclear potential nor the molecular electronic density changes sign and both vanish at infinite distance from the nuclei. The Hartree–Fock *ab initio* LCAO electronic densities, and most other representations of the electronic density $\rho(\mathbf{r})$ are continuous and differentiable functions of the position variable $\mathbf{r}$. The composite nuclear potential $V_n(\mathbf{r})$ is only an almost everywhere continuous and differentiable function of $\mathbf{r}$. Clearly, differentiability of $V_n(\mathbf{r})$ is ensured only if $\mathbf{r} \neq \mathbf{R}_i$, that is, if $\mathbf{r}$ does not coincide with any of the nuclear positions $\mathbf{R}_i$.

Chemical bonding in a molecule is not possible without electrons; the shape of the composite nuclear potential by itself describes only the relative locations of the various nuclei in the molecule. Although close proximity of two nuclei usually implies chemical bonding, this bonding is certainly not the result of the repulsive nuclear potential. Evidently, that part of the electronic density $\rho(\mathbf{r})$ that faithfully reproduces the shape of the composite nuclear potential $V_n(\mathbf{r})$ reveals little about chemical bonding; the shape of this part of the electronic density can be attributed to atomic, that is, to noninteracting, separate contributions. That part of the electronic density that is perfectly mimicked by the composite nuclear potential is not responsible for bonding; it is the *remainder* of the electronic density that is indicative of chemical bonding. In terms of shape features, it is the *difference* between the shapes of the composite nuclear potential $V_n(\mathbf{r})$ and the electronic density $\rho(\mathbf{r})$ that provides a description of chemical bonding.

Consider the following *Somoyai function*[9] $S(\mathbf{r},s)$, a special representation of the difference between the electronic density $\rho(\mathbf{r})$ and the composite nuclear potential $V_n(\mathbf{r})$, defined as

$$(3) \qquad\qquad S(\mathbf{r},s) = \rho(\mathbf{r}) - s\, V_n(\mathbf{r})$$

The *Somoyai parameter* s has the physical dimensions of bohr^{-2}. For any fixed choice of s, the Somoyai function $S(\mathbf{r},s)$ is an almost everywhere continuous and differentiable function of the position variable $\mathbf{r}$.

It is convenient to consider the shape description of all three functions within a common framework. For any molecular property P that is described by a 3D function $P(\mathbf{r})$ which is continuous in $\mathbf{r}$, such as the electronic density $\rho(\mathbf{r})$, and the composite nuclear potential $V_n(\mathbf{r})$, or the Somoyai function $S(\mathbf{r},s)$ with a constant s parameter, the *level sets* $F(a)$ for any constant value a of function $P(\mathbf{r})$ are defined as the following collection of points:

$$(4) \qquad\qquad F(a) = \{\ \mathbf{r} : P(\mathbf{r}) < a\ \}$$

The boundary surfaces $G(a)$ of level sets $F(a)$ are the isoproperty contours (IPCOs) defined as

$$(5) \qquad\qquad G(a) = \{\ \mathbf{r} : P(\mathbf{r}) = a\ \}$$

The entire 3D property function $P(\mathbf{r})$ can be represented by an infinite family of IPCOs, by taking one such surface $G(a)$ for each value of the contour parameter a, throughout the entire range $a_{\min} \leq a \leq a_{\max}$. The minimum and maximum values $a_{\min}$ and $a_{\max}$ of the contour threshold a depend on the property P and the actual molecular system studied.

Parameter s in the Somoyai function can be chosen so that $S(\mathbf{r},s)$ becomes small within a given part of the space. For example, one may select an envelope $E(a_1,a_2)$ of *a* given property $P(\mathbf{r})$, bounded by two IPCOs, $G(a_1)$ and $G(a_2)$. The envelope $E(a_1,a_2)$ is defined by the corresponding level sets, $F(a_1)$ and $F(a_2)$, as

(6) $$E(a_1,a_2) = F(a_1) \setminus F(a_2)$$

where [assuming a nonnegative 3D property $P(\mathbf{r})$] the two thresholds are ordered as

(7) $$a_1 \leq a_2$$

One may require that parameter s is chosen so that the absolute value $| S(\mathbf{r},s) |$ of the Somoyai function $S(\mathbf{r},s)$ is minimized within the envelope $E(a_1,a_2)$ of the given 3D property $P(\mathbf{r})$:

(8) $$\int_{E(a_1,a_2)} | S(\mathbf{r},s) | \, d\mathbf{r} = \text{minimum}$$

For example, property $P(\mathbf{r})$ can be chosen as the composite nuclear potential $V_n(\mathbf{r})$, and the envelope $E(a_1,a_2)$ can be taken over a finite range

(9) $$a_1 \leq a \leq a_2$$

of the potential value $V_n(\mathbf{r})$. The choice of the range is ultimately arbitrary; however, it is advantageous to select a range that involves those function values of property $P(\mathbf{r})$ that show characteristic changes during chemical reactions and conformational changes.

The *molecular subrange* of the 3D property $P(\mathbf{r})$ is of special importance: this range contains all thresholds *a* for which the level set $F(a)$ of property $P(\mathbf{r})$ is arcwise connected. For the molecular subrange level sets $F(a)$ of the composite nuclear potential this condition implies that all nuclei are contained within each such $F(a)$.

The shape of the Somoyai function $S(\mathbf{r},s)$ given for a suitably chosen parameter value s provides a 3D description of the bonding pattern within the molecule.

2.3. 3D SHAPE CHARACTERIZATIONS OF MOLECULAR FUNCTIONS $\rho(\mathbf{r})$, $V_n(\mathbf{r})$, AND $S(\mathbf{r},s)$

An important aspect of the similarities between these three functions is of special significance in our studies: the continuity properties discussed above ensure that for any nonzero contour value, the MIDCOs, the NUPCOs, and SOMCOs are closed (finite and bounded) surfaces. Hence, the topological Shape Group Methods (SGM),[10-12] originally introduced for a detailed shape analysis of electrostatic potentials and electronic charge densities, are easily applicable for all three functions. The Shape Group Methods have been developed using some of the techniques of algebraic topology and homology groups.[13] The algorithms based on the shape group approach have been inplemented as the GSHAPE 90 computer program.[14]

The topological Shape Group Methods[10-12] and their various applications have been reviewed extensively in the literature.[1,15-19] Here only a brief summary of the main features of these methods will be given.

A detailed yet rather concise description of the shapes of 3D molecular properties $P(\mathbf{r})$ represented by almost everywhere continuous and differentiable functions can be given by the topological Shape Group Methods. One family of shape groups describes the patterns of mutual arrangements of all possible local curvature domains of IPCO surfaces, relative to all possible reference curvatures b. The general scheme below lists the main steps involved in the application of the Shape Group Method:

STEP 1. *For each contour value a within a range of values for 3D property $P(\mathbf{r})$, the IPCOs $G(a)$ are partitioned into local curvature domains relative to each value b of a range of reference curvatures.*

In this step, the points of each IPCO surface $G(a)$ are classified into curvature domains of three types, denoted by $D_0(b)$, $D_1(b)$, or $D_2(b)$. The classification involves a comparison of the local canonical curvatures (the eigenvalues of the local Hessian matrices of the IPCOs) at each surface point of $G(a)$ to the curvature b of a reference tangent sphere. Each point $\mathbf{r}$ of the IPCO surface $G(a)$ is assigned to a $D_0(b)$, $D_1(b)$, or $D_2(b)$ curvature domain, depending on whether neither, only one, or both of the eigenvalues of the local Hessian matrix of the IPCO surface $G(a)$ at point $\mathbf{r}$ are smaller than the reference curvature b. In the special case of zero reference curvature, $b = 0$, that is, if the reference tangent sphere becomes a tangent plane, the three domains types, $D_0(0)$, $D_1(0)$, and $D_2(0)$, are the locally concave, saddle type, and convex domains, respectively. In general, these domains $D_0(b)$, $D_1(b)$, and $D_2(b)$ are called the *domains of local relative convexity* of the IPCO surface $G(a)$ of property $P(\mathbf{r})$, with reference to curvature b.

STEP 2. *For each pair of values of IPCO threshold parameter a and reference curvature parameter b, all curvature domains $D_\mu(b)$ of a specified type μ are formally removed from the corresponding IPCO $G(a)$.*

In this step, a *truncated surface* is generated for each IPCO $G(a)$ for each pair of parameter values a and b. Although there are infinitely many pairs of parameter values a and b, consequently, there are infinitely many such truncated surfaces, for most small changes of parameters a and b the truncated surfaces remain topologically equivalent. Consequently, there are only a finite number of classes of topologically different truncated surfaces for the entire range of parameter values a and b.

STEP 3. *The shape groups of the entire 3D molecular property are calculated by determining the algebraic homology groups for each topological equivalence class of the truncated surfaces.*

By definition, the *shape groups* of the property $P(\mathbf{r})$ are the algebraic homology groups of the family of topological equivalence classes of the truncated surfaces. The family of all of these equivalence classes involves all property thresholds a as well as all reference curvatures b. The shape groups provide a detailed shape description of the entire 3D property $P(\mathbf{r})$.

The so-called *Betti numbers* of the shape groups are the ranks of the corresponding homology groups. These Betti numbers of the shape groups are concise numerical shape descriptors, if they are determined for the entire range of physically relevant values of the two continuous parameters a and b. The Betti numbers of the 1D shape groups provide the most important shape information; in practice, the analysis is often restricted to this class of Betti numbers. The Betti numbers are topological invariants of the 3D property $P(\mathbf{r})$, describing the mutual relations of the various local shape domains of IPCOs of $P(\mathbf{r})$.

A detailed shape analysis based on the Shape Group Methods is equally applicable to 3D properties $P(\mathbf{r})$ of complete molecules and to the corresponding properties of smaller molecular fragments,[20] such as local molecular moieties, and functional groups,[21,22] as long as the same physical property is meaningful for these molecular fragments. For both molecular fragments and complete molecules, the Shape Group Method provides a collective description of the shape features of all isoproperty contours for all chemically relevant thresholds a and reference curvatures b.

The sequence of Betti numbers obtained in a shape group analysis is a concise numerical shape descriptor that can be used as a numerical shape code for the molecule or molecular fragment. The (a,b)-parameter map is the basis of a numerical shape code. After selecting ranges for both parameters a and b, a 2D map of the distribution of shape groups within the (a,b)-parameter plane is generated. A grid is defined within the parameter plane, and at each grid point the set of relevant Betti numbers is encoded into an integer. For a finite range of parameters a and b, and for any nonzero grid size, these integers form a finite numerical sequence denoted by $c(a,b)$, that can be used as a shape code.[1] Based on such numerical shape codes, nonvisual shape similarity measures, and shape complementarity measures can be determined for complete molecules and molecular fragments.

An interesting molecular descriptor is the similarity measure defined by the similarity s between the $c(a,b)$ shape codes of the electronic density and the composite nuclear potential of the same molecule A:

(10) $$d(\rho, V_n, A) = s(c_{\rho,A}(a,b),\, c_{Vn,A}(a,b))$$

A high degree of similarity $d(\rho, V_n, A)$ indicates a bonding pattern that is highly "atomic," dominated by the nuclear distribution. A lower degree of similarity $d(\rho, V_n, A)$ indicates a bonding pattern where electronic interactions within the charge cloud have features that "defy" the dominance of the coulombic effect of nuclear arrangements, for example, such "defiance" is manifested in π-bonds.

The primary role of numerical shape codes is in the comparison of different molecules. Numerical shape codes and the corresponding similarity measures and complementarity measures have important advantages when compared to visual evaluations of similarity and complementarity. For shape comparisons there is no need to reconstruct the 3D molecular properties for each comparison; the shape codes can be compared by algorithmic methods, using a computer, eliminating the subjective human bias and providing reproducible results. The numerical local and global shape similarity and shape complementarity measures are useful tools in various areas of chemistry and biochemistry, including pharmaceutical drug design, and computer-aided molecular engineering.

In the topological study and evaluation of molecular similarity, a general principle, the geometrical similarity as topological equivalence (GSTE) principle applies.[1] Two IPCOs that are very similar in a geometrical sense, have identical shape groups for many choices of the curvature parameter b. Clearly, their geometrical similarities are manifested in the topological equivalence in some of their shape groups. A similar observation can be made about small conformational changes and other small deformations of molecules: for most small changes of the nuclear arrangement K, many of the shape groups remain invariant. Again, small geometrical changes, that is, the geometrical similarities of nuclear arrangements, are manifested in a topological equivalence.

If the goal of the molecular shape description is a comparison of dynamic molecular features, that is, if one must account for the dynamic behavior and conformational flexibility of the nuclear arrangements, then one may consider various ranges of deformations of the molecule. Those ranges that preserve some of the shape descriptors, that is, the shape equivalence classes within conformational domains can be determined. The infinitely many, energetically accessible nuclear arrangements K can be classified into a finite number of topologically different shape equivalence classes. The geometrical similarities of these arrangements within each class are described by a topological equivalence. The shape invariance classes themselves can be characterized by algebraic topological techniques.[1,15,17-19]

2.4. A SHAPE REPRESENTATION OF MACROMOLECULES

An early approach to the local shape problem was based on the Shape Group Methods as applied to truncated surfaces, where the truncation was selected using essentially arbitrary criteria for deciding which part of the molecule belongs to a local region.[23] The purpose of the study was to analyze the influence of various substituents on the local shape of the rest of the molecule.

Using a more recent approach of *pseudodensities* of local molecular regions, as proposed by Walker,[24,25] the diagnosis of local shape changes is much enhanced. If one is interested in detecting the influence of one region of a molecule on the local shape of the rest of the molecule, denoted by X, then a shape change can be diagnosed by the pseudodensity of Walker, as defined in terms of the pseudodensity matrix $^*\mathbf{P}^X$ for region X:

(11) $^*P_{ij}^X = \begin{matrix} P_{ij} \\ 0 \end{matrix}$ if φ_i or φ_j is centered on a nucleus of region X
                                                                otherwise

Walker's pseudodensity $^*\rho^X(\mathbf{r})$ for the region X is obtained by replacing the density matrix elements P_{ij} with the pseudodensity matrix elements $^*P^X_{ij}$ in equation (11). The pseudodensities are nonadditive and exaggerate the actual shape changes; hence they are not suitable for building electronic densities for molecules, although they have diagnostic advantages in detecting interactions between a substituent and the rest of the molecule.

A simple, *additive* fragmentation approach to the molecular electronic density, proposed by the author, can be used for the construction of electronic densities and density-based shape representations for macromolecules. The simplest of these approaches is motivated by Mulliken's population analysis technique,[26,27] and can be regarded as a natural generalization of Mulliken's approach: a formal "population analysis without integration." This method, the Mulliken–Mezey approach, is the simplest realization of a more general, additive fuzzy density fragmentation (AFDF) principle.[18,19]

As the acronym indicates, the key properties of the AFDF schemes are *additivity* and *fuzziness* of the density fragments. If a molecular electronic density can be partitioned into *additive* fuzzy fragments, then it is natural to exploit this additivity and construct molecular electronic densities from these fragments. The fuzziness of density fragments is also essential; fuzziness ensures that no density gaps or density doublings occur when combining fragments. The appearance of density gaps or density doublings are the common failures of attempts based on nonfuzzy density fragments with boundaries.

The simplest additive, fuzzy density fragmentation method, the Mulliken–Mezey method, is the basis of the MEDLA (molecular electron density Lego assembler) technique of Walker and Mezey.[28–32] The first applications[28] involved only relatively small molecules; however, the additivity property of the fragmentation scheme has provided the first possibility of studying large systems. The simplest, Mulliken–Mezey version of the general AFDF scheme[18,19,28] has been used within the MEDLA method for the generation of *ab initio* quality electronic densities for macromolecules such as proteins.[28–32]

The same Mulliken–Mezey method, and the more elaborate alternative partitioning schemes[18,19] serve as the basis of the ALDA (adjustable local density assembler) method[33–35] and the ADMA (adjustable density matrix assembler) technique.[33,34,36] The ALDA and ADMA methods generate geometry-adjustable electronic densities and macromolecular density matrices. The ADMA macromolecular density matrices can be computed without determining a macromolecular wave function, a feature advantageous in the macromolecular application of a variety of quantum-chemical techniques which, until recently, were applicable only for small molecules. The ALDA and ADMA approaches also serve as the basis of new techniques for the computation of a variety of macromolecular properties, such as energy and the forces acting on various nuclei.

According to the Mulliken–Mezey AFDF method, the set of nuclei of the molecule are divided into m mutually exclusive families

$$(12) \qquad f_1, f_2, \ldots, f_k, \ldots, f_m$$

These nuclear families serve as reference for m fuzzy, additive density fragments, $F_1, F_2, \ldots, F_k, \ldots, F_m$, where these fuzzy molecular fragments correspond to fragment density functions

$$(13) \qquad \rho^1(\mathbf{r}), \rho^2(\mathbf{r}), \ldots, \rho^k(\mathbf{r}), \ldots, \rho^m(\mathbf{r})$$

respectively.

A formal membership function $m_k(i)$ describes whether a given AO $\varphi_i(\mathbf{r})$ belongs to a set of AOs centered on a nucleus of nuclear set f_k of molecular fragment F_k:

$$(14) \qquad m_k(i) = \begin{array}{ll} 1 & \text{if } \varphi_i(\mathbf{r}) \text{ is centered on one of the nuclei of set } f_k \\ 0 & \text{otherwise} \end{array}$$

The elements P^k_{ij} of Mezey's $n \times n$ *fragment density matrix* $\mathbf{P}^k$ of the kth fuzzy fragment F_k are defined as

$$(15) \qquad P^k_{ij} = [m_k(i)\, w_{ij} + m_k(j)\, w_{ji}]\, P_{ij}$$

where the weighting factors w_{ij} and w_{ji} are constrained by

$$(16) \qquad w_{ij} + w_{ji} = 1$$

If the simplest, Mulliken–Mezey version of this fragmentation scheme is used, then

$$(17) \qquad w_{ij} = w_{ji} = 0.5$$

whereas alternative, more general, but still exactly additive weighting schemes are discussed in Refs. 18 and 19.

The AO set and the fragment density matrix $\mathbf{P}^k$ define the fuzzy electronic density of the kth density fragment:

$$(18) \qquad \rho^k(\mathbf{r}) = \sum_{i=1}^{n} \sum_{j=1}^{n} P^k_{ij}\, \varphi_i(\mathbf{r}) \varphi_j(\mathbf{r})$$

Within the Mulliken–Mezey and the more general fuzzy fragmentation methods,[18,19] the sum of the fragment density matrices $\mathbf{P}^k$ is equal to the density matrix $\mathbf{P}$ of the complete molecule:

$$(19) \qquad\qquad P_{ij} = \sum_{k=1}^{m} P_{ij}^{k}$$

This implies that the sum of the $\rho^k(\mathbf{r})$ fragment densities is equal to the density $\rho(\mathbf{r})$ of the complete molecule:

$$(20) \qquad\qquad \rho(\mathbf{r}) = \sum_{k=1}^{m} \rho^k(\mathbf{r})$$

The above exact additivity implies that the method is applicable for the construction of approximate electronic densities for large molecules from fragment densities calculated for small molecules if the nuclear geometry and the local surroundings of the given fragment match those in the large target molecule. The fuzzy fragment densities have no boundaries. If one combines these fragments forming molecular electronic densities, then the fragments merge, and a smoothly varying electronic density charge cloud is obtained. This feature ensures that the MEDLA method is free from the errors that occur when molecular pieces with boundary surfaces are combined. In particular, neither density gaps nor regions of density doubling occur in the electronic density clouds obtained from the fuzzy MEDLA fragments. By contrast, large errors are often found when combining "atoms in molecule" fragments, defined by boundary surfaces of zero density flux condition, where density gaps (100% underestimation of density) and density doubling (100% overestimation of density) may occur. Mismatches at fragment boundaries are common if the fragments are used to build the density of a new molecule where the local surroundings are slightly different. No such accumulation of errors is possible if the MEDLA technique is used.

According to detailed tests,[28,30,32] the MEDLA method generates *ab initio* quality electronic densities at a level better than standard SCF calculations using 3-21G bases. The MEDLA electronic densities are virtually indistinguishable from electronic densities obtained by standard SCF calculations at the 6-31G** basis set level.[28,30,32]

The MEDLA method relies on an electronic density fragment data bank that stores precalculated, 6-31G** *ab initio* electronic densities of fuzzy molecular fragments obtained from direct *ab initio* calculations on various small molecules. The variations in nuclear geometries and local surroundings are accounted for by several versions of each fragment type in the data base. Various versions of each fragment type are obtained from standard *ab initio* computations of small molecules, where the local nuclear configuration and the local surroundings match those in the target molecule. Several *ab initio* quality MEDLA calculations have been carried out for large molecules [crambin,[30] the gene-5 protein (g5p) of bacteriophage M13,[30] bovine insulin,[31] taxol,[25] and the HIV-1 protease monomer[25]].

The MEDLA method, based on numerical electronic density data base, the more advanced, geometry-adjustable ALDA method,[33-35] based on a fragment density matrix data base, and the ADMA method,[33,34,36] generating macromolecular density matrices, serve as "computational microscopes" of resolution far exceeding that of current X-ray diffraction, electron diffraction, and other experimental techniques. The computer time requirements of the MEDLA, ALDA, and ADMA methods grow linearly with molecular size.

2.5. SHAPE–ENERGY RELATIONS FOR THE COMPUTATION OF FORCES AND GEOMETRY OPTIMIZATION BASED ON MACROMOLECULAR ELECTRONIC DENSITIES AND THE ELECTROSTATIC THEOREM

If reasonably accurate electronic densities are available, then the forces acting on the nuclei can be approximately determined by a simple application of the electrostatic theorem, an important variant of the Hellmann–Feynman theorem.[37–39] In turn, these forces can be used for geometry optimization.

The electrostatic theorem, as a special case of the Hellmann–Feynman theorem, has some limitations. These limitations are briefly reviewed below, following the discussion of the connections between the Hellmann–Feynman theorem and general properties of potential energy hypersurfaces.[40]

The Hellmann–Feynman theorem is a fundamental, yet rather simple theorem of quantum mechanics that describes the first derivatives of energy with respect to a parameter σ satisfying some constraints. According to the theorem, the derivative of the energy is the expectation value of the derivative of the Hamiltonian:

(21) $$\partial E/\partial\sigma = \langle\psi|\partial\hat{H}/\partial\sigma|\psi\rangle \,/\, \langle\psi|\psi\rangle$$

In the above equation E is the eigenvalue belonging to a normalizable wave function ψ,

(22) $$\hat{H}\,\psi = E\psi$$

whereas σ is a real parameter in the Hamiltonian $\hat{H}$.

Within a variational framework, a simple, sufficient condition of applicability of the Hellmann–Feynman theorem has been proposed by Hurley.[41,42] If the variation extends to a family of trial functions where the family is invariant to changes in a parameter σ, then the optimum trial function fulfills the Hellmann–Feynman theorem. In other words, if the variation in parameter σ merely interconverts the trial functions within the same family, then the Hellmann–Feynman theorem applies. One trivial case is a family of trial functions where each function is independent of σ. Hurley's condition is fulfilled in variational approaches involving Lagrange multipliers, such as the Hartree–Fock and multiconfigurational self consistent field (MCSCF) methods.[43]

A theorem by Parr[44] can be regarded as an integrated form of the Hellmann–Feynman theorem:

(23)
$$E_A - E_B = \langle \psi_A | \hat{H}_A - \hat{H}_B | \psi_B \rangle / \langle \psi_A | \psi_B \rangle$$

The Parr theorem, regarded as the finite difference formula associated with the Hellmann–Feynman theorem, has some interesting consequences. In most quantum-chemical models, the difference of two Hamiltonians on the right-hand side is a one-electron operator. Consequently, at least in principle, in the calculation of the conformational energy difference $E_A - E_B$, the evaluation of complicated two-electron integrals can be avoided.

If parameter σ is chosen as a nuclear coordinate, then the Hellmann–Feynman theorem gives the corresponding energy derivative, a formal force acting on the nucleus. The results of such force calculations are very sensitive to the quality of wave function ψ. In the integral Hellmann–Feynman theorem, the error in the wave functions ψ_A and ψ_B affects the error in the energy difference $E_A - E_B$ in the first order. The Hellmann–Feynman theorem has been applied for the calculation of rotational barriers.[45–49] By taking an alternative choice for parameter σ, the energy differences between isoelectronic molecules differing in their nuclear charges can be calculated.[50,51] After the introduction of analytic differentiation of the energy expectation value within the Hartree–Fock framework and related approximations by Pulay, and by Pulay and Meyer,[52–59] the earlier interest in force calculations based on the Hellmann–Feynman theorem has diminished.

However, a special form of the Hellmann–Feynman theorem, which can be expressed in terms of electronic density and does not require knowledge of the electronic wave function, has recently acquired additional significance.

Consider a molecule of N nuclei and k electrons, and assume that the electronic density $\rho(\mathbf{R})$ of this molecule is known. The position vectors $\mathbf{R}_a$ of nucleus a, and $\mathbf{R}_i$ of electron i are defined in terms of their components as

(24)
$$\mathbf{R}'_a = (X_a, Y_a, Z_a)$$

and

(25)
$$\mathbf{R}'_i = (X_i, Y_i, Z_i)$$

respectively. Using these notations for electronic position vectors, and the Dirac delta formalism, the electronic density $\rho(\mathbf{R})$ can be expressed in terms of the electronic wave function ψ as

(26)
$$\rho(\mathbf{R}) = \langle \psi \mid d \sum_{i=1}^{k} \delta(\mathbf{R} - \mathbf{R}_i) \mid \psi \rangle / \langle \psi \mid \psi \rangle$$

By using the nuclear and electronic position vectors $\mathbf{R}_a$ and $\mathbf{R}_i$ to express the electron–electron, electron–nucleus, and nucleus–nucleus distances, r_{ij}, r_{ia}, and r_{ab}, respectively, the roles of nuclear coordinates within the terms of the molecular Hamiltonian can be identified. Following the standard treatment within the framework of the Born–Oppenheimer approximation, the molecular Hamiltonian is given as a sum of two operators,

$$(27) \qquad \hat{H} = \hat{H}_e + \hat{H}_n$$

where the electronic electronic Hamiltonian $\hat{H}_e$ is given in terms of electronic kinetic energy operator $\hat{T}_e$ and the electronic potential energy operator $\hat{V}_e$,

$$(28) \qquad \hat{H}_e = \hat{T}_e + \hat{V}_e$$

and the nuclear Hamiltonian $\hat{H}_n$ is defined within the clamped nuclei model as

$$(29) \qquad \hat{H}_n = V_{nn}$$

Here V_{nn} is the nuclear repulsion term,

$$(30) \qquad V_{nn} = \sum_{a<b}^{N} z_a z_b \, r^{-1}{}_{ab}$$

whereas the electronic kinetic and potential energy operators are given as

$$(31) \qquad \hat{T}_e = -1/2 \sum_{i=1}^{k} \Delta_i$$

and

$$(32) \qquad \hat{V}_e = \sum_{i<j}^{k} r^{-1}{}_{ij} - \sum_{i=1}^{k} \sum_{a=1}^{N} z_a \, r^{-1}{}_{ia}$$

respectively. In terms of these operators, the molecular Hamiltonian $\hat{H}$ takes the usual form

$$(33) \qquad \hat{H} = -1/2 \sum_{i=1}^{k} \Delta_i + \sum_{i<j}^{k} r^{-1}{}_{ij} - \sum_{i=1}^{k} \sum_{a=1}^{N} z_a \, r^{-1}{}_{ia} + \sum_{a<b}^{N} z_a z_b \, r^{-1}{}_{ab}$$

If the Hamiltonian $\hat{H}$ of the form given in equation (33) is differentiated according to components of nuclear position vector $\mathbf{R}_a$, then one obtains

$$-\partial \hat{H} / \partial \mathbf{R}_a = - \sum_{i=1}^{k} z_a (\mathbf{R}_a - \mathbf{R}_i)/|\mathbf{R}_a - \mathbf{R}_i|^3$$

$$(34) \qquad\qquad + \sum_{a<b}^{N} z_a z_b (\mathbf{R}_a - \mathbf{R}_b)/|\mathbf{R}_a - \mathbf{R}_b|^3$$

This equation defines a force operator $\mathbf{F}_a$,

(35)
$$\mathbf{F}_a = -\partial \hat{\mathbf{H}}/\partial \mathbf{R}_a$$

This force operator $\mathbf{F}_a$ represents the combined forces related to all of the electrons and all of the other nuclei, acting on nucleus a. If the parameter σ of the Hellmann–Feynman theorem is identified with the components of vector $\mathbf{R}_a$, then the application of the Hellmann–Feynman theorem gives

(36)
$$-\partial E/\partial \mathbf{R}_a = \langle \psi | \mathbf{F}_a | \psi \rangle / \langle \psi | \psi \rangle$$

If the expectation value of operator $\mathbf{F}_a$ is denoted by $\langle \mathbf{F}_a \rangle$, then equation (34) can be written in the concise form

(37)
$$-\partial E/\partial \mathbf{R}_a = \langle \mathbf{F}_a \rangle$$

Although not yet obvious from the deceptively simple form of equation (37), this equation, the electrostatic Hellmann–Feynman theorem, allows one to use the electronic density and the simple internuclear Coulomb interactions to describe the forces acting on the nuclei of the molecule. A simple, classical interpretation of this theorem provides the key to the use of macromolecular electronic densities, such as those obtained within the MEDLA, ALDA, or ADMA methods, for the computation of forces within the macromolecule.

The force $-\partial E/\partial \mathbf{R}_a$, which acts on nucleus a of coordinates X_a, Y_a, and Z_a, is the negative gradient of the Born–Oppenheimer potential energy hypersurface E.[40] This force is equal to the Hellmann–Feynman force $\langle \mathbf{F}_a \rangle$. Differentiation of the Hamiltonian of equation (33) by $\mathbf{R}_a$, leading to equation (32), eliminates all of the electronic kinetic energy terms and all of the terms describing electron–electron interactions. In fact, the force operator $\mathbf{F}_a$ is a one-electron operator, and the corresponding expectation value $\langle \mathbf{F}_a \rangle$ has the actual form of

(38)
$$\langle \mathbf{F}_a \rangle = -z_a \int \rho(\mathbf{R})(\mathbf{R}_a - \mathbf{R})|\mathbf{R}_a - \mathbf{R}|^{-3} \, d\mathbf{R}$$
$$+ z_a \sum_{a \neq b}^{N} z_b(\mathbf{R}_a - \mathbf{R}_b) |\mathbf{R}_a - \mathbf{R}_b|^{-3}$$

where $\rho(\mathbf{R})$ is the electronic density, described in the Dirac delta formalism given in equation (24). That is, the expectation value $\langle \mathbf{F}_a \rangle$ of the force operator is a simple sum of a classical contribution from the internuclear repulsion and a classical contribution from the electronic charge density.

If molecular electronic densities $\rho(\mathbf{R})$ of satisfactory accuracy can be computed for large molecules, using the MEDLA, ALDA, or ADMA methods, then a 3D integration in the first term, and a trivial summation in the second term of equation (36) provides the force acting on nucleus a of the molecule. Quantum-chemical forces, of accuracy compatible with the *ab initio* quality electronic density obtained with the MEDLA, ALDA, or ADMA methods, can be calculated without requiring the computation of molecular wave functions. This combination of macromolecular electronic density methods and the electrostatic theorem suggests a promising approach for macromolecular geometry optimization and for the computation of global and local shape changes, and macromolecular conformational motions, such as protein folding.

2.6. SUMMARY

The Somoyai function is defined in terms of the electronic density function and the composite nuclear potential, providing a 3D shape representation of the bonding pattern within the molecule under study. Some of the topological techniques of molecular shape analysis have been reviewed, with special emphasis on applications to the Somoyai function. A combination of a family of recently introduced *ab initio* quality macromolecular electronic density computation methods with the electrostatic Hellmann–Feynman theorem provides a new technique for the computation of forces acting on the nuclei of large molecules. This method of force computation offers a new approach to macromolecular geometry optimization.

ACKNOWLEDGMENT

The original research work leading to the methods and results described in this report was supported by both strategic and operating research grants from the Natural Sciences and Engineering Research Council of Canada.

REFERENCES

1. P. G. Mezey, *Shape in Chemistry: An Introduction to Molecular Shape and Topology*, VCH Publishers, New York (1993).
2. R. G. Parr, S. R. Gadre, and L. J. Bartolotti, *Proc. Natl. Acad. Sci. USA 76*, 2522 (1979).
3. R. G. Parr and A. Berk, The bare-nuclear potential as harbinger for the electron density in a molecule, in: *Chemical Applications of Atomic and Molecular Electrostatic Potentials* (P. Politzer and D. G. Truhlar, eds.), pp. 51–62, Plenum Press, New York (1981).
4. D. R. Hartree, *Proc. Cambridge Philos. Soc. 24*, 111, 426 (1928); *25*, 225, 310 (1929).
5. V. Fock, *Z. Phys. 61*, 126 (1930).
6. C. C. Roothaan, *Rev. Mod. Phys. 23*, 69 (1951); *32*, 179 (1960).
7. G. G. Hall, *Proc. R. Soc. London Ser. A 205*, 541 (1951).
8. M. J. Frisch, M. Head-Gordon, G. W. Trucks, J. B. Foresman, H. B. Schlegel, K. Raghavachari, M. A. Robb, J. S. Binkley, C. Gonzalez, D. J. Defrees, D. J. Fox, R. A. Whiteside, R. Seeger, C. F. Melius, J. Baker, R. L. Martin, L. R. Kahn, J. J. P. Stewart, S. Topiol, and J. A. Pople, Program GAUSSIAN 90, Gaussian, Inc., Pittsburgh, Pennsylvania (1990).
9. F. Somoyai, unpublished results.
10. P. G. Mezey, *Int. J. Quantum Chem. Quantum Biol. Symp. 12*, 113 (1986).
11. P. G. Mezey, *J. Comput. Chem. 8*, 462 (1987).

12. P. G. Mezey, *J. Math. Chem. 2*, 325 (1988).
13. M. Greenberg, *Lectures on Algebraic Topology*, Benjamin, New York (1967).
14. P. D. Walker, G. A. Arteca, and P. G. Mezey, Program GSHAPE 90, Mathematical Chemistry Research Unit, University of Saskatchewan, Saskatoon, Canada (1990).
15. P. G. Mezey, Three-dimensional topological aspects of molecular similarity, in: *Concepts and Applications of Molecular Similarity* (M. A. Johnson and G. M. Maggiora, eds.), John Wiley & Sons, New York (1990).
16. P. G. Mezey, Molecular surfaces, in: *Reviews in Computational Chemistry* (K. B. Lipkowitz and D. B. Boyd, eds.), VCH Publishers, New York (1990).
17. P. G. Mezey, Molecular similarity measures for assessing reactivity, in: *Molecular Similarity and Reactivity: From Quantum Chemical to Phenomenological Approaches* (R. Carbó, ed.), Kluwer, Dordrecht (1995).
18. P. G. Mezey, Density domain bonding topology and molecular similarity measures, in: *Topics in Current Chemistry*, Vol. 173, *Molecular Similarity I* (K. Sen, ed.), Springer-Verlag, Berlin (1995).
19. P. G. Mezey, Methods of molecular shape-similarity analysis and topological shape design, in: *Molecular Similarity in Drug Design* (P. M. Dean, ed.), Chapman & Hall/Blackie Publishers, Glasgow (1995).
20. P. G. Mezey, *Int. J. Quantum Chem. Quantum Biol. Symp. 14*, 127 (1987).
21. J.-E. Dubois and P. G. Mezey, *Int. J. Quantum Chem. 43*, 647 (1992).
22. P. G. Mezey, *Can. J. Chem. 72*, 928 (1994).
23. G. A. Arteca, V. B. Jammal, and P. G. Mezey, *J. Comput. Chem. 9*, 608 (1988).
24. P. D. Walker, private communication (1993).
25. P. D. Walker and P. G. Mezey, *J. Math. Chem. 17*, 203 (1995).
26. R. S. Mulliken, *J. Chem. Phys. 23*, 1833, 1841, 2338, 2343 (1955).
27. R. S. Mulliken, *J. Chem. Phys. 36*, 3428 (1962).
28. P. D. Walker and P. G. Mezey, *J. Am. Chem. Soc. 115*, 12423 (1993).
29. P. D. Walker and P. G. Mezey, Program MEDLA 93, Mathematical Chemistry Research Unit, University of Saskatchewan, Canada (1993).
30. P. D. Walker and P. G. Mezey, *J. Am. Chem. Soc. 116*, 12022 (1994).
31. P. D. Walker and P. G. Mezey, *Can. J. Chem. 72*, 2531 (1994).
32. P. D. Walker and P. G. Mezey, *J. Comput. Chem. 16*, 1238 (1995).
33. P. G. Mezey, Local shape analysis of macromolecular electron densities, in: *Computational Chemistry: Reviews and Current Trends* (J. Leszczynski, ed.), World Scientific Publishers, Singapore (1996).
34. P. G. Mezey, *J. Math. Chem. 18*, 141 (1995).
35. P. G. Mezey, Program ALDA 95, Mathematical Chemistry Research Unit, University of Saskatchewan, Saskatoon, Canada (1995).
36. P. G. Mezey, Program ADMA 95, Mathematical Chemistry Research Unit, University of Saskatchewan, Saskatoon, Canada (1995).
37. H. Hellmann, *Einführung in die Quantenchemie*, Sect. 54, Deuticke and Co., Leipzig (1937).
38. R. P. Feynman, *Phys. Rev. 56*, 340 (1939).
39. S. T. Epstein, The Hellmann–Feynman theorem, in: *The Force Concept in Chemistry* (B. M. Deb, ed.), Van Nostrand–Reinhold, Princeton, New Jersey (1981).
40. P. G. Mezey, *Potential Energy Hypersurfaces*, Elsevier, Amsterdam (1987).
41. A. C. Hurley, *Proc. R. Soc. London Ser. A 226*, 170 (1954).
42. A. C. Hurley, *Proc. R. Soc. London Ser. A 226*, 179 (1954).
43. S. T. Epstein, *Theor. Chim. Acta 55*, 251 (1980).
44. R. G. Parr, *J. Chem. Phys. 40*, 3726 (1964).
45. J. Goodisman, *J. Chem. Phys. 44*, 2085 (1966).
46. J. Goodisman, *J. Chem. Phys. 45*, 4689 (1966).
47. J. Goodisman, *J. Chem. Phys. 47*, 334 (1967).
48. L. Zulicke and H. J. Spangenberg, *Theor. Chim. Acta 5*, 139 (1966).

49. M. T. Marron, Energies, energy differences and mechanisms of internal motions, in: *The Force Concept in Chemistry* (B. M. Deb, ed.), Van Nostrand–Reinhold, Princeton, New Jersey (1981).
50. C. Trindle and J. K. George, *Int. J. Quantum Chem. 10*, 21 (1976).
51. P. G. Mezey, *Mol. Phys. 47*, 121 (1982).
52. P. Pulay, *Mol. Phys. 17*, 197 (1969).
53. P. Pulay, *Mol. Phys. 18*, 473 (1970).
54. P. Pulay, *Mol. Phys. 21*, 329 (1971).
55. P. Pulay and W. Meyer, *J. Mol. Spectrosc. 40*, 59 (1971).
56. P. Pulay and W. Meyer, *J. Chem. Phys. 57*, 3337 (1972).
57. P. Pulay, Direct use of gradients for investigating molecular energy surfaces, in: *Applications of Electronic Structure Theory* (H. F. Schaefer, ed.), Plenum Press, New York (1977).
58. P. Pulay, Calculation of forces by non-Hellmann–Feynman methods, in: *The Force Concept in Chemistry* (B. M. Deb, ed.), Van Nostrand–Reinhold, Princeton, New Jersey (1981).
59. P. Pulay, TEXAS, Ab Initio Hartree–Fock Gradient Program, *Theor. Chim. Acta 50*, 299 (1979).

3

3D Molecular Design
Searching for Active Conformers in QSAR

OVANES G. MEKENYAN and GILMAN D. VEITH

3.1. INTRODUCTION

It is difficult to estimate how long drawings of chemical structures have been a part of science. Even so, it is clear that graphical representation of chemicals as two-dimensional (2D) structures has deep historical roots. Prior to 1874 when van't Hoff and Le Bel found 2D drawings inadequate to explain some chemical properties, this representation of structure was the primary language of chemists for describing chemicals and their attributes. 2D representations, or molecular topology, have immense value in education and research even though the portrayals are now known to be naive. Not only are the quantitative models that are based on topology important to QSAR, but also the application of graph theory to molecular topology using high-speed computing still provides one of the more efficient approaches for storing, sorting, and retrieving chemical information.

Mention of trade names or specific products or approaches does not constitute in any way an endorsement or official opinion on the part of the U. S. Environmental Protection Agency.

OVANES G. MEKENYAN • Department of Physical Chemistry, Bourgas University "As Zlatarov" 8010 Bourgas, Bulgaria; and Lake Superior Research Institute, University of Wisconsin–Superior, Superior, Wisconsin 54880.　GILMAN D. VEITH • National Health and Environmental Effects Research Laboratory, U. S. Environmental Protection Agency, Research Triangle Park, North Carolina 27711.

From Chemical Topology to Three-Dimensional Geometry, edited by Balaban. Plenum Press, New York, 1997

However, Edwin A. Abbott's fanciful two-dimensional world described in *Flatland* is no more the world of molecules and chemical reactions than it was the world around us. While molecular topology is adequate to explain many aspects of chemical behavior, the evolution of quantitative models for structure–activity relations is unavoidably moving into the realm of three-dimensional (3D) structure. Modern computing enables rapid manipulation of 3D chemical structures, and it is leading the way for a proliferation of models for the quantification of electronic structure. For complex behavior of chemicals including physicochemical properties, reactivity, and biological activity, 3D structures are essential.

This evolution in QSAR was slow. As in many sciences, the evolution has been driven by discoveries of chemical behavior that could not be explained using conventional concepts and models. For example, Louis Pasteur[1] recognized that optical activity (a phenomenon observed earlier[2]) was the result of the molecular dissymmetry later called *chirality* (from Greek *cheir* = hand). The concept of stereochemistry, however, was introduced by van't Hoff and Le Bel.[3] It was V. Prelog[4] who pointed out that stereochemistry is not a branch of chemistry but a point of view. Part of this point of view is the description of structure that explains relevant behavior, which necessarily leads to additional levels of taxonomic analysis of chemicals.

Molecular topology describes the 2D connectivity of molecules. Stereoisomers have the same 2D structure with different 3D geometries. Accordingly, they may be classified as either *enantiomers* (nonsuperposable mirror images, possibly due to the configuration of asymmetric atoms, termed *chiral*) or as *diastereomers* (differing in spatial array in other ways), as shown in Figure 1. Enantiomers are identical in most of their physical properties (such as boiling and melting point, refractivity index, spectra, free energy). They have identical chemical behavior toward achiral reagents as well. In fact, the enantiomers have the same distances between atoms (they are "isometric"), which explains the similarity in most of their properties. However,

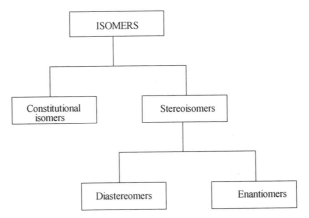

Figure 1. Isomeric versus homomeric relation.

enantiomers differ in sign with respect to physical properties depending on interactions with chiral molecules[5] or photons (e.g., optical rotation).

To enter the 3D world is also to leave behind the idea of rigid bonds and molecules and enter the world of flexible molecules, which contains limitless conformations of single chemicals. The different conformations of a molecule can be obtained by rotations around single bonds. As groups of atoms are rotated, the energy of the molecule changes, often dramatically, as a result of forces of attraction and repulsion. Conformational isomers (conformers) are those conformations corresponding to energy minima in the total energy profile. Many conformers also differ markedly in reactivity, as the case of polysubstituted cyclohexane conformers having axial and equatorial geometries. Because of the possibility of such dramatic differences in reactivity among conformers, it seems inappropriate to describe chemicals as single conformers.

This chapter reviews the basic concepts of applying 3D chemical models to the identification of new chemicals that have sought-after properties or desired biological behavior. A key development in this area is the design of new biologically active compounds. Two general approaches can be applied here. The first includes 3D search systems that identify molecules and fragments having a specific 3D shape in large chemical data bases. The second approach deals with 3D structure building from scratch (*de novo* design) where the information for receptor shape, and/or quantitative structure–activity relationship (QSAR) studies, direct the design of new chemicals.

Next, we will focus on 3D structure generation of molecules used for QSAR studies. Details for the 2D–3D builders (converters) most applied nowadays can be found in the next section. Their common feature is the capability to arrive eventually to a single good-quality stereoisomer corresponding to input molecular topology. The quality of the designed models is usually assessed by energy estimates or using similarity with X-ray crystallographic data. There are, however, many complexities inherent in the topic. One of them is the so called *multiple-minimum* problem during the search of energetic minima of the conformational space. The problem stems from the difficulty of locating the populated, low-energy conformers of molecules defined by the potential energy hyperspace. The latter may contain a remarkably large number of minima. The conformational space searching methods generate eventually a set of 3D structures, representing points distributed throughout the conformational space.

The search process can be conventionally divided into two steps. First, crude starting geometries are produced. The generated structures serve as initial geometries for optimization (by making use of force field or quantum-chemical energy minimization procedures) to nearby local minima (conformers). Thus, the energy minimization part of the process simply refines starting geometries. Another difficulty that arises, especially in dealing with flexible molecules, is the *local* versus *global* minimum problem. It is assumed[6,7] that the conformers of physical interest used in QSAR studies correspond to the best few local energy minima. Molecular modeling studies that omit significantly populated conformers are doomed to failure. This problem has prompted the development of algorithms to search for a global minimum. Molecules, however, may populate different conformations depending on the environment in which they exist. Moreover, they can shift among the conformations depending on energetic and environmental conditions or the specific interaction in which they take part. The available force field or quantum-chemical calculations are usually only applicable under the basic conditions of the isolated molecule in the gas phase. Thus, they do not take into account the neighboring effects of other molecules including those of solvents. The proper treatment of solvation effects, however, is still a research frontier.[8–10]

In spite of all efforts to reach the set of energetically favorable structures (or the global minimum conformer), it is important to recognize that the most stable conformations of a molecule are not always the active ones. For example, because of the requirement for an antiparallel arrangement of the leaving groups, it is the less stable axial conformer of cyclohexyl bromide (or p-toluenesulfonate) that undergoes the elimination (E2) reaction, instead of the more stable equatorial conformer.[5] Analogous principles hold for enzymatic reactions, where enzymes may change their conformations either by activation through external molecules or by contact with the substrate, according to the "allosteric" effect (induced fit). For example, in order for acetylcholine to bind with the nicotinic receptor, it adopts a conformation distinctly different from its conformation in solution or in the crystal state.[11] Similarly, in a recent molecular modeling study of steroidal estrogens,[12] it was found that the transition barriers between different conformers predicted by the force field (MMP2) are less than the free energy of receptor binding. This allows particular ligands to mimic the conformation of the steroid binding site (E_2-receptor) taking a more planar shape, similar to the relatively flat E_2-urea.

3.2. ALGORITHMS FOR 3D MOLECULAR DESIGN

Usually the systems providing a 3D molecular design have been created as specific subroutines in larger commercial packages. The lack of powerful software in the public sector prompted the development of a variety of algorithms for a rational 3D molecular modeling which will be discussed in forthcoming. Examples of such methods are algorithms providing the generation of all configurational stereoisomers[13] corresponding to a given chemical formula, conveying in a tabular or other symbolic form only the stereospecificity (parity) of the stereo and chiral centers or the configurational (*cis* or *trans*) orientation of double bonds. The algorithms of Wipke and Dyott,[14] Hanessian et al.,[15] Nourse et al.,[16,17] and Sasaki and his co-workers[18] can be related to that group of approaches.

3.2.1. 2D Input Geometry

The starting 2D geometry required for the 3D modeling will be defined more precisely here. In the first edition of his classical book, E. Eliel[19] wrote that "the structure of a molecule is completely defined by the number and kind of atoms and the linkages between atoms." This general statement is consistent with the fact that molecular topochemistry (topology combined with the type of atoms and bonds) is enough to completely define the conformational space of molecules. If one designs a particular stereoisomer (point of this space), however, additional information for the stereospecificity of atoms and bonds is required. Accordingly, the existing algorithms for obtaining 3D structures can either proceed from molecular topochemistry only or, depending on the generation purposes, the topochemical information is combined with the required stereospecificity.

3.2.2. Approaches to 3D Molecular Design

In this section we are aiming to classify the concepts for 3D structure design.[20]

As illustrated in Figure 2, one of the standard ways to generate the 3D geometry of a molecule is by using numerical procedures such as the conventional force field or quantum-chemical packages. Proceeding from molecular topochemistry, one attempts to produce the precise geometry of a local energy minimum conformer of a molecule. These methods, however, are extremely time-consuming and not convenient for handling large sets of molecules. Moreover, the timing and success of their perform-ance strongly depend on the initial coordinates.

In general, to solve the problem of obtaining energy-optimized 3D models of a molecule, one could define different criteria of achieving computational speed and/or structural accuracy. For example, one can reduce the time needed for the numerical calculations at the cost of simple (for the user) input or nonperfect algorithms

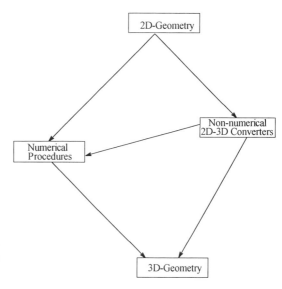

Figure 2. Numerical and nonnu-merical approaches to 3D molecular modeling.

generating strained 3D models for a subsequent optimization. Wipke and co-workers[21,22] improved the effectiveness and convergency problem of energy minimization procedures by providing some pseudo-3D coordinates (in their *SYMIN* and *PRXBLD* programs). The information input for the numerical model builder (force field geometry optimization procedure) is taken directly from the structure drawing on the graphics terminal. Pseudo-3D (screen) coordinates are generated, proceeding from the location of the atoms with respect to the plane of the screen by using the special bond symbols ("up" and "down"). The solution of Wipke and co-workers (and related approaches) does not provide quite satisfactory 2D–3D conversion since it is very sensitive to the information not included in the graphical representation of molecules. Various other algorithms convert 2D information taken from graphic screen ($x,y,0$ coordinates and configuration qualifiers) to an (x,y,z) set. For example, computer systems like MOLY,[23] MACROMODEL,[24] CAMSEQ,[25] MMMS,[26] and TUTORS[27] use similar approaches.

Other 3D modeling algorithms can sacrifice optimization accuracy for enhanced speed by applying simple potential energy functions. Levitt[28] has proposed such an empirical function as the sum of strain energies resulting from stretching, bending, interaction between nonbonded atoms, and torsional angles. Nonbonded interactions are calculated between pairs of atoms separated by at least three bonds and closer than a cutoff distance. Within the TUTORS system of Sasaki *et al.*,[27] another term has been introduced for minimization of strain energy related to steric configurations of atoms. The latter are determined by the 2D screen coordinates of the input structural diagram. Another force field scheme for energy calculation omitting the electrostatic terms in the potential energy function is that of Winter *et al.*[29] Because of the simple force fields used in the above algorithms, these should be considered as geometry strain relief procedures or pseudo-molecular mechanics (PMM), rather than molecular mechanics methods. The simplified character of the energylike function significantly accelerates the optimization procedure at the cost of the relatively crude geometries of the obtained 3D models.

Because of the above discussed drawbacks of numerical methods, rapid nonnumerical 2D–3D converters were recently developed. In general, they proceed from molecular topochemistry and produce a reasonable 3D geometry. The 3D models obtained can either be used directly for structure–property relationship studies or further refined by numerical methods. The concepts implemented in nonnumerical 2D–3D converters (or 2D–3D converters only, for brevity) are classified in Figure 3, similarly to the classification presented in Ref. 20.

The interactive building, although computerized, is still based on the ideas developed at the beginning of three-dimensional thinking in organic chemistry. The user constructs a 3D model interactively, positioning each atom and bond on a 3D graphics interface using standard bond lengths, valence and torsional angles, and joining fragments with predefined geometries. In contrast, the "automatic" methods transform 2D structure of molecules into 3D models directly without any user intervention. These methods can be divided into "template (or knowledge) based,"

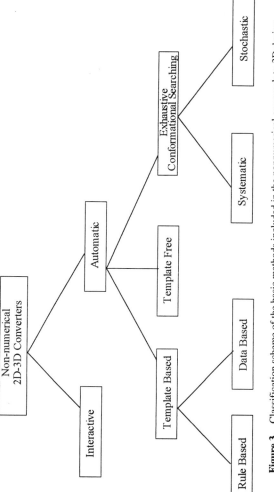

Figure 3.   Classification scheme of the basic methods included in the nonnumerical approach to 3D design.

"template-free," and "conformational searchers." In turn, the template-based method could be subdivided into "rule based" and "data based." They both generate a set of low-energy conformers of a given molecule proceeding from its atom connectivity.

The knowledge-based methods use the chemical knowledge on fragment and/or molecular geometry and energy. This can be done explicitly (e.g., by rules in rule-based methods) or implicitly (e.g., using data on allowed conformations in data-based methods). It is difficult to make a distinct separation between rule-based and data-based methods. They both use geometries of fragment units from a library, and assembling rules. The methods are specified as data-based in case of extensive use of implicit knowledge methods, usually taken from large data bases. However, explicit rules are also used here on the fragmentation of input structures, on detecting the closest analogues in the data bases, and on assembling the fragments found during the search. On the other hand, rule-based methods are dominated by knowledge rules during model generation and do not share extensively geometric information from libraries, where geometries of fragments as large as single rings are usually stored.

Similarly to the knowledge-based methods, the template-free generation algorithms proceed from connectivity information. They are also based on knowledge of molecular geometry, such as bonding and nonbonding ranges, valence and torsional angles, and constraints for configuration of specified chiral centers. However, in general, they do not use templates. Their basic goal is the generation of a conformer or a set of conformers corresponding to the geometric constraints imposed on the generation procedure.

Distinct from template-based and template-free algorithms, the methods for conformational search perform an exhaustive search of the conformational space of the molecule. Generally, they proceed from one (usually low energy) starting geometry of the molecule. The systematic conformational search methods provide a deterministic search that covers exhaustively all areas of conformational space, whereas the stochastic methods employ random elements in exploring this space. Recently, methods for conformational search were developed (e.g., 3DGEN) proceeding from molecular topochemistry as well. Usually, template-free and conformational search methods meet computational (e.g., distance geometry) and/or combinatorial (e.g., internal coordinate tree-searching procedure) difficulties, requiring a relatively large computing time.

3.2.2.1. Interactive Methods

According to the basic idea developed in these procedures, 3D molecular models are designed interactively starting from a 3D fragment (or atom). The positions of atoms and bonds are specified sequentially on a 3D graphics interface by using standard stereochemical data. Examples of such systems include MAGIC,[30] MOL-BUILD,[31] CHEMMOD,[32] ALCHEMY,[33] MACROMODEL,[24] and OASIS.[34,35] Analyzing the capabilities of these programs, one can distinguish the following more significant 3D modeling procedures: substitution of a single (terminal) atom by a mono- or polyatomic functional group; editing of a 3D model with respect to atom types and geometric parameters; adding single atoms by specifying a bond length, bond angle, and torsional angle; breaking or creating bonds; joining two molecules or fragments into a single molecule by pointing out the terminal atoms and the lengths of junction bonds; automatically including hydrogen atoms; cyclization of an acyclic fragment into a ring and vice versa; and so forth. One of the main features of these systems is the user's control of conformational and other geometric characteristics of the growing molecular structure. This allows one to construct any conformation or nontrivial (transition state) 3D models.

3.2.2.2. Template-Based Methods

a. Rule-Based Algorithms. This approach to 2D–3D conversion utilizes models and/or symbolic knowledge from a library of templates. A set of rules govern the assembly of template fragments into complete molecules.[36–42]

The first attempt to generate a molecular 3D model was made by Wipke and co-workers, introducing their program PRXBLD[22] as a module of the SECS synthesis planning system.[21] A knowledge from a heuristic part is combined here with a simplified force field method. Being interactively driven, the program has nowadays only a historical significance.

Cohen's SCRIPT algorithm[37] is one of the most typical among this group of approaches. Two main principles are applied here. The first one is the employment of a "conformational diagram" symbolism, where the sign of torsional angles for each ring and chain fragments of the molecule are described. These symbolic representations of the conformational fragments of the molecules are coded in a special template library. They provide a logical basis for the assembly rules depending on hybridization and stereochemistry of the atoms and the type of ring junction. For example, the symbolic representations of the two possible conformers of one tricyclic molecule and the allowed conformational assembly rules for the respective hybridization of the carbon–carbon junction are illustrated in Figure 4.

Once the molecule is designed on a symbolic level, then the 3D structures of the conformational surface are directly constructed by using the 3D coordinate representations of the symbolic models, which are also coded in the template library. Cohen's procedure and related ones,[36–42] however, are limited by the size and the nature of the sets of templates and assembly rules. The latter are not large enough to provide 3D building for complicated molecules such as bridged and polycyclic systems. For example, it is known that Cohen's SCRIPT system often fails in generating 3D models of polycyclic molecules with more than three cycles.

Other template-based model builders try to account for the intra- and intermolecular forces of a molecule, thus aiming to discover a set of 3D structures that would satisfactorily represent its low-energy conformers. The basic principles of these methods are embodied into the computer systems WIZARD[40] and COBRA.[42] The mole-

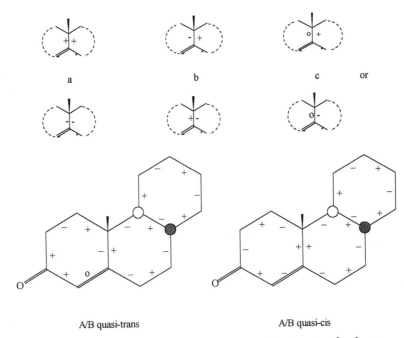

A/B quasi-trans   A/B quasi-cis

Figure 4. Symbolic representations and conformational assembly rules for sp^3–sp^2 carbon–carbon junction of a tricyclic molecule (Ref. 37).

cules are partitioned into conformational units about whose conformational behavior (e.g., energies, coordinates, flexibility) the systems have expert assessments. It is assumed that the conformational properties of each unit do not depend on the molecules in which they appear (Figure 5). Once the conformational units are recognized, a search of the conformational space is performed by examining all possible combinations of low-energy conformation units. The structural combinations are first produced at an abstract level. The symbolic suggestions are then examined to check for possible violations of assembling rules. If no problems are detected, the 3D coordinate representations of the conformational units are read in from the template library and joined in a sequential fashion. The obtained suggestions are again critically examined according to possible geometry violation of common atoms and bonds, van der Waals repulsion, and other intramolecular forces.

One of the recently developed rule-based modeling programs, CORINA[20] (see also Refs. 38–41), deserves attention being applicable to the entire range of organic chemistry including reactive intermediates, macrocycles, and organometallic compounds. The program serves as 3D model builder for the reaction prediction system EROS.[43] Bond lengths and bond angles are set to standard values parameterized for the entire periodic table. The following steps are performed in generating a 3D model.

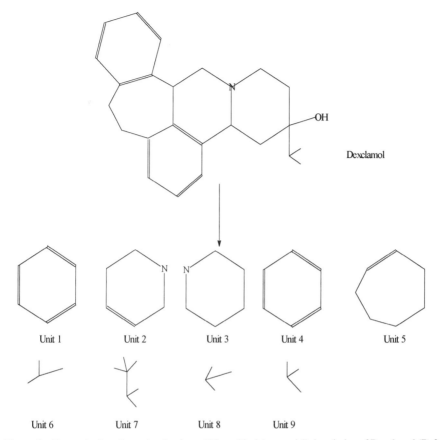

Figure 5.   Recognized conformational units and hierarchical (sequential) description of Dexclamol (Ref. 42).

First, the molecule is fragmented into cyclic and acyclic parts. The exocyclic atoms are included in the ring systems since their positions and long-range interactions are conditioned by ring conformations. Rings belong to the same ring system if they have at least one common atom. A library of allowed single-ring conformations is used to handle small-ring systems since rings of sizes three to eight have a limited number of available conformations. Symbolic representations of these templates, including ordered sequences of torsional angles, are stored in the template library. Moreover, the templates are ordered by strain energy values. To find the conformations of a cyclic part, a backtracking algorithm is used exploring the possible combinations of the conformations of the single rings. Central rings are processed first because of their highest long-range interactions with neighboring rings. Each combination of ring conformations is critically examined according to assembling rules. Finally, a list of symbolic representation of the conformations of the ring systems is generated that satisfies the requirements of these rules until exhausting all structural combinations. These conformations are ordered by determining their pseudo-energy assessments (evaluated by an energylike function). The best conformations (having lowest values of these energies) are then translated into 3D atomic coordinates using standard bond lengths, valence and torsional angles. The resulting approximate 3D models of cyclic fragments are further refined by a pseudo-force field calculation (*vide infra*). The latter seeks to optimize the geometry of generated structures instead of their energies. Acyclic chains are designed by maximizing interatomic distances, following the assumption that the longest pathways usually yield lower-energy structures.

Although CORINA provides a reduced conformational analysis while exploring the conformational space of cyclic fragments, it is basically a single low-energy conformer generator. Recently,[20] the program was tested using a large data set of X-ray structures from the Cambridge Structural Database. CORINA has modeled most of the ring structures with an average RMS deviation[44] of less than 0.3 Å.

b. Data-Based Algorithms. Wipke and Hahn[45–47] have proposed a template-based approach for building molecular models using the "knowledge" on 3D structural geometry captured by the experiment (crystallography). The basic assumption is that the large collection of experimental geometries contains the knowledge on how to build molecules. The user should be able to easily perform this reasoning by analogy on the basis of information on similar problems. The respective AIMB program uses the Cambridge Crystallographic Database as a knowledge base. The system "perceives" the query structure identifying its rings, chains, aromaticity, and stereochemistry. Close analogues are searched in the data base. If not found, the "decomposer" continues its work until the fragmentation products are detected in the data base. The cyclic units can be decomposed up to elemental single rings; chains, however, may be broken up to simple atom pairs. The subdivision is based on an assumption that interactions exist within the units only, whereas those between units could be omitted. The matching criteria during the search for analogues in the knowledge base are varied gradually. If no exact correspondence of a unit can be found, the matching tolerance is increased until an analogy is obtained. The different level of abstraction during the search guarantees early detection of the best analogues. Various attributes are included assessing similarity between a target and an analogue in order to select the best analogy, such as atom number and type, charge, valence, hybridization, and stereochemistry. The analogies found for the units are assembled to a coordinate representation of the original molecule. Although the system does not employ any energy evaluation procedure and does not take into account long-range interactions, it rapidly builds a low-energy conformer via knowledge of experimental 3D structures. Since several analogies can be selected for a given unit, a conformational search may be performed. A limitation of the AIMB program (and related ones) is that the quality of the designed model strongly depends on the quality and amount of the knowledge accumulated in the data base. The necessity of a "rich data base" requires in turn a significant amount of disk space.

Other knowledge-based methods eliminate this drawback by discarding from the data bases the redundant information related to long-range distance interactions. Thus, Ai and Wei[48] have recorded in their data bases conformational units including bond lengths, valence and torsional angles corresponding to the center atoms in the specific environments of nearby atoms. After detection of acyclic and cyclic conformational units, the latter are constructed by using the distance geometry method. The joining part of the method is also based on the knowledge base.

Another popular 3D builder is that included in the Chem-X package developed by Chemical Design Ltd., assembling fragments retrieved from a data base.[49,50] Two types of stored knowledge are used here. A relatively small library contains several hundred preoptimized common specific rings (carbocycles and heterocycles). The model builder first checks exact matches for cyclic substructures in this library. If no exact matches can be detected, generalized ring fragments with different saturation and stereospecificity are taken. Various torsional angles are used for the construction of acyclic parts in order to generate the main chain in an extended form. If more than one ring is found in the data base, a conformational space sampling can be obtained. Different stereoisomers may be generated using a special handling of stereochemistry. The program is more general than AIMB since the acyclic fragments are generated automatically instead of taking them from the library.

3.2.2.3. Template-Free Methods

The distance geometry method introduced by Crippen and Havel[51-53] and recently extended by Waldman and Havel[54] is the standard template-free procedure for molecular design. The method generates a 3D model of the molecule satisfying (1) distance constraints imposed by the distance ranges for bonded and nonbonded atoms and (2) stereo constraints expressed by the configuration designation of stereocenters. The distance geometry approach uses any geometrical information that can be translated into minimum/maximum constraints on interatomic distances, torsional angle restriction (e.g., from ^{1}H-NMR coupling constants), or imposed spatial relationships of potential pharmacophores. The distance and configuration constraints, however, do not specify a unique 3D structure, and hence a set of conformers is generated maintaining input stereochemistry.

First, an initial set of coordinates is computed from the first three largest eigenvalues and their corresponding eigenvectors of the distance matrix with respect to the mass center of the molecule (assuming unit mass for all atoms). The truncation to the subspace defined by the largest eigenvalues usually leads to coordinates that violate the distance and configuration constraints. This problem is solved by local minimization of a *penalty (error) function*, E_{err}, which consists of a sum of terms, one for each constraint:

(1) $E_{err} = F + C$

The first term reflects the violation of the distances from the constraint ranges, whereas any configuration of a particular chiral center is specified by the second term of the error function.

Although well accepted by 3D molecular modelers, the distance geometry method has specific limitations. Thus, the atoms are placed randomly within the allowed distance and configuration ranges, and this produces random sampling of the conformational space. For the same reason, the optimization procedure ends nearly always in a local minimum (greater than zero penalty function). Inconsistency in the choice of the distance matrix elements could result in severe distortion of the modeled geometry, especially when large rings or long chains are generated. Moreover, the lack of an energy penalty function of the generated structure can yield chemically erroneous models in the case of highly strained molecules. Additionally, because of the invariant character of that matrix with respect to the permutation of substituents at stereocenters, it is impossible to generate models with a prescribed geometry for chiral structures. The configuration of stereocenters is adjusted during the geometry optimization stage of model building.

Historically, the first practical realization of the distance geometry method is the EMBED system. Different variations on this system have been developed,[55,56] the most popular of which is DGEOM.[57] Linearized embedding approach[58] is another development of the distance geometry method in which all of the constraints are organized into a linearized representation of the molecule. Some other 3D modeling developments are based on distance geometry, as well.[59,60] Weiner and Profeta[61] proceed from screen coordinates applying the distance geometry approach in the next step. The MOLGEO program[60] uses only the geometry optimization part of the method, replacing the embedding method (often resulting in rather poor starting geometry) by a depth-first-search (DFS) procedure for constructing better initial coordinates. This procedure also has the advantage of being much faster since it does not require eigenvalues/eigenvectors to be evaluated. The DFS algorithm is a backtrack procedure used in MOLGEO for systematic search in conformational space imposing stereochemical constraints at the early stage of model generation. Using standard values for bond lengths ($E_{1,2}$) and valence angles ($E_{1,2,3}$) and omitting dihedral angles ($E_{1,2,3,4}$) and nonbonded interactions (E_{vdw}, having small importance), the authors assume that the approximate energylike function ($E = E_{1,2} + E_{1,2,3} + E_{1,2,3,4} + E_{vdw}$) will be conditioned only by a self-consistent set of torsional angles, satisfying stereochemical constraints, and defined by using the combinatorial enumeration technique. At every kth step of the search, the program generates a new partial conformation defined by the sequence of torsional angles $w_1, w_2, \ldots, w_k$. Next, Cartesian coordinates of the partial model are evaluated by using these torsional angles, combined with standard values of bond length and valence angles. If the geometry obtained meets the stereochemical criteria, the program recursively passes on to the next $(k + 1)$th level. Otherwise, the system scans the conformational space by perturbing w_k by an increment Δw. It was shown[60] that the MOLGEO method often results in the generation of local energy minimum conformers. Of course, the stereochemical constraints (i.e., the configurations at double bonds and stereocenters, and constraints for torsional angles) have to be specified in the input, otherwise the system generates a random stereoisomer of the molecule.

Pearlman's CONCORD system[62–64] is another elegant template-free method for rapidly generating good-quality 3D structures, proceeding from molecular topology. The building module of the CONCORD algorithm combines aspects of a knowledge-based (expert system) approach and PMM. It uses a simple data base containing a detailed list of bond lengths and valence angles, accounting for atom hybridization states and first sphere neighborhood. CONCORD identifies a gross conformation of each ring within each cyclic system in the molecule by making use of a logical analysis. A general conformation (e.g., "chair," "boat") is assigned to each ring. These general conformations reflect the constraints imposed by the presence of the other rings. Then, PMM is performed for the individual rings in such a way as to retain those gross conformations while optimizing the valence and torsion angles. The PMM optimization procedure is implemented for each ring according to a certain priority order. The PMM algorithm is similar to the energy minimization by molecular mechanics calculations. Unlike the conventional energy minimization procedures, where the torsional and bond angles vary independently, in CONCORD they are coupled in a novel composite strain function. Minimization over this single variable energylike function is fast and provides chemically reasonable geometries. In a next phase, a collection of rules based on very fast logical analysis administers the decisions regarding acyclic bonds and torsional angle values of the building algorithm. CONCORD optimizes the values of some of the dihedral angles associated with the rotatable bonds by minimizing the force field energy. Only those rotatable bonds are optimized, which are involved in paths connecting close-contact (nonbonded) atoms (e.g., minimization of the steric interactions of the largest 1,4-interactions).

CONCORD is able to produce good-quality structures for most organic compounds, including those with complex heteroatom functional groups and ring systems (bridges, cages, and fused ring systems). It is one of the most frequently used 3D model builders that is available publicly. CONCORD was used to build 3D models for Lederle (225,000 structures), a large industrial data base[65] and has been used at Chemical Abstracts Service to generate approximately 5 million models. CONCORD's strength lies in its speed, being at least an order of magnitude faster than force field algorithms.[66] Its major weakness appears in the case of flexible structures.[66] Often it simply fails. In other cases, it makes ad hoc decisions when faced with different conformational possibilities. The ultimate goal of CONCORD is the generation of a single conformer and it cannot be used for conformational search.

Recently[67] (see also Refs. 68–70), genetic algorithms (GAs) are used to find a reasonable ensemble of low-energy conformations for a given molecule (up to 12 rotatable bonds) rather than to extensively enumerate all of the conformational space proceeding from the lowest found conformation. The GAs are optimization methods based on analogies with certain strategies of the biological evolution. One of these strategies is related to a breeding population in which individuals who are better "fit" in some sense have a higher chance of producing offspring and of passing their genetic information onto succeeding generations. Another strategy is based on crossover in which a mixture of the parents' genetic material conditions that of successors. Finally, the genetic materials can occasionally be corrupted to maintain a certain level of genetic diversity in the population. We present here a brief introduction to the canonical GA method. The latter works with a population of "individuals" (3D structures) who will interact through genetic operators in order to perform an optimization process. The three principal genetic operators used in the procedure are selection of parents, crossover, and mutation, corresponding to the above three evolution strategies, respectively. The conformation energy is usually used as a fitness function. Often, GA methods start with a template conformation file for the molecule (crystallographic geometry). The dihedral angles to be varied are chosen. The energies are evaluated by a force field method. At the end of the "evolution phase," the best structures found are additionally optimized using a steepest descent gradient minimization routine.

3.2.2.4. Exhaustive Conformational Searching Methods

The algorithms of this group were developed to provide an exhaustive search of the conformational space. Most of them produce crude starting geometries, which are then optimized by any of the numerical methods for energy minimization. At any stage, the current minimum energy conformer is compared with previously found conformers to check for possible duplication. The search of the conformation space is terminated when all starting geometries have been treated. Since the numerical methods simply refine the starting geometries, the overall effectiveness of these algorithms depends on the quality of starting geometry generators. The latter, in turn, can be divided into two broad classes[71]: deterministic searchers, systematically exploring all areas of conformational space, and the stochastic (Monte Carlo) class which uses a random element in the searching procedure. The first category of methods generates all combinations of selected values for all rotatable torsional angles to produce starting geometries for the next optimization. Because of their "absolute" character, the algorithms of this group generate an extraordinarily large number of starting structures when conformationally flexible molecules are analyzed. The Monte Carlo methods do not search the conformational space in a completely random manner. Usually they start with stable conformers and limit their conformational exploration by using random or pseudorandom variation of molecular geometry. Such a random search of Cartesian space is described by Saunders.[72] His approach generates new starting geometries using random translations of the previous conformer. The most serious problem inherent in the random search is the dense population in the high-energy region. Even for small molecules, the high-energy space was found to be more crowded than expected.[73] The advantage of the approach, however, is that it allows conformational searches of molecules of any size optimizing the density of random points.

In general, to cover exhaustively the conformational space of a flexible molecule is a difficult task. Realistic approaches were recently developed that are neither totally systematic nor completely random. They are based on the idea of covering a limited portion of the conformational space, searching exhaustively only the chemically reasonable, low-energy regions. A good example of this approach is the algorithm of Goto and Osawa, introduced recently.[74] Starting from any conformation, local perturbation is systematically applied to all flexible fragments of the molecule to produce new potential conformers. The perturbations include flapping and/or flipping endocyclic bonds and stepwise rotation of acyclic bonds. The flip/flap perturbation changes three to five contiguous dihedral angles, based on the assumption that good conformations are surrounded by other compatibly good conformations differing only in local structural variations of the conformational space. Because of that a systematic search is performed varying molecular segments only ("variable segmented systematic search"). The algorithm provides, first, the identification of the global energy minimum (GEM) structure of the starting domain of conformational space by always choosing the most stable of the conformers, produced in the last perturbation cycle, as the next initial structure. Once the GEM of the domain is reached, higher-energy regions are explored by local perturbations to define all low-energy conformers therein.

Another grouping of the various conformation search methods is based on the coordinate system in which they operate. For the potential advantages of the external (Cartesian) and internal (bond lengths, bond and torsional angles) systems, the reader is referred to Refs. 71 and 75.

A different technique used for exploring conformational space is the molecular dynamics method.[76] While it is effective for searching local conformational space, it requires remarkably more computer time than the other methods for global search problems.[71]

One of the most flexible systems for exhaustive conformational space analysis is the internal coordinate tree-searching procedure introduced by Lipton and Still.[75] It generates the starting geometries (for subsequent optimization), proceeding from a local minimum of the conformational space, by rotating torsional angles around all rotatable bonds. All possible combinations of rotamers at some torsional angle resolution are thus analyzed. In an acyclic system one alters succeeding dihedral angles starting from one end and traveling down the chain. In cyclic systems, rings are temporarily opened according to certain rules, forming a 3D spanning tree which is then treated as the acyclic case but with additional constraints permitting ring closure. Only those structures that pass geometrical tests introduced specially, such as the nonbonded cutoff (distance between nonbonded atoms) and ring-closure parameters, to reject high-energy molecular geometries are retained. Thus, the number of the generated geometries is limited while retaining a set (as complete as possible) of low-energy final conformers.

We have recently proposed[77] a template-free algorithm, termed 3DGEN, which provides generation of all but the high-energy conformers, proceeding from molecular topochemistry only. More specifically, 3DGEN exhaustively searches the conformational space "escaping" from its high-energy subspaces, defined by ad hoc geometric and/or expert constraints. The system allows the identification of all possible stereoisomers, as well. First, an initial structural analysis of the molecular topochemical information is performed, including: a ranking of atoms according to their type, extended connectivity and multiplicity of incident bonds; recognition of the smallest rings incident to each pair of neighboring bonds (to specify the valence angles between these bonds, because the size of the smallest rings are crucial for the angle between two bonds); and identification of the potential stereocenters. The tertiary and quaternary carbons, as well as carbon–carbon double bonds are handled as potential stereocenters. Then, the unique topochemical naming is obtained, including the set of the "best" topochemical numberings (according to the topochemical symmetry of the molecule) by using the DFS procedure.

The core of the algorithm is the original idea named "propagating 3D spanning tree." As a 3D spanning tree (3DST) we term an acyclic 3D molecular skeleton (3DMS). Its construction starts from a particular atom (specified by the atom ranking). At any step of the algorithm one generates a subsequent bond of the 3D spanning tree. A current bond of 3DST is positioned in the space by using a recursive procedure based on the 3D information of its already constructed neighboring bonds: the type and hybridization of consisting atoms, incidence to (a)cyclic fragments, and the size of the smallest ring(s) (if any) to which they belong. Bond lengths and valence angles are adjusted in advance by using MMPI parameterization. During the propagation, 3DST loses its acyclic character by closure of rings and eventually transforms into a 3DMS. All possible rotamers at allowed torsional angle resolution are explored, retaining those 3DMSs that pass the tests for geometric constraints, including nonbonded cutoffs and ring-closure parameters. The number of generated isomers can be controlled by managing the values of geometric constraints.

The next stereochemical and conformational naming provides identification of the discernible stereoisomers, enantiomers, and conformers. DFS procedure is performed assigning n numberings of the atoms of each 3DMS, where n stands for the number of the "best" topochemical numberings. Based on the geometric configuration of each potential stereocenter in the 3DMS, one determines the real stereocenters and the "best" stereonames. The set of "best" stereonames, each related to one of the 3DMSs, is further checked for degeneracy. The redundant names (and corresponding 3DMSs) are eliminated from the list of stereonames, but preserved for the next conformational analysis. The recognition of any existing enantiomeric relationship is performed by comparing the "best" stereonames of each distinct 3DMS. The next performance of DFS provides a complete conformational specification of the 3DMSs. The values of all torsional angles are coded. Again, the "best" conformational names are chosen to represent each 3DMS. The redundant conformers (with degenerate names) are eliminated.

In the case of strained cyclic structures, the exhaustive generation of 3D isomers usually requires looser geometric constraints. Thus, some of the conformers obtained could be distorted with respect to the reference geometric parameters. In this case an original strain minimization technique (PMM) is applied based on a simple energylike function, where the electrostatic terms are omitted.

There are conformational search methods struggling with the problem of finding the GEM structure, which is an extremely difficult problem in dealing with flexible molecules of any size.[78,79] Usually one starts with a trial structure, refined to a local minimum. Then a randomly selected torsional angle is rotated by a randomly selected increment. After this random "kick," the resulting structure is optimized again by energy minimization. Thus, one can either return to the initial conformation or one can obtain a new one, which can be used as a starting point for the next random torsional angle rotation. This cycle is repeated until the global minimum is reached. The energy of each distinct conformation is usually evaluated by force field techniques.

Let us follow the above review on the approaches to the 3D molecular design with a brief summary on their ultimate goals. Most of the methods (e.g., CONCORD, MOLGEO, AIMB, EMBED, DGEOM, CORINA) produce a single low-energy conformation, which is further refined using quantum-mechanical or force field methods. The minimum energy conformations are justified in that such conformations correspond well with crystal structures. Some of the methods are even hunting for the lowest-energy conformation (GEM). Besides the cases of the simplest of molecules, however, it is unlikely that the single global minimum found by a gas-phase quantum-mechanical or force field approximation will remain as such, once the effects of entropy and solvation are included. The properties of real systems usually represent average effects resulting from a multitude of populated low-energy structures. Hence, regardless of the systems under investigation, the set of all populated minima (or a representative sampling) should be searched for structure–property studies. This requirement prompted the development of algorithms for an effective conformational space search as tools for a further selection of active conformers, which we believe are subsets of all populated minima. Moreover, these subsets are specific for particular interactions, processes, or behaviors of the modeled molecular systems. In general, they should differ within the subsets of the most significantly populated minima, because the most stable conformations might be the least likely to interact with solvents, receptors, or macromolecules. More importantly, solvation and binding interactions might more than compensate for energy differences among many conformers of one and the same compound.

3.3. ACTIVE CONFORMER SEARCHES WITH THE DYNAMIC METHOD

As described in the preceding section, the conventional translation of structures from topology to 3D geometry generally selects one or a few low-energy conformers. As shown, the literature is replete with examples of chemical reactivity involving conformers other than the lowest-energy one. In complex reaction environments (especially in biological systems) or with solvents of different polarity one should expect that the molecule can take the form of different conformers depending on the particular interaction step, such as penetration, substrate–receptor complex formation, or stereoelectronic interaction. Moreover, the specificity of the processes makes particular subsets of conformers "real" ("active") for the different end points under investigation. The identification of those conformers, however, is a difficult problem, especially when there exists an easily attained equilibrium among them. At the present time there is no general approach identifying the "real" conformers among computed ones.

A new method for the selection of active conformers was recently introduced by us,[80] named conventionally "dynamic" (or "multiple") since we attempt to find a set of "static" conformers in order to mimic the multiplicity of 3D isomers involved in different reaction stages. Implemented in the OASIS computer system for QSAR analysis,[35] the method combines a conformation generation routine with a routine for conformer screening. It begins with an exhaustive set of conformations and allows one to narrow the conformation list based on the problem being addressed, by using a hierarchical set of expert rules for screening the generated 3D models according to the chemical expertise. The stereoelectronic structure of each conformer from the initial set is assessed by quantum-chemical methods. Usually the MOPAC package is applied in a combination with the OASIS routine for the calculation of stereoelectronic indices.[34,35] The molecular descriptor array can be computed either after geometry optimization or by direct single point (1SCF) calculations of the unoptimized conformations. In many instances when highly strained structures are not involved, the stereoelectronic indices obtained without geometry optimization may be sufficient for estimating the chemical activity. Then, in an interactive mode, the user has the opportunity to introduce working hypotheses in the selection of conformations. We have often found great variation in stereoelectronic indices among different conformers of a given molecule which shows that the selection of active conformers appears to be as crucial for QSAR analysis as the selection of suitable molecular descriptors. The OASIS system contains 3D geometry-dependent physicochemical and steric indices such as the van der Waals volume and surface area of the molecule, maximum interatomic distance, sum of all interatomic distances (geometric analogue of the Wiener topological index and its informational counterparts[81]), as well as stereoelectronic reactivity indices. Some of the electronic indices include global estimates (such as the heat of formation, total electron energy, energies of frontier orbitals and the HOMO–LUMO gap, dipole moment, volume polarizability) and local parameters

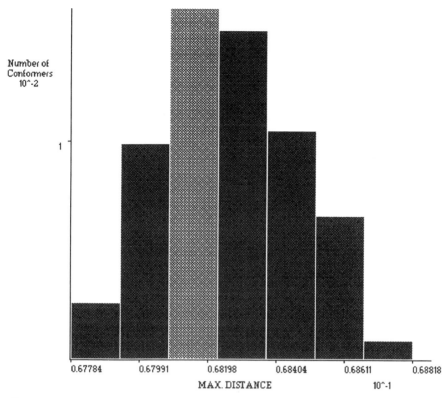

Figure 6. Conformer distribution of 2-decyn-1-ol based on the maximum geometric distances in the molecules [Lmax (Å) on the abscissa]. The number of quintals is seven.

(such as atomic charges, bond orders, frontier atomic charges, atomic self-polarizabilities, donor and acceptor superdelocalizabilities, and frontier donor and acceptor superdelocalizabilities).

Conformers can be selected based on their distribution according to specific geometric and/or physicochemical and/or stereoelectronic parameters. The conformer distribution (number of conformers belonging to a certain range of parameter values) is displayed as a selected number of parameter windows, evenly dividing the whole parameter range (Figure 6). There are a variety of selection schemes, which can be applied after examining the conformer distributions, providing:

1. *Representation of the whole range of parameter variation.* Here, the number of conformers is diminished by selecting a single conformer representing each window, thereby avoiding the loss of representation for the whole range of parameter variation.

2. *The prevailing values of the parameter.* Here, one takes all conformers belonging to the most populated window. Apparently, these conformers will provide the prevailing range of parameter values for the studied molecule.

3. *The extreme values of the parameter.* Alternatively, one can select conformers providing the extreme (maximum or minimum) values of the parameter(s) under investigation, specified by the outermost windows. Thus, for example, one can select lowest-energy conformers having minimum values of heats of formation. One can also obtain the conformers with the highest electron acceptor (donor) properties, taking those having minimum (maximum) values of E_{LUMO} (E_{HOMO}) or maximum (minimum) values of acceptor (donor) super-delocalizabilities or the respective frontier charges.

4. *Certain populations of energy level.* Another option for conformer selection provides certain percent of energy level population ($P^i,\%$), calculated by Boltzmann's distribution formulas according to the respective heats of formation ΔH_f^i (at $T = 298$ K):

$$(2) \qquad P^i[\%] = \exp[-(\Delta H_f^i - \Delta H_f^1)/RT]/\Sigma^n \exp[-(\Delta H_f^i - \Delta H_f^1)/RT] \cdot 100$$

5. *Parameter weighting.* In the first three screening methods above, one assumes that the conformers chosen are equally weighted. In an alternative approach, each compound in the studied series can be represented by a single record, where all parameters are weighted according to the values for the different conformers, V^i, and the energy level population provided by the respective conformers, $P^i/100$. Four different types of averaging methods including arithmetic, geometric, quadratic, and harmonic methods are provided. Thus, the following formula for mean arithmetic type of parameter averaging holds:

$$(3) \qquad V^{av} = \Sigma^n V^i \cdot P^i/100$$

For example, if one denotes by $E_{LUMO}^1, E_{LUMO}^2, E_{LUMO}^3, \ldots, E_{LUMO}^n$ the E_{LUMO} values for the conformers with the respective energy level population $p^1, p^2, p^3, \ldots, p^n$, then:

$$(4) \qquad E_{LUMO}^{av} = \Sigma^n E_{LUMO}^i \cdot p^i$$

3.4. APPLICATIONS OF THE DYNAMIC METHOD

3.4.1. Ah Receptor Binding

The efficiency of the dynamic approach for active conformer selection is illustrated by presenting the different schemes of deriving QSAR for aryl hydrocarbon receptor (AhR) binding [$\log(1/EC_{50})$] of 14 polychlorinated biphenyls (PCBs).[82] The 3DGEN system generated 103 conformers for the 14 compounds analyzed, at reasonably specified geometric constraints. A stacking type of interaction with AhR was

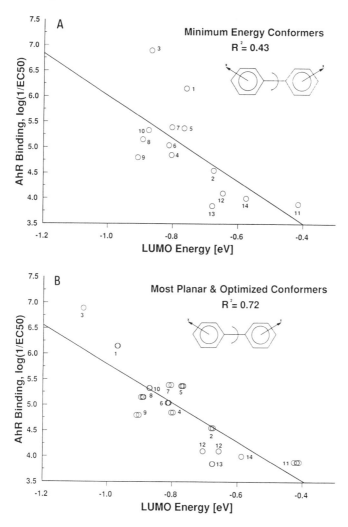

Figure 7. Plot of observed [log(1/EC$_{50}$)] AhR binding data versus E(LUMO) for optimized PCBs. In the correlation samples are included: (a) the most stable (single) conformers for each compound; (b) the most planar conformers.

hypothesized and the receptor affinity of PCBs was hypothesized to be correlated with the energies of frontier orbitals, particularly the energy of the lowest unoccupied molecular orbital (E_{LUMO}). The latter was chosen as a relevant molecular descriptor based on the experimental findings that charge-transfer complexes are obtained with a charge delocalization toward PCBs.

The geometries of all conformations were optimized by the PM3 Hamiltonian. A poor correlation between AhR binding and E_{LUMO} was found ($r^2 = 0.38$) when all conformers were included in the correlation sample. It was not significantly improved ($r^2 = 0.43$; see Figure 7a) when the most stable conformers were selected from the conformation sample (occupying 10% of E-level). If the most planar conformers were selected, however, the variance r^2 increased significantly ($r^2 = 0.72$, Figure 7b). The planarity of conformers was assessed quantitatively using the planarity index[82] obtained by a simple (normalized) summation of the dihedral angles. The conformers with minimum values of planarity index were specified as most planar.

Similar results were obtained for nonoptimized conformers whose electronic structure was assessed by single point (1SCF) calculations. The selection of the most planar conformers yielded again a significant QSAR model, $r^2 = 0.78$.

It may be summarized that the description of the most planar conformers of PCBs provides a correspondence to the binding data. One could suppose that the active conformers of PCBs taking part in the AhR receptor binding are the most planar rather than the most stable (ground state) ones, which also supports the experimental fact of charge-transfer complex formation as well as the hypothesis for a stacking type of interactions. After the elimination of the extremely unstable conformations, even the 1SCF calculations were found to be quite sufficient to assess the stereoelectronic structure of PCBs with respect to the receptor binding. The time-consuming geometry optimization (quantum-chemical) procedures did not improve the quality of the derived ultimate QSARs.

3.4.2. Semicarbazide Toxicity in Frog Embryos

The acute lethality of 36 semicarbazides [such as $(R_1)NHNHC(=O)NH(R_2)$, $(R_1)NHNHC(=O)(R_2)$] and thiosemicarbazides [such as $(R_1)NHNHC(=S)NH(R_2)$, $(R_1)NHNHC(=S)(R_2)$] evaluated using the frog embryo teratogenesis assay $Xenopus$ (FETAX) were modeled by using the dynamic method for selection of active conformers.[83] The assumed mode of action, osteolathyrism, was defined by the failure of connective tissue to polymerize properly because of interference with lysyl oxidase. The active conformer selection was based on 3D isomer distribution according to frontier orbital energies and volume polarizability, conditioning their reactivity and hydrophobicity, respectively. The best modeling results were obtained after selection of conformers ($n = 110$) providing prevailing values of electron acceptor ability, which appears to be the active 3D isomers for the studied end points. The best two-parameter QSARs, encompassing all of the evaluated compounds, incorporate a steric parameter, the geometric analogue of the Wiener topological index, and the local electronic characteristics of the C=O or C=S group, superdelocalizabilities, and charges. The best QSAR model (Figure 8) for FETAX acute lethality has the following statistics: $n = 110$; $r^2 = 0.805$; $s^2 = 0.21$; $F = 220.8$; $\alpha = 99$; $r^2_{cros} = 0.812$; $s^2_{cros} = 0.27$ (the conventional statistics were extended by the cross-validation test estimates: cross-validated r^2 and s^2). It was found that toxicity increased with the increase of the size of substituents (probably related to the penetration factor), increase in the electron acceptor properties of the carbonyl oxygen or sulfur, and the positive charge at the

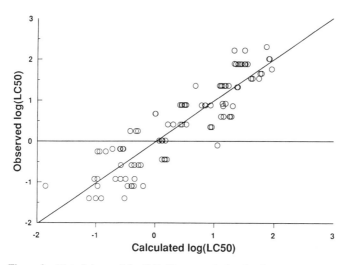

Figure 8. Plot of observed [log(LC$_{50}$)] versus calculated embryo lethality data.

carbonyl carbon. The results support the hypotheses that the acute lethality of semi-carbazides and thiosemicarbazides is conditioned by a single mechanism of action. The QSAR study was obtained without geometry optimization of the generated conformations.

3.4.3. Photoinduced Toxicity of α,α′-Terthienyls

The generality of the multilinear (parabolic) relationship between phototoxicity and electronic gap that has been established for rigid polycyclic aromatic hydrocarbons (PAHs)[84,85] was examined for flexible aromatic chemicals such as naturally occurring and synthetic tricyclic α,α′-terthienyl compounds.[86] An initial set of 812 conformers was generated to represent 41 α,α′-terthienyl compounds. It was found that the ranges of calculated HOMO–LUMO energy gaps for different conformers of a single chemical are significantly large (in some cases up to 0.5 eV). This unambiguously showed the necessity of applying the dynamic approach to active conformer selection. The screening of the initial set of 812 conformers was performed according to: (1) heats of formation (formation enthalpies); (2) volume polarizability, VolP, which is directly related to the octanol/water partition coefficient; and (3) HOMO–LUMO gap, the ground-state electronic parameter that dictates the light absorbance and stability of PAHs relative to their phototoxicity. The conformers found to develop a parabolic phototoxicity/electronic gap model were the energetically most favorable ones and those having minimum electronic gaps (the most "red-shifted"; Figure 9).

The range of the phototoxic region along a HOMO–LUMO energy gap axis for thiophene derivatives (within the PM3 Hamiltonian) coincides almost exactly with

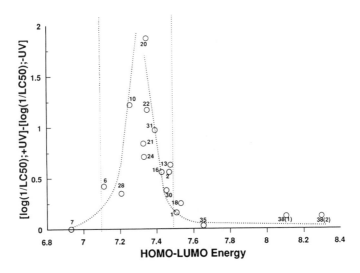

Figure 9. Variation of photoenhanced toxicity [$\Delta\log(1/LC_{50})$] for mosquito larvae with HOMO–LUMO gap of α-terthienyls, for geometry-optimized conformers occupying 10% of the energy levels.

that established for PAHs. Its left-side boundary was around 7.1 eV, whereas the right side was around 7.5 eV.

3.5. CONCLUDING REMARKS

The basic approaches to conversion of the connection information into 3D molecular models were classified and illustrated by the most representative methods. The basic features of those methods and the resulting 3D outcomes were discussed. Most methods attempt to generate the GEM structures, or a set of energetically favorable conformers, proceeding from the presumption that the active conformers are the most stable ones. Although convenient, this principle has limited many QSAR studies. On the contrary, molecular behavior can be expected to be variable so that different conformers should be expected to be active for different interactions and end points. Moreover, even for a given end point, different conformers may be taking part in the various successive interactions requiring a multiple representation of molecules in QSAR studies with flexible molecules.

The dynamic method for selection of active conformers appears to be useful in this venture. Its first step is a procedure for generation of 3D isomers, representing the conformational space of molecules. Although the 3DGEN algorithm is used in the OASIS system, most of the other algorithms for conformational space search (after minor modifications) could be used for that purpose. Next, a method for conformer screening is used, based on a hierarchical set of rules, in an attempt to find the active conformers.

The implementation of the method showed that, in general, the searching for molecules that can assume a specific conformation may be more important than determining the geometry of the most stable 3D conformers. If poor correlations are found between chemical activity and a mechanistically plausible stereoelectronic index, it seems as important to consider models with conformations other than that of lowest energy as to search for different stereoelectronic indices. This is reasonable after the finding that the variation in indices among different conformers of a given molecule is often larger than that between conformers of different molecules. The indices of electronic structures without geometry optimization may in many cases be sufficient for modeling a chemical property when the active conformers are selected properly. If so, delaying geometry optimization until necessary would save a significant amount of computing time and facilitate the modeling process.

ACKNOWLEDGMENTS

This work was supported, in part, by U. S. EPA Cooperative Agreement CR 822306-01-0 with the Higher Institute of Chemical Technology (Bulgaria), as well as by the Bulgarian Science Foundation, Grant X-409. The participation of Dr. Stoyan Karabunarliev and Dr. Julian Ivanov in the development of the OASIS system, and fruitful discussions with Dr. Steve Bradbury (U. S. EPA) and Dr. Jose Kanetti (Bulgarian Academy of Sciences) are gratefully acknowledged.

REFERENCES

1. L. Pasteur, "Researches in Molecular Asymmetry of Natural Organic Products" (1860); Alembic Club Reprint No. 14, Edinburgh, 1905. [Note the erroneous translation from "Recherches sur la Dissymetrie Moleculaire. . . "]
2. J. B. Biot, *Bull. Soc. Philomath. Paris 190* (1815).
3. J. H. van't Hoff, Voorsteel tot uidbreiding der tegenwoordig in de scheikunde gebruikte structuurformules in de ruimte, *Arch. Neerl. Sci. Exactes Nat. 9*, 445–454 (1874); J. A. Le Bel, Sur les relations qui existent entre les formules atomiques des corps organiques et le pouvoir rotatoire de leurs dissolutions, *Bull. Soc. Chim. Fr. (2) 22*, 337–347 (1874).
4. V. Prelog and G. Helmchen, *Angew. Chem. Int. Ed. Engl. 21*, 567–583 (1982).
5. E. L. Eliel, Chemistry in three dimensions, in: *Chemical Structures* (W.A. Warr, ed.), Vol. 1, pp. 1–8, Springer, Berlin (1993).
6. G. M. Crippen and T. F. Havel, *J. Chem. Inf. Comput. Sci. 30*, 222–227 (1990).
7. M. Saunders, *J. Am. Chem. Soc. 109*, 3150–3152 (1987).
8. C. J. Cramer and D. G. Truhlar, *J. Am. Chem. Soc. 113*, 8305–8311 (1991).
9. C. J. Cramer and D. G. Truhlar, *J. Comput. Chem. 13*, 1089–1097 (1992).
10. G. Rauhut, T. Clark, and T. Steinke, *J. Am. Chem. Soc. 115*, 9174–9181 (1993).
11. R. W. Behling, T. Yamane, G. Navon, and L. W. Jelinski, *Proc. Natl. Acad. Sci. USA 85*, 6721 (1988).
12. E. W. Thomas and S. C. Brooks, *J. Steroid Biochem. Mol. Biol. 50*, 61–73 (1994).
13. J. G. Nourse, D. H. Smith, R. E. Carhart, and C. Djerassi, *J. Am. Chem. Soc. 102*, 6289 (1980).
14. W. T. Wipke and T. M. Dyott, *J. Am. Chem. Soc. 96*, 4825, 4834 (1974).

15. S. Hanessian, J. Franco, G. Gagnon, D. Laramee, and B. Larouche, *J. Chem. Inf. Comput. Sci. 30*, 413–425 (1990); S. Hanessian, *Pure Appl. Chem. 65*, 1189–1204 (1993).

16. J. G. Nourse, *J. Am. Chem. Soc. 101*, 1210 (1979).

17. J. G. Nourse, R. E. Carhart, D. H. Smith, and C. Djerassi, *J. Am. Chem. Soc. 101*, 1216 (1979).

18. H. Abe, H. Hayasaka, Y. Miyashita, and S. Sasaki, *J. Chem. Inf. Comput. Sci. 24*, 216 (1984).

19. E. L. Eliel, *Stereochemistry of Carbon Compounds*, p. 1, McGraw–Hill, New York (1962).

20. J. Sadowski and J. Gasteiger, *Chem. Rev. 93*, 2567–2581 (1993).

21. W. T. Wipke, H. Braun, G. Smith, F. Choplin, and W. Sieber, *ACS Symp. Ser. 61*, 97–127 (1977).

22. W. T. Wipke, J. Verbalis, and T. Dyott, Three-dimensional interactive model building, presented at the 162nd National Meeting of the American Chemical Society, Los Angeles (1972).

23. T. M. Dyott, A. J. Stuper, and G. S. Zander, *J. Chem. Inf. Comput. Sci. 20*, 28 (1980).

24. F. Mohamadi, N. G. J. Richards, W. C. Guida, R. Liskamp, M. Lipton, C. Caufield, G. Chang, T. Hendrickson, and W. C. Still, *J. Comput. Chem. 11*, 440 (1990).

25. R. Potenzone, Jr., E. Cavicchi, H. J. R. Weintraub, and A. J. Hopfinger, *Comput. Chem. 1*, 187 (1977).

26. P. Gund, J. D. Andose, J. B. Rhodes, and G. M. Smith, *Science 208*, 1425 (1981).

27. S. Sasaki, H. Abe, Y. Hirota, Y. Ishida, Y. Kudo, S. Ochiai, K. Saito, and T. Yamasaki, *J. Chem. Inf. Comput. Sci. 18*, 211 (1978); H. Abe, I. Fujiwara, T. Nishimura, T. Okuyama, T. Kida, and S. Sasaki, *Comput. Enhanced Spectrosc. 1*, 55 (1983); S. Sasaki, H. Abe, and Y. Takahashi, TUTORS - system.

28. M. Levitt, *J. Mol. Biol. 82*, 393 (1974).

29. J. G. Winter, A. Davis, and M. R. Saunders, *J. Comput. Aided Mol. Des. 1*, 31 (1987).

30. J. Bauer and W. Schubert, *Comput. Chem. 2*, 61–65 (1982).

31. T. Liljefors, *J. Mol. Graphics 1*, 111 (1983).

32. D. N. J. White, J. K. Tyler, and M. R. Lindley, *Comput. Chem. 3*, 193 (1986).

33. D. Turk and J. Zupan, *J. Chem. Inf. Comput. Sci. 28*, 116 (1988).

34. O. Mekenyan, S. Karabunarliev, and D. Bonchev, *Comput. Chem. 14*, 193 (1990).

35. O. Mekenyan, S. Karabunarliev, J. Ivanov, and D. Dimitrov, *Comput. Chem. 18*, 173–187 (1994).

36. E. J. Corey and N. F. Feiner, *J. Org. Chem. 45*, 757 (1980).

37. N. Cohen, P. Colin, and G. Lemoin, *Tetrahedron 37*, 1711 (1981).

38. P. J. DeClercq, *Tetrahedron 40*, 3717 (1981). [A program limited to joining ring templates]

39. W. T. Wipke and M. Hahn, *ACS Symp. Ser. 306*, 136 (1986). [A template assembly program that obtains its templates from an abbreviated X-ray data base]

40. D. P. Dolata, A. R. Leach, and K. Prout, *J. Comput. Aided Mol. Des. 1*, 73 (1987).

41. H. Bogel and J. Sadowski, 2nd Workshop "Computer in Chemistry," Magdeburg, October (1990).

42. A. R. Leach and K. Prout, *J. Comput. Chem. 11*, 1193 (1990).

43. J. Gasteiger, M. G. Hutchings, B. Christoph, L. Gann, C. Hiller, P. Low, M. Marsili, H. Saller, and K. Yuki, *Top. Curr. Chem. 137*, 19 (1987).

44. M. J. Sippl and H. Stegbuchner, *Computers Chem. 15*, 73–78 (1991).

45. W. T. Wipke and M. A. Hahn, *ACS Symp. Ser. 306*, 136–146 (1986).

46. W. T. Wipke and M. A. Hahn, *Tetrahedron Comput. Method. 2*, 141 (1988).

47. M. A. Hahn and W. T. Wipke, in: *Chemical Structures* (W.E. Warr, ed.), Vol. 1, pp. 269–278, Springer, Berlin (1988).

48. Z. Ai and Y. Wei, *J. Chem. Inf. Comput. Sci. 33*, 635–638 (1993).

49. K. Davies and R. Upton, *R. Soc. Chem. (Spec. Publ.) 142*, 51–55 (1994).

50. M. C. Nicklaus, G. W. A. Milne, and D. Zaharevitz, *J. Chem. Inf. Comput. Sci. 33*, 639–646 (1993).

51. G. M. Crippen and T. F. Havel, *Acta Crystallogr. Sect. A A34*, 282 (1978).

52. T. F. Havel, I. D. Kuntz, and G. M. Crippen, *Bull. Math. Biol. 45*, 665 (1983).

53. G. M. Crippen and T. F. Havel, *J. Math. Chem. 6*, 307–324 (1991).

54. M. Waldman and T. F. Havel, "The Sketcher," available from BIOSYM Technologies Inc., San Diego (1992).

55. G. M. Crippen, *J. Math. Chem. 6*, 307 (1991).

56. A. S. Smellie, CONSTRUCTOR, Oxford Molecular Ltd., Terrapin House, University Science Area, South Parks Road, Oxford, England (1989).
57. J. M. Blaney, DGEOM, Quantum Chemistry Program Exchange, Department of Chemistry, Indiana University, Bloomington, Indiana 47405 (1989).
58. G. M. Crippen, A. S. Smellie, and W. W. Richardson, *J. Comput. Chem. 13*, 1262 (1992).
59. J. C. Wenger and D. H. Smith, *J. Chem. Inf. Comput. Sci. 22*, 29 (1982).
60. E. V. Gordeeva, A. R. Katritzky, V. V. Shcherbukhin, and N. S. Zefirov, *J. Chem. Inf. Comput. Sci. 33*, 102 (1993).
61. P. K. Weiner and S. Profeta, Jr., Abstracts, 181st National Meeting of the American Chemical Society, Atlanta, American Chemical Society, Washington, DC, COMP Division, Abstract No. 15 (1981).
62. A. Rusinko, III, Tools for Computer Assisted Drug Design; Ph.D. thesis, University of Texas at Austin (1988).
63. R. S. Pearlman, *Chem. Des. Auto. News 2*, 1 (1987).
64. M. A. Hendrickson, M. C. Nicklaus, G. W. A. Milne, and D. Zaharevitz, *J. Chem. Inf. Comput. Sci. 33*, 155–163 (1993).
65. A. T. Brint and P. Willett, *J. Mol. Graphics 49*, 5 (1987).
66. M. A. Hendrickson, M. C. Nicklaus, and G. W. A. Milne, *J. Chem. Inf. Comput. Chem. 33*, 155–163 (1993).
67. R. S. Judson, E. P. Jaeger, A. M. Treasurywala, and M. L. Peterson, *J. Comput. Chem. 14*, 1407–1414 (1993).
68. R. S. Judson, M. E. Colvin, A. Huffer, and D. Gutierrez, *Int. J. Quantum Chem. 44*, 277 (1992).
69. S. Legrand and K. Merz, *J. Global Opt. 3*, 49–66 (1993).
70. P. Tuffery, C. Etchebest, S. Hazout, and R. Levery, *J. Biomol. Struct. Dyn. 8*, 1267 (1991).
71. M. Saunders, K. N. Houk, Y. D. Wu, W. C. Still, M. Lipton, G. Chander, and W. C. Guida, *J. Am. Chem. Soc. 112*, 1419–1427 (1990).
72. M. Saunders, *J. Am. Chem. Soc. 109*, 3150 (1987).
73. H. Goto, E. Osawa, and M. Yamato, *Tetrahedron 49*, 387–396 (1992).
74. H. Goto and E. Osawa, *J. Chem. Soc. Perkin Trans. 2*, 189–198 (1993).
75. M. Lipton and W. C. Still, *J. Comput. Chem. 4*, 343 (1988).
76. J. A. McCammon and M. Karplus, *Acc. Chem. Res. 16*, 187 (1983).
77. J. Ivanov, S. Karabunarliev, and O. G. Mekenyan, *J. Chem. Inf. Comput. Chem. 34*, 234–243 (1994).
78. M. Saunders, *J. Am. Chem. Soc. 109*, 3150–3152 (1987).
79. Z. Li and H. A. Scheraga, *Proc. Natl. Acad. Sci. USA 84*, 6611 (1987).
80. O. G. Mekenyan, J. M. Ivanov, G. D. Veith, and S. P. Bradbury, *Quant. Struct. Act. Relat. 13*, 302–307 (1994).
81. A. T. Balaban, I. Motoc, D. Bonchev, and O. Mekenyan, *Top. Curr. Chem. 114*, 21–71 (1984).
82. O. G. Mekenyan, G. T. Ankley, G. D. Veith, and D. J. Call, *Environ. Health Perspect.* (in press).
83. O. G. Mekenyan, T.-W. Schultz, G. D. Veith, and V. B. Kamenska, *J. Appl. Toxicol. 16*, 355–363 (1996).
84. O. G. Mekenyan, G. T. Ankley, G. D. Veith, and D. J. Call, *Chemosphere 28*, 567–582 (1994).
85. O. G. Mekenyan, G. T. Ankley, G. D. Veith, and D. J. Call, *SAR and QSAR Environ. Res. 2*, 237–247 (1994).
86. G. D. Veith, O. G. Mekenyan, G. T. Ankley, and D. J. Call, *Environ. Sci. Technol. 29*, 1267–1272 (1995).

4

Use of Graph-Theoretic and Geometrical Molecular Descriptors in Structure–Activity Relationships

SUBHASH C. BASAK, GREGORY D. GRUNWALD, and GERALD J. NIEMI

> *Ostensibly there is color, ostensibly sweetness, ostensibly bitterness, but actually only atoms and the void.*
>
> GALEN
>
> (*Nature and the Greeks*, Erwin Schrödinger, 1954)

4.1. INTRODUCTION

One of the current interests in pharmaceutical drug design,[1-20] chemistry,[21-40] and toxicology[41-53] is the prediction of physicochemical, biomedicinal, and toxicological

SUBHASH C. BASAK, GREGORY D. GRUNWALD, and GERALD J. NIEMI • Center for Water and the Environment, Natural Resources Research Institute, University of Minnesota at Duluth, Duluth, Minnesota 55811.

From Chemical Topology to Three-Dimensional Geometry, edited by Balaban. Plenum Press, New York, 1997

properties of molecules from nonempirical structural parameters which can be calculated directly from their structure. Both in drug design[3,4,31,33,54] and in hazard assessment of chemicals,[31,33,46–53,55] one has to evaluate therapeutic or toxic potential of a large number of compounds, many of which have not even been synthesized. Drug design usually begins with the discovery of a "lead" compound which has the particular therapeutic activity of interest. The lead is altered through molecular modifications and the analogues thus produced are tested until a compound of desirable activity and toxicity profile is found. The combination of possibilities in such a process is almost endless. For example, let us assume the compound in Figure 1 is a lead. The medicinal chemist can carry out numerous manipulations on the lead in terms of substitution. On a very limited scale, if one carries out 50 substitutions in each of the aromatic positions, 10 modifications for esterification, 10 substitutions for the aliphatic carbon and 10 substitutions for the nitrogen, the total number of possible analogues comes to $50^5 \times 10 \times 10 \times 10 = 312.5$ billion structures. This astronomical number is reached by considering only a small fraction of the possible substituents that the medicinal chemist has in his repertoire.[54]

A similar situation exists for the hazard assessment of environmental pollutants. More than 15 million distinct chemical entities have been registered with the Chemical Abstract Service and the list is growing by nearly 775,000 per year. About 1000 of these chemicals enter into societal use every year.[56] Few of these chemicals have experimental properties needed for risk assessment. Table 1 gives a partial list of properties necessary for a reasonable risk assessment of a chemical.[31,33] In the United States, the Toxic Substances Control Act Inventory has about 74,000 entries and the list is growing by nearly 3000 per year. Of the approximately 3000 chemicals

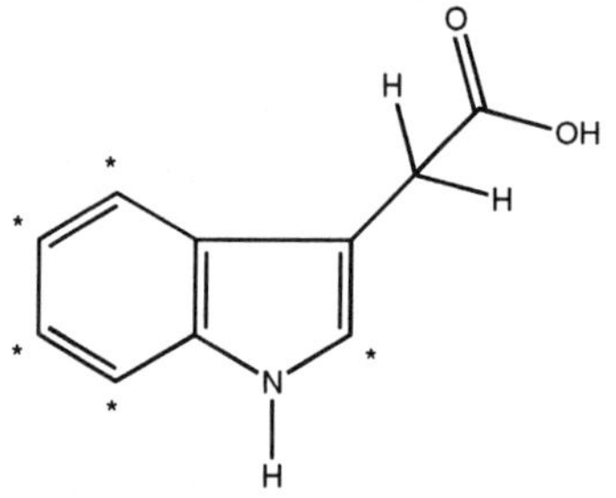

- 50 groups for each aromatic position (*)
- 10 groups for esterification
- 10 groups for aliphatic C
- 10 groups for ring N

Total analogs = 50^5 x 10 x 10 x 10 = 312.5 billion

Figure 1. Probable number of derivatives from a lead via molecular modification.

Table 1. Properties Necessary for Risk Assessment of Chemicals

Physicochemical	Biological
Molar volume	Receptor binding (K_D)
Boiling point	Michaelis constant (K_m)
Melting point	Inhibitor constant (K_i)
Vapor pressure	Biodegradation
Aqueous solubility	Bioconcentration
Dissociation constant (pKa)	Alkylation profile
Partition coefficient	Metabolic profile
Octanol–water (log P)	Chronic toxicity
Air–water	Carcinogenicity
Sediment–water	Mutagenicity
Reactivity (electrophile)	Acute toxicity
	LD_{50}
	LC_{50}
	EC_{50}

submitted yearly to the U.S. Environmental Protection Agency for the premanufacture notification process, more than 50% have no experimental data, less than 15% have empirical mutagenicity data, and only about 6% have experimental ecotoxicological and environmental fate data.[55] Also, limited data are available for many of the over 700 chemicals found on the Superfund list of hazardous substances.

In the face of this massive unavailability of experimental data for the vast majority of chemicals, practitioners in drug discovery and hazard assessment have developed the use of nonempirical parameters to estimate molecular properties.[1,3,4,20,31–33] By *nonempirical*, we mean those parameters that can be calculated directly from molecular structure without any other input of experimental data. Topological indexes (TIs), substructural parameters defined on chemical graphs, geometrical (3D or shape) parameters, and quantum-chemical parameters fall in this category.[3,4,21–40,46–55,57–61]

A large number of quantitative structure–activity relationships (QSARs) pertaining to chemistry, pharmacology, and toxicology have used these nonempirical parameters. QSARs are mathematical models that relate molecular structure to their physicochemical, biomedicinal, and toxic properties. Two distinct processes are involved in the derivation of nonempirical parameters for a chemical: (1) defining the model object called "structure" which represents the salient features of the architecture of the chemical species and (2) calculating structural quantifiers from a selected set of critical features of the model object.[31,62] Figure 2 depicts the process of experimental determination of properties vis-à-vis prediction of properties using descriptors.

Figure 2 represents an empirical property as a function $\alpha:C \to R$ which maps the set C of chemicals into the real line R. A nonempirical QSAR may be regarded as a composition of a description function, $\beta_1:C \to D$, mapping each chemical structure of C into a space of nonempirical structural descriptors (D) and a prediction function, $\beta_2:D \to R$, which maps the descriptors into the real line. When $[\alpha(C) - \beta_2\beta_1(C)]$ is within the range of experimental errors, we say that we have a good nonempirical

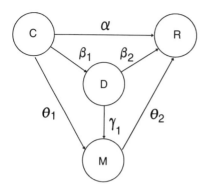

Figure 2. Composition functions for quantitative structure–activity relationship (QSAR) and property– activity
relationship (PAR).

predictive model. On the other hand, a property–activity relationship (PAR) is the
composition of θ_1:C → M, which maps the set C into the molecular property space M,
and θ_2:M → R, mapping those molecular properties into the real line R. PAR seeks to
predict one property (usually a complex property) of a molecule in terms of another
(usually simpler or available) property. The latter group of properties may consist
either of a number of experimentally determined quantities (e.g., melting point, boiling
point, vapor pressure, partition coefficient) or substituent constants or solvatochromic
parameters (e.g., steric, electronic, hydrophobic, charge transfer substituent constants,
hydrogen bond donor acidity, hydrogen bond acceptor basicity).[54,60] PAR using a
calculated property, e.g., calculated partition coefficient (log P, octanol–water), may
be looked on as a mapping $\theta_2\gamma_1\beta_1$:C → R, which is a composition of β_1:C → D, γ_1:D
→ M mapping the descriptor space into the molecular property space (e.g., calculation
of log P from fragments using additivity rule), and θ_2:M → R.

   Graph invariants have been used in a large number of QSARs.[1–53] A graph
invariant is a graph-theoretic property that is preserved by isomorphism.[63,64] A graph
invariant may be a polynomial, a sequence of numbers, or a single numerical index.
Numerical indexes derived from the topological characteristics of molecular graphs
are called topological indexes. Molecular structures can be symbolized by graphs
where the atomic cores are represented by vertices and covalent chemical bonds are
depicted by edges of the graph. Such a graph depicts the connectivity of atoms in a
chemical species irrespective of the metric parameters (e.g., equilibrium distance
between nuclei, valence angles) associated with the molecular structure. It is in this
sense that molecular graphs can be seen as topological, rather than geometrical,
representations of molecular structure.[65] TIs are numerical quantifiers of molecular
topology and are sensitive to such structural features of molecules as size, shape,
symmetry, branching, and cyclicity. Two nonisomorphic graphs may have the same
set of graph invariants. In that sense, TIs do not uniquely characterize molecular
topology. Yet, it has to be emphasized that TIs quantify many salient aspects of
molecular structure. As a result, different graph invariants have been successfully used
in characterizing the structural similarity/dissimilarity of molecules,[1–4 28,29,47,49,50,66]

quantifying the degree of molecular branching,[34,35,67] and developing structure–activity relationships in chemistry, biomedical sciences, and environmental toxicology.[5–53,64,67–81]

4.2. TOPOLOGICAL INDEXES AND QSAR

TIs have been used in developing QSAR models for predicting various properties. We give below some examples of successful QSARs using TIs. Definitions of the TIs used in the following equations and throughout this chapter may be found in Table 2.

Table 2. Symbols for Topological Indexes, Geometrical Parameters, and Hydrogen Bonding Parameter and Their Definitions

Index symbol	Definition
I_D^W	Information index for the magnitudes of distances between all possible pairs of vertices of a graph
$\overline{I}_D^W$	Mean information index for the magnitude of distances
$^v\overline{I}_D^E$	Mean information index for the equality of distances
W	Wiener index = half-sum of the off-diagonal elements of the distance matrix of a graph
I^D	Degree complexity
H^V	Graph vertex complexity
H^D	Graph distance complexity
$\overline{IC}$	Information content of the distance matrix partitioned by frequency of occurrences of distance h
O	Order of neighborhood when IC_r reaches its maximum value for the hydrogen-filled graph
I_{ORB}	Information content or complexity of the hydrogen-suppressed graph at its maximum neighborhood of vertices
M_1	A Zagreb group parameter = sum of square of degree over all vertices
M_2	A Zagreb group parameter = sum of cross-product of degrees over all neighboring (connected) vertices
IC_r	Mean information content or complexity of a graph based on the rth ($r = 0$–6) order neighborhood of vertices in a hydrogen-filled graph
SIC_r	Structural information content for r^{th} ($r = 0$–6) order neighborhood of vertices in a hydrogen-filled graph
CIC_r	Complementary information content for r^{th} ($r = 0$–6) order neighborhood of vertices in a hydrogen-filled graph
TIC_r	Total information content for r^{th} order neighborhood of vertices in a hydrogen-filled graph
$^h\chi$ or $^h\chi_p$	Path connectivity index of order $h = 0$–6
$^h\chi_C$	Cluster connectivity index of order $h = 3$–6
$^h\chi_{Ch}$	Chain connectivity index of order $h = 3$–6
$^h\chi_{PC}$	Path–cluster connectivity index of order $h = 4$–6
$^h\chi^b$	Bonding path connectivity index of order $h = 0$–6
$^h\chi_C^b$	Bonding cluster connectivity index of order $h = 3$–6
$^h\chi_{Ch}^b$	Bonding chain connectivity index of order $h = 3$–6
$^h\chi_{PC}^b$	Bonding path–cluster connectivity index of order $h = 4$–6

(continued)

Table 2.   (Continued)

Index symbol	Definition
$^h\chi^v$	Valence path connectivity index of order $h = 0–6$
$^h\chi_C^v$	Valence cluster connectivity index of order $h = 3–6$
$^h\chi_{Ch}^v$	Valence chain connectivity index of order $h = 3–6$
$^h\chi_{PC}^v$	Valence path–cluster connectivity index of order $h = 4–6$
χ_t	Total structure index
Ω-MCI	Orthogonal molecular connectivity indexes
τ	Branchedness indexes
κ	Shape indexes
ϕ	Flexibility indexes
A_3	Half-sum of the cube of the adjacency matrix
p_3	Polarity number: number of third neighbors
N_2	Gordon–Scantlebury index: number of second neighbors
P_h	Number of paths of length $h = 0–10$
J	Balaban's J index based on distance
J^B	Balaban's J index based on multigraph bond orders
J^X	Balaban's J index based on relative electronegativities
J^Y	Balaban's J index based on relative covalent radii
U,V,X,Y	Balaban's information-based indexes on distance sums
AZV	Local vertex invariant based on the adjacency matrix, atomic numbers, and vertex degrees
D	Mean distance topological index for any graph
D_1	Mean distance topological index for acyclic graphs
Z	Hosoya index
HB_1	Hydrogen bonding potential of molecule
ID	Molecular identification numbers
V_W	Volume of molecule
$^{3D}W_H$	3D Wiener number including hydrogens
^{3D}W	3D Wiener number without hydrogens

4.2.1. Physicochemical Properties

4.2.1.1. Boiling Point of Alkanes

Needham *et al.*[21] used TIs to develop a regression equation to predict the normal boiling point (BP) for 74 alkanes:

(1) $BP = -9.6 + 38.1(^1\chi) - 49.0(1/^0\chi) + 5.7(^4\chi_{PC}) - 94.5(\chi_t) + 8.4(^6\chi_p)$
$(N = 74, r = 0.999, s = 1.86, F = 9030)$

Subsequently, Basak and Grunwald[78] derived the following equation:

(2) $BP = -263 + 237(^1\chi) + 18.6(CIC_2)$
$(N = 74, r = 0.997, s = 3.83, F = 5287)$

4.2.1.2. Boiling Point of Chlorofluorocarbons (CFCs)

Balaban et al.[70] were able to model the boiling points of a large set of CFCs using TIs with the following equation:

$$(3) \quad BP = -73.65 + 33.21(^1\chi^v - {}^0\chi^v) - 64.06(^D\chi^0) + 94.46(^1\chi) - 20.65(N_{Br}) - $$

$$22.18(N_I) + 6.36(^2\chi^v - {}^1\chi^v)$$

$$(N = 532, r = 0.98, s = 10.94, F = 2953)$$

Using a backpropagation neural network (NN), Balaban et al.[25] successfully predicted BP for 276 CFCs. As inputs to the NN, the following parameters were used: J index, Wiener index (W), number of carbon atoms (N_c), number of chlorine atoms (N_{Cl}), and number of fluorine atoms (N_F). This NN resulted in a correlation $(r) = 0.992$ of observed BP with predicted BP, with a standard error (s) of 8.5°C. The data set used for NN model development consisted of 276 CFCs with, at most, four carbon atoms.

4.2.1.3. Lipophilicity of Diverse Sets of Compounds

Basak et al.[27] derived the following equation to predict lipophilicity (log P, octanol–water):

$$(4) \quad \log P = 1.76 - 0.50(HB_1) - 5.28(IC_0) - 1.48(CIC_1) + 3.75(^0\chi^v) + 0.41(P_6)$$

$$(N = 382, r = 0.95, s = 0.27, F = 1186)$$

where HB_1, is a theoretically calculated hydrogen bonding parameter.

Basak et al.[31] developed a refined model for chemicals with HB_1 equal to zero:

$$(5) \quad \log P = -3.13 - 1.64(IC_0) + 2.12(^5\chi_C) - 2.91(^6\chi_{Ch}) + 4.21(^0\chi^v)$$

$$+ 1.06(^4\chi^v) - 1.02(^4\chi_{PC}^v)$$

$$(N = 137, r = 0.98, s = 0.26, F = 446)$$

4.2.1.4. Chromatographic Retention Time of Alkanes, Alkylbenzenes

Bonchev and Trinajstić[79] derived the following correlation for alkylbenzenes:

$$(6) \quad RI = 683 + 2.97(^V T_D^E) + 2.71(P_0 - 6)$$

$$N = 28, r = 0.99, s = 0.58)$$

For alkanes, Kier and Hall[77] found the following relationship:

$$(7) \quad RI = -0.242 + 0.719(^1\chi) + 0.125(^3\chi_p)$$

$$(N = 18, r = 0.998, s = 0.045, F = 1702)$$

4.2.2. Biomedicinal Properties

4.2.2.1. Anesthetic Dose (AD_{50}) of Barbiturates

Basak et al.[13] predicted AD_{50} of barbiturates using various TIs:

(8)
$$AD_{50} = -49.1 + 200(SIC_1) - 190(SIC_1)^2$$
$$(N = 13, r = 0.76, s = 0.20, F = 6.6)$$

(9)
$$AD_{50} = -200 + 153(IC_1) - 28.3(IC_1)^2$$
$$(N = 13, r = 0.74, s = 0.21, F = 6.1)$$

(10)
$$AD_{50} = -41.2 + 11.5(^1\chi) - 0.740(^1\chi)^2$$
$$(N = 13, r = 0.72, s = 0.21, F = 5.4)$$

4.2.2.2. Analgesic Potency (A-ED_{50}) of Barbiturates

Basak et al.[13] correlated A-ED_{50} of barbiturates using graph-theoretic parameters:

(11)
$$\text{A-}ED_{50} = 4700 - 26300(SIC_1) + 36700(SIC_1)^2$$
$$(N = 7, r = 0.97, s = 6.5, F = 29)$$

(12)
$$\text{A-}ED_{50} = 5280 - 2800(CIC_1) + 372(CIC_1)^2$$
$$(N = 7, r = 0.96, s = 7.4, F = 27)$$

(13)
$$\text{A-}ED_{50} = 2400 - 444(^1\chi) + 20.4(^1\chi)^2$$
$$(N = 7, r = 0.94, s = 9.1, F = 17)$$

4.2.2.3. Enzymatic Acetyl Transfer Reaction

Several TIs have been found to correlate with the enzymatic acetyl transfer reaction,[12] as shown by the following equations:

(14)
$$A_\chi = 3.20 - 0.62(^1\chi)$$
$$(N = 9, r = 0.88, s = 0.24, F = 23)$$

(15)
$$A_\chi = 2.67 - 0.83(IC_1)$$
$$(N = 9, r = 0.91, s = 0.20, F = 35)$$

(16)
$$A_\chi = 3.13 - 4.07(\text{SIC}_1)$$
$$(N = 9, r = 0.92, s = 0.20, F = 36)$$

4.2.2.4. Hill Reaction Inhibitory Potency of Triazinones[6]

(17)
$$\text{pI}_{50} = -13.36 + 71.15(\text{SIC}_1) - 63.64(\text{SIC}_1)^2$$
$$(N = 11, r = 0.937, s = 0.316, F = 28.6)$$

4.2.2.5. Complement Inhibition by Benzamidines[80]

(18)
$$1/\log_{10} C = -1.125 + 0.487(\text{ID}) + 0.011(O)$$
$$(N = 105, r = 0.941, s = 0.020, F = 391)$$

4.2.2.6. Binding of Barbiturates to Cytochrome P_{450}

Basak[43] used several TIs to correlate the binding of barbiturates to cytochrome P_{450}:

(19)
$$K_s = 27.79 - 36.78(\text{IC}_0) + 12.17(\text{IC}_0)^2$$
$$(N = 10, r = 0.99, s = 0.01, F = 156.1)$$

(20)
$$K_s = 5.94 - 41.26(\text{SIC}_0) + 71.84(\text{SIC}_0)^2$$
$$(N = 10, r = 0.99, s = 0.01, F = 224.3)$$

(21)
$$K_s = 35.74 - 18.45(H^D) + 2.38(H^D)^2$$
$$(N = 10, r = 0.98, s = 0.01, F = 94.6)$$

4.2.3. Toxicological Properties

4.2.3.1. Nonspecific Narcotic Activity of Alcohols

Basak and Magnuson[81] correlated the nonspecific narcotic activity (LC_{50}) of alcohols using TIs:

(22)
$$\log \text{LC}_{50} = 1.979 - 1.896(\text{CIC}_1)$$
$$(N = 10, r = 0.989, s = 0.323, F = 355.3)$$

4.2.3.2. Nonspecific Toxicity of Esters to *Pimephales promelas*[41]

(23)
$$\log \text{LC}_{50} = -0.774 - 0.364(\text{CIC}_1) - 0.774(^1\chi^v)$$
$$(N = 15, r = 0.965, s = 0.194, F = 81.1)$$

$$\log LC_{50} = 1.012 - 0.774(CIC_1) - 0.615(I_D^W)$$

(24)
$$(N = 15, r = 0.961, s = 0.204, F = 72.7)$$

4.2.3.3. Mutagenicity of Nitrosamines

Basak et al.[42] correlated information- or complexity-based parameters with mutagenic potency of nitrosamines:

(25)
$$\ln R = 61.0 - 86.8(IC_0) + 29.2(IC_0)^2$$
$$(N = 15, r = 0.96, s = 1.17, p < 0.001)$$

(26)
$$\ln R = 12.0 - 15.3(IC_1) + 3.84(IC_1)^2$$
$$(N = 15, r = 0.98, s = 0.86, p < 0.001)$$

4.2.3.4. Mutagenicity of Diverse Structures

Basak et al.[46] used six TIs and four substructure (subgraph) indicator variables to develop a linear model to classify a set of 520 diverse chemicals as mutagens or nonmutagens as defined by the Ames mutagenicity test.[82] The data set used in their study consisted of 260 mutagens and 260 nonmutagens. The TIs included three information-based indexes: information content of the graph orbits (I_{ORB}), information content at sixth order (IC_6), and structural information content at zeroth order (SIC_0). A fourth index included number of paths of length 10 (P_{10}). The remaining two indexes were connectivity type: third-order bond-corrected cluster connectivity ($^3\chi_C^b$) and third-order valence-corrected chain connectivity ($^3\chi_{Ch}^v$). The four substructure indicators were: (1) nitroso chemicals, (2) halogen-substituted mustard, sulfur mustard, or oxygen mustard, (3) organic sulfate or sulfonate, and (4) a biphenyl amine, benzidine, or 4,4'-methylene dianiline derivative.

Using these parameters, a 74.8% overall correct classification rate was achieved. Jackknifed classification tests showed a 74.6% overall correct classification rate.

4.2.3.5. Toxicity of Monoketones

Basak et al.[83] derived the following correlations between TIs and the toxicity (LD_{50}) of monoketones:

(27)
$$LD_{50}(control) = 620.0 - 448.0(CIC_1) + 83.5(CIC_1)^2$$
$$(N = 13, r = 0.95, s = 9.62, F = 48.9)$$

(28)
$$LD_{50}(CCl_4) = 407.0 - 235.0(CIC_0) + 35.1(CIC_0)^2$$
$$(N = 13, r = 0.97, s = 4.76, F = 74.0)$$

4.2.3.6. Inhibition of p-Hydroxylation of Aniline by Alcohols[84]

(29) $pIC_{50} = -13.85 + 25.17(IC_0) - 27.89(SIC_1) - 1.87(CIC_2)$

$$(N = 20, r = 0.96, F = 62.7)$$

Table 3 gives more exhaustive information about the list of properties of different chemical classes that have been successfully correlated using TIs.

Table 3. Summary of QSARs Using Topological Parameters

Property	Chemical class	Variables[b]	Method	Citation	Ref. No.
BP	Aliphatic alcohols	MCI,ID,J, κ, Elec.	LR	Smeeks and Jurs	85
BP	Alkanes	LOVI/LOIS	NLR	Filip *et al.*	73
BP	Alkanes	MCI	LR	Needham *et al.*	21
BP	Alkanes	^1X	LR	Randić	35
BP	Haloalkanes	$W, J, N_{Cl}, N_{Br},$ N_F, N_I	NN	Balaban *et al.*	25
BP	Haloalkanes	MCI, N_X	LR	Balaban *et al.*	70
BP	Haloalkanes	N, MCI, κ, ϕ, J	LR	Balaban *et al.*	70
BP	Nonanes–dodecanes	Z, W, p_3, N_2, A_3	LR	Gao and Hosoya	86
BP	Paraffins	Platt's No.	LR	Platt	88
BP	Paraffins	W	LR	Wiener	87
CD	α-Amino acids	MCI	LR	Pogliani	76
CNDO/2 charge	Alkanes	MCI	LR	Hall and Kier	89
Cavity SA	Alcohols	LOVI/LOIS	NLR	Filip *et al.*	73
d	Nonanes–dodecanes	Z, W, p_3, N_2, A_3	LR	Gao and Hosoya	86
d	Alkanes	LOVI/LOIS	NLR	Filip *et al.*	73
d_4^{20}	Infinite linear polymers	W	LR,NLR	Mekenyan *et al.*	69
d_4^{20}	Organophosphorus	MCI	LR	Pogliani	90
d_C	Nonanes–dodecanes	Z, W, p_3, N_2, A_3	LR	Gao and Hosoya	86
ΔG_f	Nonanes–dodecanes	Z, W, p_3, N_2, A_3	LR	Gao and Hosoya	86
ΔH_{vap}	Alkanes	LOVI/LOIS	NLR	Filip *et al.*	73
ΔS	Nonanes–dodecanes	Z, W, p_3, N_2, A_3	LR	Gao and Hosoya	86
Diverse profile	Diverse	TI	CCA	Boecklen and Niemi	91
E_s	Diverse	LOVI/Sub.-TI	LR	Balaban and Catana	92
E_s	Hydrocarbons	$^1\chi^V$	LR	Gupta and Singh	93
ΔH_F	Nonanes–dodecanes	Z, W, p_3, N_2, A_3	LR	Gao and Hosoya	86
ΔH_F	Paraffins	Platt's No.	LR	Platt	88

(continued)

Table 3. (Continued)

Property	Chemical class	Variables[b]	Method	Citation	Ref. No.
MON	Alkanes	J, D, D_1	LR	Balaban	23
MON	Alkanes	LOVI/LOIS	NLR	Filip *et al.*	73
MON	Alkanes	MCI	LR	Pogliani	90
MP	Alkanes	MCI	LR	Pogliani	90
MP	Caffeine homologues	MCI	LR	Pogliani	90
MP	Infinite linear polymers	W	LR,NLR	Mekenyan *et al.*	69
MR	Alkylbenzenes	Ω-MCI	LR	Randić	94
MR	Alkylgermanes	1st-order MCI	LR	Kupchik	75
MR	Heptanes	Ω-MCI	LR	Randić	94
MR	Nonanes–dodecanes	Z, W, p_3, N_2, A_3	LR	Gao and Hosoya	86
MR	Organophosphorus	MCI	LR	Pogliani	90
MR	Paraffins	Platt's No.	LR	Platt	88
MR	Nonanes–dodecanes	Z, W, p_3, N_2, A_3	LR	Gao and Hosoya	86
MV	Paraffins	Platt's No.	LR	Platt	88
MW	α-Amino acids	MCI	LR	Pogliani	76
n_D	Nonanes–dodecanes	Z, W, p_3, N_2, A_3	LR	Gao and Hosoya	86
n_D^{20}	Organophosphorus	MCI	LR	Pogliani	90
P_C	Alkanes	$J, X, Y, V, U,$ $^1\chi, AZV$	LR	Balaban and Feroiu	74
P_C	Nonanes–dodecanes	Z, W, p_3, N_2, A_3	LR	Gao and Hosoya	86
R_1	α-Amino acids	MCI	LR	Pogliani	76
RI	Alkanes	MCI	LR	Kier and Hall	77
RI	Alkylbenzenes	I^D, P_0	LR	Bonchev and Trinajstić	79
RI	Diverse drugs	MCI, P_n, κ, Elec.	LR	Rohrbaugh and Jurs	95
RI	Organophosphorus	MCI	LR	Pogliani	90
RON	Alkanes	τ	LR	Pal *et al.*	96
S	α-Amino acids	MCI	LR	Pogliani	76
S	Caffeine homologues	MCI	LR	Pogliani	90
T_C	Alkanes	$J, X, Y, V, U,$ $^1\chi, AZV$	LR	Balaban and Feroiu	74
T_C	Nonanes–dodecanes	Z, W, p_3, N_2, A_3	LR	Gao and Hosoya	86
Ultrasonic sound	Alkanes, alcohols	W, J, MCI, ID	LR	Rouvray and Tatong	98
VP	α-Amino acids	MCI	LR	Pogliani	76
VP	Polychlorinated biphenyls	W, J, MCI, N_{Cl}	LR	Rouvray and Tatong	97
V_C	Alkanes	$J, X, Y, V, U,$ $^1\chi, AZV$	LR	Balaban and Feroiu	74
V_C	Nonanes–dodecanes	Z, W, p_3, N_2, A_3	LR	Gao and Hosoya	86

(continued)

Table 3. (Continued)

Property	Chemical class	Variables[b]	Method	Citation	Ref. No.
$\alpha_D 22$	Infinite linear polymers	W	LR,NLR	Mekenyan et al.	69
log P	Diverse	TI, HB_1	LR	Basak et al.	27
log P	Diverse	TI	LR	Niemi et al.	45
log P	Diverse	MCI	LR,NLR, PCR	Niemi et al.	99
log P	Diverse, $HB_1 = 0$	TI	LR	Basak et al.	31
pI	α-Amino acids	MCI	LR	Pogliani	76
Biomedicinal	Bioactive	Inf.	LR	Ray et al.	9
Pharmacological	Bioactive agents	Inf.	LR	Basak et al.	5
1/log$C$	Benzamidines	TI	LR	Basak et al.	80
A-ED$_{50}$	Barbiturates	MCI, Inf.	LR	Basak et al.	13
AD$_{50}$	Barbiturates	MCI, Inf.	LR	Basak et al.	13
AD$_{50}$	Barbiturates	MCI, Inf., W	NLR	Basak et al.	16
A_X	Anilines	MCI, Inf.	LR	Basak et al.	12
Antihistaminic	2-(Piperidin-4-ylamino)-1H-benzimidazoles	Sub.-W	LR	Lukovits	100
BOD	Diverse	MCI	Clustering, DA	Niemi et al.	99
BOD	Diverse	MCI, log$P$	Clustering, DA	Niemi et al.	101
Biodegradation	Diverse	MCI, κ, Sub.	DA	Gombar and Enslein	102
Carcinogenicity	Diverse	σ, κ, MCI, Sub.	LR,DA	Blake et al.	103
Cytostatic activity	1 H-Isoindolediones	Sub.-W	LR	Lukovits	100
Estrogen binding	2-Phenylindoles	Sub.-W	LR	Lukovits	100
K_S	Barbiturates	TI	LR	Basak	43
LC$_{50}$	Alcohols	CIC	LR	Basak and Magnuson	81
LC$_{50}$	Esters	W, $^1\chi$, $^1\chi^V$, Inf.	LR	Basak et al.	41
LD$_{50}$	Monoketones	TIC$_0$, TIC$_1$, CIC$_0$, CIC$_1$	NLR	Basak et al.	83
Mutagenicity	Diverse	σ, κ, MCI, Sub.	LR,DA	Blake et al.	103
Taste	Sulfamates	Wt-paths	SIMCA	Okuyama et al.	104
Therapeutic type	Therapeutics	Wt-paths	Clustering	Randić	40
ln$R$	Nitrosamines	IC$_0$, IC$_1$	NLR	Basak et al.	42
pIC$_{50}$	N-alkylnorketo-bemidones/triazinones	Inf.	LR	Ray et al.	6
pIC$_{50}$	Alcohols	Inf.	LR	Magnuson et al.	84

[a]Property: BP = boiling point; CD = crystal density; SA = surface area; d = liquid state density; d_4^{20} = density; d_C = critical density; ΔG_f = free energy of formation; ΔH_{vap} = vaporization enthalpy; ΔS = entropy; E_S = Taft's steric parameter; ΔH_F = heat of formation; MON = motor octane number; MP = melting point; MR = molar refractivity; MV = molar volume; n_D = refractive index; n_D^{20} = refractivity index; P_C = critical pressure; R_1 = relaxation rate; RI = retention index; RON = research octane number; S = solubility; T_C = critical temperature; VP = vapor pressure; V_C = critical volume; α_D^{22} = specific rotation; log$P$ = logarithm of the octanol–water partition coefficient; pI = isoelectric points; C = molar concentration of inhibitor required for 50% inhibition of complement; A-ED$_{50}$ = analgesic effective dose; AD$_{50}$ = anesthetic dose; A_χ = enzymatic acetyl transfer reaction rate; BOD = biological oxygen demand; K_S = binding constant; LC$_{50}$ = lethal concentration; LD$_{50}$ = lethal dose; ln$R$ = natural logarithm of the number of revertants per nanomole; pIC$_{50}$ = negative logarithm of the inhibition concentration;
[b]Variables: MCI = molecular connectivity indexes; Inf. = information indexes; LOVI = local vertex invariant; LOIS = local invariant set; Elec. = electronic variables; TI = diverse set of topological indexes; Sub. = substructure.

4.3. TOPOLOGICAL APPROACHES TO MOLECULAR SIMILARITY

One important application of TIs and substructural parameters has been in the quantification of molecular similarity. In practical drug design and risk assessment, good-quality QSARs of specific classes of chemicals, if available, are the best option. However, class-specific QSARs are often not available. In such cases, one selects analogues of the chemical of interest (lead or toxicant), and uses the property of

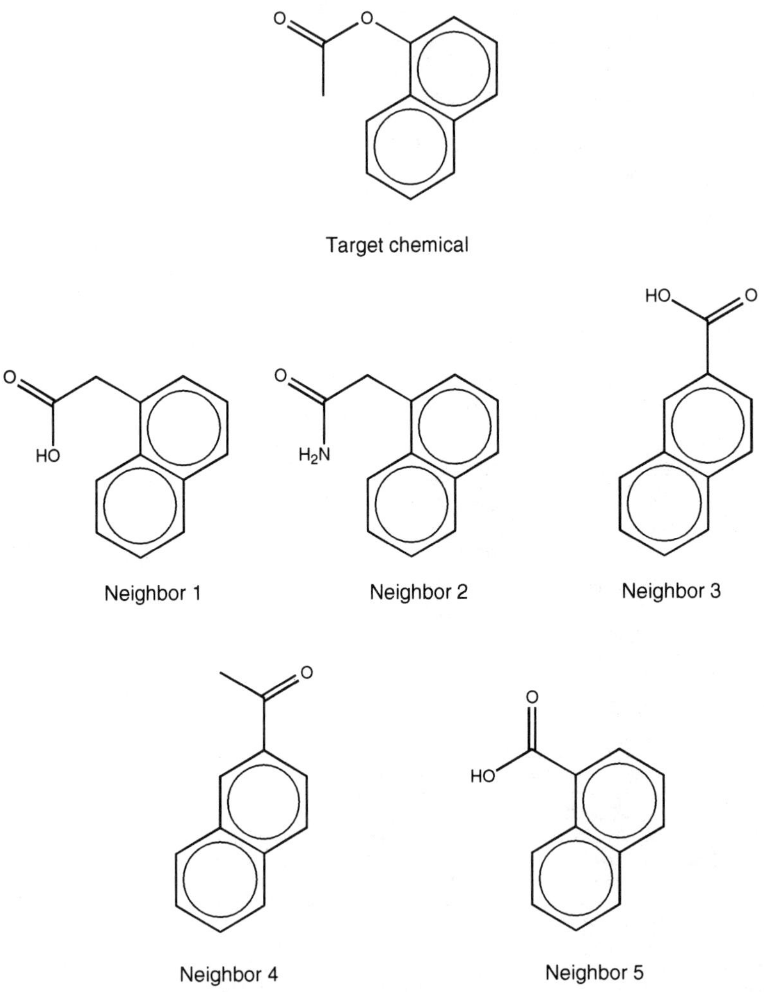

Target chemical

Neighbor 1   Neighbor 2   Neighbor 3

Neighbor 4   Neighbor 5

Figure 3. Target chemical and five selected analogues using ED method from the set of 3692 chemicals.

selected analogues for the estimation of the biomedicinal/toxic potential of the chemical.

4.3.1. Quantification of Similarity Using Path Numbers

Path numbers P_h ($h = 1, 2, \ldots$) and weighted paths have been used by Randić and co-workers in determining partial orderings relating dopamine agonist properties for 2-aminotetralins,[105] physicochemical properties of decanes,[106] therapeutic potential of diverse compounds,[40] and antitumor activity of phenyldialkyltriazines.[107] Randić[66] has also reviewed the use of path numbers and weighted paths as they are applied in molecular similarity approaches to property optimization. The results show that the ordering of molecules by path numbers reflects the pattern of activity reasonably well.

4.3.2. Quantification of Similarity Using Topological Indexes

Basak et al.[2] used TIs to compute intermolecular similarity of chemicals. Ninety TIs were calculated for a set of 3692 chemicals with diverse structures. Principal component analysis (PCA) was used to reduce the 90-dimensional space to a 10-dimensional subspace which explained 93% of the variance. In the 10-dimensional PC space, the intermolecular similarity of chemicals were quantified in terms of their Euclidean distance (ED). Ten chemicals were then chosen at random from the set of 3692 structures and their analogues were selected using the Euclidean distance as the criterion for nearest-neighbor selection. Figure 3 gives one example of a probe chemical and its five chosen neighbors using this method. The results show that the probe and its selected analogues have a reasonable degree of structural similarity.

4.3.3. Quantification of Intermolecular Similarity Using Substructural Parameters

4.3.3.1. Atom Pairs (APs)

Carhart et al.[4] developed the AP method of measuring molecular similarity. An AP is defined as a substructure consisting of two nonhydrogen atoms i and j and their interatomic separation:

$$\langle \text{atom descriptor}_i \rangle\text{-}\langle \text{separation}\rangle\text{-}\langle \text{atom descriptor}_j \rangle$$

where $\langle \text{atom descriptor}_i \rangle$ encodes information about the element type, number of nonhydrogen neighbors, and number of π electrons. Interatomic separation of two atoms is the number of atoms traversed in the shortest bond-by-bond path containing both atoms.

For two molecules, M_i and M_j, AP-based similarity is defined as:

(30)
$$S_{ij} = 2C/(T_i + T_j)$$

where C is the number of APs common to molecule i and j. T_i and T_j are the total number of APs in chemicals i and j, respectively. The numerator is multiplied by 2 to reflect the presence of shared APs in both molecules.

The Lederle group has used the AP similarity method to compare chemicals in their data base. Basak *et al.*[28,29,46,47,49,50,53,108] have used the AP method in selecting analogues of chemicals in different and diverse data bases. The relative effectiveness of the AP and ED methods in selecting analogues of chemicals in the STARLIST[109] data base containing more than 4000 chemicals are shown in Figure 4.[108]

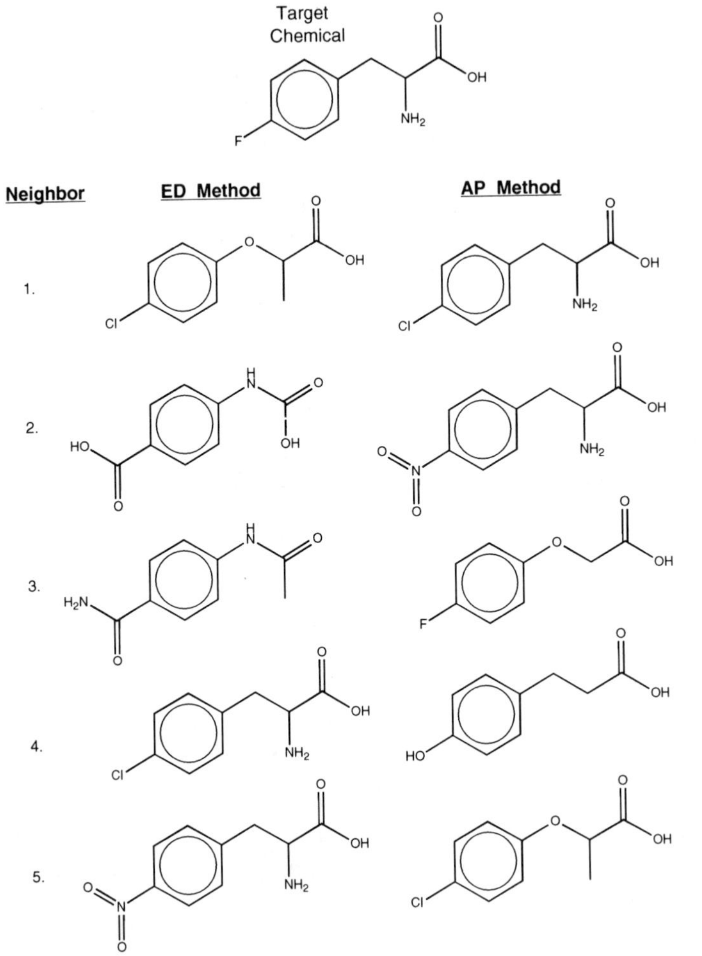

Figure 4. Target chemical and five selected analogues using ED and AP methods from the STARLIST data base of chemicals.

4.3.3.2. Similarity Methods Based on Substructures

Willett and co-workers[110–115] have developed several novel and useful techniques in molecular similarity based on substructural fragments. These approaches are based on the frequency of occurrence of generated fragment descriptors within the molecular graph. Success of these methods has been shown in 2D and 3D matchings of chemical structure, classification of chemical data bases, as well as property estimation.

4.3.4. *K* Nearest-Neighbor (KNN) Method of Estimating Properties

Basak and co-workers also used K ($K = 1–10, 15, 20, 25$) nearest neighbors of compounds in predicting properties like lipophilicity,[29] boiling point,[28,47,49,116] and mutagenicity[28,47,49,50] of diverse data bases. For a structurally diverse set of 76 compounds, lipophilicity (log P, octanol–water) could be reasonably estimated using AP ($r = 0.85$) and ED ($r = 0.85$) methods for $K = 5$.[29]

Four topologically based methods were used by Basak and Grunwald[47] in estimating the boiling point of a set of 139 hydrocarbons and a group of 15 nitrosamines using the nearest neighbor ($K = 1$).

Basak and Grunwald[50] carried out a comparative study of five molecular similarity techniques, four topologically and one physicochemically based, in estimating the mutagenicity of a set of 73 aromatic and heteroaromatic amines. Of the five methods, two measures of molecular similarity were calculated using topological descriptors, two were derived using physical properties, and the fifth was based on a combination of both topological and physicochemical parameters. The best estimated values were obtained with $K = 4–5$.

Basak and Grunwald[49] also used topologically based similarity for KNN estimation of the mutagenicity of a set of 95 aromatic amines and the boiling point of a group of over 2900 chemicals with good results.

4.4. GEOMETRICAL/SHAPE PARAMETERS IN SAR

Geometrical parameters, such as molecular shape parameters,[117] sterimol descriptors,[59] volume,[61] bulk parameters,[60,118] and 3D Wiener index,[119] have been developed and used in SARs. Such parameters are derived from the relative distances of atoms in the 3D Euclidean space. We give below some examples of QSARs using 3D descriptors.

4.4.1. van der Waals Volume (V_W)

4.4.1.1. Physicochemical Properties

Bhatnagar et al.[120] studied the relationship of boiling point with V_W for several classes of chemicals, including saturated alcohols, primary amines, and alkyl halides:

(31)
$$BP_{alcohols} = 5.019 + 127.969(V_W)$$
$$(N = 48, r = 0.964, s = 8.25, F = 605)$$

(32)
$$BP_{amines} = -60.175 + 166.419(V_W)$$
$$(N = 21, r = 0.995, s = 5.13, F = 2061)$$

(33)
$$BP_{alkyl\ halides} = -108.431 + 226.874(V_W)$$
$$(N = 24, r = 0.896, s = 16.35, F = 90)$$

Correlation of water solubility (molality) with V_W has also been determined for the saturated alcohols[120]:

(34)
$$logS = 6.908 - 8.596(V_W)$$
$$(N = 48, r = 0.974, s = 0.464, F = 860)$$

4.4.1.2. Biomedicinal Properties

Moriguchi and Kanada[61] developed a regression equation modeling the effective concentration (C) of penicillins against *Staphylococcus aureus* in mice:

(35)
$$log\,(1/C) = 5.911 - 1.692(V_W)$$
$$(N = 18, r = 0.927, s = 0.18)$$

4.4.1.3. Toxicological Properties

For tadpole narcosis of a diverse set of chemicals, the following equation has been developed[61]:

(36)
$$log\,(1/C) = -2.022 + 2.940(V_W)$$
$$(N = 53, r = 0.969, s = 0.29)$$

Correlation of nonspecific toxicity on the Madison 517 fungus, expressed as $log(1/C)$ (C is the minimum toxic dose), with V_W was found to be[61]:

(37)
$$log\,(1/C) = -1.236 + 2.645(V_W)$$
$$(N = 45, r = 0.982, s = 0.19)$$

4.4.2. Comparative Molecular Field Analysis (CoMFA) Approach

In the CoMFA method developed by Cramer *et al.*,[121] a molecule is described using electrostatic, steric, and, sometimes, hydrogen bonding fields calculated at the intersections of a 3D lattice. The partial least-squares method is used to describe statistical relationships between these fields and biological activity.

4.5. COMPARATIVE STUDY OF TOPOLOGICAL VERSUS GEOMETRICAL DESCRIPTORS IN QSARs

It is clear from the above that both topological and 3D descriptors have been extensively used in QSARs of large sets of molecules. However, no systematic work has been carried out on the relative effectiveness of TIs versus 3D parameters in the prediction of properties using QSAR models. We summarize below the results of our recent studies on the utility of graph-theoretic indexes and geometrical parameters such as 3D Wiener index and volume in estimating: (1) normal boiling point of a set of 140 hydrocarbons, (2) lipophilicity ($\log P$, octanol–water) of a diverse set of 254 molecules, and (3) mutagenic potency ($\ln R$, R being the number of revertants per nanomole in the Ames test) of a set of 95 aromatic and heteroaromatic amines.

4.5.1. Property Data Bases

4.5.1.1. Boiling Point

All normal BP data for the hydrocarbons were found in the literature. The hydrocarbons analyzed include 74 alkanes,[21] 29 alkyl benzenes,[122] and 37 polycyclic aromatic hydrocarbons.[123] Table 4 presents a list of the hydrocarbon compounds with their normal BP (°C).

Table 4. Normal Boiling Point (°C) for 140 Hydrocarbons and Predicted Boiling Point Using Equations (44) and (45)

			Predicted BP	
No.	Chemical name	Obsd. BP	Eq. (44)	Eq. (45)
1	ethane	−88.6	−108.1	−94.7
2	*n*-propane	−42.1	−61.3	−47.7
3	*n*-butane	−0.5	−16.1	−2.3
4	2-methylpropane	−11.7	−17.9	−9.3
5	*n*-pentane	36.1	26.3	36.8
6	2-methylbutane	27.8	21.6	27.6
7	2,2-dimethylpropane	9.5	15.8	22.0
8	*n*-hexane	68.7	64.6	70.5
9	2-methylpentane	60.3	51.8	59.9
10	3-methylpentane	63.3	57.6	64.1
11	2,2-dimethylbutane	49.7	51.9	53.9
12	2,3-dimethylbutane	58.0	61.6	65.0
13	*n*-heptane	98.4	99.0	99.8
14	2-methylhexane	90.0	83.6	88.9

(continued)

Table 4.   (Continued)

| No. | Chemical name | Obsd. BP | Predicted BP | |
			Eq. (44)	Eq. (45)
15	3-methylhexane	91.8	86.5	91.5
16	3-ethylpentane	93.5	91.6	98.3
17	2,2-dimethylpentane	79.2	75.6	80.3
18	2,3-dimethylpentane	89.8	90.9	91.1
19	2,4-dimethylpentane	80.5	83.6	89.3
20	3,3-dimethylpentane	86.1	81.8	83.6
21	2,2,3-trimethylbutane	80.9	88.4	88.4
22	*n*-octane	125.7	129.9	124.7
23	2-methylheptane	117.7	113.4	113.9
24	3-methylheptane	118.9	115.1	116.2
25	4-methylheptane	117.7	114.3	115.8
26	3-ethylhexane	118.5	119.2	123.0
27	2,2-dimethylhexane	106.8	102.7	104.3
28	2,3-dimethylhexane	115.6	113.7	114.5
29	2,4-dimethylhexane	109.4	114.9	112.2
30	2,5-dimethylhexane	109.1	108.8	110.4
31	3,3-dimethylhexane	112.0	105.2	106.4
32	3,4-dimethylhexane	117.7	119.0	117.9
33	2-methyl-3-ethylpentane	115.6	115.0	116.7
34	3-methyl-3-ethylpentane	118.3	111.8	115.6
35	2,2,3-trimethylpentane	109.8	111.7	109.1
36	2,2,4-trimethylpentane	99.2	106.5	105.1
37	2,3,3-trimethylpentane	114.8	115.0	112.3
38	2,3,4-trimethylpentane	113.5	120.6	120.6
39	2,2,3,3-tetramethylbutane	106.5	119.0	112.7
40	*n*-nonane	150.8	158.0	147.3
41	2-methyloctane	143.3	140.8	136.7
42	3-methyloctane	144.2	142.2	138.6
43	4-methyloctane	142.5	140.7	138.4
44	3-ethylheptane	143.0	145.4	145.8
45	4-ethylheptane	141.2	144.7	145.5
46	2,2-dimethylheptane	132.7	128.7	126.2
47	2,3-dimethylheptane	140.5	138.9	136.8
48	2,4-dimethylheptane	133.5	136.7	133.3
49	2,5-dimethylheptane	136.0	137.5	133.8
50	2,6-dimethylheptane	135.2	135.3	132.3
51	3,3-dimethylheptane	137.3	130.4	128.5
52	3,4-dimethylheptane	140.6	141.3	139.1
53	3,5-dimethylheptane	136.0	142.8	138.8
54	4,4-dimethylheptane	135.2	130.2	127.9
55	2-methyl-3-ethylhexane	138.0	139.6	138.9
56	2-methyl-4-ethylhexane	133.8	137.2	136.7
57	3-methyl-3-ethylhexane	140.6	135.9	136.5
58	3-methyl-4-ethylhexane	140.4	145.9	146.1

Table 4. (Continued)

No.	Chemical name	Obsd. BP	Predicted BP Eq. (44)	Predicted BP Eq. (45)
59	2,2,3-trimethylhexane	133.6	131.7	130.0
60	2,2,4-trimethylhexane	126.5	130.2	125.4
61	2,2,5-trimethylhexane	124.1	129.0	124.1
62	2,3,3-trimethylhexane	137.7	134.8	132.9
63	2,3,4-trimethylhexane	139.0	144.9	140.7
64	2,3,5-trimethylhexane	131.3	138.5	137.4
65	2,4,4-trimethylhexane	130.6	133.0	128.0
66	3,3,4-trimethylhexane	140.5	137.5	133.2
67	3,3-diethylpentane	146.2	145.0	144.5
68	2,2-dimethyl-3-ethylpentane	133.8	132.9	130.8
69	2,3-dimethyl-3-ethylpentane	142.0	140.0	136.7
70	2,4-dimethyl-3-ethylpentane	136.7	144.9	139.2
71	2,2,3,3-tetramethylpentane	140.3	141.3	132.8
72	2,2,3,4-tetramethylpentane	133.0	139.3	134.7
73	2,2,4,4-tetramethylpentane	122.3	130.9	128.2
74	2,3,3,4-tetramethylpentane	141.6	145.0	139.2
75	benzene	80.1	99.2	76.0
76	toluene	110.6	121.3	112.2
77	ethylbenzene	136.2	152.6	137.5
78	*o*-xylene	144.4	142.3	146.7
79	*m*-xylene	139.1	137.3	135.1
80	*p*-xylene	138.4	137.3	134.5
81	*n*-propylbenzene	159.2	182.0	163.6
82	1-methyl-2-ethylbenzene	165.2	170.1	169.2
83	1-methyl-3-ethylbenzene	161.3	166.8	158.7
84	1-methyl-4-ethylbenzene	162.0	166.0	157.9
85	1,2,3-trimethylbenzene	176.1	162.7	175.6
86	1,2,4-trimethylbenzene	169.4	161.4	166.5
87	1,3,5-trimethylbenzene	164.7	162.1	167.5
88	*n*-butylbenzene	183.3	209.0	190.4
89	1,2-diethylbenzene	183.4	195.2	192.8
90	1,3-diethylbenzene	181.1	193.0	189.2
91	1,4-diethylbenzene	183.8	194.4	188.1
92	1-methyl-2-*n*-propylbenzene	184.8	194.8	193.5
93	1-methyl-3-*n*-propylbenzene	181.8	191.0	183.5
94	1-methyl-4-*n*-propylbenzene	183.8	190.4	182.5
95	1,2-dimethyl-3-ethylbenzene	193.9	187.4	196.2
96	1,2-dimethyl-4-ethylbenzene	189.8	186.6	187.5
97	1,3-dimethyl-2-ethylbenzene	190.0	187.2	193.5
98	1,3-dimethyl-4-ethylbenzene	188.4	186.4	190.4
99	1,3-dimethyl-5-ethylbenzene	183.8	187.0	188.6
100	1,4-dimethyl-2-ethylbenzene	186.9	187.2	190.7
101	1,2,3,4-tetramethylbenzene	205.0	185.2	202.9
102	1,2,3,5-tetramethylbenzene	198.2	185.1	198.8

(continued)

Table 4.   (Continued)

No.	Chemical name	Obsd. BP	Predicted BP	
			Eq. (44)	Eq. (45)
103	1,2,4,5-tetramethylbenzene	196.8	185.7	198.5
104	naphthalene	218.0	228.5	209.9
105	acenaphthalene	270.0	234.4	267.6
106	acenaphthene	279.0	—	272.0
107	fluorene	294.0	299.2	295.7
108	phenanthrene	338.0	346.4	329.8
109	anthracene	340.0	344.8	330.6
110	4*H*-cyclopenta(*def*)phenanthrene	359.0	326.9	349.7
111	fluoranthene	383.0	416.0	378.4
112	pyrene	393.0	392.2	384.3
113	benzo(*a*)fluorene	403.0	386.0	406.4
114	benzo(*b*)fluorene	398.0	390.8	406.4
115	benzo(*c*)fluorene	406.0	386.3	404.5
116	benzo(*ghi*)fluoranthene	422.0	443.4	427.0
117	cyclopenta(*cd*)pyrene	439.0	—	435.7
118	chrysene	431.0	445.8	439.2
119	benz(*a*)anthracene	425.0	440.1	439.7
120	triphenylene	429.0	454.6	433.6
121	naphthacene	440.0	445.2	444.2
122	benzo(*b*)fluoranthene	481.0	503.9	476.0
123	benzo(*j*)fluoranthene	480.0	492.2	474.8
124	benzo(*k*)fluoranthene	481.0	501.1	489.9
125	benzo(*a*)pyrene	496.0	485.9	487.9
126	benzo(*e*)pyrene	493.0	490.8	484.4
127	perylene	497.0	484.0	479.7
128	anthanthrene	547.0	527.7	539.2
129	benzo(*ghi*)perylene	542.0	526.9	529.7
130	indeno(1,2,3-*cd*)fluoranthene	531.0	547.3	541.2
131	indeno(1,2,3-*cd*)pyrene	534.0	540.1	534.5
132	dibenz(*a,c*)anthracene	535.0	535.2	533.7
133	dibenz(*a,h*)anthracene	535.0	528.9	546.2
134	dibenz(*a,j*)anthracene	531.0	529.6	543.6
135	picene	519.0	531.1	545.1
136	coronene	590.0	575.6	591.6
137	dibenzo(*a,e*)pyrene	592.0	574.9	581.6
138	dibenzo(*a,h*)pyrene	596.0	569.8	591.8
139	dibenzo(*a,i*)pyrene	594.0	569.2	591.1
140	dibenzo(*a,l*)pyrene	595.0	573.9	590.5

4.5.1.2. Lipophilicity ($\log P$, Octanol–Water)

The 254 chemicals used to model $\log P$ are presented in Table 5. These chemicals were a subset of 382 chemicals studied by us previously[27] and consist of only those compounds with measured $\log P$ available in STARLIST,[109] a selected subset of data deemed to be of very high quality by experts in the field. The $\log P$ values are provided in Table 5.

Table 5. Observed $\log P$ and Estimated $\log P$ from Equations (46) and (47) for 254 Diverse Chemicals

| No. | Chemical name | Obsd. $\log P$ | Estimated $\log P$ | |
			Eq. (46)	Eq. (47)
1	butane	2.89	1.61	1.69
2	pentane	3.39	2.08	2.16
3	cyclopentane	3.00	2.61	2.46
4	cyclohexane	3.44	2.27	2.46
5	1-butene	2.40	1.65	1.57
6	1-hexene	3.39	2.64	2.63
7	cyclohexene	2.86	2.44	2.48
8	1-pentyne	1.98	1.66	1.63
9	ethylchloride	1.43	1.05	1.20
10	1-chloropropane	2.04	1.43	1.65
11	1-chlorobutane	2.64	1.81	2.04
12	1-chloroheptane	4.15	3.09	3.32
13	carbon tetrachloride	2.83	2.25	2.49
14	1,2-dichloroethane	1.48	1.60	2.03
15	1,1,1-trichloroethane	2.49	1.88	2.19
16	1,1,2,2-tetrachloroethane	2.39	2.95	3.36
17	trichloroethylene	2.42	2.86	3.16
18	tetrachloroethylene	3.40	3.79	4.04
19	trichlorofluoromethane	2.53	2.66	2.46
20	benzene	2.13	2.21	2.14
21	toluene	2.73	2.59	2.43
22	*o*-xylene	3.12	3.09	2.98
23	*m*-xylene	3.20	3.08	3.00
24	*p*-xylene	3.15	2.89	2.79
25	1,3,5-trimethylbenzene	3.42	3.61	3.58
26	1,2,4-trimethylbenzene	3.78	3.62	3.57
27	1,2,3-trimethylbenzene	3.66	3.60	3.50
28	1,2,3,4-tetramethylbenzene	4.11	3.96	3.84
29	1,2,3,5-tetramethylbenzene	4.17	4.25	4.14
30	1,2,4,5-tetramethylbenzene	4.00	3.98	3.85
31	pentamethylbenzene	4.56	4.70	4.73
32	hexamethylbenzene	5.11	5.15	5.13
33	ethylbenzene	3.15	2.94	2.83

(continued)

Table 5.   (Continued)

No.	Chemical name	Obsd. log P	Estimated log P	
			Eq. (46)	Eq. (47)
34	propylbenzene	3.72	3.37	3.30
35	isopropylbenzene	3.66	3.35	3.30
36	butylbenzene	4.26	3.82	3.82
37	*t*-butylbenzene	4.11	3.76	3.71
38	*p*-cymene	4.10	3.85	3.78
39	fluorobenzene	2.27	2.37	2.00
40	chlorobenzene	2.84	2.37	2.42
41	bromobenzene	2.99	2.37	2.65
42	iodobenzene	3.25	2.37	3.15
43	*o*-dichlorobenzene	3.38	2.97	3.12
44	1,3-dichlorobenzene	3.60	2.97	3.15
45	*p*-dichlorobenzene	3.52	2.89	3.03
46	1,2,3-trichlorobenzene	4.05	3.71	3.92
47	1,2,4-trichlorobenzene	4.02	3.56	3.74
48	1,3,5-trichlorobenzene	4.15	3.66	3.95
49	1,2,3,4-tetrachlorobenzene	4.64	4.34	4.55
50	1,2,3,5-tetrachlorobenzene	4.92	4.32	4.57
51	1,2,4,5-tetrachlorobenzene	4.82	4.31	4.56
52	pentachlorobenzene	5.17	5.15	5.40
53	hexachlorobenzene	5.31	6.05	6.27
54	*o*-dibromobenzene	3.64	2.97	3.52
55	*p*-dibromobenzene	3.79	2.89	3.48
56	*o*-chlorotoluene	3.42	2.87	2.88
57	*m*-chlorotoluene	3.28	2.87	2.90
58	*p*-chlorotoluene	3.33	2.79	2.83
59	ethyl ether	0.89	1.23	1.21
60	dipropyl ether	2.03	2.12	2.15
61	dibutyl ether	3.21	2.91	2.96
62	tetrahydrofuran	0.46	1.82	1.50
63	ethyl vinyl ether	1.04	1.25	1.09
64	anisole	2.11	2.08	1.91
65	*o*-methylanisole	2.74	2.70	2.58
66	*m*-methylanisole	2.66	2.69	2.59
67	*p*-methylanisole	2.81	2.59	2.50
68	4-chloroanisole	2.78	2.36	2.39
69	phenetole	2.51	2.50	2.41
70	phenyl propyl ether	3.18	2.95	2.88
71	formic acid, propyl ester	0.83	1.16	0.92
72	acetic acid, methyl ester	0.18	0.72	0.18
73	acetic acid, ethyl ester	0.73	1.04	0.70
74	propionic acid, ethyl ester	1.21	1.49	1.23
75	acrylic acid, methyl ester	0.80	1.03	0.50
76	methacrylic acid, methyl ester	1.38	1.35	0.91
77	benzoic acid, methyl ester	2.12	2.29	2.08

Table 5. (Continued)

No.	Chemical name	Obsd. log P	Estimated log P	
			Eq. (46)	Eq. (47)
78	ethyl benzoate	2.64	2.68	2.46
79	*o*-toluic acid, methyl ester	2.75	2.87	2.68
80	acetic acid, benzy lester	1.96	2.70	2.58
81	acetic acid, β-phenylethyl ester	2.30	3.00	2.95
82	phenylacetic acid, methy lester	1.83	2.69	2.53
83	β-phenylpropionic acid, ethyl ester	2.73	3.36	3.32
84	benzyl benzoate	3.97	3.68	3.70
85	acetic acid, phenyl ester	1.49	2.32	2.09
86	*o*-tolylacetate	1.93	2.89	2.72
87	*m*-tolylacetate	2.09	2.81	2.61
88	*p*-tolylacetate	2.11	2.81	2.66
89	2-chlorophenyl acetate	2.18	2.69	2.59
90	3-chlorophenyl acetate	2.32	2.62	2.56
91	2-bromophenyl acetate	2.20	2.69	2.76
92	propionaldehyde	0.59	0.87	0.64
93	butyraldehyde	0.88	1.25	1.16
94	hexaldehyde	1.78	2.16	2.13
95	benzaldehyde	1.48	2.11	1.87
96	acetone	−0.24	0.47	0.17
97	2-butanone	0.29	1.20	0.94
98	2-pentanone	0.91	1.63	1.46
99	2-hexanone	1.38	2.20	2.10
100	2-heptanone	1.98	2.58	2.50
101	cyclohexanone	0.81	1.85	1.87
102	acetophenone	1.58	2.52	2.36
103	*m*-chloroacetophenone	2.51	2.89	2.86
104	*p*-chloroacetophenone	2.32	2.76	2.77
105	*p*-bromoacetophenone	2.43	2.76	2.95
106	*p*-fluoroacetophenone	1.72	2.76	2.42
107	*p*-methylacetophenone	2.10	2.99	2.86
108	propiophenone	2.19	2.92	2.76
109	1-phenyl-2-propanone	1.44	2.95	2.77
110	ethylamine	−0.13	−0.66	−0.51
111	propylamine	0.48	−0.13	0.23
112	butylamine	0.97	0.29	0.70
113	amylamine	1.49	0.81	1.22
114	hexylamine	2.06	1.23	1.64
115	heptylamine	2.57	1.64	2.06
116	diethylamine	0.58	0.72	1.00
117	dipropylamine	1.67	1.64	1.93
118	dibutylamine	2.83	2.43	2.70
119	trimethylamine	0.16	0.11	0.13
120	triethylamine	1.45	1.99	1.95
121	tripropylamine	2.79	3.30	3.25

(continued)

Table 5. (Continued)

No.	Chemical name	Obsd. log P	Estimated log P	
			Eq. (46)	Eq. (47)
122	aniline	0.90	0.71	0.89
123	*o*-toluidine	1.32	1.31	1.51
124	*m*-toluidine	1.40	1.31	1.54
125	*p*-toluidine	1.39	1.23	1.43
126	*m*-chloroaniline	1.88	1.12	1.46
127	*p*-chloroaniline	1.83	1.04	1.38
128	*m*-bromoaniline	2.10	1.12	1.69
129	*p*-bromoaniline	2.26	1.04	1.62
130	*m*-fluoroaniline	1.30	1.12	1.05
131	*p*-fluoroaniline	1.15	1.04	0.97
132	benzidine	1.34	1.37	1.94
133	α-naphthylamine	2.25	2.20	2.39
134	β-naphthylamine	2.28	2.16	2.31
135	*N,N*-dimethylaniline	2.31	2.45	2.47
136	*N,N*-dimethyl-*p*-toluidine	2.81	2.95	2.96
137	*N,N*-diethylaniline	3.31	3.40	3.44
138	*N,N*-dimethylbenzylamine	1.98	2.90	2.94
139	pyridine	0.65	1.45	1.46
140	3-methylpyridine	1.20	1.69	1.65
141	3-chloropyridine	1.33	1.67	1.81
142	3-bromopyridine	1.60	1.67	2.07
143	4-bromopyridine	1.54	1.67	2.08
144	acetonitrile	−0.34	0.45	0.24
145	propionitrile	0.16	0.93	0.84
146	butyronitrile	0.53	1.27	1.21
147	benzonitrile	1.56	2.17	2.01
148	phenylacetonitrile	1.56	2.54	2.40
149	benzylacetonitrile	1.72	2.96	2.93
150	acrylonitrile	0.25	1.19	0.99
151	nitromethane	−0.35	−0.37	−1.01
152	nitroethane	0.18	0.47	0.10
153	1-nitropropane	0.87	0.75	0.52
154	1-nitrobutane	1.47	1.21	1.03
155	1-nitropentane	2.01	1.57	1.45
156	nitrobenzene	1.85	1.64	1.38
157	*m*-nitrotoluene	2.45	2.13	1.99
158	*p*-nitrotoluene	2.37	2.02	1.87
159	2-chloro-1-nitrobenzene	2.24	2.12	2.02
160	3-chloro-1-nitrobenzene	2.41	2.12	2.07
161	4-chloro-1-nitrobenzene	2.39	2.01	1.98
162	3-bromo-1-nitrobenzene	2.64	2.12	2.30
163	4-bromo-1-nitrobenzene	2.55	2.01	2.20
164	*m*-dinitrobenzene	1.49	1.71	1.43
165	*p*-dinitrobenzene	1.46	1.63	1.38

Table 5. (Continued)

No.	Chemical name	Obsd. log P	Estimated log P Eq. (46)	Estimated log P Eq. (47)
166	dimethylformamide	−1.01	0.19	−0.03
167	N,N-dimethylacetamide	−0.77	0.61	0.44
168	diethylacetamide	0.34	1.74	1.60
169	benzamide	0.64	0.75	0.85
170	dimethylsulfoxide	−1.35	0.07	0.36
171	diethylsulfide	1.95	1.72	2.06
172	methanol	−0.77	−0.13	−0.66
173	ethanol	−0.31	−0.07	−0.30
174	propanol	0.25	0.39	0.43
175	butanol	0.88	0.80	0.97
176	isobutanol	0.76	0.64	0.67
177	pentanol	1.56	1.30	1.50
178	isopentanol	1.42	1.10	1.21
179	hexanol	2.03	1.71	1.89
180	octanol	2.97	2.45	2.70
181	allyl alcohol	0.17	0.38	0.26
182	isopropanol	0.05	−0.00	−0.05
183	s-butanol	0.61	0.78	0.82
184	3-pentanol	1.21	1.04	1.07
185	cyclohexanol	1.23	1.49	1.75
186	t-butanol	0.35	0.26	0.22
187	2-ethyl-2-propanol	0.89	1.02	1.02
188	benzyl alcohol	1.10	1.59	1.54
189	m-methylbenzyl alcohol	1.60	2.20	2.23
190	p-methylbenzyl alcohol	1.58	2.10	2.12
191	m-chlorobenzyl alcohol	1.94	1.97	2.13
192	p-chlorobenzyl alcohol	1.96	1.87	2.02
193	2-phenylethanol	1.36	2.01	2.08
194	3-phenylalcohol	1.88	2.46	2.60
195	cinnamyl alcohol	1.95	2.36	2.38
196	phenol	1.46	1.28	1.19
197	m-methylphenol	1.96	1.84	1.83
198	p-methylphenol	1.94	1.75	1.73
199	m-chlorophenol	2.50	1.70	1.83
200	p-chlorophenol	2.39	1.62	1.75
201	m-bromophenol	2.63	1.70	2.08
202	p-bromophenol	2.59	1.62	1.99
203	m-fluorophenol	1.93	1.70	1.40
204	p-fluorophenol	1.77	1.62	1.27
205	acetic acid	-0.17	-0.19	-0.74
206	propionic acid	0.33	0.23	-0.13
207	butyric acid	0.79	0.55	0.39
208	valeric acid	1.39	1.07	1.01
209	hexanoic acid	1.92	1.43	1.43

(continued)

Table 5.   (Continued)

No.	Chemical name	Obsd. log P	Estimated log P	
			Eq. (46)	Eq. (47)
210	decanoic acid	4.09	2.69	2.86
211	benzoic acid	1.87	1.47	1.32
212	*m*-toluic acid	2.37	2.01	1.95
213	*p*-toluic acid	2.27	1.88	1.78
214	*m*-chlorobenzoic acid	2.68	1.89	1.93
215	*p*-chlorobenzoic acid	2.65	1.76	1.84
216	*m*-bromobenzoic acid	2.87	1.89	2.13
217	*p*-bromobenzoic acid	2.86	1.76	2.01
218	*m*-fluorobenzoic acid	2.15	1.89	1.55
219	*p*-fluorobenzoic acid	2.07	1.76	1.48
220	phenylacetic acid	1.41	1.83	1.72
221	*m*-chlorophenylacetic acid	2.09	2.13	2.23
222	*p*-chlorophenylacetic acid	2.12	2.12	2.23
223	*m*-bromophenylacetic acid	2.37	2.13	2.42
224	*o*-fluorophenylacetic acid	1.50	2.20	1.93
225	*m*-fluorophenylacetic acid	1.65	2.13	1.91
226	*p*-fluorophenylacetic acid	1.55	2.12	1.89
227	β-phenylpropionic acid	1.84	2.21	2.27
228	4-phenylbutyric acid	2.42	2.51	2.61
229	1-naphthoic acid	3.10	2.79	2.76
230	naphthalene	3.30	3.59	3.36
231	1-methylnaphthalene	3.87	4.07	3.91
232	2-methylnaphthalene	3.86	4.03	3.83
233	1,3-dimethylnaphthalene	4.42	4.53	4.39
234	1,4-dimethylnaphthalene	4.37	4.56	4.40
235	1,5-dimethylnaphthalene	4.38	4.46	4.28
236	2,3-dimethylnaphthalene	4.40	4.50	4.37
237	2,6-dimethylnaphthalene	4.31	4.46	4.32
238	1-nitronaphthalene	3.19	2.95	2.82
239	anthracene	4.45	4.80	4.59
240	9-methylanthracene	5.07	5.15	5.03
241	phenanthracene	4.46	4.83	4.66
242	pyrene	4.88	5.49	5.24
243	fluorene	4.18	3.69	3.77
244	acenaphthene	3.92	3.68	3.69
245	quinoline	2.03	2.69	2.60
246	isoquinoline	2.08	2.69	2.60
247	2,2'-biquinoline	4.31	4.49	4.63
248	biphenyl	4.09	4.10	3.97
249	2-chlorobiphenyl	4.38	4.27	4.24
250	2,4'-dichlorobiphenyl	5.10	4.56	4.62
251	2,5-PCB	5.16	4.59	4.68
252	2,6-PCB	4.93	4.67	4.72
253	2,4,6-PCB	5.47	4.99	5.09
254	bibenzyl	4.79	4.55	4.52

4.5.1.3. Mutagenicity (ln*R*)

The set of compounds used to model mutagenic potency consisted of 95 aromatic and heteroaromatic amines available from the literature.[124] A list of these chemicals and their mutagenic potency is presented in Table 6. The mutagenic potency of the aromatic amines in *S. typhimurium* TA98 + S9 microsomal preparation is expressed by the natural logarithm of the number of revertants per nanomole.

Table 6. Mutagenicity (ln*R*)[a] of 95 Aromatic and Heteroaromatic Amines and Predicted Mutagenicity by Equations (48) and (49)

			Predicted ln*R*	
No.	Chemical name	Obsd. ln*R*	Eq. (48)	Eq. (49)
1	2-bromo-7-aminofluorene	2.62	2.14	2.66
2	2-methoxy-5-methylaniline	−2.05	−2.57	−2.07
3	5-aminoquinoline	−2.00	−1.60	−1.71
4	4-ethoxyaniline	−2.30	−3.75	−3.40
5	1-aminonaphthalene	−0.60	−0.93	−0.86
6	4-aminofluorene	1.13	0.73	1.02
7	2-aminoanthracene	2.62	1.26	1.22
8	7-aminofluoranthene	2.88	1.60	2.27
9	8-aminoquinoline	−1.14	−1.79	−2.02
10	1,7-diaminophenazine	0.75	0.18	0.23
11	2-aminonaphthalene	−0.67	0.21	−0.42
12	4-aminopyrene	3.16	2.99	2.89
13	3-amino-3′-nitrobiphenyl	−0.55	−0.26	0.19
14	2,4,5-trimethylaniline	−1.32	−1.20	−0.55
15	3-aminofluorene	0.89	1.35	1.37
16	3,3′-dichlorobenzidine	0.81	0.24	0.95
17	2,4-dimethylaniline	−2.22	−2.88	−2.34
18	2,7-diaminofluorene	0.48	0.85	1.02
19	3-aminofluoranthene	3.31	2.88	3.06
20	2-aminofluorene	1.93	1.75	1.74
21	2-amino-4′-nitrobiphenyl	−0.62	0.06	0.20
22	4-aminobiphenyl	−0.14	0.10	−0.04
23	3-methoxy-4-methylaniline	−1.96	−3.21	−2.55
24	2-aminocarbazole	0.60	0.61	0.25
25	2-amino-5-nitrophenol	−2.52	−2.65	−3.16
26	2,2′-diaminobiphenyl	−1.52	−0.42	−0.46
27	2-hydroxy-7-aminofluorene	0.41	1.29	1.44
28	1-aminophenanthrene	2.38	1.06	1.19
29	2,5-dimethylaniline	−2.40	−2.41	−2.34
30	4-amino-2′-nitrobiphenyl	−0.92	0.06	0.09
31	2-amino-4-methylphenol	−2.10	−3.59	−2.88
32	2-aminophenazine	0.55	0.83	0.66

(continued)

Table 6.   (Continued)

No.	Chemical name	Obsd. ln$R$	Predicted ln$R$ Eq. (48)	Eq. (49)
33	4-aminophenyl sulfide	0.31	0.32	0.18
34	2,4-dinitroaniline	−2.00	−0.59	−1.54
35	2,4-diaminoisopropylbenzene	−3.00	−1.79	−1.35
36	2,4-difluoroaniline	−2.70	−1.95	−2.68
37	4,4′-methylenedianiline	−1.60	−1.23	−1.19
38	3,3′-dimethylbenzidine	0.01	−0.65	−0.76
39	2-aminofluoranthene	3.23	2.51	2.79
40	2-amino-3′-nitrobiphenyl	−0.89	−0.33	0.04
41	1-aminofluoranthene	3.35	2.61	2.92
42	4,4′-ethylenebis(aniline)	−2.15	−1.79	−1.59
43	4-chloroaniline	−2.52	−2.54	−2.52
44	2-aminophenanthrene	2.46	2.02	1.59
45	4-fluoroaniline	−3.32	−2.85	−3.01
46	9-aminophenanthrene	2.98	1.33	1.26
47	3,3′-diaminobiphenyl	−1.30	−1.28	−1.14
48	2-aminopyrene	3.50	3.43	3.01
49	2,6-dichloro-1,4-phenylenediamine	−0.69	−1.50	−1.78
50	2-amino-7-acetamidofluorene	1.18	1.09	1.46
51	2,8-diaminophenazine	1.12	0.17	0.34
52	6-aminoquinoline	−2.67	−1.31	−1.51
53	4-methoxy-2-methylaniline	−3.00	−3.07	−2.55
54	3-amino-2′-nitrobiphenyl	−1.30	−0.22	−0.10
55	2,4′-diaminobiphenyl	−0.92	0.08	−0.31
56	1,6-diaminophenazine	0.20	0.41	0.54
57	4-aminophenyl disulfide	−1.03	−0.12	0.43
58	2-bromo-4,6-dinitroaniline	−0.54	−1.05	−1.33
59	2,4-diamino-n-butylbenzene	−2.70	−2.93	−3.09
60	4-aminophenyl ether	−1.14	−0.50	−0.56
61	2-aminobiphenyl	−1.49	0.02	−0.40
62	1,9-diaminophenazine	0.04	0.13	0.26
63	1-aminofluorene	0.43	0.99	1.11
64	8-aminofluoranthene	3.80	2.77	3.01
65	2-chloroaniline	−3.00	−3.16	−2.95
66	3-amino-$\alpha\alpha\alpha$-trifluorotoluene	−0.80	−0.78	−1.16
67	2-amino-1-nitronaphthalene	−1.17	−0.24	−0.47
68	3-amino-4′-nitrobiphenyl	0.69	−0.29	0.28
69	4-bromoaniline	−2.70	−2.23	−2.04
70	2-amino-4-chlorophenol	−3.00	−2.56	−2.68
71	3,3′-dimethoxybenzidine	0.15	−0.50	−0.43
72	4-cyclohexylaniline	−1.24	−1.97	−1.89
73	4-phenoxyaniline	0.38	−0.28	−0.80
74	4,4′-methylenebis(o-ethylaniline)	−0.99	−0.11	−1.62
75	2-amino-7-nitrofluorene	3.00	1.56	2.36
76	Benzidine	−0.39	−0.98	−0.94

Table 6. (Continued)

No.	Chemical name	Obsd. ln$R$	Predicted ln$R$	
			Eq. (48)	Eq. (49)
77	1-amino-4-nitronaphthalene	−1.77	−0.21	−0.36
78	4-amino-3′-nitrobiphenyl	1.02	−0.36	0.08
79	4-amino-4′-nitrobiphenyl	1.04	−0.27	0.28
80	1-aminophenazine	−0.01	0.67	0.51
81	4,4′-methylenebis(*o*-fluoroaniline)	0.23	0.46	0.66
82	4-chloro-2-nitroaniline	−2.22	−2.31	−2.79
83	3-aminoquinoline	−3.14	−1.50	−1.82
84	3-aminocarbazole	−0.48	0.84	0.51
85	4-chloro-1,2-phenylenediamine	−0.49	−1.50	−1.68
86	3-aminophenanthrene	3.77	1.64	1.30
87	3,4′-diaminobiphenyl	0.20	−0.74	−1.03
88	1-aminoanthracene	1.18	1.32	1.33
89	1-aminocarbazole	−1.04	0.14	−0.11
90	9-aminoanthracene	0.87	1.65	1.45
91	4-aminocarbazole	−1.42	0.33	0.06
92	6-aminochrysene	1.83	2.49	3.31
93	1-aminopyrene	1.43	2.91	2.88
94	4,4′-methylenebis(*o*-isopropylaniline)	−1.77	0.71	−1.29
95	2,7-diaminophenazine	3.97	1.42	1.10

[a]ln$R$ = log revertants per nanomole, *S. typhimurium* TA98 with metabolic activation.

4.5.2. Calculation of Parameters

4.5.2.1. Computation of Topological Indexes

The first TI reported in the chemical literature, the Wiener index W,[87] may be calculated from the distance matrix $D(G)$ of a hydrogen-suppressed chemical graph G as the sum of the entries in the upper triangular distance submatrix. The distance matrix $D(G)$ of a nondirected graph G with n vertices is a symmetric $n \times n$ matrix (d_{ij}), where d_{ij} is equal to the topological distance between vertices v_i and v_j in G. Each diagonal element d_{ii} of $D(G)$ is zero. We give below the distance matrix $D(G_1)$ of the labeled hydrogen-suppressed graph G_1 of isobutane (Figure 5):

$$D(G_1) = \begin{array}{c} \\ 1 \\ 2 \\ 3 \\ 4 \end{array} \begin{array}{cccc} (1) & (2) & (3) & (4) \\ \left[\begin{array}{cccc} 0 & 1 & 2 & 2 \\ 1 & 0 & 1 & 1 \\ 2 & 1 & 0 & 2 \\ 2 & 1 & 2 & 0 \end{array}\right] \end{array}$$

W is calculated as:

(38)
$$W = {}^1\!/_2 \sum_{i,j} d_{ij} = \sum_h h \cdot g_h$$

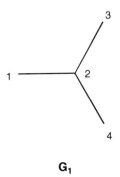

G₁

Figure 5.   Labeled hydrogen-suppressed graph of isobutane.

where g_h is the number of unordered pairs of vertices whose distance is h.

Randić's connectivity index,[35] higher-order connectivity indexes, and path, cluster, path–cluster, and chain types of simple, bond and valence connectivity parameters developed by Kier and Hall[77] were calculated by a computer program POLLY 2.3 developed by Basak, Harriss, and Magnuson[125] at the University of Minnesota. Also, P_h parameters, the number of paths of length h ($h = 0$–10) in the hydrogen-suppressed graph, are calculated using standard algorithms.

Balaban[22–24] defined a series of indexes based on distance sums within the distance matrix for a molecular graph which he designated as J indexes. Unlike W, these indexes are independent of molecular size and have low degeneracy.

Information-theoretic TIs are calculated by the application of information theory on molecular graphs. An appropriate set A of n elements is derived from a molecular graph G depending on certain structural characteristics. On the basis of an equivalence relation defined on A, the set A is partitioned into disjoint subsets A_i of order n_i ($i = 1$, $2, \ldots, h$; $\Sigma_i n_i = n$). A probability distribution is then assigned to the set of equivalence classes:

$$A_1, A_2, \ldots, A_h$$

$$p_1, p_2, \ldots, p_h$$

where $p_i = n_i/n$ is the probability that a randomly selected element of A will occur in the ith subset.

The mean information content of an element of A is defined by Shannon's relation[126]:

(39)
$$IC = -\sum_{i=1}^{h} p_i \log_2 p_i$$

The logarithm base 2 is used to measure the information content in bits. The total information content of the set A is then n times IC.

To account for the chemical nature of vertices, as well as their bonding pattern, Sarkar, Roy, and Sarkar[127] calculated information content of molecular graphs on the basis of an equivalence relation where two atoms of the same element are considered equivalent if they possess an identical first-order topological neighborhood. Since properties of atoms or reaction centers are often modulated by physicochemical characteristics of distant neighbors, i.e., neighbors of neighbors, it was deemed essential to extend this approach to account for higher-order neighbors of vertices. This can be accomplished by defining open spheres for all vertices of a molecular graph. If r is any nonnegative real number, and v is a vertex of the graph G, then the open sphere $S(v,r)$ is defined as the set consisting of all vertices v_i in G such that $d(v,v_i) < r$. Obviously, $S(v,0) = \phi$, $S(v,r) = v$ for $0 < r < 1$, and if $1 < r < 2$, then $S(v,r)$ is the set consisting of v and all vertices v_i of G situated at unit distance from v.

One can construct such open spheres for higher integral values of r. For a particular value of r, the collection of all such open spheres $S(v,r)$, where v runs over the whole vertex set V, forms a neighborhood system of the vertices of G. A suitably defined equivalence relation can then partition V into disjoint subsets consisting of topological neighborhoods of vertices up to r^{th} order neighbors. Such an approach has already been developed and the information-theoretic indexes calculated are called indexes of neighborhood symmetry.[128]

In this method, chemical species are symbolized by weighted linear graphs. Two vertices u_0 and v_0 of a molecular graph are said to be equivalent with respect to r^{th} order neighborhood if and only if corresponding to each path $u_0, u_1, \ldots, u_r$ of length r, there is a distinct path $v_0, v_1, \ldots, v_r$ of the same length such that the paths have similar edge weights, and both u_0 and v_0 are connected to the same number and type of atoms up to the rth order bonded neighbors. The detailed equivalence relation is described in our earlier studies.[128]

Once partitioning of the vertex set for a particular order of neighborhood is completed, IC_r is calculated by equation (39). Basak, Roy, and Ghosh[129] defined another information-theoretic measure, structural information content (SIC_r), which is calculated as:

$$(40) \qquad\qquad SIC_r = IC_r / \log_2 n$$

where IC_r is calculated from equation (39) and n is the total number of vertices of the graph.

Another information-theoretic invariant, complementary information content (CIC_r), is defined as[81]:

$$(41) \qquad\qquad CIC_r = \log_2 n - IC_r$$

CIC_r represents the difference between maximum possible complexity of a graph (where each vertex belongs to a separate equivalence class) and the realized topological information of a chemical species as defined by IC_r. Figure 6 provides an example of the first order ($r = 1$) calculations of IC, SIC, and CIC.

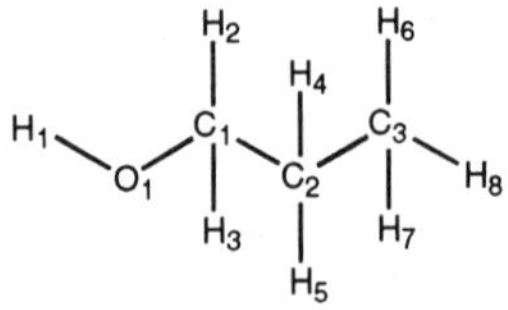

$$
\begin{array}{c}
H_2 \quad\quad H_6 \\
| \quad H_4 \quad | \\
H_1{-}O_1{-}C_1{-}C_2{-}C_3{-}H_8 \\
| \quad | \quad | \\
H_3 \quad H_7 \\
H_5
\end{array}
$$

First order neighbors:

I	II	III	IV	V	VI

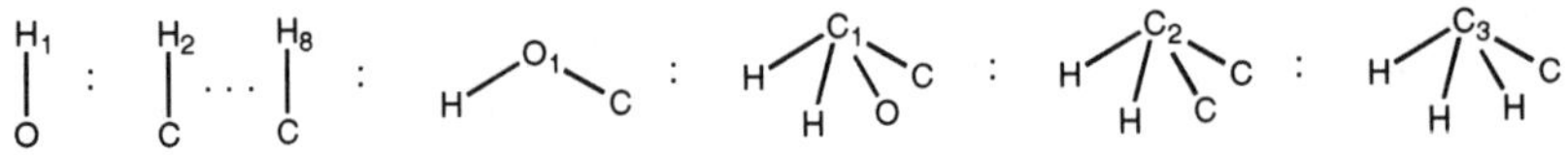

$$
\begin{array}{ccccc}
H_1 & H_2 \quad H_8 \\
| & | \quad\cdots\quad | \\
O & C \quad\quad C
\end{array}
$$

III. $H{-}O_1{-}C$: IV. $\overset{C_1}{H{-}\underset{H}{C}{-}O}$: V. $\overset{C_2}{H{-}\underset{H}{C}{-}C}$: VI. $\overset{C_3}{H{-}\underset{H}{C}{-}H}$

Subsets:

I	II	III	IV	V	VI
(H_1)	$(H_2{-}H_8)$	(O_1)	(C_1)	(C_2)	(C_3)

Probability:

I	II	III	IV	V	VI
1/12	7/12	1/12	1/12	1/12	1/12

$$
\begin{aligned}
IC_1 &= 5*1/12*\log_2 12 + 7/12*\log_2 12/7 = 1.950 \text{ bits} \\
SIC_1 &= IC_1/\log_2 12 = 0.544 \text{ bits} \\
CIC_1 &= \log_2 12 - IC_1 = 1.635 \text{ bits}
\end{aligned}
$$

Figure 6. Derivation of first-order neighborhoods and calculation of complexity indexes (IC_1, SIC_1, and CIC_1) for *n*-propanol.

The information-theoretic index on graph distance, I_D^W is calculated from the distance matrix $D(G)$ of a molecular graph G as follows[34]:

$$
I_D^W = W \log_2 W - \sum_h g_h \cdot h \log_2 h
$$

The mean information index, $\overline{I}_D^W$, is found by dividing the information index I_D^W by W. IC_r, SIC_r, CIC_r, I_D^W, and $\overline{I}_D^W$ were calculated by POLLY 2.3.[125] The information-theoretic parameters defined on the distance matrix, H^D and H^V, were calculated by the method of Raychaudhury et al.[36]

4.5.2.2. Computation of Geometrical Parameters

Volume (V_W) was calculated using the SYBYL[130] package from Tripos Associates, Inc. The 3D Wiener numbers were calculated using SYBYL with an SPL (Sybyl Programming Language) program. Calculation of 3D Wiener numbers consists of the sum of entries in the upper triangular submatrix of the topographic Euclidean distance matrix for a molecule. The 3D coordinates for the atoms were determined using CONCORD 3.0.1.[131] Two variants of the 3D Wiener number were calculated. For $^{3D}W_H$, hydrogen atoms are included in the computations and for ^{3D}W, hydrogen atoms are excluded from the computations.

4.5.2.3. Computation of HB_1

The hydrogen bonding parameter HB_1 was calculated using a program developed by Basak.[132] This program is based on the ideas of Ou et al.[133]

The list of parameters used in this chapter is given in Table 2.

4.5.3. Statistical Methods

4.5.3.1. Index Selection

Since the scale of the TIs vary by several orders of magnitude, each TI was transformed by the natural log of the index plus one.

The large number of TIs, and the fact that many of them are highly correlated, confounds the development of predictive models. Therefore, we attempted to reduce the number of TIs to a smaller set of relatively independent variables. Variable clustering[134] was used to divide the TIs into disjoint subsets (clusters) that are essentially unidimensional. These clusters form new variables which are the first principal component derived from the members of the cluster. From each cluster of indexes, a single index was selected. The index chosen was the one most correlated with the cluster variable. In some cases, a member of a cluster showed poor group membership relative to the other members of the cluster, i.e., the correlation of an index with the cluster variable was much lower than the other members. Any variable showing poor cluster membership was selected for further studies as well. A correlation of a TI with the cluster variable less than 0.7 was used as the definition of poor cluster membership.

4.5.3.2. Regression Analysis

The variables used to model each of the properties in this study were TIs, HB_1 and three geometry-related parameters, volume (V_W) and the two 3D Wiener numbers

(^{3D}W and $^{3D}W_H$). The TIs were restricted to those selected by the variable clustering procedure described previously.

All subsets regression was used for the development of the models. The criteria used for defining the "best" model were R^2 and Mallow's C^P.[135] For each of the properties examined, initial models used only the TIs and HB_1 as potential variables. Subsequently, we added the three geometric variables to examine the improvement provided by the addition of geometric information.

4.5.4. Results

4.5.4.1. Boiling Point

HB_1 is zero for all hydrocarbons and, therefore, was deleted from analyses of BP. Twelve of the TIs were deleted for the analysis of the 140 hydrocarbons as well. These indexes included the third- and fourth-order chain connectivity indexes, which were zero for all chemicals, the fourth- and sixth-order bond and valence corrected cluster connectivity indexes, which were perfectly correlated with the simple cluster connectivity indexes ($r = 1.0$), and J^X and J^Y, which were perfectly correlated with J^B for hydrocarbons.

Variable clustering of the remaining 89 TIs resulted in ten clusters. These clusters explained 89.7% of the total variation. In Table 7, we present the indexes selected from each cluster for subsequent use in modeling the BP of hydrocarbons. O, IC_0, IC_1, IC_2, SIC_0, and SIC_1 were selected because of their poor relationship with their clusters ($r < 0.7$).

With the 16 TIs, all subsets regression resulted in a seven-parameter model as follows:

$$BP = -322.86 + 5.46(P_4) - 45.76(IC_1) - 53.23(IC_2) + 799.94(^6\chi_{Ch}) +$$

Table 7. Topological Indexes Selected by Variable Clustering 89 Indexes for the Set of 140 Hydrocarbons with Measured Boiling Point

Cluster	Indices selected	Correlation
1	P_4	0.992
2	CIC_4, IC_0	0.968, 0.338
3	IC_5, O, IC_2, SIC_0	0.969, 0.400, 0.581, 0.456
4	$^5\chi_C^v$	0.964
5	$^5\chi_{Ch}^v$	0.998
6	$^0\chi^v$, SIC_1	0.986, 0.537
7	$^3\chi_C^v$	0.990
8	$^6\chi_{Ch}$, IC_1	0.916, 0.292
9	CIC_2	0.986
10	$^5\chi_{PC}^v$	0.973

(43)
$$288.26(^0\chi^v) - 32.76(^3\chi_C^v) - 2518.52(^5\chi_{Ch}^v)$$
$$(N = 140, r = 0.9956, s = 15.5, F = 2114)$$

Two chemicals, acenaphthene and cyclopenta(*cd*)pyrene (Nos. 106 and 117 of Table 4, respectively) had rather large residuals (> 60° C). The Cook's distance[136] for these two chemicals indicated they were influential cases. Given these circumstances, an outlier test[136] was performed and both chemicals had a significant result. After the removal of these chemicals, the following model was developed:

$$BP = -349.11 - 0.71(P_4) - 31.93(IC_1) - 44.70(IC_2) + 884.75(^6\chi_{Ch}) +$$

(44)
$$291.24(^0\chi^v) - 33.10(^3\chi_C^v) - 3327.61(^5\chi_{Ch}^v)$$
$$(N = 138, r = 0.9976, s = 11.4, F = 3876)$$

With the inclusion of the geometric parameters, an eight-parameter model was developed, which included two of the geometric parameters:

$$BP = -626.4 + 1050.8(SIC_0) - 204.0(SIC_1) - 249.8(^6\chi_{Ch}) + 364.0(^0\chi^v) -$$

(45)
$$32.3(^3\chi_C^v) + 833.4(^5\chi_{Ch}^v) + 20.4(^{3D}W)^{1/2} - 8.0(^{3D}W_H)^{1/2}$$
$$(N = 140, r = 0.9994, s = 6.1, F = 12246)$$

Table 4 presents the predicted normal BP for the hydrocarbons when using equations (44) and (45).

4.5.4.2. Lipophilicity (log P)

Twelve of the TIs were dropped from the study of the log P data set. The third- and fourth-order chain connectivity indexes were zero for all chemicals and the fifth-order chain connectivity index was nonzero for only one chemical. The sixth-order cluster connectivity indexes were nonzero for only one compound as well. Therefore, 89 indexes were used for the variable clustering.

There were 12 clusters generated by variable clustering for the 89 TIs used for the log P data set. The total variation explained by these clusters was 87.8% of the total. Table 8 presents the indexes selected from each of the clusters. The indexes O, SIC_0, J, IC_0, IC_1, and SIC_2 showed poor membership ($r < 0.7$) within the clusters and were retained as well.

Using all subsets regression with the selected TIs and HB_1 as independent variables resulted in a nine-parameter model:

$$\log P = -3.64 + 4.81(P_0) - 0.54(\overline{IC}) - 9.30(IC_0) + 13.65(SIC_0) +$$

(46)
$$3.88(SIC_4) - 7.68(^6\chi_{Ch}) + 0.63(^6\chi_{PC}^b) - 1.52(J^B) - 0.49(HB_1)$$
$$(N = 254, r = 0.912, s = 0.56, F = 134.1)$$

Table 8.   Topological Indexes Selected by Variable Clustering 89
Indexes for the Set of 254 Chemicals with Measured log P

Cluster	Indices selected	Correlation
1	P_0	0.981
2	SIC_4	0.944
3	$^5\chi_C^b$	0.929
4	IC_5, SIC_0, O	0.972, 0.469, 0.681
5	$^6\chi_{PC}^b$	0.976
6	$^4\chi_C$	0.980
7	$^6\chi_{CH}$, J	0.855, 0.406
8	SIC_1, IC_0, IC_1, SIC_2	0.910, 0.503, 0.585, 0.689
9	J^B	0.996
10	$^3\chi_C^b$	0.968
11	$\overline{IC}$	0.843
12	$^4\chi$	0.963

Inclusion of the geometric parameters resulted in the following 11-parameter model:

(47)
$$\log P = -12.06 - 0.68(\overline{IC}) - 8.13(IC_0) + 2.25(IC_5) + 12.62(SIC_0) -$$
$$5.65(^6\chi_{Ch}) + 0.66(^6\chi_{PC}^b) - 2.22(J) - 0.37(HB_1) +$$
$$4.23 \log(V_W) + 0.60 \log(^{3D}W) - 0.75 \log(^{3D}W_H)$$
$$(N = 254, \, r = 0.932, \, s = 0.50, \, F = 129.1)$$

Predicted values of log P using equations (46) and (47) are presented in Table 5.

4.5.4.3. Mutagenicity

Twelve TIs were dropped from the analyses of the 95 aromatic and heteroaromatic amines. The indexes dropped included the third- and fourth-order chain connectivity indexes, which were zero for all chemicals, and the fourth- and sixth-order cluster connectivity indexes, which were nonzero for only three compounds.

There were eight clusters generated by variable clustering the 89 TIs used for the aromatic amine data set. The total variation explained by these clusters was 88.6% of the total. In Table 9, we present the indexes selected from each of the clusters. Indexes O, $\overline{IC}$, IC_2, SIC_3, CIC_3, and $^3\chi_C^v$ were retained because of their poor cluster membership ($r < 0.7$).

Using all subsets regression with the selected TIs and HB_1 as independent variables resulted in an eight-parameter model:

$$\ln R = 9.308 + 5.141(\overline{IC}) - 3.018(O) - 23.814(IC_3) - 15.050(SIC_1) +$$

Table 9. Topological Indexes Selected by Variable Clustering 89 Indexes for the Set of 95 Aromatic and Heteroaromatic Amines with Measured $\ln R^a$

Cluster	Index selected	Correlation
1	$^4\chi$	0.982
2	SIC_4, SIC_3, CIC_3	0.969, 0.698, 0.665
3	$^6\chi_{PC}^b$, $^3\chi_C^v$	0.951, 0.631
4	SIC_1, O	0.922, 0.478
5	$^6\chi_{Ch}^b$	0.944
6	P_0	0.990
7	$^4\chi_{PC}^b$	0.919
8	IC_3, $\overline{IC}$, IC_2	0.909, 0.698, 0.537

^{a}Natural log of number of revertants per nanomole.

$$(48) \quad 41.572(SIC_4) + 2.636(^4\chi) + 3.728(^6\chi_{PC}^b) + 3.018(^3\chi_C^v)$$
$$(N = 95, r = 0.872, s = 0.98, F = 34.2)$$

Addition of the geometric parameters resulted in the following model:

$$\ln R = 15.785 + 3.883(\overline{IC}) - 1.374(O) - 14.152(SIC_1) + 2.878(^4\chi) +$$

$$(49) \quad 3.409(^6\chi_{PC}^b) + 4.625(^3\chi_C^v) - 7.867(P_0) - 0.0021(^{3D}W_H) + 0.0096(^{3D}W)$$
$$(N = 95, r = 0.893, s = 0.91, F = 37.2)$$

The predicted mutagenicity values for each of the aromatic amine chemicals from equations (48) and (49) are presented in Table 6.

4.6. DISCUSSION

The objectives of this chapter were to review the utility of TIs and 3D parameters in QSARs as well as to report recent results on the relative effectiveness of TIs versus geometrical parameters in the development of QSARs for estimating properties. A large number of QSAR models summarized here show that graph-theoretic invariants correlate reasonably well with physicochemical, biomedicinal, toxicological, and biochemical properties of diverse congeneric sets of molecules. It is also clear that TIs and substructures have found successful applications in the quantification of molecular similarity, selection of analogues, and molecular similarity-based estimation of properties. Of special interest is the fact that the molecular similarity method developed by Basak et al.[2] has been successfully used in the discovery of a novel class of human immunodeficiency virus reverse transcriptase (HIV-RT) inhibitors, showing the utility of such nonempirically based methods in practical drug discovery. Examples of

QSARs using 3D descriptors also demonstrate that geometrical parameters alone can predict properties of congeneric molecules quite satisfactorily.

In this context, it was of interest to compare the capabilities of TIs and 3D descriptors in QSAR analysis. To this end, we reported the QSAR studies developed on three different properties, viz., normal boiling of 140 hydrocarbons, lipophilicity (log P, octanol–water) of a diverse set of 254 chemicals, and mutagenicity (ln$R$) of a group of 95 aromatic and heteroaromatic amines. Results of these QSARs show that TIs contain important structural information sufficient to develop useful predictive models for these properties. However, in the case of BP, the addition of geometrical parameters, viz., ^{3D}W and $^{3D}W_H$, to the list of independent variables resulted in improved models in the sense that, while the TI-based QSARs had two outliers, the addition of geometrical variables gave well-behaved models including all of the hydrocarbons in the set. Also, estimate errors were significantly smaller for the regression equation using geometrical descriptors [equation (45) versus (44)]. This indicates that for BP of hydrocarbons, 3D or geometrical parameters encode some pertinent information relevant to BP which are not included in TIs. For log P and mutagenicity, however, the addition of geometrical descriptors resulted in only a slight improvement in the quality of the QSAR models over those derived from TIs only.

For log P, we also used HB_1, an algorithmically derived hydrogen bonding parameter, in addition to TIs, V_W, ^{3D}W, and $^{3D}W_H$. This is because the magnitude of log P of a molecule is known to depend significantly on the strength of hydrogen-bonding ability of solutes with solvents.[60,133] Our earlier studies on the correlation of log P using algorithmically derived parameters show that HB_1 is an important parameter in predicting log P.[27,31,45] QSARs of log P reported in this chapter provide evidence that the role of HB_1 cannot be carried out by a combination of TIs and 3D parameters.

In many practical situations of drug design and risk assessment, one has to estimate physical/biomedical/toxicological properties of chemicals without access to any empirical data.[55] Similarity-based models[1–3,28,29,47,49,50,66,111] and estimated values based on nonempirical structural parameters[25,27,31,32,45,70] are two viable alternatives for deriving property values under such data-poor situations. The QSAR models reported here based on TIs and geometrical parameters may find applications in selecting analogues and in estimating properties of chemicals in such cases.

ACKNOWLEDGMENTS

This chapter is contribution No. 159 from the Center for Water and the Environment of the Natural Resources Research Institute. Research reported in this chapter was supported, in part, by grant F49620-94-1-0401 from the United States Air Force, Exxon Corporation, and the Structure–Activity Relationship Consortium (SARCON) of the Natural Resources Research Institute at the University of Minnesota.

REFERENCES

1. M. A. Johnson, S. C. Basak, and G. Maggiora, *Math. Comput. Modeling 11*, 630 (1988).
2. S. C. Basak, V. R. Magnuson, G. J. Niemi, and R. R. Regal, *Discrete Appl. Math. 19*, 17 (1988).
3. M. S. Lajiness, in: *Computational Chemical Graph Theory* (D. H. Rouvray, ed.), pp. 299–316, Nova, New York (1990).
4. R. E. Carhart, D. H. Smith, and R. Venkataraghavan, *J. Chem. Inf. Comput. Sci. 25*, 64 (1985).
5. S. C. Basak, C. Raychaudhury, A. B. Roy, and J. J. Ghosh, *Indian J. Pharmacol. 13*, 112 (1981).
6. S. K. Ray, S. C. Basak, C. Raychaudhury, A. B. Roy, and J. J. Ghosh, *Arzneim. Forsch. 32*, 322 (1982).
7. S. C. Basak, S. K. Ray, C. Raychaudhury, A. B. Roy, and J. J. Ghosh, *IRCS Med. Sci. 10*, 145 (1982).
8. A. K. Samanta, S. K. Ray, S. C. Basak, and S. K. Bose, *Arzneim. Forsch. 32*, 1515 (1982).
9. S. K. Ray, S. C. Basak, C. Raychaudhury, A. B. Roy, and J. J. Ghosh, *Indian J. Chem. 20B*, 894 (1981).
10. S. C. Basak, D. P. Gieschen, V. R. Magnuson, and D. K. Harriss, *IRCS Med. Sci. 10*, 619 (1982).
11. S. K. Ray, S. C. Basak, C. Raychaudhury, A. B. Roy, and J. J. Ghosh, *Arzneim. Forsch. 33*, 352 (1983).
12. S. C. Basak, D. P. Gieschen, D. K. Harriss, and V. R. Magnuson, *J. Pharm. Sci. 72*, 934 (1983).
13. S. C. Basak, D. K. Harriss, and V. R. Magnuson, *J. Pharm. Sci. 73*, 429 (1984).
14. A. B. Roy, C. Raychaudhury, S. K. Ray, S. C. Basak, and J. J. Ghosh, in: *Proceedings of the Fourth European Symposium on Chemical Structure–Biological Activity: Quantitative Approaches* (J. C. Deardon, ed.), pp. 75–76, Elsevier, Amsterdam (1983).
15. S. K. Ray, S. Gupta, S. C. Basak, C. Raychaudhury, A. B. Roy, and J. J. Ghosh, *Indian J. Chem. 24B*, 1149 (1985).
16. S. C. Basak, L. J. Monsrud, M. E. Rosen, C. M. Frane, and V. R. Magnuson, *Acta Pharm. Jugosl. 36*, 81 (1986).
17. S. C. Basak, B. D. Gute, and L. R. Drewes, *Pharm. Res. 13*, 775 (1996).
18. S. C. Basak, *Med. Sci. Res. 15*, 605 (1987).
19. S. C. Basak, in: *Proceedings of the NATO Advanced Study Institute (ASI) on Pharmacokinetics*, Erice, Sicily, April 4–17, 1994, Plenum, New York.
20. R. Nilakantan, N. Bauman, and R. Venkataraghavan, *J. Chem. Inf. Comput. Sci. 31*, 527 (1991).
21. D. E. Needham, I. C. Wei, and P. G. Seybold, *J. Am. Chem. Soc. 110*, 4186 (1988).
22. A. T. Balaban, *Chem. Phys. Lett. 89*, 399 (1982).
23. A. T. Balaban, *Pure Appl. Chem. 55*, 199 (1983).
24. A. T. Balaban, *MATCH 21*, 115 (1986).
25. A. T. Balaban, S. C. Basak, T. Colburn, and G. D. Grunwald, *J. Chem. Inf. Comput. Sci. 34*, 1118 (1994).
26. A. T. Balaban, *J. Chem. Inf. Comput. Sci. 35*, 339 (1995).
27. S. C. Basak, G. J. Niemi, and G. D. Veith, in: *Computational Chemical Graph Theory* (D. H. Rouvray, ed.), p. 235, Nova, New York (1990).
28. S. C. Basak and G. D. Grunwald, *J. Chem. Inf. Comput. Sci. 35*, 366 (1995).
29. S. C. Basak and G. D. Grunwald, *New J. Chem. 19*, 231 (1995).
30. A. T. Balaban, S. Bertelsen, and S. C. Basak, *Math. Chem. 30*, 55 (1994).
31. S. C. Basak, G. J. Niemi, and G. D. Veith, *J. Math. Chem. 4*, 185 (1990).
32. S. C. Basak, G. J. Niemi, and G. D. Veith, *Math. Comput. Modelling 14*, 511 (1990).
33. S. C. Basak, G. J. Niemi, and G. D. Veith, *J. Math. Chem. 7*, 243 (1991).
34. D. Bonchev and N. Trinajstić, *J. Chem. Phys. 67*, 4517 (1977).
35. M. Randić, *J. Am. Chem. Soc. 97*, 6609 (1975).
36. C. Raychaudhury, S. K. Ray, J. J. Ghosh, A. B. Roy, and S. C. Basak, *J. Comput. Chem. 5*, 581 (1984).
37. D. H. Rouvray and R. B. Pandey, *J. Chem. Phys. 85*, 2288 (1986).
38. D. H. Rouvray, *New Sci. May*, 35 (1993).
39. K. Balasubramanian, *SAR QSAR Environ. Res. 2*, 59 (1994).
40. M. Randić, *Int. J. Quantum Chem. Quant. Biol. Symp. 11*, 137 (1984).
41. S. C. Basak, D. P. Gieschen, and V. R. Magnuson, *Environ. Toxicol. Chem. 3*, 191 (1984).

42. S. C. Basak, C. M. Frane, M. E. Rosen, and V. R. Magnuson, *IRCS Med. Sci. 14*, 848 (1986).
43. S. C. Basak, *Med. Sci. Res. 16*, 281 (1988).
44. G. J. Niemi, S. C. Basak, and G. D. Veith, in: *Envirotech Vienna: Proceedings of the First Conference of the International Society of Environmental Protection* (K. Zirm and J. Mayer, eds.), pp. 57–68. W. B. Druck Gmbh and Co., Reiden, Austria (1989).
45. G. J. Niemi, S. C. Basak, G. D. Veith, and G. D. Grunwald, *Environ. Toxicol. Chem. 11*, 893 (1992).
46. S. C. Basak, S. Bertelsen, and G. D. Grunwald, *Toxicol. Lett. 79*, 239 (1995).
47. S. C. Basak and G. D. Grunwald, *SAR QSAR Environ. Res. 2*, 289 (1994).
48. S. C. Basak and G. D. Grunwald, in: *Proceeding of the XVI International Cancer Congress* (R. S. Rao, M. G. Deo, and L. D. Sanghvi, eds.), pp. 413–416, Monduzzi, Bologna, Italy (1995).
49. S. C. Basak and G. D. Grunwald, *SAR QSAR Environ. Res. 3*, 265 (1995).
50. S. C. Basak and G. D. Grunwald, *Chemosphere 31*, 2529 (1995).
51. T. Colburn, D. Axtell, and S. C. Basak, *Mutat. Res.* (in preparation).
52. S. C. Basak, in: *Practical Applications of Quantitative Structure–Activity Relationships (QSAR) in Environmental Chemistry and Toxicology* (W. Karcher and J. Devillers, eds.), pp. 83–103, Kluwer Academic, Dordrecht (1990).
53. S. C. Basak, G. D. Grunwald, G. E. Host, G. J. Niemi, and S. Bradbury, *Environ. Toxicol. Chem.* (in preparation).
54. C. Hansch, *Adv. Pharmacol. Chemother. 13*, 45 (1975).
55. C. M. Auer, J. V. Nabholz, and K. P. Baetcke, *Environ. Health Perspect. 87*, 183 (1990).
56. J. C. Arcos, *Environ. Sci. Technol. 21*, 743 (1987).
57. U. Burkert and N. L. Allinger, *Molecular Mechanics, ACS Monograph 177*, American Chemical Society, Washington, DC (1982).
58. W. G. Richards, *Quantum Pharmacology*, Butterworths, London (1977).
59. A. Verloop, W. Hoogenstraaten, and J. Tipker, in: *Drug Design*, Vol. VII (E. J. Ariens, ed.), pp. 165–207, Academic Press, New York (1976).
60. M. J. Kamlet, R. M. Doherty, G. D. Veith, R. W. Taft, and M. H. Abraham, *Environ. Sci. Technol. 20*, 690 (1986).
61. I. Moriguchi and Y. Kanada, *Chem. Pharm. Bull. 25*, 926 (1977).
62. M. Bunge, *Methods, Models and Matter*, Reidel, Dordrecht (1973).
63. F. Harary, *Graph Theory*, Addison–Wesley, Reading, Massachusetts (1969).
64. N. Trinajstić, *Chemical Graph Theory*, CRC Press, Boca Raton, Florida (1983).
65. I. S. Dmitriev, *Molecules Without Chemical Bonds*, Mir Publishers, Moscow (1981).
66. M. Randić, in: *Concepts and Applications of Molecular Similarity* (M. A. Johnson and G. M. Maggiora, eds.), pp. 77–145, John Wiley & Sons, New York (1990).
67. C. Raychaudhury, S. K. Ray, J. J. Ghosh, A. B. Roy, and S. C. Basak, *J. Comput. Chem. 5*, 581 (1984).
68. A. Sabljic and N. Trinajstić, *Acta Pharm. Jugosl. 31*, 189 (1981).
69. O. Mekenyan, S. Dimitrov, and D. Bonchev, *Eur. Polym. J. 19*, 1185 (1983).
70. A. T. Balaban, N. Joshi, L. B. Kier, and L. H. Hall, *J. Chem. Inf. Comput. Sci. 32*, 233 (1992).
71. M. V. Duudea, O. Minailiuc, and A. T. Balaban, *J. Comput. Chem. 12*, 527 (1991).
72. A. T. Balaban, *Theor. Chim. Acta 53*, 355 (1979).
73. P. A. Filip, T. S. Balaban, and A. T. Balaban, *J. Math. Chem. 1*, 61 (1987).
74. A. T. Balaban and V. Feroiu, *Rep. Mol. Theory 1*, 133 (1990).
75. E. J. Kupchik, *Quant. Struct. Act. Relat. 7*, 57 (1988).
76. L. Pogliani, *J. Phys. Chem. 97*, 6731 (1993).
77. L. B. Kier and L. H. Hall, *Molecular Connectivity in Structure–Activity Analysis*, Research Studies Press, New York (1986).
78. S. C. Basak and G. D. Grunwald, *Math. Modelling Sci. Comput. 2*, 735 (1993).
79. D. Bonchev and N. Trinajstić, *Int. J. Quantum Chem. 12*, 293 (1978).
80. S. C. Basak, B. D. Gute, and S. Ghatak, *J. Chem. Inf. Comput. Sci.* (submitted for publication).
81. S. C. Basak and V. R. Magnuson, *Arzneim. Forsch. 33*, 501 (1983).

82. J. V. Soderman, *CRC Handbook of Identified Carcinogens and Noncarcinogens: Carcinogenicity–Mutagenicity Database*, Vol. I, CRC Press, Boca Raton, Florida (1982).

83. S. C. Basak, C. M. Frane, M. E. Rosen, and V. R. Magnuson, *Med. Sci. Res. 15*, 887 (1987).

84. V. R. Magnuson, D. K. Harriss, and S. C. Basak, in: *Studies in Physical and Theoretical Chemistry* (R. B. King, ed.), pp. 178–191, Elsevier, Amsterdam (1983).

85. F. C. Smeeks and P. C. Jurs, *Theor. Chim. Acta 233*, 111 (1990).

86. Y. Gao and H. Hosoya, *Bull. Chem. Soc. Jpn. 61*, 3093 (1988).

87. H. Wiener, *J. Am. Chem. Soc. 69*, 17 (1947).

88. J. R. Platt, *J. Chem. Phys. 15*, 419 (1947).

89. L. H. Hall and L. B. Kier, *Tetrahedron 33*, 1953 (1977).

90. L. Pogliani, *J. Phys. Chem. 99*, 925 (1995).

91. W. J. Boecklen and G. J. Niemi, *SAR QSAR Environ. Res. 2*, 79 (1994).

92. A. T. Balaban and C. Catana, *SAR QSAR Environ. Res. 2*, 1 (1994).

93. S. P. Gupta and P. Singh, *Bull. Chem. Soc. Jpn. 52*, 2745 (1979).

94. M. Randić, *New J. Chem. 15*, 517 (1991).

95. R. H. Rohrbaugh and P. C. Jurs, *Anal. Chem. 60*, 2249 (1988).

96. D. K. Pal, S. K. Purkayaastha, C. Sengupta, and A. U. De, *Indian J. Chem. 31*, 109 (1992).

97. D. H. Rouvray and W. Tatong, *Int. J. Environ. Stud. 33*, 247 (1989).

98. D. H. Rouvray and W. Tatong, *Z. Naturforsch. 41*, 1238 (1986).

99. G. J. Niemi, R. R. Regal, and G. D. Veith, *ACS Symp. Ser. 292*, 148 (1985).

100. I. Lukovits, *J. Chem. Soc. Perkin Trans. 2*, 1667 (1988).

101. G. J. Niemi, G. D. Veith, R. R. Regal, and D. D. Vaishnav, *Environ. Toxicol. Chem. 6*, 515 (1987).

102. V. K. Gombar and K. Enslein, in: *Applied Multivariate Analysis in SAR and Environmental Studies* (J. Devillers and W. Karcher, eds.), pp. 377–414, Kluwer Academic, Dordrecht (1991).

103. B. W. Blake, K. Enslein, V. K. Gombar, and H. H. Borgstedt, *Mutat. Res. 241*, 261 (1990).

104. T. Okuyama, Y. Miyashita, S. Kanaya, H. Katsumi, S. Sasaki, and M. Randić, *J. Comput. Chem. 9*, 636 (1988).

105. C. L. Wilkins and M. Randić, *Theor. Chim. Acta 58*, 45 (1980).

106. M. Randić and N. Trinajstić, *MATCH 13*, 271 (1982).

107. M. Randić, in: *Molecular Basis of Cancer, Part A: Macromolecular Structure, Carcinogens, and Oncogenes* (R. Rein, ed.), pp. 309–318, Alan R. Liss, New York (1985).

108. S. C. Basak, S. Bertelsen, and G. D. Grunwald, *J. Chem. Inf. Comput. Sci. 34*, 270 (1994).

109. A. Leo and D. Weininger, *CLOGP Version 3.2 User Reference Manual*, Medicinal Chemistry Project, Pomona College, Claremont, California (1984).

110. P. Willett, *J. Chem. Inf. Comput. Sci. 23*, 22 (1983).

111. P. Willett and V. Winterman, *Quant. Struct. Act. Relat. 5*, 18 (1986).

112. P. Willett, in: *Concepts and Applications of Molecular Similarity* (M. A. Johnson and G. M. Maggiora, eds.), pp. 43–63, John Wiley & Sons, New York (1990).

113. G. M. Downs and P. Willett, in: *Applied Multivariate Analysis in SAR and Environmental Studies* (J. Devillers and W. Karcher, eds.), pp 247–279, Kluwer Academic, Dordrecht (1991).

114. P. A. Bath, A. R. Andrew, and P. Willett, *J. Chem. Inf. Comput. Sci. 34*, 141 (1994).

115. R. D. Brown, G. Jones, and P. Willett, *J. Chem. Inf. Comput. Sci. 34*, 63 (1994).

116. S. C. Basak, B. D. Gute, and G. D. Grunwald, *Croat. Chem. Acta 69* (1996) (in press).

117. J. E. Amoore, *Nature 214*, 1095 (1967).

118. M. Charton. *Top. Curr. Chem. 114*, 107 (1983).

119. B. Bogdanov, S. Nikolić, and N. Trinajstić, *J. Math. Chem. 3*, 299 (1989).

120. R. P. Bhatnagar, P. Singh, and S. P. Gupta, *Indian J. Chem. 19B*, 780 (1980).

121. R. D. Cramer III, D. E. Patterson, and J. D. Bunce, *J. Am. Chem. Soc. 110*, 5959 (1988).

122. O. Mekenyan, D. Bonchev, and N. Trinajstić, *Int. J. Quantum Chem. 18*, 369 (1980).

123. W. Karcher, *Spectral Atlas of Polycyclic Aromatic Hydrocarbons*, Vol. 2, pp. 16–19, Kluwer Academic, Dordrecht (1988).

124. A. K. Debnath, G. Debnath, A. J. Shusterman, and C. Hansch, *Environ. Mol. Mutagen. 19*, 37 (1992).
125. S. C. Basak, D. K. Harriss, and V. R. Magnuson, POLLY 2.3, copyright by the University of Minnesota (1988).
126. C. E. Shannon, *Bell Syst. Tech. J. 27*, 379 (1948).
127. R. Sarkar, A. B. Roy, and R. K. Sarkar, *Math. Biosci. 39*, 379 (1978).
128. A. B. Roy, S. C. Basak, D. K. Harriss, and V. R. Magnuson, in: *Mathematical Modelling in Science and Technology* (X. J. R. Avula, R. E. Kalman, A. I. Liapis, and E. Y. Rodin, eds.), pp. 745–750, Pergamon Press, Elmsford, New York (1984).
129. S. C. Basak, A. B. Roy, and J. J. Ghosh, in: *Proceedings of the Second International Conference on Mathematical Modelling*, Vol. II (X. J. R. Avula, R. Bellman, Y. L. Luke, and A. K. Rigler, eds.), pp. 851–856, University of Missouri, Rolla (1980).
130. Tripos Associates, Inc., SYBYL Version 6.1, Tripos Associates, Inc., St. Louis, Missouri (1994).
131. Tripos Associates, Inc., CONCORD Version 3.0.1, Tripos Associates, Inc., St. Louis, Missouri (1993).
132. S. C. Basak, H-BOND: A Program for Calculating Hydrogen Bonding Parameter, University of Minnesota, Duluth (1988).
133. Y.-C. Ou, Y. Ouyang, and E. J. Lien, *J. Mol. Sci. 4*, 89 (1986).
134. SAS Institute, Inc., *SAS/STAT User's Guide, Release 6.03 Edition*, SAS Institute, Inc., Cary, North Carolina (1988).
135. R. R. Hocking, *Biometrics 32*, 1 (1976).
136. S. Weisberg, *Applied Linear Regression*, John Wiley & Sons, New York (1980).

5

Recognition of Membrane Protein Structure from Amino Acid Sequence

BONO LUČIĆ, NENAD TRINAJSTIĆ, and
DAVOR JURETIĆ

5.1. INTRODUCTION

The fundamental prerequisite for the existence of life is the existence of a boundary between the living unit (such as the cell) and the environment. This boundary must possess such a structure that divides but also connects the living unit and the environment. Therefore, it must be highly selective. The membrane (the lipid bilayer including fixed proteins) is situated on the very boundary between the living cell and the surrounding medium, and with its function it controls and provides conditions necessary for all of the numerous processes in the space that it encloses. The communication between the living and nonliving worlds through the membrane influences the increase of the order in the space that it encloses by using chemical potential in this space. Chemical potential in the living unit originates by selectively bringing into it the free energy from the nonliving world through the function of the membrane and its proteins. The entropy of the universe increases continuously, but in this unavoidable process, via the function of proteins (and related molecules) it is possible to increase

BONO LUČIĆ and NENAD TRINAJSTIĆ • The Rugjer Bošković Institute, P. O. Box 1016, HR-10001 Zagreb, The Republic of Croatia. DAVOR JURETIĆ • Department of Physics, Faculty of Science, HR-21001 Split, The Republic of Croatia.

From Chemical Topology to Three-Dimensional Geometry, edited by Balaban. Plenum Press, New York, 1997

order locally (that is, the local increase in negentropy) and because of this fact, life is also possible.

The membrane consists of lipids and proteins. Lipids in membranes are organized in a structure consisting of two layers called the lipid bilayer. Lipids in one layer are in contact with lipids in the other layer through their hydrophobic parts. Hydrophilic parts of lipids in both layers are on the outer side of the membrane and turned toward water molecules. Proteins are formed from amino acids. The activity of proteins is defined by their 3D structures. Membrane proteins characteristically contain hydrophobic parts that tend to be located in the inside of the membrane, and parts that are predominantly hydrophilic and can be found in either the cytoplasmic or periplasmic regions. Membrane proteins are classified as surface proteins (since they appear on the surface of the membrane) and integral membrane proteins (since they are integral parts of the membrane). The surface membrane proteins function attached to the surface of the membrane. The integral membrane proteins are more numerous than the surface proteins and they pass once or more through the membrane. In this chapter a few characteristics of their structures will be discussed.

The membrane influences the conformation of a protein by its electrostatic properties.[1] The membrane is a medium of low dielectric constant for every protein that is attached to it or is in its proximity. This is a consequence of the fact that membranes consist mostly of hydrophobic long-chain fatty acids. Therefore, for a protein within the membrane it is not possible to possess the global structure of an oil drop surrounded by charges as in soap micelles because the membrane is an energetically unfavorable environment for electrical charges. If polar groups of peptides are not linked to each other by hydrogen bonds, then the lipid bilayer is an energetically unfavorable environment for polypeptides.[2] Because of this and because of the tendency of hydrophobic amino acid side chains to partition into lipid (because there is no effect of the oil drop as in water), the interior of the membrane represents a suitable environment for the formation of an α-helix that passes through the membrane with that part of the polypeptide chain that contains the hydrophobic amino acid residues (hydrophobic side chains). In such regular structures the hydrogen bond potential is best satisfied if α-helices are perpendicular to the plane of the lipid bilayer. In order to pass through the membrane, α-helices must contain between 17 and 22 amino acid residues.[3-5] An antiparallel pair of helices is also a very common structure in membranes because of the stabilization of helices by creation of opposite electric dipoles in vis-à-vis helices.

The structure of the protein part in the membrane may also be arranged in several β-strands, organized in β-barrels.[6,7] These structures create pores in the membrane. Near the surface of the membrane or in the inner part of the pore there are charged hydrophilic amino acid residues that are in contact with water. The membrane proteins that possess such a structure are thought to be less frequent than the membrane proteins containing transmembrane segments in the structure of an α-helix.

Helices that form pores will be amphiphilic because it is more favorable to have situated in the inner side of the pore hydrophilic amino acid side chains, while the outer side of the pore represents a more favorable environment for hydrophobic amino acid side chains since these are in contact with lipids. Some authors point to the possibility that such a structure contains hydrogen bonds between amino acid residues and the main chain in order to compensate opposite charges and oppositely oriented dipoles.[8] A comparison between the strength of different interactions in the structure of soluble and membrane proteins leads to the conclusion that because of the decreased strength of hydrophobic interactions and increased strength of electrostatic interactions (because of the reduced dielectric constant), the electrostatic interactions play the main role in stabilizing the structure of membrane proteins.[9]

5.2. INVESTIGATIONS OF THE STRUCTURE OF MEMBRANE PROTEINS

5.2.1. Experimental Approaches

5.2.1.1. The X-Ray Diffraction Method

Well-ordered 3D crystals are the basic prerequisite for a successful analysis of biological macromolecules by the X-ray diffraction method. Until recently the crystallization of membrane proteins seemed impossible because membrane proteins consist of hydrophobic parts that are integrated with the membrane, polar parts that are integrated with the water phase on both sides of the membrane, and the polar heads of lipids. Membrane proteins are soluble neither in the water buffer nor in organic solvents with low dielectric constants. This is the reason why detergents must be added to these solutions. Micelles of detergents link hydrophobic parts of a protein and thus protect it from contact with water. Membrane proteins with large parts sticking out of the membrane crystallize much more easily. The most important factor is the size of micelles. If the crystallization of the protein is successful, then structure determination can be carried out. In this way the 3D structures of the three subunits of the photosynthetic reaction center from *Rhodopseudomonas viridis*[10] and from *Rhodobacter sphaeroides*[11] have been solved. The phase angle is determined using the method of isomorphic exchange by introducing a group with a heavy atom. Ordinarily, several exchanges of this kind are made.[12] However, the introduced group with the heavy atom should not change the conformation of the protein.

5.2.1.2. Electron Microscopy of 2D Crystals

The spatial structure of membrane proteins may also be elucidated by electron microscopy of the so-called 2D crystal. This method has been used to solve the 3D structure of bacteriorhodopsin[13] with a resolution of 3.5 Å. Some membrane proteins form ordered lattices in the membrane plane creating 2D crystals (ordered monomolecular layers). Bacteriorhodopsin is present in large quantities in special parts of the membrane in the bacterium *Halobacterium halobium* that is found in briny water. It can yield a large crystal with a diameter of up to 1 mm that contains up to 20,000 molecules of bacteriorhodopsin. Because of this circumstance, it is possible to obtain a picture with a very weak electron beam. This is important because strong beams of electrons can damage biological samples. One takes several thousand pictures and those that are the best are selected by optical diffraction and then stored in the computer and later serve for assembling the final picture of the structure. It is necessary to take 20 pictures for one projection at the temperature of liquid nitrogen or liquid helium. Similarly the phase problem is also solved by analysis of many pictures.[13]

5.2.1.3. Experimental Methods for the Determination of the Topography and Topology of Membrane Protein

Membrane proteins may be classified into monotopic proteins (proteins that are associated with the membrane and do not pass across the bilayer), bitopic proteins (proteins that pass through the membrane once), and polytopic proteins (proteins that pass through the membrane several times).[2,14] Topography provides information about topology (the number of times that the protein passes through the membrane), and also about the surface of the protein. Both topography and topology are commonly used to describe the arrangement of membrane proteins relative to the bilayer. In this chapter, we will present an algorithm by which we can build topological (not topographical) models of membrane proteins. Different experimental methods can yield the number of protein segments that are embedded in the membrane. If information about the number of transmembrane segments is available, then various theoretical methods can be used to determine the position of these segments in the primary structure of proteins.

One can place in some positions on the protein radioactive, fluorescent, or spin-labeled markers.[15] For the same purpose one may also use antibodies and proteolytic enzymes. With these procedures one can determine those parts of the protein that are outside the membrane. Next, efforts are made to determine the position of the membrane part in the sequence. A complementary method is the hydrophobic marking of the protein by building into it hydrophobic photolabeled parts from the class of phospholipid compounds in order to determine those parts that are situated in the membrane.[16] Another method uses proteolytic enzymes that split the protein in a specific position in the segment that is not located in the membrane.[17]

By using the gene fusion technique one can generate hybrid proteins in which the amino-terminal portion of the hybrid is derived from the investigated proteins and the carboxyl-terminal portion of the hybrid is from a sensor protein, either alkaline phosphatase or β-galactosidase. High alkaline phosphatase activity is expected if the fusion junction is located in or near a periplasmic domain of the membrane protein. In contrast, if the fusion junction is in or near a cytoplasmic domain of the membrane protein, the alkaline phosphatase moiety will be located on the cytoplasmic surface of the membrane, where it is inactive. High β-galactosidase activity is expected when the fusion junction is located in or near a cytoplasmic domain of the membrane protein. If the β-galactosidase fusions are within a periplasmic domain, the specific activity will be low because the β-galactosidase moiety is usually unable to translocate fully across the membrane and can become embedded in the cytoplasmic membrane as an inactive species.[18,19]

Immunological methods produce antibodies that will bind to target positions in the protein sequence.[20] The target segment in the protein will bind antibodies if this segment is outside the membrane or if it is positioned only on one side of the membrane. From this basis one then attempts to reconstruct the topology of proteins.

Using genetic approaches it is possible to work with targeted mutations in order to determine which members in the amino acid sequence are key in the functioning of the protein. In this approach, helpful theoretical analyses are used to compare a specific protein with similar proteins, that is, proteins that are similar to the investigated protein and that perform equal functions in different organisms. In this way one can study the conservation of particular parts in the sequence in different proteins. Usually the conservation is highest in the active position and its surroundings.[21–23]

The method of spin labeling is also used in the investigation of the structure and dynamics of membrane proteins.[15] This method is based on the substitution of a given amino acid residue at a selected position by an amino acid residue that contains a nitroxide group that is then analyzed by EPR spectroscopy. In most cases one chooses cysteine in which the nitroxide group (as a side chain) is linked to sulfur. One may also use other synthetic amino acids whose side chain contains a nitroxide group. EPR spectroscopy can determine the topography of the polypeptide chain and the electrostatic potential on a given surface. These analyses can be used for the identification of a particular secondary structure and the determination of its orientation in the protein. They can also be used for the determination of the distance between two groups with spin labels, for the investigation of interactions in the tertiary protein structure, interactions of protein with the membrane, the time dependence of all of these interactions and for the determination of structural parameters.

5.2.2. Theoretical Approaches

Why is it necessary to predict the protein secondary structure? Research on the project related to the human genome, in which scientists are determining the sequences of bases in the entire human genome, produces on a daily basis new genes that serve for the synthesis of new proteins for which one must determine the structure. Thus, today we know the amino acid sequence for more than 30,000 proteins but the structure for only 3000 proteins. This disproportion is increasing daily, since it is expected that by the end of this century we will know the amino acid sequence for 100,000 proteins. On the other hand, the experimental determination of a protein structure is a rather slow process that is likely to remain so for the next decade. Thus, theoretical approaches to the determination of protein structures are especially important. The most reliable theoretical prediction of protein structure exists for cases in which one already knows the structure of some protein with a very similar amino acid sequence.[24,25] Unfortunately, only one-seventh of new protein sequences have a known protein with a similar amino acid sequence.[26]

Attempts to predict the 3D structure of proteins by molecular dynamic simulation directly from the amino acid sequence have shown to be futile,[27] because they cannot reveal the hidden connection between the primary and tertiary structures. However, the complexity of the problem may be reduced because in the tertiary structures there appear parts that possess a regular periodic structure[28,29] which is called the secondary structure. Therefore, the problem of predicting the 3D protein structures can be projected into one dimension, that is, into the amino acid string where each amino acid residue in the primary structure is associated with one kind of the secondary structure. The large gap between the primary and tertiary protein structures is bridged by defining the secondary structure. In this way the problem of understanding and predicting the protein structure is no more just the direct computation of the structure but the processing and interpretation of available data contained in the protein for which the structure is determined.

In the case of membrane proteins one distinguishes (1) the part that is present in the membrane and that mostly possesses the secondary structure of an α-helix and (2) the part that is outside of the membrane. This part is taken in all theoretical approaches as undefined structure although in reality it may possess the structure of an α-helix or β-strand equally to those in soluble proteins. But, because of the limited number of membrane proteins with known 3D structures, it is not possible to train the algorithm to predict the secondary structure for the soluble part of membrane proteins. Therefore, remedy algorithms for predicting the secondary structures of soluble proteins can be used. Among the available algorithms the most reliable is the PHD method.[30–33]

It is generally believed that the structure of membrane proteins is simpler than that of soluble proteins because the lipid bilayer in which the membrane proteins are immersed diminishes the degrees of freedom.[34] Thus, the prediction of the membrane protein structures is expected to be more accurate and the obtained models should be of considerable help when the activity of the membrane proteins is analyzed.

Using only experimental methods, one cannot obtain enough information about the structure of new proteins. For this reason the best results are usually achieved in modeling the structure of membrane proteins when experimental results are combined with theoretical models obtained via one of the many available algorithms. Among the theoretical models, most advances are achieved by developing algorithms that predict the membrane protein structures on the basis of analysis of hydrophobicity,[3–5,35,36] rules of grouping the positively charged amino acid side chains (Arg, Lys) in the cytoplasmic region of the sequence, known as the "positive inside" rule,[37,38] statistical procedures that produce much better results when combined with multiple sequence alignments,[39] statistical procedures along with prediction of topology,[40] neural network algorithms,[41] and neural network algorithms along with the utilization of the multiple sequence alignments as the entry instead of the sequence itself.[42]

The best known and most used procedure is the Kyte–Doolittle method[3] which computes within the sliding window of specific width the hydrophobicity/amphiphilicity of the segment. This represents a certain probability that the specific segment will or will not be present in the membrane.

Eisenberg *et al.* have developed a method by which one can compute the hydrophobicity moment.[43] This moment is calculated from the hydrophobicity and orientation of successive side chains arranged in an α-helix. Sequences in which hydrophobic and hydrophilic side chains tend to orient in opposite directions have a high hydrophobic moment. A hydrophobicity profile for a protein is a graph of the average hydrophobicity of each residue (which is calculated over several neighbors) against position in the sequence. Beforehand one expects a periodicity in the hydrophobicity, because of the periodicity of the secondary structure and because of regular alternation of hydrophobic and hydrophilic amino acid side chains in the helical segment or in the segment that is in the β-sheet structure and that passes through the membrane. In that position where the hydrophobic moment is the largest, the probability is also largest for the existence of the amphiphilic segment. Eisenberg *et al.* have combined this with the amount of hydrophobicity of the segment itself.[43] The largest hydrophobic moments are exhibited by surface membrane proteins while transmembrane proteins possess the largest hydrophobicity and somewhat smaller hydrophobic moment.[43,44]

5.3. THE PREFERENCE FUNCTIONS METHOD

5.3.1. Training and Testing Sets of Membrane Proteins

The preference functions method[45] extracts conformational preferences that are functions of the local sequence environment. The method combines in one algorithm (PREF) the sliding window procedure that extracts a running average of the local amino acid attributes with a very simple training procedure. To test a protein of unknown structure, sequence environments are collected from its primary structure and used for evaluation and comparison of preference functions for different folding motifs. A more or less accurate location of transmembrane segments, presumably in the α-helix conformation, is known in many integral membrane proteins and can be used to extract statistically significant folding parameters. In this chapter only two folding motifs are distinguished: the transmembrane segments and the extramembrane residues of integral membrane proteins.

As the set of amino acid attributes, any hydrophobicity scale or any statistical, physical, chemical, or mathematical property of the 20 natural amino acid types may be used. Many amino acid scales either are obtained experimentally (by determining properties of individual amino acids) or are computed by statistical analysis on a certain set of membrane proteins.[5] However, it is not clear in advance which one among these scales is the best, that is, which one reflects most closely the actual behavior of amino acid residues in structuring membrane proteins.

The advantage of using Chou–Fasman[46]-type statistical conformational preferences is twofold. First, it enhances the strength of the training procedure because both amino acid attributes and preference functions are extracted from the same training data base of proteins. In the two-state model, when the only folding motifs considered in integral membrane proteins are transmembrane helix conformations, transmembrane helix preferences are the natural and unique choice of hydrophobicity scale associated with the training data set of proteins. The training set of well-known membrane protein structures defines the united set of constant conformational preferences. The choice of the training set is at the same time the choice of the amino acid attributes. The second advantage of such a choice is that it is not an arbitrary and subjective choice of a hydrophobicity scale favored by the authors.

The subjective choice of attributes can be eliminated also by the procedure of testing all published amino acid scales with respect to the prediction accuracy for transmembrane helices in the chosen training data set of proteins. Our PREF suite of algorithms can indeed perform such a task. Such results are beyond the scope of this chapter and will be published elsewhere.

Our earlier publications have explored secondary structure predictions for membrane proteins by using the PREF method and by varying the choice of the training data set of proteins and of the hydrophobicity scale. All such results either included soluble proteins in the training data set,[47–49] or used only such proteins.[45,50,51] Up to this work, secondary structure predictions by PREF did not include the explicit prediction of transmembrane helical segments although such segments could be recognized from the α-helix preference profiles. Therefore, the specificity of the present work is to change the prediction goal from a three-state model (α-helix, β-sheet, and undefined conformation) to a two-state model: transmembrane α-helix conformations and non-transmembrane (extramembrane) conformation. Another simplification made possible by the reduction from three-state to two-state model is that we now use only integral membrane proteins in the training data set of proteins. In the present work we do not attempt to predict the β-sheet structure of membrane proteins either in transmembrane or in extramembrane regions.

The set of proteins used in this work contains two lists: the training list with 95 proteins (list-train), 43,336 residues (Table 1), and the testing list with 71 proteins (list-test), 28,487 residues (Table 2).

The primary structure, without signal (FT SIGNAL) and propeptide (FT PROPEP) sequence, and the locations of the transmembrane segments for both protein lists were extracted from Release 31.0 of the SWISS-PROT data bank.[52] There are 8840 and 5562 residues in the transmembrane region of training and testing lists, respectively. In order to enable a straightforward comparison with other methods, we decided to use codes from the SWISS-PROT data bank. For all of these proteins it is assumed that the secondary structure of transmembrane regions is α-helical. All residues found in

Table 1. Training Data Subset of Proteins Used for Optimization of All Parameters Needed for Construction of Preference Functions Method[a]

Observed[b]	Predicted[c]	Observed[b]	Predicted[c]	Observed[b]	Predicted[c]
4f2_human		**aqp1_human**		118–138	123–144
82–104	82–103	18–35	13–33	195–215	196–215
5ht3_mouse		49–67	52–71	221–241	223–241
223–249	224–244	—	77–95	308–328	305–324
255–273	259–277	94–115	99–117	331–350	328–346
283–301	284–305	136–156	138–156	860–880	858–883
442–461	439–459	165–184	168–188	912–932	911–932
a1aa_human		211–232	208–229	991–1011	986–1006
—	43–61	**athb_rat**		1014–1034	1010–1029
54–79	65–82	37–63	41–62	1103–1123	1103–1121
92–117	91–112	**athp_neucr**		1129–1149	1129–1147
128–150	128–148	116–136	116–138	**cgcc_bovin**	
172–196	171–189	141–160	143–163	163–183	164–183
210–233	215–235	292–313	290–311	197–215	—
307–331	308–329	325–347	325–348	240–259	—
339–363	340–361	—	694–712	298–320	298–320
a4_human		720–738	718–736	373–392	376–398
683–706	683–705	755–774	755–775	477–497	—
aa1r_canfa		827–847	824–844	**cp5a_cantr**	
11–33	10–34	860–876	855–878	29–48	29–47
47–69	46–68	**atm1_yeast**		**cyda_ecoli**	
81–102	81–103	116–136	115–134	23–42	16–38
124–146	124–144	156–176	154–174	—	55–73
177–201	180–203	233–253	232–251	95–114	95–116
236–259	235–256	258–278	259–278	130–149	128–149
268–292	268–290	352–372	351–370	188–207	185–205
adt2_yeast		379–399	374–393	220–239	219–238
27–44	26–44	**atpl_ecoli**		393–412	391–411
89–107	—	11–31	9–31	471–490	471–493
131–148	128–148	53–73	53–74	**cydb_ecoli**	
192–211	192–210	**bacr_halha**		9–28	9–29
231–248	222–240	10–29	10–30	—	61–79
287–305	—	44–63	45–68	80–99	83–99
adt_ricpr		82–101	83–101	123–142	121–142
34–54	28–45	108–127	106–127	165–184	164–185
68–88	67–85	135–154	135–156	206–225	206–226
93–113	93–113	178–197	176–196	263–282	259–280
148–168	148–169	204–223	201–224	293–312	291–312
185–205	184–204	**cb21_pea**		337–356	337–359
219–239	217–237	62–81	—	**cyoa_ecoli**	
280–300	277–297	114–134	114–134	27–45	20–41
321–341	321–341	182–198	182–198	69–87	66–84
349–369	349–370	**cek2_chick**		**cyob_ecoli**	
380–400	378–399	346–370	348–370	17–35	18–39
439–459	444–462	**cftr_human**		58–76	57–76
466–486	466–482	81–103	76–94	102–121	106–127

(continued)

Table 1. (Continued)

Observed[b]	Predicted[c]	Observed[b]	Predicted[c]	Observed[b]	Predicted[c]
144–162	141–163	88–107	88–107	gmcr_human	
195–213	190–213	113–134	116–136	299–324	302–324
232–250	231–252	153–176	153–175	gpbb_human	
277–296	273–298	183–203	183–203	122–146	125–146
320–339	315–334	223–243	223–243	gpt_crilo	
348–366	342–361	255–281	254–274	7–32	9–32
382–401	379–404	dsbb_ecoli		58–79	60–79
410–429	413–434	15–31	16–37	95–114	95–115
457–476	456–477	50–65	49–65	126–145	129–147
494–513	493–516	72–89	69–87	165–184	160–182
588–607	589–607	145–162	143–163	195–211	190–211
614–634	611–629	edg1_human		222–240	223–241
cyoc_ecoli		47–71	48–70	253–269	252–270
32–50	27–50	79–107	83–106	275–294	274–292
67–85	68–88	122–140	122–142	379–397	379–397
102–120	99–119	160–185	160–181	hema_cdvo	
143–161	138–162	202–222	202–222	35–55	38–58
185–203	180–200	256–277	256–277	—	310–328
cyod_ecoli		294–314	294–312	hema_pi4ha	
18–36	18–38	egf_mouse		28–47	26–46
46–64	46–66	—	9–29	—	301–321
81–99	81–102	1039–1058	1039–1059	hg2a_human	
cyoe_ecoli		egfr_human		46–72	49–69
10–28	12–30	622–644	624–644	hly4_ecoli	
38–56	36–56	fce2_human		60–80	61–81
79–97	80–98	22–47	24–45	hoxn_alceu	
108–126	106–124	fixl_rhime		20–40	20–40
—	160–180	21–40	22–40	52–72	44–60
198–216	207–226	47–67	49–67	95–115	91–113
229–247	230–249	71–86	71–89	129–149	128–149
269–287	264–282	97–118	98–117	200–220	197–220
dhg_ecoli		ftsh_ecoli		244–264	233–258
11–37	12–31	5–24	4–22	270–290	273–292
41–58	39–57	96–120	100–120	il2a_human	
63–81	65–85	g2lf_human		220–238	218–238
96–110	94–110	9–25	9–28	il2b_human	
119–141	114–136	—	304–324	215–239	218–239
—	709–729	glp_pig		imma_citfr	
dhsc_bacsu		63–85	63–85	14–37	17–37
12–31	12–32	glpc_human		69–89	76–91
60–79	57–80	58–81	59–81	107–123	106–123
93–113	95–115	glra_rat		143–165	144–165
135–155	135–167	521–540	520–540	ita5_mouse	
178–196	180–199	567–585	563–581	356–381	359–381
dmsc_ecoli		596–614	597–618	kdgl_ecoli	
10–32	10–32	788–808	785–808	33–48	35–53
44–66	45–66			52–68	57–75

Table 1. (Continued)

Observed[b]	Predicted[c]	Observed[b]	Predicted[c]	Observed[b]	Predicted[c]
96–118	97–116	486–504	484–504	**myp0_human**	
kgtp_ecoli		**mas6_yeast**		125–150	127–150
26–54	35–54	99–128	100–118	**mypr_human**	
62–80	62–84	146–166	146–164	10–35	12–34
96–116	96–114	175–189	—	59–87	64–87
120–136	120–140	197–215	197–216	151–177	154–177
163–185	164–184	**mdr3_human**		238–267	231–249
196–214	199–214	58–78	57–77	—	253–267
244–261	244–264	123–143	121–141	**nep_human**	
275–301	277–298	192–211	193–211	28–50	29–49
312–330	312–330	216–235	217–235	**ngfr_human**	
337–360	336–359	301–320	299–320	223–244	224–244
369–392	371–391	336–354	330–350	**nk11_mouse**	
403–423	402–422	712–732	714–736	43–62	43–63
lacy_ecoli		756–776	753–776	**nqoc_parde**	
11–33	9–30	832–851	832–851	4–24	4–20
47–67	45–65	854–873	855–873	30–50	30–50
75–99	75–95	937–956	936–956	79–99	79–99
103–125	103–123	976–993	973–991	116–135	119–138
145–163	146–165	**melb_ecoli**		139–158	142–160
168–187	169–187	8–28	15–47	179–199	178–197
212–234	219–237	33–53	—	224–244	221–240
260–281	264–284	76–96	76–94	256–276	258–276
291–310	290–312	103–123	100–119	290–310	290–308
315–334	—	146–166	146–164	325–344	324–342
347–366	349–370	172–192	174–194	347–366	346–364
380–399	377–398	231–251	234–255	381–401	386–407
lech_human		263–283	263–283	415–435	421–443
40–60	40–60	293–313	294–316	475–495	473–492
lep_ecoli		320–340	324–346	580–600	580–598
4–22	4–23	370–390	373–393	679–699	677–698
58–76	63–81	408–428	403–426	**nqod_parde**	
lspa_ecoli		**mota_ecoli**		3–23	5–23
12–22	12–32	2–21	4–22	34–54	36–54
70–88	70–88	34–51	30–48	81–101	81–101
96–113	96–114	171–191	167–187	112–131	114–133
139–153	135–155	201–222	200–223	134–153	137–156
magl_mouse		**motb_ecoli**		164–184	165–185
498–517	495–514	28–49	30–50	211–231	209–231
malf_ecoli		**mpi2_yeast**		250–270	251–270
17–35	17–36	12–32	15–35	277–297	283–303
40–58	40–58	59–79	58–76	312–332	317–337
73–91	71–92	88–108	88–108	340–360	341–360
277–295	279–305	113–133	115–133	383–403	379–398
319–337	319–340	**mprd_human**		418–438	415–438
371–389	371–392	160–184	162–183	463–483	463–479
418–436	425–447				

(continued)

Table 1.　(Continued)

Observed[b]	Predicted[c]	Observed[b]	Predicted[c]	Observed[b]	Predicted[c]
och1_yeast		**ppb_yeast**		443–463	443–461
16–30	16–34	34–59	36–56	**ptma_ecoli**	
oec6_spiol		**psaa_pinth**		25–44	19–39
18–40	19–40	73–96	75–93	51–69	51–69
oppb_salty		162–182	159–179	—	82–103
10–30	9–27	198–222	198–216	135–154	132–153
100–121	99–119	294–312	294–312	166–184	160–183
138–158	134–158	349–372	349–369	—	219–241
173–190	170–190	388–414	390–412	—	245–263
227–250	233–253	436–458	439–459	274–291	269–291
272–293	275–296	534–552	535–557	314–333	313–334
oppc_salty		592–613	592–612	**rceh_rhovi**	
38–59	38–59	667–689	668–689	12–32	10–30
103–122	103–125	727–747	727–747	**rcel_rhovi**	
140–160	140–158	**psab_pinth**		32–55	27–49
164–180	162–180	46–69	48–68	84–112	82–102
216–236	215–240	135–157	135–153	115–140	114–134
268–290	270–290	175–199	178–196	170–199	176–196
ops1_calvi		273–291	273–291	225–251	230–250
48–72	51–74	330–353	329–349	**sece_ecoli**	
85–110	85–105	369–395	374–394	19–36	16–36
125–144	123–143	417–439	421–441	45–63	41–61
164–187	163–184	517–535	517–535	93–111	94–116
212–239	215–237	575–596	575–596	**spg1_strsp**	
275–298	275–295	643–665	644–665	390–410	388–406
306–330	—	707–727	707–728	**suis_human**	
opsb_human		**ptgb_ecoli**		13–32	11–33
34–58	33–55	15–35	22–43	—	617–636
71–96	74–96	51–71	51–71	**tcb1_rabit**	
111–130	111–131	80–100	79–100	292–313	291–311
150–173	149–169	112–132	115–133	**trbm_human**	
200–227	201–221	152–172	153–174	495–518	493–514
250–273	252–272	191–211	—	**trsr_human**	
282–306	—	250–270	248–270	63–88	67–87
pigr_human		280–300	280–298	**vmt2_iaann**	
621–643	622–641	310–330	305–327	25–42	28–48
		356–376	354–374	**vnb_inbbe**	

[a]Each protein is identified by its SWISS-PROT identifier (Ref. 52).
[b]The positions for the transmembrane helices observed according to SWISS-PROT counted from the first residue or from the first residue after the signal or propeptide sequence.
[c]The prediction by preference functions method achieved in the training procedure.

Table 2. Testing Data Set of Proteins[a]

Observed[b]	Predicted[c]	Observed[b]	Predicted[c]	Observed[b]	Predicted[c]
1b14_human		785–806	785–803	140–155	140–161
285–308	283–305	—	833–850	177–195	—
ach1_xenla		844–865	854–869	215–234	208–231
211–235	210–233	**c561_bovin**		327–351	327–350
243–261	245–265	38–60	38–60	448–466	448–466
277–296	276–296	75–97	74–93	482–501	478–496
409–427	405–427	107–129	109–127	510–528	516–539
acm5_human		145–167	145–168	539–557	—
30–53	32–53	185–207	184–204	577–596	573–596
67–87	72–93	219–241	221–241	652–675	652–676
105–126	106–126	**cadn_mouse**		816–834	814–832
147–169	149–169	566–587	561–584	851–870	849–869
444–464	193–213	**car1_dicdi**		883–901	883–901
479–498	—	14–33	14–36	909–927	—
ag22_mouse		48–68	—	947–966	938–962
46–71	48–69	84–109	79–105	—	1011–1029
81–102	81–101	121–139	124–145	1057–1081	1057–1081
120–140	118–138	163–181	161–183	1135–1153	1131–1151
161–179	158–176	206–224	208–228	1169–1188	1168–1188
209–234	210–233	236–260	—	1197–1215	1198–1216
257–278	257–281	—	260–282	1253–1271	—
286–313	289–312	**cb2_human**		1291–1310	1290–1311
atn1_human		34–59	37–59	1378–1402	1377–1402
91–112	91–111	72–92	75–96	**cik1_drome**	
125–144	123–141	105–129	109–129	186–204	184–204
286–308	285–305	150–172	150–170	229–250	228–246
315–343	309–328	189–214	190–211	261–282	260–278
782–805	782–805	247–267	246–269	290–308	—
844–869	850–871	280–301	281–300	324–345	324–344
911–931	911–929	**cd2_human**		—	354–372
948–973	—	186–211	186–208	385–406	385–406
atp9_wheat		**cd72_human**		**cox2_parli**	
8–32	9–30	96–116	95–116	27–48	26–49
45–72	47–71	**cd7_human**		63–82	65–84
b3at_human		156–176	154–175	**cox9_yeast**	
404–424	403–423	**cd81_human**		15–33	14–34
437–456	445–466	12–35	12–33	**cxb5_rat**	
460–479	—	58–78	59–81	20–40	19–39
491–510	491–512	90–115	92–114	76–96	76–94
523–541	520–538	202–226	202–227	137–159	129–151
569–588	570–588	**cd8a_human**		185–205	187–206
604–624	606–624	162–187	162–183	**cyf_brara**	
661–680	661–684	**cic1_cypca**		251–270	251–270
699–719	702–724	71–86	71–90	**divb_bacsu**	
763–780	761–779	108–127	110–131	32–50	32–52

(continued)

Table 2. (Continued)

Observed[b]	Predicted[c]	Observed[b]	Predicted[c]	Observed[b]	Predicted[c]
fmlr_rabit		**hmdh_human**		515–535	515–535
28–50	21–53	10–39	17–37	553–573	556–574
62–83	—	57–78	60–80	579–599	579–597
101–121	97–121	90–114	90–112	603–623	603–621
141–162	145–162	124–149	—	629–648	630–648
208–228	199–225	160–187	167–187	659–679	659–679
245–268	245–268	192–220	195–218	697–717	697–718
288–307	284–304	315–339	319–337	735–754	735–754
frdd_provu		—	744–764	758–776	761–779
14–41	19–43	—	841–863	790–810	800–818
55–80	59–79	**isp6_yeast**		814–836	822–838
99–119	96–116	32–52	35–54	**nram_iabda**	
frsl_ecoli		**itb1_human**		7–36	12–33
38–57	38–57	709–731	712–733	**psbi_horvu**	
furi_human		**lha1_rhosh**		6–27	6–24
609–631	611–631	16–36	19–39	**secy_ecoli**	
gaa1_bovin		**lhb4_rhopa**		23–42	24–44
225–246	225–244	24–46	28–46	75–95	75–98
252–273	256–273	**ly49_mouse**		122–139	122–140
286–307	285–306	45–66	45–66	154–174	151–172
395–416	398–416	**m49_strpy**		183–203	183–202
gasr_human		324–342	321–339	217–237	215–236
58–79	58–81	**malg_ecoli**		274–294	273–294
88–109	91–117	19–39	16–37	316–335	315–335
132–150	125–150	82–102	87–107	376–395	373–391
171–189	174–192	124–144	124–144	399–414	400–420
220–242	220–241	151–171	156–177	**spc2_canfa**	
334–355	331–350	205–225	208–229	11–32	8–28
374–394	369–389	260–280	262–282	**spir_spine**	
gesr_human		**mepa_mouse**		165–184	—
604–626	602–622	650–677	652–674	—	3–21
ghr_human		**mpcp_rat**		**stub_drome**	
247–270	247–269	14–36	14–34	59–80	57–77
grhr_human		72–91	—	**sy65_drome**	
39–58	39–58	112–133	—	108–134	113–134
78–97	—	169–188	—	**syb1_human**	
116–137	118–138	212–234	213–233	97–116	97–115
165–184	158–178	265–283	264–284	**synp_rat**	
213–232	213–235	**nals_bovin**		2–25	4–22
282–300	—	25–44	25–43	81–105	81–102
307–326	—	**nntm_bovin**		113–136	116–136
ha21_human		—	187–207	180–198	177–197
195–217	197–217	432–450	435–453	**ta16_human**	
hb23_human		458–478	457–475	10–30	11–29
199–219	200–219	484–503	486–506	46–70	47–67

Table 2. (Continued)

Observed[b]	Predicted[c]	Observed[b]	Predicted[c]	Observed[b]	Predicted[c]
89–116	93–114	111–129	108–128	—	138–159
162–192	161–188	140–162	142–162	—	168–190
tat2_yeast		165–185	166–185	tsa4_giala	
88–108	87–105	201–221	199–217	663–691	667–688
114–134	112–134	223–240	221–239	ucp_rat	
161–181	163–184	256–276	256–279	10–31	16–34
193–213	194–214	297–317	292–315	73–95	
227–247	224–244	324–344	325–343	116–132	113–133
267–287	267–285	346–365	347–365	178–194	—
307–327	307–328	—	393–413	212–231	211–229
359–379	356–376	432–451	428–448	266–288	—
405–425	408–426	tee6_strpy		va34_vaccc	
429–449	432–451	492–510	489–509	1–20	15–34
473–493	476–498	thas_human		vcal_human	
515–535	517–535	10–30	11–31	675–696	675–695
tca_human		139–164	—	vglg_hrsva	
118–137	116–137	209–239	224–240	38–66	41–63
tcc1_mouse		331–364	332–353	vs10_rotbn	
135–155	133–154	480–498	—	29–44	29–47
tcrb_bacsu		tnr1_human		wapa_strmu	
12–33	15–35	191–213	192–212	394–412	392–410
—	50–68	tolq_ecoli			
81–100	80–100	23–43	10–33		

[a]Each protein is identified by its SWISS-PROT identifier (Ref. 52).
[b]The positions for the transmembrane helices observed according to SWISS-PROT counted from the first residue or from the first residue after the signal or propeptide sequence.
[c]The prediction by preference functions method achieved in the testing procedure.

extramembrane regions are assumed to be in undefined conformations. This procedure was used even for membrane proteins of crystallographically known structure. Because the exact locations of transmembrane segments are often controversial, the selection of these 166 proteins was based on hydrophobicity analysis, experimental, and topological data collected by the cited authors in the SWISS-PROT data bank for each protein. All of these proteins were chosen such that pairwise sequence identity was less than, or equal to, 30%.

A comparison between our set of proteins and the sets used by other authors leads to the following conclusions: (1) Our set of proteins is the largest; (2) the similarity between any 2 proteins among our 166 proteins is less than 30% (the amount of similarity is taken from the HSSP data bank[24]); (3) we decided to divide our set of proteins into a training subset and a testing subset rather than use the cross-validation procedure (this will be explained below). It is especially important that proteins in the training subset and the testing subset, as well as proteins from the training subset in relation to proteins in the testing subset, be as nonhomologous as possible. Among the papers published in this area, this datum is given by Rost *et al.*[42] where the pairwise identity in the set of 69 proteins was smaller than 25% and by Jones *et al.*[40] where the pairwise identity was smaller than 60%. A pairwise identity greater than 25–30% is too high and indicates similarities in the structure of the proteins in the set.[24] Thus, a method that is trained on such a set may not give a good structure prediction for a new protein that is not homologous to proteins used for optimization of the method employed.

It is important to mention that probably for a few proteins from the SWISS-PROT data bank the location and number of transmembrane segments are wrong because they were determined by combining experimental data that are not entirely reliable with theoretical methods for predicting the location of transmembrane segments, most often with the Kyte–Doolittle method that also is not quite accurate. This possibility of an error also exists in works of other authors. Errors of this kind represent a hindrance to comparison of the accuracy of different methods. Besides, these errors limit the accuracy of all theoretical methods. We assumed that these errors can be considered as random noise that will on average tend to cancel out. This means that the method will not predict structures inaccurately because there is an error in the location and number of transmembrane segments in the training subset, since such errors cancel each other, but errors will be caused by the difficulty of the problem of structuring membrane proteins and by the inability of the algorithm to solve the problem completely accurately.

5.3.2. Performance Measures

In predicting the transmembrane segments, we distinguish the numbers of residues that are associated with positive correct prediction p, negative correct prediction n, underprediction u, and overprediction o. Out of these four data it is possible to define and compute many performance parameters.[53] We consider as the best parameter the performance parameter A_2 defined by Ponnuswamy and Gromiha[54]:

(1)
$$A_2 = \frac{p - o}{p + u}$$

However, since we want to compare it with the best method[42] we will also compute the Q_2 parameter:

(2)
$$Q_2 = \frac{p + n}{p + n + u + o}$$

Parameter A_2 can assume values smaller than zero, while its highest value is 1.0 in the case of a perfect prediction. This parameter is more sensitive than the Q_2 parameter. Since there are far greater numbers of residues in extramembrane segments, the value of n is considerably greater than the value of p. This causes the value of the

Q_2 parameter to be very large and equally good for both good and bad predictions and thus it shows that there are no significant differences in the accuracy of the prediction between various existing theoretical methods. Therefore, one artificially increases the accuracy of the larger set of proteins, measured by either Q_2 or A_2 parameters, by putting for proteins with only one transmembrane segment all residues into the extramembrane region. Hence, we will also compute the Q_2 parameter for random prediction[30,55]:

$$(3) \qquad Q_{2,\text{random}} = \frac{(p + u)(p + o) + (n + u)(n + o)}{(p + n + u + o)^2}$$

In the above equation $p + u$ is the number of residues in the actual location of transmembrane segments according to the SWISS-PROT data bank, while $p + o$ is the number of residues predicted to be in the transmembrane segments; $n + o$ and $n + u$ have analogous meanings for the extramembrane segments. The true contribution of a certain theoretical method for predicting the location of transmembrane segments and, more generally, for predicting the protein secondary structure, is given by the following difference equation:

$$(4) \qquad \Delta Q_{2,\text{method}} = Q_2 - Q_{2,\text{random}}$$

5.3.3. Propensity Values

Propensity values of 20 amino acid residues were computed on the training subset with the equation

$$(5) \qquad p_i = \frac{f_{i,\text{mem}}}{f_{i,\text{tot}}}$$

where p_i is the propensity value of residue type i, $f_{i,\text{mem}}$ is the frequency of occurrence of residue i in the transmembrane regions, and $f_{i,\text{tot}}$ is the frequency of occurrence of residue i along the entire sequences of the training protein data subset. We have also added two values for spaces before the beginning of the sequence at the NH_2-terminus and at the COOH-terminus of the sequence, similarly to other authors.[30,42,55] In the procedure for optimization predictions on the training set, we also obtained propensity values for the NH_2-terminus and for the COOH-terminus for which the prediction was the best. Abbreviations for the amino acid residues and for the two end spaces, together with their propensity value, are as follows: A, Ala (1.327); C, Cys (0.926); L, Leu (1.621); M, Met (1.451); E, Glu (0.026); Q, Gln (0.507); H, His (0.599); K, Lys (0.142); V, Val (1.579); I, Ile (1.804); F, Phe (1.688); Y, Tyr (1.096); W, Trp (1.319); T, Thr (0.823); G, Gly (0.996); S, Ser (0.732); D, Asp (0.163); N, Asn (0.689); P, Pro (0.106); R, Arg (0.057); space(NH_2-end) (−0.435), space(COOH-end) (−0.07).

5.3.4. Normal Approximation for Frequency Distribution of Amino Acid Residues over Local Sequence Environment

The initial assumption of the propensity functions method[48,50,56,57] is that the propensity of each amino acid residue is a function of properties of neighboring amino acid residues in the primary structure. It is well known that the external environment affects the amino acid residue preference for secondary structure.[58] In addition, the protein 3D environment has an important influence on preferences of amino acid residues for certain folding motifs. For instance, heme binding increases the preference of neighboring amino acid residues for the α-helix conformation.[59] Local sequence environment must be taken into account as far as eight neighbors away from the residue considered in order to obtain good prediction for its secondary conformation.[60,61] A markedly increased probability of α-helix conformation in soluble proteins has been found in bulkier primary structure environment.[56] Not surprisingly, increased probability of α-helix conformation is also found in the case when the Chou–Fasman conformational parameters for α-helix are used as sequence attribute.[46,57] Sequence environment influence is even more important for membrane proteins that must achieve the partition of their sequence between transmembrane and extramembrane regions.

In this work the sequence environment of a residue, found in transmembrane or extramembrane regions, is defined as an arithmetic average of the above-defined propensities for its six left and six right neighbors, without the central residue. Differing from earlier configuration preference functions methods, the sliding window width parameter is optimized from the beginning starting with width 3 to width 21. The best prediction is obtained for a sliding window of width 13.

The frequency distributions of all environments X from the training data subset were collected for each amino acid residue type and for two conformations: transmembrane and extramembrane. There are often cases of membrane proteins that have very long extramembrane segments. Thus, there are nearly four times more amino acid residues in the extramembrane state than in the transmembrane state in the training subset of proteins. In an early version of the PREF method[45,48,50] for obtaining the frequency distributions, we used entirely soluble proteins or a mixture of soluble and membrane proteins. When only membrane proteins are used, as in the present two-state model, there is an imbalance in the number of transmembrane and nontransmembrane environments, the latter being much more abundant. This can decrease the resolution of prediction between transmembrane and nontransmembrane conformation. However, the prediction in the region of the wide extramembrane segments does not become worse by significant reduction in the number of environments for the extramembrane state (not shown).

Because of this we used in the procedure of training for the frequency distributions of environments for the extramembrane state only those environments that are removed from the NH_2- and COOH-terminal transmembrane segments for at most 16 amino acid residues. By using this methodology we obtained approximately the same number of environments in the transmembrane state (8840 cases) as in the extramembrane state (8730 cases).

We stated the null hypothesis, that is, that the frequency distributions are approximately normal,[50,56] since all environments X were obtained by the averaging procedure. In order to test the null hypothesis, the χ^2 test of goodness of fit was used on all 20 amino acid residues on transmembrane and extramembrane states. In order to obtain the best possible approximation to the frequency distribution, we have optimized the number of classes of the equal sliding window width that divide the set of environments X starting from X_{min} and ending with X_{max} for each amino acid residue. The number of classes that divide the set of environments X is varied from 4 to 15.

This is a change from PREF version 1.0, where we always used a distribution into nine classes in the procedure of approximating the frequency distribution by the normal distribution for each amino acid residue and for each kind of secondary structure.[45,48,50,56] The optimum χ^2 values, the number of classes for which the χ^2 value is obtained, and the probabilities p that allow higher χ^2 values to be found are given in Table 3.

Table 3. Normality Tests for the Frequency Distribution of Environments Computed from Propensities Obtained Using Equation (5)[a]

	Transmembrane state			Extramembrane state		
AA	n	χ^2	p	n	χ^2	p
A	8	9.718	> 0.05	5	3.950	> 0.10
C	5	0.005	> 0.95	5	0.275	> 0.80
L	4	1.418	> 0.20	13	10.060	> 0.30
M	5	2.905	> 0.20	7	56.420	< 0.001
E	5	0.258	> 0.80	13	3.036	> 0.98
Q	4	5.875	> 0.01	4	1.919	> 0.20
H	7	0.991	> 0.90	7	2.578	> 0.50
K	7	8.388	> 0.05	7	10.260	> 0.02
V	8	19.580	> 0.001	13	8.887	> 0.50
I	7	15.490	> 0.001	8	7.822	> 0.10
F	11	9.441	> 0.30	6	4.097	> 0.20
Y	9	5.937	> 0.30	11	11.140	> 0.10
W	11	7.898	> 0.30	5	0.067	> 0.95
T	5	4.440	> 0.10	9	7.680	> 0.20
G	4	3.919	> 0.02	5	1.332	> 0.50
S	7	19.720	< 0.01	4	5.678	> 0.01
D	5	0.066	> 0.95	7	4.661	> 0.30
N	6	5.813	> 0.10	9	7.084	> 0.30
P	5	5.873	> 0.05	13	15.650	> 0.10
R	4	3.860	> 0.05	7	11.520	> 0.02

[a] n represents the number of classes. χ^2 values were determined using our computer program. p values are the probabilities that higher χ^2 values can be found. The probabilities were extracted from the standard table (Ref. 79) for the $n-3$ degrees of freedom.

We point out that in the case of the frequency distribution approximation for environments, we have optimized the number of classes only with the aim to obtain the best possible approximation according to the criterion of the minimum χ^2 test. This means that the optimal number of classes is not optimized with respect to final prediction accuracy of the PREF method.

In Table 3 it can be seen that in almost all cases studied the probability $p(\chi^2 > \chi_0^2)$ is very high. Therefore, the approximation to the frequency distribution of environments X by normal distribution is justified.

Frequency distributions for methionine and glutamic acid residues are shown in Figure 1.

Of all amino acid residues, methionine in the extramembrane state appears most often at the beginning of the sequence because the codon for this amino acid is at the same time the start codon. However, methionine, similarly to other amino acid residues, also appears inside of the sequence. As the first amino acid residue in the sequence, methionine does not possess left-hand neighbors. They are mimicked by the NH_2-terminal space with a constant low propensity value. The lowest propensity value for the local environment X is that for methionine in the extramembrane state ($p = 0.0771$). For this reason, the frequency distributions for methionine in the extramembrane state deviate most from the regular frequency distributions.

Glutamic acid is an electrically charged amino acid. Charged amino acid residues in membrane proteins possess an environment that is very well defined, that is, they are very rarely present in the membrane and are predominantly located in extramembrane regions. Glutamic acid may be found in the membrane only if its environment possesses a high propensity for finding it in the membrane.

As a result of approximating the frequency distributions by normal distribution, one obtains the average deviation μ_{ij}, the sample standard deviation σ_{ij}, and the number of amino acid residues i found in the transmembrane ($j = 1$) and extramembrane ($j = 2$) state N_{ij}. All of these data are shown in Table 4.

5.3.5. The Definition of Preference Functions in the PREF Method (Version 2.0)

After we have approximated frequency distributions of environments X by normal distribution, that is, by analytical function, we can define preference functions $P_{ij}(X)$ as:

$$(6) \qquad P_{ij}(X) = \frac{(N_{ij}/\sigma_{ij}) \exp[-(X - \mu_{ij})^2/2\sigma_{ij}^2]}{\sum\limits_{k=1,2} (N_{ik}/\sigma_{ik}) \exp[-(X - \mu_{ik})^2/2\sigma_k^2]}$$

In the normal approximation mentioned above, the preference function is expressed as a ratio of one Gaussian function of X for state j to the sum of all Gaussian functions of X for all (in this case 2) conformations. By defining preference functions according to equation (6) one secures their normalization, that is, the sum of preference functions

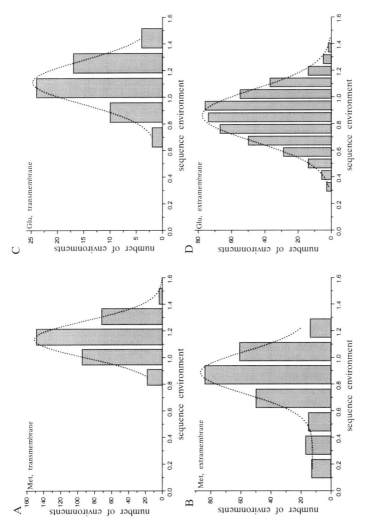

Figure 1. Approximation of frequency distributions by normal distribution for the case of the worst approximation (according to p values in Table 3) for Met (A) in the transmembrane state and (B) in the extramembrane state; and the case of the best approximation for Glu (C) in the transmembrane state and (D) in the extramembrane state.

Table 4.   Parameters for the Construction of Gaussian Curves[a]

AA	Transmembrane state			Extramembrane state		
	N_{ij}	μ_{ij}	σ_{ij}	N_{ij}	μ_{ij}	σ_{ij}
A	1007	1.1487	0.1394	673	0.8540	0.1894
C	147	1.1582	0.1468	101	0.8272	0.1789
L	1520	1.1267	0.1583	731	0.8377	0.1817
M	338	1.1327	0.1302	254	0.8033	0.2595
E	57	1.1071	0.1720	432	0.8543	0.1899
Q	94	1.1111	0.1953	335	0.8923	0.1949
H	118	1.1130	0.1239	214	0.8629	0.1874
K	49	1.0825	0.1631	594	0.8576	0.1821
V	1031	1.1405	0.1415	467	0.8598	0.1672
I	984	1.1287	0.1525	415	0.8305	0.1800
F	800	1.1229	0.1356	415	0.8675	0.1781
Y	324	1.1143	0.1383	303	0.8697	0.1689
W	217	1.1105	0.1414	205	0.8766	0.1828
T	420	1.1641	0.1437	491	0.8767	0.1698
G	796	1.1534	0.1504	628	0.8887	0.1883
S	473	1.1483	0.1479	606	0.8580	0.1969
D	57	1.1328	0.1305	423	0.8722	0.1975
N	133	1.1094	0.1392	388	0.8935	0.1793
P	227	1.1591	0.1679	406	0.8738	0.1776
R	48	1.0597	0.1773	649	0.8822	0.1835

[a]The average μ_{ij} and standard deviation σ_{ij} of sequence environments X are given together with the total number of environments N_{ij} in the training protein data set for amino acid residue types i in transmembrane ($j = 1$) and in extramembrane ($j = 2$) regions.

for each residue for the transmembrane state and for the extramembrane state is equal to 1.0. In the process of prediction, preference functions are computed for each amino acid residue in the sequence for the transmembrane state and for the extramembrane state. The results are preference profiles for transmembrane and extramembrane states. The accuracy of prediction in this version of the algorithm is the best if the smoothing procedure is performed on a sliding window of width 7 for the transmembrane state and if for the extramembrane state the smoothing procedure is not performed. The preference profiles smoothed in such a way are used further for deciding in which state which amino acid residue will be found. The amino acid residue i will be in that state for which after the smoothing procedure the preference function value is larger.

5.3.6. Filtering the Prediction

The preference functions method needs some mending because some potential transmembrane segments are too short, others are too long, and very often two transmembrane segments are joined into one very long segment. The mending of the preference functions method is done by means of filtering the prediction. The direction of the segment enlarging or shortening is determined by the difference of preference functions for transmembrane or extramembrane segments, respectively, which is defined by $\Delta P(X) = P_{tm}(X) - P_{extm}(X)$. It should be noted that: (1) Potential transmembrane segments longer than 8 and shorter than 12 residues are enlarged in the direction of larger values of $\Delta P(X)$ until a length of 17 to 22 residues is reached and if $\Delta P(X) > -0.9$ only if the maximum value of $\Delta P_{max}(X)$ on segment is larger than 0.82. (2) Transmembrane segments with a length of 12 to 20 residues are enlarged in the direction of larger values of $\Delta P(X)$ until a length of 17 to 22 residues is reached and if $\Delta P(X) > -0.3$ only if the maximum value of $\Delta P_{max}(X) > 0.76$. (3) Transmembrane segments with a length of 24 to 33 residues are shortened in the direction of smaller values of $\Delta P(X)$ until a length of up to 22 residues is reached and if $\Delta P(X) < 0.49$. (4) Transmembrane segments with a length of 34 to 38 residues are enlarged in the direction of larger values of $\Delta P(X)$ until a length of up to 40 residues is reached and if $\Delta P(X) < 0.90$ and after that they are broken in the middle with three residues that are in the extramembrane state. (5) Transmembrane segments longer than 40 residues break in the middle with 7 residues which are in the extramembrane state.

The process of filtering was carried out following the above order on the training data set (list-train, 95 proteins). Because of the size of the problem, the optimization of parameters in the process of filtering was carried out in such a way that during the search for the optimum value for one parameter all other parameters were fixed. The values of parameters connected with the limiting values of $\Delta P(X)$ were optimized with steps of 0.1 and when the best value was found for this step the optimization was continued with step 0.01. The length of segments was changed with steps of 1.0.

The process of filtering is necessary because the initial prediction was reached by means of a two-state digitalization of the $\Delta P(X)$ profile which is a nearly analogous function that depends on the order of amino acid residues in the sequence. The window widths used to compute average propensities as well as the window widths used in the smoothing procedure are too small for some parts of the sequence and in some cases they are too long and this leads to prediction of too short segments or to joining the neighboring transmembrane segments into one too long segment. The digitalization cannot describe all details of the $\Delta P(X)$ profile and thus it gives the arrangement of transmembrane segments that can be improved by the process of filtering.

Considerably more information about the arrangement of actual and potential transmembrane segments is contained in the $\Delta P(X)$ profile than in digitalized form which tells us only whether a certain residue is in the membrane segment (=1) or in the extramembrane segment (=0). However, the digitalization is necessary in the optimization procedure as well as in comparing other competing methods. In predicting the structure of a certain new membrane protein for which the location of transmembrane segments is not known, we consider the whole profile and combine the prediction with the methods for determining the topology of the protein, such as the tendency of positively charged amino acid side chains to appear more often in the extramembrane part which is in the cytoplasm.[37,38,62] With such an analysis one can predict more reliably whether a certain potential transmembrane segment [that was left out according to rules of the digitalization of the $\Delta P(X)$ profile] is after all a transmembrane segment. Likewise one can detect proteins and protein parts for which the location of transmembrane segments and their topology is not quite certain, and one can also indicate the need for additional experimental verification.

5.4. RESULTS AND EXAMPLES ILLUSTRATING THE PREDICTION ACCURACY

We will first show the prediction accuracy of the preference functions method on the training set of proteins. We do this in the following way. In order to investigate the accuracy of the method relative to other methods, we will compare the preference functions method with the best method published so far for predicting the location of transmembrane segments that is based on the neural network (NN) algorithm.[42] Then we will analyze predictions for individual proteins from the training set and from the testing set in order to present examples where the method is deficient and where it gives predictions that are good and that in accordance with other methods suggest that the published structures of some proteins are inaccurate. Finally, we will describe the procedure for modeling the structure and topology of transmembrane proteins by presenting examples of very reliable predictions for proteins for which the location of transmembrane segments is not known. We consider as very reliable those predictions for which the maxima in the $\Delta P(X)$ profile, corresponding to transmembrane segments, are clear, high, and wide for about 20 amino acid residues and also if the number of positively charged amino acid side chains in extramembrane segments is distributed in accordance with the "positive inside" rule.

5.4.1. Prediction Accuracy of the Transmembrane Segments on the Training and Testing Data Sets

We obtained the following parameter values which are used for indicating the prediction accuracy on the list-train (95 proteins), when the preference functions are extracted from the same data set of proteins (see Section 5.3.2 for definitions):

$p = 7747, n = 33,404, u = 1078, o = 1107$
$A_2 = 0.7524, Q_2 = 94.96\%$
$Q_{2,random} = 67.53\%$
$\Delta Q_{2,method} = 27.43\%$

Observed and predicted transmembrane segments for 95 proteins are shown in Table 1.

The method was checked on the "never-seen data set" of testing proteins in which the pairwise identity between any two proteins within the testing data set (list-test) as well as the pairwise identity between any protein from the testing set (list-test) and any protein from the training set (list-train) is smaller than 30%. The following parameter values are obtained for indicating the prediction accuracy on the testing data set of proteins (list-test, 71 proteins), when the preference functions are extracted from the training data set of proteins (list-train, 95 proteins):

$$p = 4502, n = 22,246, u = 1060, o = 679$$
$$A_2 = 0.6873, Q_2 = 93.90\%$$
$$Q_{2,random} = 69.39\%$$
$$\Delta Q_{2,method} = 24.51\%$$

Observed and predicted transmembrane segments for 71 proteins are shown in Table 2. The values of parameters used in the process of filtering are optimized on the list-train.

Some authors have also reported the prediction accuracy on soluble proteins.[42,63] It is expected that a good method will predict a number of transmembrane segments in soluble proteins as small as possible. If so, such a method could be used to predict whether a given protein is soluble or is a membrane protein before the experiment. However, we consider that different rules are needed by which the method could be supplemented in order to decide after the prediction whether a given protein is a soluble protein or a membrane protein, because in soluble proteins there appear segments of high hydrophobicity that in the actual 3D protein structure appear in its interior. These segments may be predicted by the PREF method which, in our opinion, should not be considered as a shortcoming of the method. In the set of 131 soluble proteins taken from the paper by Rost and Sander,[30] whose 3D structures have been determined with the highest resolution, for 15 proteins it is predicted that they contain one transmembrane segment. All of these proteins were less than 25% similar among themselves.

We tested PREF 2.0 in predicting transmembrane segments for β-strand proteins on the data set consisting of six porin (ompf_ecoli, pori_rhoca, ompa_ecoli, om32_comac, phoe_ecoli, lamb_ecoli) and one defensin (defn_human), and found that PREF 2.0 predicts no single transmembrane segment.

5.4.2. A Comparison of the Preference Functions Method with the Neural Network (NN) Method

The prediction accuracy comparison between different methods is rather difficult to carry out correctly.[42] Some reasons for this are as follows: (1) Each method was trained on a different set of proteins; (2) most authors allow a high pairwise identity in the training set and in the testing set of proteins; (3) certain methods are not accessible to other authors for testing. A significant step to remove the last obstacle was made by Rost et al., who made their method available to everyone over the user-friendly E-mail server (predictprotein@embl-heidelberg.de) and with the help of the World Wide Web (http://www.embl-heidelberg.de/predictprotein/predictprotein.html). Among all methods available, the best performance measure parameters were obtained by the method of Rost et al.[42] An additional advantage of this method is that it has been trained on a set of proteins with the lowest pairwise identity. Here we compare the method of Rost *et al.* and our PREF method (version 2.0). To carry out the comparison as correctly as possible, the testing data set (list-test, 71 proteins) is composed in such a way that the pairwise identities between any two proteins from the training data used by Rost *et al.*,[42] and any protein from the testing data set (list-test, 71 proteins), are similar for an amount less than 30%. The following parameter performance measures are obtained by the method of Rost *et al.*[42] on the testing data set of proteins (list-test, 71 proteins):

$p = 4431, n = 21,970, u = 1131, o = 955$
$A_2 = 0.6250, Q_2 = 92.68\%$
$Q_{2,\text{random}} = 68.95\%$
$\Delta Q_{2,\text{method}} = 23.73\%$

It is easily seen that our method is comparable, if not better, to the method based on the NN algorithm. The prediction method of Rost *et al.*[42] has been trained using rigorous cross-validation test on only 69 proteins and yielded $A_2 = 0.7298$, $Q_2 = 94.72\%$. We can see that the difference between the prediction accuracy on never-seen proteins and that on the training data set of proteins for the method of Rost *et al.*[42] is considerably higher.

5.4.3. Predictions for Proteins with Known 3D Structure

It is especially important to check whether the method gives accurate predictions for membrane proteins with known 3D structures. We consider three subunits of the photosynthetic reaction center of *Rhodopseudomonas viridis*,[10] three subunits of the reaction center of *Rhodobacter sphaeroides*,[11] bacteriorhodopsin[13] (bacr_halha in Table 1), chlorophyll A-B binding protein of LHCII type I precursor[64] (cb22_pea), and light-harvesting protein B-800/820, alpha chain[65] (lha2_rhoac). The number and location of transmembrane segments are accurately predicted in all of these proteins, apart from one error in the first transmembrane segment in the L subunit of *Rhodobacter sphaeroides* where instead of one long transmembrane segment, two transmembrane segments are predicted. In real predictions we would use, besides the $\Delta P(X)$ profile, the distribution of the positively charged amino acid side chains in the extramembrane parts as well as the predictions on similar proteins. If there were two shorter transmembrane segments instead of the first long transmembrane segment of 32 to 55 amino acid residues, then the NH_2-terminus with positively charged amino acid side chains would be in the extracytoplasmic region instead of in the cytoplasm. Also accurate is the prediction for 1b14_human protein (Table 2) whose 3D structure is almost completely known. The M and L subunits from the photosynthetic reaction center of *Rhodopseudomonas viridis* are more than 30% similar. Thus, in the training procedure we used only the L and H subunits. The H, M, and L subunits of *Rhodopseudomonas viridis* and of *Rhodobacter sphaeroides* are also more than 30% similar, but for these proteins the prediction of the transmembrane segment location is also similar.

The preference functions profiles $\Delta P(X)$ with marked predicted and actual location of transmembrane segments, obtained from the 3D structures, are shown for two proteins (Table 1)—M subunit (rcem_rhovi) from photosynthetic reaction center[10] and bacteriorhodopsin (bacr_halha) from *Halobacterium halobium*[13]—in Figure 2A and B, respectively. It is seen that the number and length of transmembrane segments in both cases are correctly predicted. A somewhat larger difference is obtained between predicted and actual lengths of transmembrane segments in the case of the rcem_rhovi subunit. In general, the largest inaccuracy is also obtained in the experimental determination of the location and length of transmembrane segments at both ends where transmembrane segments are in contact with the solvent. Thus, for example, in the case of photosynthetic reaction center, in two papers[10,12] the ends of almost every transmembrane segment are reported to be different for at least one amino acid residue. It is clearly seen from the comparison with preference profiles for these proteins, obtained with the PREF method (1.0 version),[50] that one can achieve much higher prediction resolution with the present version of the PREF method (2.0 version).

In Figure 3A are shown the $\Delta P(X)$ profiles for lactose permease (lacy_ecoli).[21,62,66,67] We see that one can easily detect regions that correspond to transmembrane segments in the $\Delta P(X)$ profile for lacy_ecoli. Table 1 reveals that the preference functions method in digitalized form did not predict one transmembrane segment in the region 315–334. However, we see in the $\Delta P(X)$ profile in Figure 3A in the region 315–334 one more peak which corresponds to yet another transmembrane segment. This peak was removed in the digitalization process because the maximum $\Delta P(X)$ value for this potential transmembrane segment is a little lower than the boundary value in the filter which is obtained by the optimization on the set of proteins used for training. If the $\Delta P(X)$ profile is combined with the "positive inside" rule, it appears that the maximum in the $\Delta P(X)$ between 315 and 334 amino acid residues corresponds to the transmembrane segment. This agrees with the experimental and theoretical results for the prediction of the structure and topology of this well-investigated protein.[21,67]

5.4.4. Insufficiently Reliable SWISS-PROT Structures of Membrane Proteins

We have shown for the example of protein lacy_ecoli that sometimes errors can be made when using the digitalized profile of the PREF method in predicting the location of transmembrane segments. These errors can be corrected by considering the entire $\Delta P(X)$ profile in combination with the "positive inside" rule. However, we observed in the case of some less investigated proteins in Table 1 that there are also possible errors in the location of transmembrane segments in the SWISS-PROT data bank. With such an analysis one can detect segments in the sequence for which it is necessary to carry out additional experimental and theoretical research with the aim to establish the actual structure and topology of these segments and the whole protein.

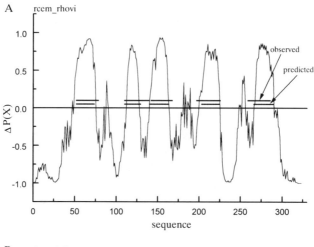

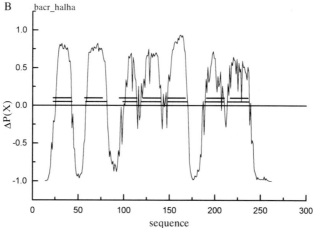

Figure 2. $\Delta P(X)$ profiles for proteins for which the 3D structures are known. (A) Photosynthetic reaction center M subunit (rcem_rhovi) and (B) bacteriorhodopsin (bacr_halha). The $\Delta P(X)$ profile is computed as the preference profiles difference: $P(X)_{\text{transmembrane}} - P(X)_{\text{extramembrane}}$. The preference profile for the transmembrane state is smoothed in a window of width 7, while the preference profile for the extramembrane state is not smoothed. Each observed (extracted from the SWISS-PROT data bank) and predicted (based on digitalized form of the $\Delta P(X)$ profile—see Section 5.3.6) transmembrane segment is denoted by a horizontal line.

In some cases, the prediction that can now be obtained by means of the PREF method (2.0 version) is sufficient to correct the error in the number and location of transmembrane segments and consequently the error in the topology of the membrane protein. However, for some other proteins it is necessary to use the "positive inside" rule, similarity with related proteins as well as predictions obtained by other methods.

Figure 3B shows the $\Delta P(X)$ profile of the cytochrome o ubiquinol oxidase (cyoe_ecoli) operon protein.[68] In the preference functions $\Delta P(X)$ profile there appear one smaller and one distinct maximum between 130 and 180 amino acid residues. Table 1 predicts one transmembrane segment in the sequence region between 160 and 180 amino acid residues. The method of Rost $et\ al.$[42] predicts two, somewhat shorter transmembrane segments in regions 142–158 and 166–181. Analysis of topological models, obtained by the method of alkaline phosphatase and β-galactosidase fusions, indicates that in the above-mentioned region there must be either two transmembrane segments or none. If there were only one transmembrane segment in this region, then the topology of the protein would have to be altered, a fact that is unfavorable according to the "positive inside" rule. On the other hand, it is known that transmembrane segments come in pairs leading to the "helical hairpins" structure.[69] The PREF method gives a prediction that is very reliable for one transmembrane segment and less reliable for the other. The present topological model is obtained without any alkaline phosphatase or β-galactosidase fusions in this sequence region,[68] but in the case that the quoted sequence contains two transmembrane segments, then the topology of the sequence residue would be in agreement with the experimental determination of its topology. In agreement with this and in regard to the $\Delta P(X)$ profile, we consider that it is necessary to correct the prediction obtained using only the PREF method and prognosticate that there is also one transmembrane segment between 135 and 154 residues. For this final arrangement of transmembrane segments we computed, following the "positive inside" rule,[37,38,62] the number of positively charged amino acid residues (Arg and Lys) in the extramembrane segments. According to the topological model,[68] the NH_2- and COOH-termini are in the cytoplasmic region. Predictions obtained by our method and by the method of Rost $et\ al.$[42] give the same topology for the larger part of the sequence as well as the SWISS-PROT model with the addition that between residues 130 and 180 there exist two more transmembrane segments which alter the topology of only this segment while the topology of the rest of the sequence remains as in the previous model.[68] The sum of the positively charged amino acid side chains in all cytoplasmic and periplasmic regions in our model is 20 and 1, respectively, while the prediction made by Rost $et\ al.$[42] gives 17 and 0. The topological model of Chepuri and Gennis[68] gives 17 and 1. In the next section we will show predictions for the protein ubia_ecoli which is similar to protein cyoe_ecoli; their similarity is 9.2% according to BLITZ-server, when we used the default values of the MultiPle search (MPsrch) program. BLITZ is an automatic electronic mail server (E-mail address: blitz@ebi.ac.uk) for the MPsrch program[70] which uses the well-known Smith and Waterman algorithm[71] for searching the data base. The PREF method predicts for

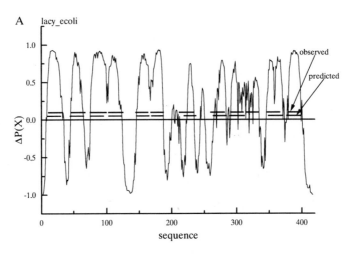

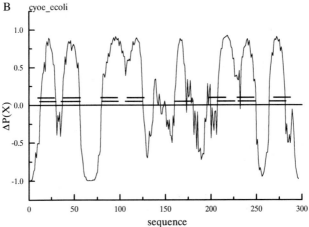

Figure 3. $\Delta P(X)$ profiles for proteins for which the 3D structures are unknown. (A) The very well investigated membrane protein lactose permease from *E. coli* (lacy_ecoli) and (B) cytochrome *o* ubiquinol oxidase from *E. coli* (cyoe_ecoli). The $\Delta P(X)$ profile is computed as the preference profiles difference: $P(X)_{transmembrane} - P(X)_{extramembrane}$. The preference profile for the transmembrane state is smoothed in a window of width 7, while the preference profile for the extramembrane state is not smoothed. Each observed (extracted from the SWISS-PROT data bank) and predicted transmembrane segment is denoted by a horizontal line.

ubia_ecoli nine transmembrane segments. On the basis of everything mentioned above, it could be concluded that our model gives predictions for cyoe_ecoli close to the real model and that it best satisfies the "positive inside" rule. Figure 4 shows a new topological model for the protein cyoe_ecoli which is based on the preference functions method.

The number of transmembrane segments for the membrane protein atpi_pea according to the SWISS-PROT data bank is probably incorrect. The location predictions of transmembrane segments for the membrane protein atpi_pea by the method of Rost et al.[42] and the method of Persson and Argos[39] indicate that there are five transmembrane segments in protein atpi_pea instead of four as listed in the SWISS-PROT data bank.[52] The preference functions method also predicts five transmembrane segments. The four transmembrane segments—(1) 20–40, (2) 78–98, (3) 118–136, (4) 200–220—are in excellent agreement with the locations from the SWISS-PROT data bank: (1) 21–40, (2) 79–97, (3) 116–135, (4) 203–222. All locations of transmembrane segments are denoted starting from residue 19 because protein atpi_pea in the SWISS-PROT data bank has an 18-residue-long signal reference. Equally large $\Delta P(X)$ values (profile not shown) as those for the four transmembrane segments appear in segment 172–192. This clearly indicates the existence of the fifth transmembrane segment.

The case of tolq_ecoli is also instructive. The topology proposed by Müller et al.[72] in the SWISS-PROT version 31.0 is: 1–22 cytoplasmic, 23–43 potential transmembrane, and 44–230 periplasmic domain. We have a strong prediction for three transmembrane helices (Table 2). Interestingly, the exbb_ecoli biopolymer transport protein is reported to have a strong similarity to the colicin transport protein tolq_ecoli, but its topology assigned by Eick-Helmerich and Braun[73] in the same version of the SWISS-PROT data base is with three transmembrane segments!

5.4.5. Predictions for Proteins for Which, According to the SWISS-PROT Data Bank, Structure and Topology Are Not Known

In this section we will demonstrate the use of the preference functions method in combination with the "positive inside" rule for determining the location and number of transmembrane segments, as well as the topology in proteins using as examples 4-hydroxybenzoate octaprenyltransferase (EC 2.5.1-) (4-HB polyprenyltransferase) (ubia_ecoli) and hypothetical 39.2-kDa protein in rhsb-pit intergenic region (yhim_ecoli) for which in the SWISS-PROT Release 31.0 there is no information about their structure or topology. The primary structure for yhim_ecoli has been determined by Sofia et al.[74] The primary structure of ubia_ecoli has also been determined.[75] It is predicted using the Kyte–Doolittle method that this protein contains six transmembrane segments,[76] but this piece of information is not included in the SWISS-PROT data bank. The preference functions $\Delta P(X)$ profile for ubia_ecoli is shown in Figure 5A.

The profile contains nine very clear transmembrane segments with very high maxima which also correspond by length fully to transmembrane segments. The distribution of positively charged amino acid side chains in the extramembrane part

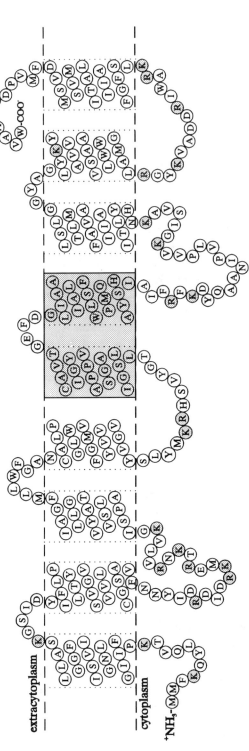

Figure 4. Proposed topological model of cytochrome *o* ubiquinol oxidase from *E. coli* (cyoe_ecoli). The topological model of protein cyoe_ecoli is obtained only on the basis of the $\Delta P(X)$ profile acquired by means of the PREF method (version 2.0) and the "positive inside" rule. Two transmembrane segments that are predicted in this report and that have not been predicted earlier (Ref. 68) are shown in boldface. Each amino acid residue is given in one-letter code (see Section 5.3.3).

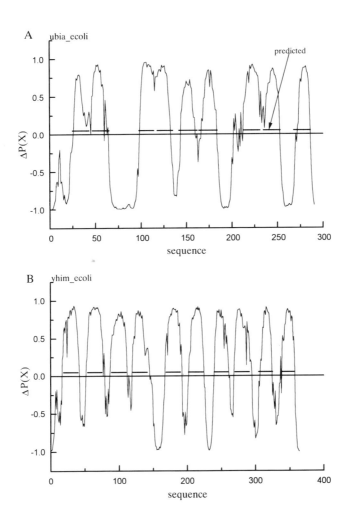

Figure 5. $\Delta P(X)$ profiles for proteins for which, according to the SWISS-PROT Release 31.0, either the location of transmembrane segments or topology is not known. (A) 4-Hydroxybenzoate octaprenyltrans-ferase (EC 2.5.1-) (4-HB polyprenyltransferase) (ubia_ecoli) and (B) hypothetical 39.2-kDa protein in rhsb-pit intergenic region (yhim_ecoli). The $\Delta P(X)$ profile is computed as the preference profiles difference: $P(X)_{\text{transmembrane}} - P(X)_{\text{extramembrane}}$. The preference profile for the transmembrane state is smoothed in a window of width 7, while the preference profile for the extramembrane state is not smoothed. Each predicted transmembrane segment is denoted by a horizontal line.

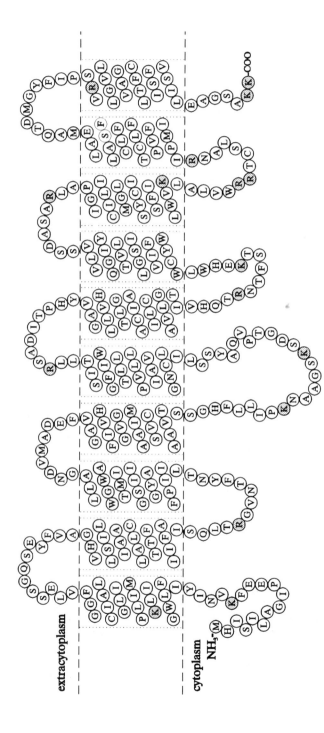

Figure 6. Proposed topological model of the hypothetical 39.2-kDa protein in rhsb–pit intergenic region (yhim_ecoli). This is the first published topological model for the protein yhim_ecoli obtained by the PREF method (version 2.0) and the "positive inside" rule. Each amino acid residue is given in one-letter code (see Section 5.3.3).

of the sequence agrees perfectly with the "positive inside" rule; 21 positively charged amino acid residues are, according to this model, in the cytoplasmic side, while in the extracytoplasmic side of the sequence there are no positively charged amino acid residues. In accordance with this we predict that the first extramembrane segment is at the NH_2-terminus of the sequence in the cytoplasm, while all other extramembrane segments enter alternatively the extracytoplasmic and cytoplasmic regions, respectively. There is no information in the literature about the number or location of transmembrane segments for the protein yhim_ecoli. According to the $\Delta P(X)$ profile, given in Figure 5B, yhim_ecoli has ten transmembrane segments with very high and almost identical maxima. Positively charged amino acid residues are distributed in the ratio 11:2, but regions with more positively charged amino acid residues are in the cytoplasm. Thus, we predict that the NH_2-terminus is in the cytoplasm while other extramembrane segments enter alternatively extracytoplasm and cytoplasm, respectively. The topological model, which agrees with our predictions and the "positive inside" rule, is shown in Figure 6.

5.5. CONCLUDING REMARKS

It is most important to know the number, location, and length of transmembrane segments in, and the topology of, membrane proteins that give information about which part of the sequence is turned toward the cytoplasm and which toward the extracytoplasmic region. Since transmembrane segments pass through the membrane almost perpendicularly, knowledge regarding the above data allows the modeling of the spatial structure of membrane proteins. For the prediction of the secondary structure of the membrane protein part that is not in the membrane, we can use algorithms that predict the secondary structure of soluble proteins, the best being the Rost and Sander method.[30] On the other hand, even if we know for some soluble protein its secondary structure, it is still very hard or almost impossible to model its spatial structure. However, it is much easier to isolate, crystallize, and determine by X-ray diffraction the 3D structure of the soluble protein. The same process for membrane proteins is much more difficult. This is why the development of reliable methods for predicting the location of transmembrane segments in membrane proteins is of utmost importance.

We have described the preference functions method (version 2.0) and have shown that it is comparable to, if not better than, the method based on the NN algorithm. The PREF method gives, for a large number of proteins, predictions in which transmembrane segments are clearly distinguished from the protein part that is not in the membrane. Thus, we obtain predictions that are in very good agreement with the "positive inside" rule. In the case that the prediction is not reliable enough [e.g., predictions for which the maxima in the $\Delta P(X)$ profile, corresponding to transmembrane segments, are not clear, high, and wide for about 20 amino acid residues and also if the number of positively charged amino acid residues in extramembrane segments is not distributed in accordance with the "positive inside" rule] to lead to the reliable topological model, we use predictions obtained for similar proteins for which it is not necessary to know the location of transmembrane segments and their topology, because similar proteins have similar $\Delta P(X)$ profiles. If after this it is still not possible to obtain a reliable model, one can use predictions obtained by other methods though they still may not be as good as our method. Each method is optimized on a different set of proteins and for predictions it always employs a special algorithm that utilizes different properties of amino acid residues and membrane proteins. Thus, different methods may give similar or identical predictions for some proteins while for other proteins some methods will produce more accurate predictions than others. However, no method is entirely exact. We can rank them according to predictions for a larger set of proteins. Individual applications are useful for the comparison of methods because in this way more reliable predictions are obtainable.

The main shortcoming of the presented algorithm is that it cannot predict structural motifs in membrane proteins besides transmembrane helices. This deficiency is shared with many other published methods. However, the PREF suite of algorithms has been designed to overcome this drawback, and it is capable of predicting all other secondary structure motifs in membrane proteins when an appropriate data base of both soluble and membrane proteins is used for training.[47,49,50] With the appropriate choice of hydrophobicity scale it is also capable of predicting transmembrane β-strands although with lower accuracy than transmembrane α-helices.[77] With the choice of the Kyte–Doolittle hydrophobicity scale as amino acid attributes, PREF version 3.0 superimposes the transmembrane helix prediction on the standard three-state model prediction of secondary structure in membrane proteins and the prediction accuracy is very much competitive with the neural network results,[42] and with the results presented here (in preparation).

Another shortcoming of the PREF method (version 2.0) follows from the procedure that we call the digitalization of the $\Delta P(X)$ profile (which is the deficiency of all other methods too). It is caused by the use of filters that make the method dependent on the set of proteins used for the optimization of parameters that are employed in filters.

It can be seen from the $\Delta P(X)$ profiles shown for several proteins in this report that the profiles in the present version of the PREF method are much clearer than in the earlier version.[45] These profiles are far clearer than those obtained by the Kyte–Doolittle method on the example of cyoe_ecoli.[68]

One important advantage of our method is that it avoids topological frustration related to fixed length prediction of transmembrane segments. The distribution of predicted lengths of transmembrane segments (Figure 7A) is similar to the observed distribution as found in the training and testing data set (Figure 7B). The allowance for variable length and the explicit prediction of short and long transmembrane helical segments increase the prediction accuracy and flexibility in constructing topological models that would satisfy the "positive inside" rule and experimental observations. It may seem strange that a procedure that does not use homology or the "positive inside" rule in the digital part of the algorithm can be as accurate as the much more

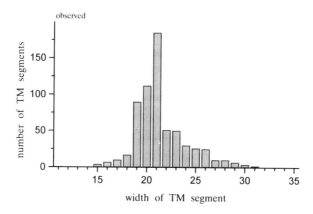

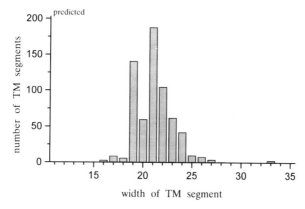

Figure 7. The number of transmembrane (TM) segments of given width in the testing and training data set (Tables 1 and 2): (A) observed (extracted from the SWISS-PROT data bank); (B) predicted with PREF 2.0.

sophisticated neural network method.[42] It is fair to point out that only a small number of objects are difficult to predict when the goal is to predict transmembrane helices, even when close to 100 proteins are used for training and for testing. Therefore, overtraining is quite easy to achieve, and it is also very easy to select proteins to be tested for prediction accuracy that belong to the easy-to-predict category with respect to transmembrane segments. We tried to avoid as much as possible both of these dangers. The drop in performance quality parameters is indeed somewhat less in our case when 71 never-before-seen proteins are tested than in the case of neural network algorithms.[42] Optimization of different parameters in the predictor can indeed lead to overtraining but it is less dangerous when a larger data set of proteins is used for training (95 in our case).

False-positive predictions of transmembrane segments are not found in porins (no transmembrane helical segments predicted) and are rare in soluble proteins (11.5%), but still more frequent than with the neural network method (4.35%).[42] With respect to overprediction of transmembrane helical segments in soluble proteins and with respect to the capability of distinguishing signal sequences from transmembrane sequences, our PREF method is weaker than the neural network method. In the present version 2.0, the PREF method does not give accurate predictions for other classes of membrane proteins besides integral membrane proteins with transmembrane helices. Weak predictions are associated with self-inserting membrane toxins such as melittin and beetle delta endotoxin, and colicin A (not shown) and with membrane proteins that are thought to enter only one side of the bilayer, such as prostaglandin synthase,[78] and that contain very amphiphilic transmembrane segments (not shown).

It has been pointed out that the SWISS-PROT data bank includes a certain number of membrane proteins for which the structure is not quite reliably determined, in part because older methods were used to determine the location and number of transmembrane segments. These older methods are less reliable because they were trained on a smaller sets of proteins. Hence, we suggest that it is necessary to use new and better methods for detecting these defective topologies in order to carry out additional research on them with the aim to obtain their more accurate structures and topologies. In the present report we have shown one of possible ways to carry out such a procedure and this has been illustrated using several examples. In this we also used the "positive inside" rule, predictions on similar proteins, and predictions obtained by other methods.

We have also shown how it is possible to obtain by the PREF method (version 2.0) very reliable predictions for the structure and topology of proteins with unknown structural and topological features. With this we have demonstrated the usefulness of the development of such methods and especially the quality of the PREF method which in many cases gives very reliable predictions and thus considerably shortens the time-consuming and expensive experimental research in the determination of membrane protein structures.

ACKNOWLEDGMENTS

We are grateful to Burkhard Rost from EMBL, Heidelberg, Germany, for data bases of membrane proteins kindly provided for our use. The authors thank Vladimir Braus, Tomislav Došlić, and Robert Manger for computer support. This work was supported by the Ministry of Science and Technology of the Republic of Croatia through Grants 1-07-159 and 1-03-171.

REFERENCES

1. G. D. Fasman, The role of electrostatic interactions in the structure of globular proteins, in: *Prediction of Protein Structure and the Principles of Protein Conformations* (G. D. Fasman, ed.), pp. 359–389, Plenum Press, New York (1989).
2. M. L. Jennings, Topography of membrane proteins, *Annu. Rev. Biochem. 58*, 999–1027 (1989).

3. J. Kyte and R. F. Doolittle, A simple method for displaying the hydropathic character of a protein, *J. Mol. Biol. 157*, 105–132 (1982).

4. D. M. Engelman, T. A. Steitz, and A. Goldman, Identifying nonpolar transbilayer helices in amino acid sequences of membrane proteins, *Annu. Rev. Biophys. Biophys. Chem. 15*, 321–353 (1986).

5. J. L. Cornette, K. B. Cease, H. Margalit, J. L. Spouge, J. A. Berzofsky, and C. DeLisi, Hydrophobicity scales and computational techniques for detecting amphipathic structures in proteins, *J. Mol. Biol. 195*, 659–685 (1987).

6. M. S. Weiss and G. E. Schulz, Structure of porin refined at 1.8 Å resolution, *J. Mol. Biol. 227*, 493–509 (1992).

7. S. W. Cowan and J. P. Rosenbusch, Folding pattern diversity of integral membrane proteins, *Science 264*, 914–916 (1994).

8. T. M. Gray and B. W. Matthews, Intrahelical hydrogen bonding of serine, threonine and cysteine residues within alpha-helices and its relevance to membrane-bound proteins, *J. Mol. Biol. 175*, 75–81 (1984).

9. D. M. Engelman, An implication of the structure of bacteriorhodopsin. Globular membrane proteins are stabilized by polar interactions, *Biophys. J. 37*, 187–188 (1982).

10. J. Deisenhofer, O. Epp, K. Miki, R. Huber, and H. Michel, Structure of the protein subunits in the photosynthetic reaction centre of Rhodopseudomonas viridis at 3A resolution, *Nature 318*, 618–624 (1985).

11. J. P. Allen, G. Feher, T. O. Yeates, H. Komiya, and D. C. Rees, Structure of the reaction center from Rhodobacter sphaeroides R-26: The protein subunits, *Proc. Natl. Acad. Sci. USA 84*, 6162–6166 (1987).

12. J. Deisenhofer and H. Michel, The photosynthetic reaction center from the purple bacterium Rhodopseudomonas viridis, *Science 245*, 1463–1473 (1989).

13. R. Henderson, J. M. Baldwin, T. A. Ceska, F. Zemlin, E. Beckmann, and K. H. Downing, Model for the structure of bacteriorhodopsin based on high-resolution cryo-microscopy, *J. Mol. Biol. 213*, 899–929 (1990).

14. G. Blobel, Intracellular protein topogenesis, *Proc. Natl. Acad. Sci. USA 77*, 1496–1500 (1980).

15. W. L. Hubbell and C. Altenbach, Investigation of structure and dynamics in membrane proteins using site-directed spin labeling, *Curr. Opin. Struct. Biol. 4*, 566–573 (1994).

16. J. Brunner and F. M. Richards, Analysis of membranes photolabeled with lipid analogues, *J. Biol. Chem. 255*, 3319–3329 (1980).

17. M. L. Jennings, M. P. Anderson, and R. Monaghan, Monoclonal antibodies against human erythrocyte band 3 protein, *J. Biol. Chem. 261*, 9002–9010 (1986).

18. C. Manoil and J. Beckwith, TnphoA: A transposon probe for protein export signals, *Proc. Natl. Acad. Sci. USA 82*, 8129–8133 (1985).

19. C. Manoil and J. Beckwith, A genetic approach to analyzing membrane protein topology, *Science 233*, 1403–1408 (1986).

20. J. A. Berzofsky, Intrinsic and extrinsic factors in protein antigenic structure, *Science, 229*, 932–940 (1985).

21. E. Bibi and H. R. Kaback, In vivo expression of the LacY gene in two segments leads to functional *lac* permease, *Proc. Natl. Acad. Sci. USA 87*, 4325–4329 (1990).

22. J. Soppa, J. Duschl, and D. Oesterhelt, Bacterioopsin, haloopsin, and sensory opsin I of the halobacterial isolate *Halobacterium sp.* strain SG1: Three new members of a growing family, *J. Bacteriol. 175*, 2720–2726 (1993).

23. H. G. Khorana, Two light-transducing membrane proteins: Bacteriorhodopsin and the mammalian rhodopsin, *Proc. Natl. Acad. Sci. USA 90*, 1166–1171 (1993).

24. C. Sander and R. Schneider, Database of homology-derived structures and the structural meaning of sequence alignment, *Proteins Struct. Funct. Genet. 9*, 56–68 (1991).

25. C. Sander and R. Schneider, The HSSP database of protein structure–sequence alignments, *Nucleic Acids Res. 22*, 3597–3599 (1994).

26. P. Bork, C. Ouzonis, C. Sander, R. Scharaf, R. Schneider, and E. Sonnhammer, What's in a genome? *Nature 358*, 287 (1992).

27. F. Jähning and O. Edolm, Can the structure of proteins be calculated? *Z. Phys. B 78*, 137–143 (1990).

28. L. Pauling and R. B. Corey, Configuration of polypeptide chains with favored orientations around single bonds: Two new pleated sheets, *Proc. Natl. Acad. Sci. USA 37*, 729–740 (1951).

29. L. Pauling, R. B. Corey, and H. R. Branson, The structure of proteins: Two hydrogen-bonded helical configurations of the polypeptide chain, *Proc. Natl. Acad. Sci. USA 37*, 205 (1951).

30. B. Rost and C. Sander, Prediction of protein secondary structure at better than 70% accuracy, *J. Mol. Biol. 232*, 584–599 (1993).

31. B. Rost and C. Sander, Improved prediction of protein secondary structure by use of sequence profiles and neural networks, *Proc. Natl. Acad. Sci. USA 90*, 7558–7562 (1993).

32. B. Rost and C. Sander, Secondary structure prediction of all-helical proteins in two states, *Protein Eng. 6*, 831–836 (1993).

33. B. Rost and C. Sander, Combining evolutionary information and neural networks to predict protein secondary structure, *Proteins Struct. Funct. Genet. 20*, 216–226 (1994).

34. W. R. Taylor, D. T. Jones, and N. M. Green, A method for α-helical integral membrane protein fold prediction, *Proteins Struct. Funct. Genet. 18*, 281–294 (1994).

35. P. Argos, J. K. M. Rao, and P. A. Hargrave, Structural prediction of membrane-bound proteins, *Eur. J. Biochem. 128*, 565–575 (1982).

36. M. Degli Esposti, M. Crimi, and G. Venturoli, A critical evaluation of the hydropathy profile of membrane proteins, *Eur. J. Biochem. 190*, 207–219 (1990).

37. G. von Heijne, Membrane proteins—The amino acid composition of membrane-penetrating segments, *Eur. J. Biochem. 120*, 275–278 (1981).

38. G. von Heijne, Membrane protein structure prediction. Hydrophobicity analysis and the positive- inside rule, *J. Mol. Biol. 225*, 487–494 (1992).

39. B. Persson and P. Argos, Prediction of transmembrane segments in proteins utilising multiple sequence alignments, *J. Mol. Biol. 237*, 182–192 (1994).

40. D. T. Jones, W. R. Taylor, and J. M. Thornton, A model approach to the prediction of all-helical membrane protein structure and topology, *Biochemistry 33*, 3038–3049 (1994).

41. R. Lohmann, G. Schneider, D. Behrens, and P. Wrede, A neural network model for the prediction of membrane-spanning amino acid sequences, *Protein Science 3*, 1597–1601 (1994).

42. B. Rost, R. Casadio, P. Fariselli, and C. Sander, Transmembrane helices predicted at 95% accuracy, *Protein Sci. 4*, 521–533 (1995).

43. D. Eisenberg, E. Schwartz, M. Komaromy, and R. Wall, Analysis of membrane and surface protein sequences with the hydrophobic moment plot, *J. Mol. Biol. 179*, 125–142 (1984).

44. D. Eisenberg, Three-dimensional structure of membrane surface proteins, *Annu. Rev. Biochem. 53*, 595–623 (1984).

45. D. Juretić, B. K. Lee, N. Trinajstić, and R. W. Williams, Conformational preference functions for predicting helices in membrane proteins, *Biopolymers 33*, 255–273 (1993).

46. P. Y. Chou and G. D. Fasman, Conformational parameters for amino acids in helical, β-sheets and random coil regions calculated from proteins, *Biochemistry 13*, 211–222 (1974).

47. D. Juretić, Conformational preference functions and secondary structure prediction for membrane proteins, *Acta Pharm. 43*, 223–226 (1993).

48. D. Juretić, N. Trinajstić, and B. Lučić, Protein secondary structure conformations and associated hydrophobicity scales, *J. Math. Chem. 14*, 35–34 (1993).

49. D. Juretić and R. Pešić, A scale of β-preferences for structure–activity predictions in membrane proteins, *Croat. Chem. Acta 68*, 215–232 (1995).

50. D. Juretić, B. Lučić, and N. Trinajstić, Predicting membrane protein secondary structure: Preference functions method for finding optimal conformational parameters, *Croat. Chem. Acta 66*, 201–208 (1993).

51. D. Juretić, Secondary structure of membrane proteins: Prediction with conformational preference functions of soluble proteins, *Croat. Chem. Acta 65*, 921–932 (1992).

52. A. Bairoch and B. Boeckmann, SWISS-PROT protein sequence data bank: Current status, *Nucleic Acids Res. 22*, 3578–3580 (1994).

53. D. Juretić, B. Lučić, and N. Trinajstić, Secondary structure prediction quality for naturally occurring amino acids in soluble proteins, *J. Mol. Struct. (Teochem) 338*, 43–50 (1995).

54. P. K. Ponnuswamy and M. M. Gromiha, Prediction of transmembrane helices from hydrophobic characteristics of proteins, *Int. J. Peptide Protein Res. 42*, 326–341 (1993).

55. D. G. Kneller, F. E. Cohen, and R. Langridge, Improvements in protein secondary structure prediction by an enhanced neural network, *J. Mol. Biol. 214*, 171–182 (1990).

56. D. Juretić and R. W. Williams, Protein secondary structure preferences, *J. Math. Chem. 8*, 229–242 (1991).

57. G. E. Arnold, A. K. Dunker, S. J. Johns, and R. J. Douthart, Use of conditional probabilities for determining relationships between amino acid sequence and protein secondary structure, *Proteins 12*, 382–399 (1992).

58. L. Zhong and W. C. Johnston, Jr., Environment affects amino acid preference for secondary structure, *Proc. Natl. Acad. Sci. USA 89*, 4462–4465 (1992).

59. S. M. Muskal and S. H. Kim, Predicting protein secondary structure content. A tandem neural network approach, *J. Mol. Biol. 225*, 713–727 (1992).

60. J.-F. Gibrat, J. Garnier, and B. Robson, Further developments of protein secondary structure predictions using information theory. New parameters and consideration of residue pairs, *J. Mol. Biol. 198*, 425–443 (1987).

61. J.-F. Gibrat, B. Robson, and J. Garnier, Influence of the local amino acid sequence upon the zones of the torsional angles ϕ and ψ adopted by residues in proteins, *Biochemistry 30*, 1578–1586 (1991).

62. G. von Heijne, The distribution of positively charged residues in bacterial inner membrane proteins correlates with the trans-membrane topology, *EMBO J. 5*, 3021–3027 (1986).

63. J. Edelman, Quadratic minimization of predictors for protein secondary structure: Application to membrane alpha-helices, *J. Mol. Biol. 232*, 165–191 (1993).

64. W. Kühlbrandt, D. N. Wang, and Y. Fujiyoshi, Atomic model of plant light-harvesting complex by electron crystallography, *Nature 367*, 614–621 (1994).

65. G. McDermott, S. M. Prince, A. A. Freer, A. M. Hawthornthwaite-Lawless, M. Z. Papiz, R. J. Cogdell, and N. W. Isaacs, Crystal structure of an integral membrane light-harvesting complex from photosynthetic bacteria, *Nature 374*, 517–521 (1995).

66. S. C. King, C. L. Hansen, and T. H. Wilson, The interaction between aspartic acid 237 and lysine 358 in the lactose carrier of *Escherichia coli*, *Biochim. Biophys. Acta 1062*, 177–186 (1991).

67. J. Calamia and C. Manoil, *Lac* permease of *Escherichia coli*: Topology and sequence elements promoting membrane insertion, *Proc. Natl. Acad. Sci. USA 87*, 4937–4941 (1990).

68. V. Chepuri and R. B. Gennis, The use of gene fusions to determine the topology of all of the subunits of cytochrome o terminal oxidase complex of *Escherichia coli*, *J. Biol. Chem. 265*, 12978–12986 (1990).

69. G. Gafvelin and G. von Heijne, Topological 'frustration' in multispanning *E. coli* inner membrane proteins, *Cell 77*, 401–412 (1994).

70. S. S. Sturrock and J. F. Collins, *MPsrch version 1.3*, Biocomputing Research Unit, University of Edinburgh (1993).

71. T. F. Smith and M. S. Waterman, Identification of common molecular subsequences, *J. Mol. Biol. 147*, 195–197 (1981).

72. M. M. Müller, A. Vianney, J.-C. Lazzarony, R. E. Webster, and R. Portalier, Membrane topology of the *Escherichia coli* TolR protein required for cell envelope integrity, *J. Bacteriol. 175*, 6059–6061 (1993).

73. K. Eick-Helmerich and V. Braun, Import of biopolymers into *Escherichia coli*: Nucleotide sequences of the *exbB* and *exbD* genes are homologous to those of the *tolQ* and *tolR* genes, respectively, *J. Bacteriol. 171*, 5117–5126 (1989).

74. H. J. Sofia, V. Burland, D. L. Daniels, G. Plunkett, III, and F. R. Blattner, Analysis of the *Escherichia coli* genome. V. DNA sequence of the region from 76.0 to 81.5 minutes, *Nucleic Acids Res. 22*, 2576–2586 (1994).

75. F. R. Blattner, V. Burland, G. Plunkett III, H. J. Sofia, and D. L. Daniels, Analysis of the *Escherichia coli* genome. IV. DNA sequence of the region from 92.8 minutes, *Nucleic Acids Res. 21*, 5408–5417 (1993).

76. M. Melzer and L. Heide, Characterization of polyprenyldiphosphate: 4-hydroxybenzoate polyprenyl-transferase from *Escherichia coli*, *Biochim. Biophys. Acta 1212*, 93–102 (1994).

77. D. Juretić and B. Lučić, Predicting the secondary structure of membrane channel proteins: The performance of preference functions compared to other statistical methods, *HB93 Proceedings*, Zagreb (1993).

78. D. Picot, P. J. Loll, and M. Garavito, The X-ray structure of the membrane protein prostaglandin H_2 synthase-1, *Nature 367*, 243–249 (1994).

79. F. E. Croxton and D. J. Cowden, *Applied General Statistics*, Prentice–Hall, Englewood Cliffs, New Jersey (1948).

6

On Characterization of 3D Molecular Structure

MILAN RANDIĆ and MARKO RAZINGER[†]

6.1. THE NATURE OF THE CHEMICAL STRUCTURE

One of the central themes in chemistry is the study of structure–property relationships. In its early days quantum chemistry was mostly concerned, preoccupied one could say, with questions relating to the nature of the chemical bond.[1] Before the arrival of quantum theory, chemists believed in special "chemical forces" that had short range, directional properties, and exhibited saturation—all three characteristics so different from well-known forces of physics.

The first simple calculations of Heitler and London[2] on the H_2 molecule resulted in the most important conceptual contribution of quantum theory to chemistry: reduction of the special "chemical forces" to Coulomb interactions between electrons and atomic nuclei governed by the laws of quantum mechanics. The emphasis in theoretical chemistry shifted immediately to solving the Schrödinger equation for molecular systems.

The difficulties of the application of quantum mechanics to chemistry were voiced already in 1928 by Dirac in the opening paragraph of his paper on many-electron systems[3]:

[†]Deceased

MILAN RANDIĆ • Department of Mathematics and Computer Science, Drake University, Des Moines, Iowa 50311.   MARKO RAZINGER • National Institute of Chemistry, 61115 Ljubljana, P. O. Box 30, Republic of Slovenia.

From Chemical Topology to Three-Dimensional Geometry, edited by Balaban. Plenum Press, New York, 1997

> The general theory of quantum mechanics is now almost complete . . . . The underlying physical laws necessary for the mathematical theory of large part of physics and the whole of chemistry are thus completely known, and the difficulty is only that the exact application of these laws leads to equations much too complicated to be soluble.

Today, thanks to the dramatic technological opportunities offered by computers the "predictions" of Dirac are somewhat outdated. S. F. Boys[4] was the pioneer of the early *ab initio* calculations. He outlined the mathematical approach to molecular integrals that require no use of empirical parameters. Hence, not only do we understand the nature of the chemical bond, but we can also in practice find out all about chemical bonds in sufficient detail, at least for relatively small molecules.

While theoretical chemistry is moving toward larger molecules and the consideration of simple chemical reactions, the topic that falls within the computational capabilities of the modern *ab initio* calculations, i.e., the questions relating to *the nature of chemical structure*, however, remain mostly unanswered[5,6]:

- What is chemical structure?
- How can one define chemical structures?
- What is molecular shape?
- How can one define molecular shapes?

To these questions we could add questions that directly or indirectly reflect the ambiguities and the uncertainties concerning a substructure of a molecule. Consider, for example, questions like:

- What is molecular complexity?
- How can we characterize the complexity?
- What is molecular surface?
- How can we characterize the molecular surface?
- How can we measure molecular similarity?
- How can we measure the degree of branching?
- How can we measure the degree of chirality?

We will try to answer some of these questions in this chapter. The emphasis will be on representation and characterization of molecules rather than the questions concerning molecular similarity and structure–property relationships. We will differentiate between the *representation* and the *characterization*. The former is based on molecular codes and the latter on structural invariants. The codes allow one to represent a structure without loss of information, hence they allow one to reconstruct the object fully from a given code. Invariants represent mathematical properties of a structure and are therefore useful in structure–property analysis, as well as in comparisons of different structures. The list of invariants need not be unique, hence, invariants do not generally allow reconstruction. Figuratively speaking, characterization depicts various

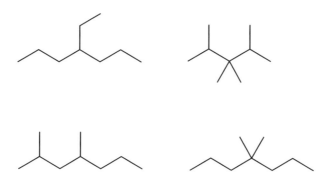

Figure 1. A few isomers of nonane C_9H_{20} that show similar boiling points.

"algebraic projections" of the structure, i.e., gives lists of selected algebraic properties of a structure. Just as different objects can have the same or similar geometric projections and yet be vastly different, so different chemical structures can have the same or similar "algebraic projections." There are molecules that are vastly different and yet may have the same collection of structural invariants. The same is true not only for selected mathematical properties of such molecules but also for their selected physicochemical and biological properties. For example, sugar and saccharin, and other sweeteners, are vastly different, yet show a similar specific property.

Consider another example: 4-ethylheptane and 2,3,3,4-tetramethylpentane have similar boiling points (BP; 141.3 and 141.5°C, respectively) (Figure 1). So also do 2,4-dimethylheptane and 4,4-dimethylheptane (136.0 and 135.2°C, respectively). Is this accidental or is there some structural reason for the similarity of their BP values? Clearly, this kind of question is not to be answered by quantum chemistry or *ab initio* calculations, even if they were possible today. In this context we may say that quantum chemistry generates questions, but not the answers. Quantum-chemical calculations are but a source of novel data on molecules. In order to rationalize data one has to resort to traditional data-reduction methods. Such methods include the statistical analysis for numerical solutions and molecular modeling, including chemical graph theory, for conceptual advances.

In the above case it happens that the similarity in BP values for 4-ethylheptane and 2,3,3,4-tetramethylpentane is accidental, while that for 2,4-dimethylheptane and 4,4-dimethylheptane is of structural origin.[7,8] The conclusion follows from an examination of the shortest paths in these molecules. It happens that the paths of length two and three in the molecular graphs dominate many bond additive properties of molecules.[9,10]

Sometimes molecules showing similar properties have similar *shape* but differ in their *structure*. This is well illustrated by macrocyclic musk civetone (**I**) and steroids

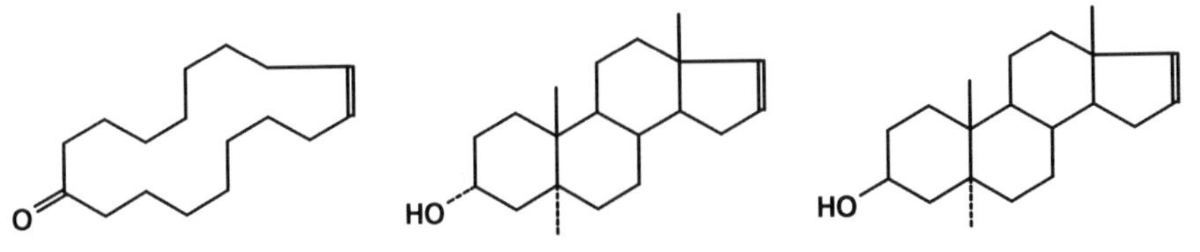

Figure 2.   Illustration of structurally different molecules having similar shapes and similar odor.

of similar shape (**II** and **III**), 3α-hydroxy-5α-androst-16-ene and 3β-hydroxy-5α-androst-16-ene, respectively (Figure 2).[11] These molecules have similar odor. The finding agrees with the theory of Ružička[12] that the character of odoriferous substances is determined by their molecular shape. The importance of the shape of molecules in biochemical activities was recognized over 100 years ago by Emil Fischer[13] who proposed the "lock and key" model for the action of enzymes.

The art in the quantitative structure–property relationship studies (QSPR), in the quantitative structure–activity relationship studies (QSAR),[14,15] in the quantitative shape–property relationship studies (QShPR),[16] and in the quantitative structure–function relationship studies (QSFR) is to find the relevant molecular "projections" that have captured the essential structural features responsible for the considered property.

6.2. MOLECULAR CODES AND MOLECULAR REPRESENTATION

Apparently chemical compounds multiply faster than chemists! Not long ago there were 5 million compounds registered and today the number approaches 15 million—but to double the human population requires more than 20 years. This fast growth of analytical and preparative chemistry has been accompanied by emergence of compounds that have novel structural features that have not been anticipated. Often such structurally novel compounds pose novel difficulties in finding their systematic names. Apparently there is no end to the wealth of combinatorial forms in nature. Our inability to cope systematically with such diversity of structures has resulted in a proliferation of trivial names in the chemical nomenclature. Moreover, some systematic names not only are difficult to derive, but are too cumbersome for practical use. An illustration is the systematic name for buckminsterfullerene.[17–19]

There are several systems of chemical nomenclature outlined in the literature.[20–41] Generally it is not fully recognized by many chemists that naming of a chemical compound is tantamount to solving the problem of graph isomorphism. One may appreciate the difficulties inherent in this problem by examining some of the early attempts to solve the graph isomorphism problem.[42,43] This problem continues to attract attention.[44–52] In its simplest form, when confined to molecular graphs, rather than chemical structures in 3D space, the problem requires one to define a unique code for a structure. This always implies precise rules for labeling of the vertices in a graph.

The abundance of trivial chemical names clearly shows that the naming of the chemical compounds is far from trivial. One may summarize the situation by saying that it is often easier to make a compound than to name the compound. A good example is buckminsterfullerene, which can now be prepared in gram quantities, while finding its systematic IUPAC name proved to be full of pitfalls. The naming of complex chemical compounds ought to be left to nomenclature experts, and even they may find the task rather difficult. This does not suggest that experimental chemists lack imagination, on the contrary. The sophistication involved in many multiple-stage syntheses on one side and the plethora of trivial chemical names, often well chosen, on the other side illustrates the imagination of organic chemists. The warning merely suggests that many experimental chemists may not be familiar with inherent difficulties of the graph isomorphism problem. The problem is not in designing a single name, or a name for a single type of compound. To recognize the inherent complexity of developing the systematic chemical names one has to consider all molecules! Naming the chemical compounds as objects in 3D space is even more involved. This has become apparent by the need for revisions of existing nomenclature systems. For example, some of the limitations of the Cahn–Ingold–Prelog system[53,54] have recently been discussed.[55–57]

The rules for labeling vertices (atoms) in a structure embedded in 2D or 3D space are still more involved than the rules for labeling vertices in graphs that are by definition devoid of a fixed geometry. An attempt was made recently to develop a scheme for labeling the vertices (carbon atoms) of planar benzenoids (embedded on a graphite lattice). The approach is based on the average distances of atoms from other atoms in the molecule.[58] It is premature to speculate if this scheme is sufficiently general and not free from degeneracy, but in its current form it offers a route to discriminate nonequivalent atoms among those that are equivalent if a molecule is viewed as a graph.

6.3. MOLECULAR GRAPHS

We will briefly review some aspects of the chemical graph theory[59,60] that are of interest in developing analogous models for 3D molecular structures. Let us start by examining types of molecular graphs. Besides molecular graphs in which atoms are represented as vertices and bonds as edges, there are graphs in which other molecular features are emphasized. For example, one can consider only π-electrons of planar conjugated hydrocarbons. In this case the adjacency matrix of the molecular graph becomes the Hückel matrix of the π-electron system.[61-64] Because of that parallel, many early critics of chemical graph theory erroneously identified graph theory with the Hückel molecular–orbital (HMO) theory. However, by the late 1960s and early 1970s, the time when chemical graph theory started to grow, the HMO theory had been replaced by more advanced MO models. Therefore, the graph theoretical approach to HMO did not receive adequate attention. Nevertheless, the graph theoretical viewing of π-electron systems has brought important novelties to structural chemistry. For example, the graph theoretical approach led to the definition of the topological

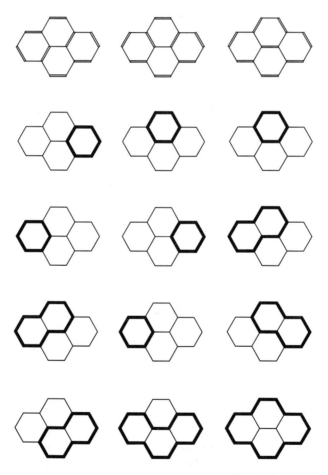

Figure 3. Conjugated circuits for three symmetry nonequivalent Kekule valence structures of pyrene.

reference for the molecular resonance energy of polycyclic aromatic compounds.[65-67] Equally, chemical graph theory has revived an interest in the valence bond method. The conjugated circuits, which are shown in Figure 3 for three symmetry nonequivalent Kekule valence structures of pyrene, illustrate an approach that has resulted from graph theoretical views of molecular structure.[68-77] The approach is computationally equivalent to the resonance theory formulated initially by Simpson,[78,79] and revived by Herndon.[80-83]

The important role of graph theory here is not only in deriving reliable results by simple computations (although that also has its significance), but also in the clarification of conceptual notions about the chemical structure. For example, one arrives at a classification of polycyclic aromatic hydrocarbons into aromatic, antiaromatic, and "mixed" solely based on the structural formula of a compound.[84,85] If Kekule valence structures of a molecule contain only $4n + 2$ conjugated circuits, then the molecule is aromatic. This is the case with pyrene that has only 6-, 10-, and 14-membered conjugated circuits despite the fact that it has 16 carbon atoms. The case of pyrene well illustrates the irrelevance of the perimeter model. If there are only $4n$ conjugated circuits, then the molecule is antiaromatic (the idea was introduced by Breslow[86,87]). Finally, if the set of Kekule valence structures has both types of conjugated circuits, the molecule is classified as "mixed." Such molecules have diminished aromatic character. The first such molecules were synthesized and identified as unusual the first time by Hafner and co-workers.[88]

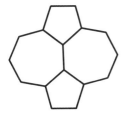 

We will now briefly illustrate other "forms" of molecular graphs depicting conjugated polycyclic hydrocarbons. Smith introduced a simplified representation of polycyclic benzenoid system by replacing each of the fused hexagonal rings by a single vertex (that one can imagine in the center of the rings).[89] The simplified graph is obtained by connecting vertices corresponding to the adjacent rings. This pure notational device was found later by Balaban and Harary to reflect useful mathematical properties of benzenoids that lead to simple molecular codes, enumeration, and even construction and benzenoid graphs.[90-93]

The so-called "factor" graphs (Figure 4) depict the individual Kekule valence structures by considering C=C bonds as the vertices of a graph that are connected only when separated from other C=C bonds by a single CC bond.[94-99] This approach can be viewed as a "translation" of the familiar Kekule valence structures to an

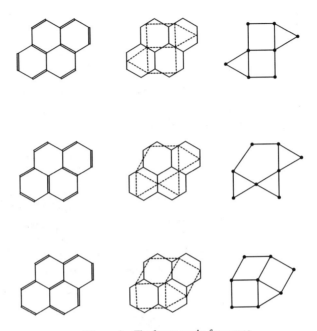

Figure 4.   The factor graphs for pyrene.

alternative representation that better illustrates the differences in the conjugation within the individual Kekule valence structures.

An interesting graph representation of benzenoid hydrocarbons, though limited in applications, is illustrated in Figure 5. Here the individual rings of *cata*-condensed benzenoids are represented by vertices and the vertices are connected only if the corresponding rings are "resonant."[100–102] The resonance is here defined in terms of the concept of the π-sextet of Clar.[103–105] The resulting graphs are known in the mathematical literature as "caterpillars," i.e., trees in which the side branches can at most be of length one.

Clar's approach amounts to use of some Kekule valence structures and neglect of others. Clar justified his structures of conjugated polycyclic hydrocarbons having π-sextets and "migrating" sextets by interpretation of the shifts in the UV spectra of benzenoids and the splitting in the NMR spectra caused by coupling of protons in neighboring CH bonds. In many cases Clar's structural formula can be derived from a mathematical scheme in which one assigns to individual Kekule valence structures different weights.[106–108] The weights are determined by the smallest Pauling bond order in a Kekule valence structure for any of the CC double bonds that appear in the Kekule structure considered. If one superimposes the Kekule structures that have so assigned largest weight, one obtains the so-called maximal valence structure, which in many instances coincides with the Clar structure, or is closely related to the empirical Clar structure.

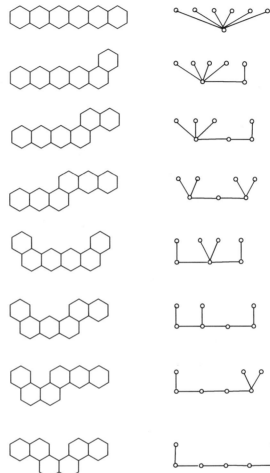

Figure 5. Caterpillar graphs representing resonance pattern for π-sextets for several *cata*-condensed benzenoid hydrocarbons.

Pauling bond orders can also lead to an alternative weighting procedure for individual Kekule valence structures. Instead of assigning to a valence structure the weight given by the smallest Pauling bond order for any CC double bonds in a structure, one simply adds all Pauling bond orders in a structure for bonds that formally appear as CC double bonds. When all so derived weights for all Kekule valence structures are added, one obtains a single descriptor, the Pauling sum, for a molecule. The Pauling sum correlates with the molecular resonance energy.[109]

In developing his resonance theory, Herndon introduced molecular graphs (Figure 6) that depict the couplings among distinct Kekule valence structures of polycyclic aromatic hydrocarbons (PAH).[80-83] When one considers only the couplings between the six-membered conjugated circuits, one obtains resonance graphs depicted in

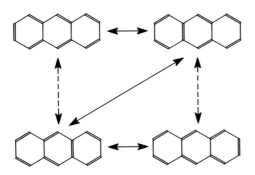

Figure 6. Herndon's interaction graph for the four Kekule valence structures of anthracene.

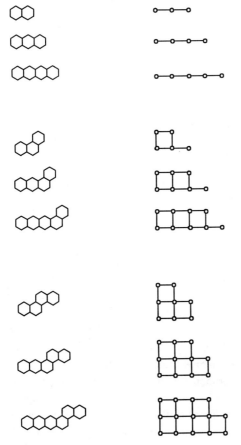

Figure 7. The resonance graphs showing the coupling between the six-member conjugated circuits in different families of benzenoids.

Figures 7 and 8 for families of benzenoid compounds and selected larger benzenoids, respectively. Resonance graphs show interesting regular shapes, while the leading eigenvalue of the graphs represent an index of the benzenoid character of PAH.[110-113]

There are still other types of molecular graphs. For example, Herndon and Bertz considered so-called bond graphs.[114,115] In these graphs one considers CC bonds as the vertices of a graph and connects only those vertices that correspond to incident edges.

Let us outline yet another kind of molecular graph designed for saturated hydrocarbons. This type, as we will see, depicts different conformers of *n*-alkanes. We will refer to these graphs that represent CC chains superimposed on a diamond grid as augmented graphs. The augmented graph for all-*transoid* *n*-pentane is illustrated in Figure 9. This particular kind of molecular graph allows one to enumerate conformers of long chains in 3D space.[116] This illustrates an extension of graph theory to solving a problem relating to a structure in 3D space. Thus, interestingly, though strictly

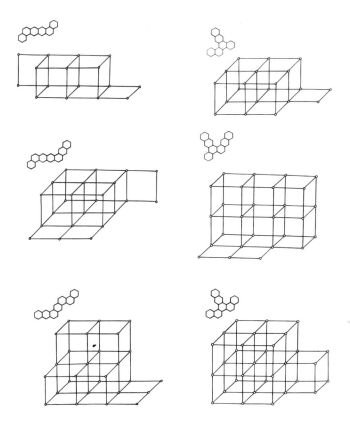

Figure 8. The resonance graphs showing the coupling between the six-member conjugated circuits for selected larger benzenoids.

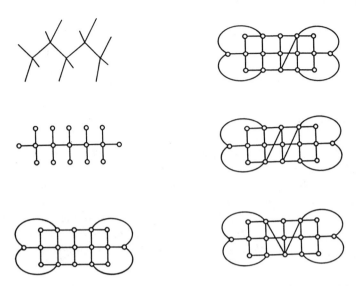

Figure 9. The augmented graph for all-*trans* n-pentane and augmented graphs for the remaining rotational isomers of *n*-pentane.

speaking graphs are devoid of 3D geometrical information character, augmented graphs (which are mathematically speaking simple graphs) allow one to reconstruct the 3D form of a conformer of *n*-alkane. Because of our interest in the characterization of 3D molecular structure, we will examine augmented graphs in more detail.

The augmented graphs are constructed in the following way: First we draw the molecular graph suppressing hydrogen atoms. In the case of *n*-pentane this results in a graph depicted as a chain of length four. In the next step we add the missing hydrogen atoms; hence, we obtain a molecular graph in which hydrogen atoms have not been suppressed. To obtain the augmented graph we introduce *additional* edges to the molecular graph by connecting those hydrogen atoms on adjacent carbon atoms if they are in an *anti* conformation. In the case of all-*transoid n*-pentane we obtain the graph shown at the bottom of the left column of Figure 9. The other three possible conformers of *n*-pentane produce augmented graphs illustrated on the right side of Figure 9. Observe that graphs for other conformers are similar to that for the all-*transoid* conformer, except that they have an *additional* line or lines. These additional edges connect *gauche* hydrogens on adjacent carbon atoms. The augmented graphs are not difficult to construct if one uses molecular models or if one can trace various conformations of a chain on a diamond grid.

From Figure 9 we see that in the case of *n*-pentane each conformer is represented by a different augmented graph. Hence, by constructing all possible augmented graphs we can enumerate conformers (or rotational isomers) of *n*-alkanes. In Figure 10 this is illustrated for *n*-hexane. Since the "end" parts of all augmented graphs are the same, we can simplify and transform augmented graphs into ternary codes defined by the

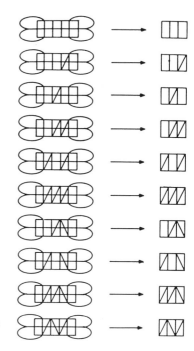

Figure 10. The augmented graphs for the rotational isomers of *n*-hexane and their contraction to ternary codes.

central parts of the augmented graphs. These abbreviated ternary codes are depicted at the right in Figure 10. The procedure has reduced the problem of enumerating all 3D conformers of *n*-alkanes to the construction of corresponding ternary codes. Mathematically speaking, we have reduced the dimensionality of the problem. The approach can be verified by comparing these results with the ternary algebraic codes for *n*-alkanes considered earlier by Balaban.[117]

The diversity of molecular graphs based on the connectivity among atoms, bonds, or selected molecular components offers a source for many structural descriptors. Applications to inorganic compounds have led to a further increase in the type of molecular graphs that are of interest in chemistry for modeling molecules.[118] We should add that not only molecules but also other mathematical objects in 3D space can be successfully modeled by graphs. Thus, recently Liang, Jiang, and Mislow[119,120] used graphs with two kinds of vertices to characterize knots—mathematical objects characteristic of 3D space.

6.4. MOLECULAR INVARIANTS AND MOLECULAR CHARACTERIZATION

In contrast to the representation of molecules by codes that are unique for each molecule, molecular characterization based on ordered sets of invariants need not be unique. If a single invariant is used to characterize a structure, one finds already among small graphs nonisomorphic structures having the same (degenerate) value for an invariant.[121-130] In Figure 11 we illustrate graphs that show degeneracy with respect to a single invariant. That different graphs can have the same invariant is not surprising since single numbers, integer or real, cannot parallel the ever-growing variations in the connectivity as the number of graphs increases with their size. In fact, it is surprising that a single invariant can in many cases differentiate graphs so well.

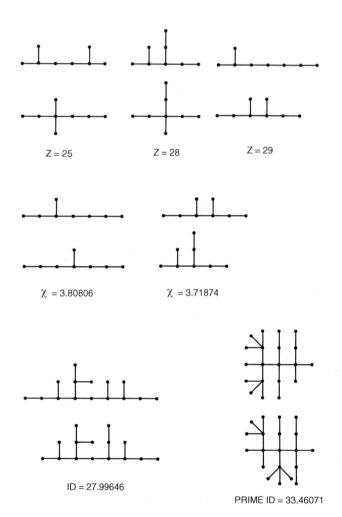

Figire 11. Graphs showing degeneracy with respect to a selected invariant. Z = Hosoya Z topological index, χ = the connectivity index, ID = identification number, and prime ID is the identification number based on prime number weights.

If we select an ordered set of descriptors (invariants), we should expect a better discrimination among graphs than when based on a single variable. We are particularly interested in descriptors that are themselves structurally related. Examples are the connectivity index and the higher-order connectivity indices.[131,132] For such descriptors there is a *natural* ordering of invariants based on the length of paths used in their construction. Several additional structurally related descriptors were recently considered. For example, the Wiener index W[133] has been generalized to higher-order Wiener indices.[134–138] The Hosoya Z topological index[139] was generalized to two distinct higher-order ^kZ topological indices.[140–143] Other novel structurally related descriptors include the extended connectivity,[144] the augmented path numbers which besides paths also count subgraphs of disjoint paths in a graph,[145] and an infinite set of descriptors based on powers of the adjacency matrix.[146,147] In addition, several new matrices for graphs were proposed.[148–151] One hopes in this way to arrive at alternative bases for characterization of structures that may better discriminate between similar molecules and be better suited in applications to structure–property/activity relationships.

It may be of interest to mention here a historical note. Generalized matrices were considered before World War II in Russia by Balandin. He used different molecular properties that allowed bond partitioning and constructed the corresponding generalized matrices based on the entries of thus partitioned molecular properties.[152,153]

6.4.1. Molecular Structure Basis

The comparison between molecules is of interest to structure–property/activity studies. An ordered set of descriptors (invariants, topological indices) can be viewed as a basis for characterization of a molecule. Different bases will in different ways discriminate between similar molecules. We are particularly interested in the use of the descriptors in multivariate regression analysis for the purpose of studying structure–property/activity correlation. The use of a single descriptor in a regression has some advantages but clearly also serious limitations. The advantage of the simple regression (if the correlation is very good) is in the simplicity of the interpretation of the results. Use of several descriptors increases the accuracy of a correlation and reduces the standard error of predictions but the traditional multivariate regression analyses, including the principal components analysis, suffer from the limited possibility to interpret the role of the individual descriptors used. The limitations of a single descriptor have recently been illustrated for some 20 properties of octanes using over 40 molecular descriptors.[154] Use of two and three descriptors often suffices to dramatically improve the predictability of the correlation.[155,156] When one wants to compare the regressions for the same set of compounds and different properties, it will be easier if one uses the same descriptors throughout, rather than each time use an ad hoc combination of descriptors. Recent studies have clearly shown that about half a dozen descriptors offer the best regressions.[154–161] According to Katrizky and Gordeeva,[159] the optimal descriptors are not the quantum-chemical parameters and the traditional QSAR descriptors (log P,[14,15] Taft's steric parameters[162]) but rather the "classical" graph theoretical indices such as the connectivity indices, the Wiener index, and other such indices.

The question can be raised: Is it possible to design a set of descriptors that will be useful for diverse applications? In other words, can we design a basis that will serve for the characterization of diverse structure–property/activity relationships?

The use of structurally related descriptors that can be viewed as a basis, in analogy with vectors in linear algebra, has important advantages. It allows one to follow the role of individual descriptors. This became possible by disengaging the parallel contributions of strongly interrelated descriptors.[163–167] The troublesome interdependence of most descriptors can be eliminated through the process of orthogonalization. By considering residuals of mutual correlation among descriptors, one can, as outlined in the literature, derive noncorrelated descriptors.[163–167] In view of the relatively long history and wide use of the multivariate regression, it is somewhat surprising that this important development was not discovered earlier. The success of the principal component method of Hotelling[168] was precisely in that it produced mutually independent combinations of descriptors. In doing this, however, one constructs linear combinations that, while independent, at the same time are difficult to interpret, in part because the components of which the combinations are made are interrelated. With orthogonalized descriptors one not only can derive stable regression equations (with satisfactory statistical parameters for the coefficients of the regression equation), but also can interpret the role of the individual descriptors. Moreover, one can arrive at regression equations corresponding to the orthogonalized descriptors *without* necessarily constructing the orthogonalized descriptors themselves. One continues with stepwise regression using selected interdependent descriptors and one extracts the contributions of individual orthogonalized descriptors from the successive coefficients of the equation corresponding to a nonorthogonalized basis.

It should be recognized that the regression equation associated with orthogonalized descriptors is numerically stable and its coefficients show improved statistical significance.[167] It is not surprising that this approach has been well received and utilized by several investigators.[169–174] Because the orthogonalization of descriptors extends to descriptors used for characterization of 3D structures, we will briefly outline the salient features of the approach in a later section. The correlation among the residuals has been considered in statistics.[175] Recently, it was shown that the residuals describing the orthogonal descriptors can also be obtained directly by considering various multiple regressions between the variables, in lieu of stepwise simple regressions.[176]

6.5. ORTHOSIMILARITY

Similarity has played an important role in chemistry in the discussion of structure–property relationships, particularly in discussions of biological activity of chemicals. The early recognition of the role of molecular similarity in structure–activity studies may be traced to Emil Fischer's "lock and key" model for the interaction of proteins and drugs.[13] Much of the work in QSAR studies rests on the paradigm that similar structures have similar properties. Recently, attempts have been made to express quantitatively the degree of mutual similarity.[177,178] In quantitative applications of molecular similarity, three distinct problems have to be tackled and resolved[177–180]:

1. Choice of molecular descriptors
2. Selection of the measure of similarity
3. Choice of the clustering technique

Hansch[14,15] and his followers use log P in their QSAR studies and the Hammett sigma constant as the principal molecular descriptors, resorting also to selected molecular properties (e.g., molar refraction). Kier and Hall[181–184] use the connectivity indices[131,132] and the valence connectivity indices.[185] The connectivity index is an illustration of a mathematical descriptor, to be contrasted with descriptors related to physicochemical properties of molecules used by the traditional QSAR school of Hansch. The earliest mathematical descriptors for characterization of molecules were proposed in the mid-1940s by Platt.[186] He suggested the use of path numbers as molecular descriptors. At the same time, Wiener introduced his W index, the descriptor determined by the sum of interatomic distances when these are measured by the number of the bonds separating atoms. Since the early 1970s there has been revived interest in the design of mathematical molecular descriptors, usually referred to as topological indices.[59,187,188]

A quantitative measure of similarity is closely related to the comparison of sequences. Each molecule is represented by a sequence $\mathbf{M} = (M_1, M_2, M_3, \ldots)$. One views $\mathbf{M}$ as a vector in n-dimensional space and then one uses the Euclidean distance

$$D = \{ (M_1 - M'_1)^2 + (M_2 - M'_2)^2 + (M_3 - M'_3)^2 + \cdots \}^{1/2}$$

as the measure of similarity. Other measures of similarity are outlined in the literature.[177,178] Once the similarity/dissimilarity table is constructed, one extracts the ranking among the compounds using one of several available clustering patterns.

Each step in deriving the quantitative measure of the similarity depends on the previous step. Hence, the selection of molecular descriptors is the most critical step. Here we will discuss the bias that is introduced into the standard approach to molecular similarity by the hitherto mostly overlooked interdependence of molecular descriptors. We will illustrate the problem using molecular path numbers as descriptors by considering the similarity among the 18 octane isomers.[189] However, the approach equally applies to other structures and other descriptors, mathematical or not.

In Table 1 we list the path numbers for 18 isomers of octane. All isomers have the same number of atoms (paths of length zero, p_0) and the same number of bonds (paths of length one, p_1). Shorter paths, particularly paths of length two (p_2) and paths of length three (p_3), dominate over many molecular properties.[9,10] Each molecule will therefore be characterized by a sequence of path numbers

$$P_k = (p_0, p_1, p_2, p_3, p_4, \dots)$$

Such characterization is associated with a loss of the information, since already the path counts for individual atoms give only the number of neighbors at a certain distance away from that atom, not their relative distribution. Thus, the atomic path count 1, 2, 2 represents two distinct distributions of the second neighbors:

Therefore, it is not surprising that two molecules can have the same count of paths of different length. For example, 2,3,4-trimethylhexane and 3-methyl-3-ethylhexane, (two isomers of nonane) have the same path counts (9, 8, 10, 10, 6, 2).[145]

Table 1. The Count of Paths for 18 Octane Isomers

		P_0	P_1	P_2	P_3	P_4	P_5	P_6	P_7
1	*n*-octane	8	7	6	5	4	3	2	1
2	2-M	8	7	7	5	4	3	2	0
3	3-M	8	7	7	6	4	3	1	0
4	4-M	8	7	7	6	5	2	1	0
5	3-E	8	7	7	7	5	2	0	0
6	2,2-MM	8	7	9	5	4	3	0	0
7	2,3-MM	8	7	8	7	4	2	0	0
8	2,4-MM	8	7	8	6	5	2	0	0
9	2,5-MM	8	7	8	5	4	4	0	0
10	3,3-MM	8	7	9	7	4	1	0	0
11	3,4-MM	8	7	8	8	4	1	0	0
12	2-M, 3-E	8	7	8	8	5	0	0	0
13	3-M, 3-E	8	7	9	9	3	0	0	0
14	2,2,3-MMM	8	7	10	8	3	0	0	0
15	2,2,4-MMM	8	7	10	5	6	0	0	0
16	2,3,3-MMM	8	7	10	9	2	0	0	0
17	2,3,4-MMM	8	7	9	8	4	0	0	0
18	2,2,3,3-MMMM	8	7	12	9	0	0	0	0

We will take the paths of Table 1 as molecular descriptors to obtain a quantitative measure of molecular similarity. In Table 2 we show the similarity/dissimilarity table for the octane isomers using the Euclidean distance as the measure of similarity. The smaller entries in Table 2 indicate molecules found similar under the procedure adopted, while the larger entries point to the least similar structures.

To select the most similar structures by inspection is somewhat risky. Are n-octane and 2-methylheptane the most similar, or 2-methyloctane and 3-methylheptane, or some other such pair? The use of the quantitative approach is therefore important when the similarity is not apparent, as is frequently the case. However, in this particular quantitative application, we see somewhat disappointingly from Table 2 that in all there are 20 pairs of isomers emerging as equally most similar! The smallest entry in the similarity/dissimilarity table (1.4142) occurs whenever path sequences differ by a single entry. Though the count of paths is distinct for each isomer, clearly the characterization by paths fails to discriminate the structures sufficiently well. Is this

Table 2. The Similarity/Dissimilarity Matrix for Octane Isomers
Based on Path Characterization

	1	2	3	4	5	6	7	8	9
1	0	1.414	2	2.449	3.464	3.742	3.464	3.162	4.690
2		0	1.414	2	3.162	2.828	3.162	2.828	2.449
3			0	1.414	2	2.449	2	2	2
4				0	1.414	2.828	2	1.414	2.828
5					0	3.162	1.414	1.414	1.414
6						0	2.449	2	2.414
7							0	1.414	2.828
8								0	2.449
9									0

	10	11	12	13	14	15	16	17	18
1	4.690	4.690	5.292	6.325	6.325	5.831	7.071	5.657	9.022
2	4	4.243	4.899	5.831	5.657	5.099	6.481	5.099	8.367
3	3.162	3.162	4	4.899	4.899	4.899	5.657	4.243	7.746
4	2.828	2.828	2.828	3.162	4.690	4.690	4	5.657	3.742
5	2.449	2	2.449	4	4.243	4.243	5.099	3.162	7.616
6	2.828	3.742	4.472	5.099	4.472	3.742	5.477	4.243	7.071
7	1.414	1.414	2.449	3.162	3.162	4	4	2.449	6.325
8	2	2.449	2.828	4.243	4	3.162	5.099	3.162	7.348
9	3.742	4.242	5.099	5.831	5.477	4.899	6.325	5.099	8
10	0	1.414	2	2.449	2	3.162	3.162	1.414	5.477
11		0	1.414	2	2.449	4.243	3.162	1.414	5.831
12			0	2.449	2.828	3.742	3.742	1.414	6.481
13				0	1.414	5.099	1.414	1.414	4.243
14					0	4.243	1.414	1.414	3.742
15						0	5.657	3.742	7.483
16							0	2.449	2.828
17								0	5.099
18									0

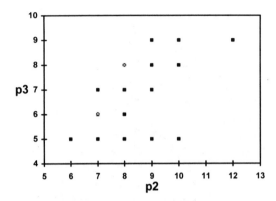

Figure 12. A regression of p_3 against p_2 for the 18 octane isomers (⊠ indicates the sites occupied by two isomers).

indeed so? It is hard to believe that the 20 pairs of isomers have the same degree of similarity. We can introduce additional descriptors to increase discrimination among isomers, but that avoids the direct answer to the question raised.

The high degeneracy seen in the similarity/dissimilarity table of octanes is the consequence of the *bias* introduced through the mutual interdependence of the path numbers. For many isomers a large p_2 is often paralleled by a large p_3, and a large p_3 is paralleled by a large p_4. This gives path p_2 greater weight as its role appears again with p_3, and so on. In Figure 12 we show the simple regression of p_3 against p_2. Although the correlation coefficient R is low, the pattern of dots in Figure 12 clearly

Table 3. Orthogonalized Path Numbers

	Average p_2 Ω_0	(Av. p_2) $-p_2$ Ω_1	Residual p_3/p_2 Ω_2	Residual p_4/p_3 Ω_3	Residual p_5/p_4 Ω_4	Residual p_6/p_5 Ω_5	Residual p_7/p_6 Ω_6
1	8.444	2.444	−0.4695	−1.3954	−1.0780	−0.3765	0
2	8.444	1.444	−1.0274	−1.0308	−0.5203	0.3932	0
3	8.444	1.444	−0.0274	−0.6952	0.2075	0.1628	0
4	8.444	1.444	−0.0274	0.3048	−0.2174	0.1804	0
5	8.444	1.444	0.9726	0.6404	0.5105	−0.0500	0.0001
6	8.444	−0.556	−2.1433	−0.3016	0.5948	−0.0674	0
7	8.444	0.444	0.4146	0.0050	0.4929	−0.0463	−0.0001
8	8.444	0.444	−0.5854	0.6694	0.3402	−0.0499	0
9	8.444	0.444	−1.5854	−0.6661	1.0372	−0.0887	0
10	8.444	−0.556	−0.1433	0.3696	0.0505	−0.0250	−0.0001
11	8.444	0.444	1.4146	0.3406	0.2208	−0.0251	0
12	8.444	0.444	1.4146	1.3406	−0.2041	−0.0074	0
13	8.444	−0.556	1.8567	0.0408	−0.0688	−0.0001	0
14	8.444	−1.556	0.2988	0.0698	−0.2391	0.0000	0
15	8.444	−1.556	−2.7012	2.0630	−0.6974	−0.0109	0
16	8.444	−1.556	1.2988	−0.5947	−0.0863	0.0036	0.0001
17	8.444	−0.556	0.8567	0.7052	−0.2216	−0.0036	0.0001
18	8.444	−3.556	0.1829	−1.8654	−0.1213	0.0110	0

indicates the relatedness of p_2 and p_3. In fact, the pattern seen in Figure 12 is the basis for ordering of isomers and it gives a template from which one can recognize regular variations in properties of isomers.[7–10,145,190–194]

The degeneracy in the similarity table can be reduced *without* introducing additional descriptors. We can eliminate the repeated influence of the shorter paths when calculating the contributions from longer paths. This is accomplished by use of orthogonalized path numbers.[7,163–167] The orthogonalized descriptors are constructed in the following way: The first descriptor, here p_2, is selected as the first orthogonal descriptor Ω_1. To find the second orthogonal descriptor we first consider the regression of p_3 against p_2. The residual of that regression, Res p_3/p_2, listed in Table 3, is taken as the second orthogonal descriptor Ω_2. Clearly, the part of p_3 that correlates with p_2 is given by the regression equation, while the parts of p_3 that cannot be determined from p_2 are the residuals. Hence, by definition the residuals make the orthogonal component. The process of orthogonalization continues by considering the regression

Table 4. The Similarity/Dissimilarity Matrix for Octane Isomers Based on Orthogonalized Path Characterization

	1	2	3	4	5	6	7	8	9
1	0	1.532	1.905	2.266	3.139	3.986	3.053	3.224	3.215
2		0	1.302	1.709	2.838	2.689	2.318	2.241	2.025
3			0	1.087	1.709	2.972	1.346	1.799	2.044
4				0	1.302	3.093	1.357	1.345	2.453
5					0	3.821	1.310	1.859	3.087
6						0	2.765	2.106	1.281
7							0	1.210	2.179
8								0	1.089
9									0

	10	11	12	13	14	15	16	17	18
1	3.690	3.518	3.992	4.199	4.425	5.764	4.571	4.005	6.141
2	2.691	3.093	3.584	3.718	3.493	4.644	3.865	3.288	5.241
3	2.282	2.046	2.724	2.863	3.149	4.960	3.299	2.637	5.152
4	2.033	1.821	2.046	2.770	3.032	4.417	3.408	2.231	5.450
5	2.352	1.170	1.483	2.341	3.216	5.097	3.315	2.134	5.684
6	2.179	3.770	4.123	4.070	2.793	2.928	3.661	3.269	4.168
7	1.281	1.010	1.809	1.844	2.134	4.400	2.340	1.483	4.465
8	1.170	2.031	2.179	2.744	2.341	3.391	3.055	1.844	4.820
9	2.265	3.269	3.818	3.818	3.119	3.963	3.687	3.231	4.682
10	0	1.859	2.106	2.031	1.170	3.312	2.007	1.090	3.759
11		0	1.087	1.170	2.352	4.975	2.232	1.281	4.744
12			0	1.704	2.619	4.659	2.788	1.310	5.273
13				0	1.859	5.124	1.310	1.210	3.929
14					0	3.631	1.210	1.310	2.788
15						0	4.841	3.966	5.299
16							0	1.704	2.619
17								0	4.009
18									0

Table 5.   The Most Similar Pairs of Octane Isomers (Illustrated in Figure 13)

Rank			Magnitude
1	3,4-dimethylhexane	2-methyl-3-ethylpentane	1.086671
2	3-methylheptane	4-methylheptane	1.086669
3	2,3-dimethylhexane	3,4-dimethylhexane	1.089548
4	3,3-dimethylhexane	2,3,4-trimethylpentane	1.089552
5	3-ethylhexane	3,4-dimethylhexane	1.170380

of p_4 against p_2, which gives Res p_4/p_2, the part of original p_4 count that does not parallel p_2. The residual Res p_4/p_2, however, may show a correlation with the second orthogonal descriptor Res p_3/p_2. We have therefore to regress Res p_4/p_2 against Res p_3/p_2 and use the residual of that correlation as our third orthogonal descriptor Ω_3.

The orthogonalized path counts are listed in Table 3. As seen, the magnitudes of the entries in Table 3 decrease after each successive step of the orthogonalization. The last column essentially has all entries zero. This indicates that p_7 does not introduce useful information to the existing set of path numbers p_2–p_6.

When we use the orthogonalized path numbers from Table 3 to characterize isomers and construct the similarity/dissimilarity table, we will obtain an unbiased similarity measure for the isomers of octane. The orthogonalization procedure, which is analogous to the Gram-Schmidt orthogonalization of vectors, requires an ordering of descriptors. In various applications one can follow different ordering of descriptors.[163–167,195–197] Here we have assumed the "natural" ordering of path numbers according to the length of the paths involved. We use the term *unbiased* to signify that the descriptors used in calculating the similarity/dissimilarity are unrelated, hence no duplication of information occurs.

In Table 4 we show the similarity/dissimilarity matrix based on orthogonal descriptors of Table 3. Observe how the degeneracy of the similarity/dissimilarity table for path numbers has almost completely disappeared. The most similar pairs of isomers are listed in Table 5 and are illustrated in Figure 13.

6.6. BEYOND GRAPH THEORY

Graph theory offers many useful characterizations of molecules. If the molecular property is bond additive, the modeling by graphs is quite adequate. Graph theory, even when not explicitly mentioned, has been behind many successful mathematical or quantum-chemical models. For example, Hameka studied the magnetic susceptibility of alkanes from a quantum-chemical point of view.[198] However, the very same quantum-chemical model can be "translated" without difficulties in the graph theoretical terms when it leads to even simpler expressions for the same magnetic susceptibilities.[199–201]

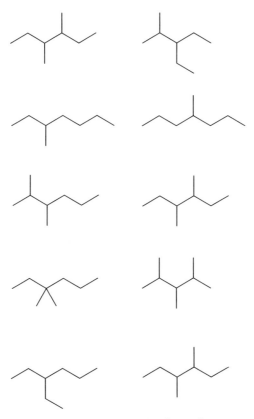

Figure 13. The most similar pairs of octane isomers.

There are still several open problems concerning the use of graph theory in the study of structure–property relationships. For example, Poshusta and McHughes pointed to an apparent difficulty associated with cluster expansions. Apparently, when applied to alkanes such expansions are associated with inherent divergence.[202–204] Different kinds of difficulties arise from inherent limitations of graph theory to cope with heteroatoms and different types of chemical bonds in a compound. Finally, even greater difficulties are associated with the limitation of graph theory to represent the three-dimensionality of chemical compounds. Because graphs have been very useful in modeling many structure–property relationships of molecules, one would like to extend methodologies developed for graphs to graphs rigidly embedded in 3D space. The question is: Can graphs be generalized so that useful properties of graphs already recognized are preserved and at the same time more realistic models for molecules are obtained?

In this review we will focus attention on the construction of models that reflect the 3D structure of molecules. Here we will only briefly consider the problem of heteroatoms, i.e., how to modify simple graphs so that they can better reflect the differences among different kinds of atoms present. Clearly molecules with heteroatoms require additional molecular descriptors. One way to generate such descriptors is to use the diagonal elements in the adjacency matrix and assign to them suitable numerical weights. The resulting matrices will generate novel invariants that may give the sought additional descriptors. A way to determine the numerical weights for different atoms has already been outlined[205-208]: Use multivariate regression analysis and minimize the standard error by varying the entries for atoms of different kind on the diagonal of the adjacency matrix. The variables are extracted from thus weighted matrices using already defined algorithms, such as those for construction of the connectivity index and the higher-order connectivity indices. Alternatively, one can define standard parameters for different heteroatoms analogous to the empirical Slater rules of early quantum chemistry,[209] or computationally derived rules for the form of the optimal Gaussian orbitals,[210] or double-ζ-type orbitals.[211] This is essentially the approach of Kier and Hall in the construction of their valence connectivity indices.[185] Similarly, Trinajstić, Balaban, and others considered a set of standard parameters for heteroatoms.[212-214] Other physicochemical properties of atoms can be used to determine the relative weights for heteroatoms, such as van der Waals radii and electronegativities.[215,216] Hence, there are ways to discriminate heteroatoms even though today we do not know the optimal parameters for different heteroatoms.

The problem of capturing the 3D nature of chemical structure appears even harder, but as we will see there are ways to generalize graphs so as to give a more realistic model of molecules as 3D objects. Most often a graph is defined by a list of neighbors, i.e., by the set of elements and the list of the binary relations, or by the adjacency matrix, as illustrated in Table 6 for methane, CH_4. In each case the given information relates to the connectivity of the graph, i.e., we know which element (vertex) is connected (related) to which other element.

Methane as a 3D structure is defined by the coordinates of all of its atoms. The precise form of the coordinates will depend on the orientation of the molecule relative to the coordinate system, just as the form of the adjacency matrix depends on the choice

Table 6. Alternative Ways of Defining a Graph Illustrated on Methane CH_4

List of neighbors:		Elements (vertices):	Adjacency matrix
1	2, 3, 4, 5	{1, 2, 3, 4, 5}	$\begin{pmatrix} 0 & 1 & 1 & 1 & 1 \\ 1 & 0 & 0 & 0 & 0 \\ 1 & 0 & 0 & 0 & 0 \\ 1 & 0 & 0 & 0 & 0 \\ 1 & 0 & 0 & 0 & 0 \end{pmatrix}$
2	1		
3	1	**Binary relations:**	
4	1		
5	1	(1,2), (1,3), (1,4), (1,5)	

Table 7. The Grid Coordinates, the Cartesian Coordinates, the Adjacency and the Geometric Distance Matrix for Methane CH_4

Grid coordinates		Cartesian coordinates	Distance matrix
1	0, 0, 0, 0	0, 0, 0	$\begin{pmatrix} 0 & 1 & 1 & 1 & 1 \\ 1 & 0 & 0 & 0 & 0 \\ 1 & 0 & 0 & 0 & 0 \\ 1 & 0 & 0 & 0 & 0 \\ 1 & 0 & 0 & 0 & 0 \end{pmatrix}$
2	1, 0, 0, 0	0, $2\sqrt{2}/\sqrt{3}$, $-1/3$	
3	0, 1, 0, 0	$\sqrt{2}$, $\sqrt{3}$, $-\sqrt{2}/3$, $-1/3$	
4	0, 0, 1, 0	$-\sqrt{2}/\sqrt{3}$, $-\sqrt{2}/3$, $-1/3$	
5	0, 0, 0, 1	0, 0, 1	

of labels for individual atoms. We can supply information on the 3D structure by grid coordinates (when appropriate), by Cartesian coordinates of the individual vertices, or by a matrix in which the interatomic distances have been recorded. If we assume the length of C–H bond to be equal to 1 and if we suitably orient the CH_4 tetrahedron relative to the x, y, z coordinates, we can represent methane by the data in Table 7.

Although the interatomic distance matrix (which we will designate by G and will refer to as the geometry matrix of the molecule) is formally similar to the adjacency matrix, it cannot be viewed as a weighted adjacency matrix. For graphs, the adjacency matrix and the distance matrix (in which the distances are measured through bonds, i.e., by the number of bonds between vertices) give information on the connectivity within the molecule. The geometry matrix G, in which the distances are measured through space, does not possess information on connectivity. One may try to infer the bonding pattern of a molecule from the G matrix but that requires *additional* considerations. For example, the adjacency matrix for a six-membered (benzene) ring and its spanning tree are different, but the geometry G matrices for the two molecules (when suppressing hydrogen atoms) are the same (Table 8).

Sometimes it is overlooked that characterization of molecules as 3D objects is *deficient*, just as is the characterization of molecules based on the molecular adjacency or the graph-distance matrix. In the former there is no information on connectivity, in the latter there is no information on spatial geometry. In a way the two approaches complement each other, one having the information that is absent in the other. Therefore, it should not be surprising that combining such matrices will yield a matrix that has information on adjacency as well as on interatomic separations.

6.7. D/D MATRICES

A way to combine the information on the connectivity and the geometry in a single matrix is to construct elements of a new matrix given as the quotient of the corresponding elements of the graph distance matrix and the molecular geometry matrix,[151] hence the term distance/distance (D/D) matrix. In Table 9 we illustrate D/D matrices for the

Table 8.   The Adjacency and the Geometric Distance Matrix
for the Benzene Ring and for Its Spanning Tree

Adjacency matrix

Benzene ring   Acyclic structure

$$\begin{pmatrix} 0 & 1 & 0 & 0 & 0 & 1 \\ 1 & 0 & 1 & 0 & 0 & 0 \\ 0 & 1 & 0 & 1 & 0 & 0 \\ 0 & 0 & 1 & 0 & 1 & 0 \\ 0 & 0 & 0 & 1 & 0 & 1 \\ 1 & 0 & 0 & 0 & 1 & 0 \end{pmatrix} \qquad \begin{pmatrix} 0 & 1 & 0 & 0 & 0 & 0 \\ 1 & 0 & 1 & 0 & 0 & 0 \\ 0 & 1 & 0 & 1 & 0 & 0 \\ 0 & 0 & 1 & 0 & 1 & 0 \\ 0 & 0 & 0 & 1 & 0 & 1 \\ 0 & 0 & 0 & 0 & 1 & 0 \end{pmatrix}$$

Geometry matrix

$$\begin{pmatrix} 0 & 1 & \sqrt{3} & 2 & \sqrt{3} & 1 \\ 1 & 0 & 1 & \sqrt{3} & 2 & \sqrt{3} \\ \sqrt{3} & 1 & 0 & 1 & \sqrt{3} & 2 \\ 2 & \sqrt{3} & 1 & 0 & 1 & \sqrt{3} \\ \sqrt{3} & 2 & \sqrt{3} & 1 & 0 & 1 \\ 1 & \sqrt{3} & 2 & \sqrt{3} & 1 & 0 \end{pmatrix}$$

six-membered ring and its spanning tree, respectively. As seen, the two matrices are different. Therefore, one should expect at least some of the matrix invariants to differ for the two systems. In order to gain some insight into the interpretation of the invariants derived from D/D matrices, we will review a few applications.

6.7.1. Linear Chains

The D/D matrices for strictly linear chains do not bring novelty, since for strictly linear chains the geometrical and the topological distances between any pair of vertices are the same. Consequently, all nondiagonal elements of D/D matrix are equal to 1, as illustrated for a straight chain of five atoms:

$$\begin{pmatrix} 0 & 1 & 1 & 1 & 1 \\ 1 & 0 & 1 & 1 & 1 \\ 1 & 1 & 0 & 1 & 1 \\ 1 & 1 & 1 & 0 & 1 \\ 1 & 1 & 1 & 1 & 1 \end{pmatrix} \qquad \text{o—o—o—o—o}$$

The above is in fact the adjacency matrix for the complete graphs K_5.

Can we interpret the first eigenvalue of a D/D matrix? The first eigenvalue of the adjacency matrix of trees has already been suggested as an index of molecular branching.[217,218] Recently it was suggested that the first eigenvalue of the resonance

Table 9.   The Distance/Distance Matrix for the Benzene Ring and for Its Spanning Tree

Benzene ring

$$
\begin{pmatrix}
0 & 1 & \sqrt{3}/2 & 2/3 & \sqrt{3}/2 & 1 \\
1 & 0 & 1 & \sqrt{3}/2 & 2/3 & \sqrt{3}/2 \\
\sqrt{3}/2 & 1 & 0 & 1 & \sqrt{3}/2 & 2/3 \\
2/3 & \sqrt{3}/2 & 1 & 0 & 1 & \sqrt{3}/2 \\
\sqrt{3}/2 & 2/3 & \sqrt{3}/2 & 1 & 0 & 1 \\
1 & \sqrt{3}/2 & 2/3 & \sqrt{3}/2 & 1 & 0
\end{pmatrix}
$$

The spanning tree

$$
\begin{pmatrix}
0 & 1 & \sqrt{3}/2 & 2/3 & \sqrt{3}/4 & 1/5 \\
1 & 0 & 1 & \sqrt{3}/2 & 2/3 & \sqrt{3}/4 \\
\sqrt{3}/2 & 1 & 0 & 1 & \sqrt{3}/2 & 2/3 \\
2/3 & \sqrt{3}/2 & 1 & 0 & 1 & \sqrt{3}/2 \\
\sqrt{3}/4 & 2/3 & \sqrt{3}/2 & 1 & 0 & 1 \\
1/5 & \sqrt{3}/4 & 2/3 & \sqrt{3}/2 & 1 & 0
\end{pmatrix}
$$

graphs for PAH can be interpreted as an index of benzenoid character.[110–113] Here we will see that the first eigenvalue of the D/D matrix offers a useful index that measures the degree of molecular folding. As is known, the largest eigenvalue of K_n is $n - 1$.[219–221] Thus, the normalized eigenvalue for D/D matrices of a strictly linear chain having n atoms is: $\lambda/n = 1 - 1/n$. We immediately see that as n increases, λ/n tends toward 1. We may interpret λ/n, for which we use the label ϕ_n, to measure the departure of a finite segment of a straight line (the chain of n atoms) from a line (geometrical object of infinite length). The larger the segment of the line, the smaller the departure from an idealized infinite line, and ϕ is closer to 1. Although the interpretation is somewhat straightforward, we will see later that such an interpretation of the normalized leading eigenvalue is of interest when considering structures that are not linear.

Let us consider chains superimposed on the graphite lattice. Similar considerations apply to chains superimposed on the diamond lattice. In Table 10 we list the leading eigenvalue of D/D matrices and the normalized index ϕ. As expected we see that as the length of the chain increases, the value of the index ϕ increases. It appears that ϕ approaches 1 as n increases indefinitely; however, without a rigorous mathematical analysis we can only say that the limit L is less than or equal to 1. The leading eigenvalue of D/D matrices apparently indicates the "degree of departure" of a system from strict linearity. The difference $(1 - 1/n)$ may be viewed as a measure of the departure of a molecule from strict linearity.

In order to strengthen the interpretation of ϕ as an index for the folding of a chain, we will examine the other extreme, the chain superimposed on a graphite lattice in an all-*cisoid* conformation. We allow superposition of a chain on itself in order to see how ϕ changes with increased folding. In a thus folded chain the geometrical distances

Table 10. The Normalized Leading Eigenvalue for "Linear" Chains of Length n (Superimposed on Graphite Lattice) Corresponding to all-*trans* Conformers and all-*cis* Conformers

| | all-*transoid* | | all-*cisoid* | |
n	λ_n	ϕ_n	λ_n	ϕ_n
4	2.8079	0.7019	2.7968	0.6767
5	3.6920	0.7384	3.3690	0.6738
6	4.5719	0.7620	3.8974	0.6496
7	5.4479	0.7783	4.3020	0.6146
8	6.3220	0.7903	4.6388	0.5798
9	7.1994	0.7994	4.9350	0.5483
10	8.0659	0.8066	5.2046	0.5204
11	8.9363	0.8124	5.4520	0.4956
12	9.8063	0.8172	5.6783	0.4732
13	10.6756	0.8212	5.8806	0.4524

between any pair of vertices never exceed 2, while the topological distances increase with the size of the chain. The last two columns of Table 10 give λ_n and ϕ_n for folded chains up to length $n = 13$. We see that as n increases, the index ϕ_n steadily decreases. This confirms the suitability of ϕ_n as a measure of the "degree of folding" of the structure. Clearly $\phi_{n+1} < \phi_n$ now holds, which in itself is a justification to view ϕ_n as an index of chain folding.

6.7.2. Conformations of n-Chains in a Plane

We will illustrate the use of index ϕ on some of the conformers of a chain of length six embedded on the graphite grid (e.g., 1,3,5-heptatriene). In Figure 14 and Table 11 we show all conformers of the chain of length six and the corresponding ϕ_n values. We see that the all-*transoid* TTTT conformer has the largest $\phi = 0.7783$ while the apparently most folded TCCC has the smallest ϕ ($\phi = 0.6558$). Here T for *trans* and C for *cis* describe the relative orientation of three successive bonds. The hypothetical isomer CCCC ($\phi = 0.6496$, see Table 10) that would overlap itself is not considered, but in helicenes such folding of a molecule is possible as the overcrowded atoms are displaced out of plane.[222] By inspection one can estimate approximately the relative folding for different structures to some degree. With ϕ_n, however, we can have a quantitative characterization of the degree of folding. Let us consider all of the conformers of the chain of length six. It is usually easy to identify the extreme forms of conformers (TTTT and TCCC). The usefulness of a quantitative index to characterize the relative degree of folding becomes apparent when considering other than the extreme cases. For the in-between cases we can now determine the degree of similarity. From Table 11 we see that after the TTTT conformer, the least bent conformers are

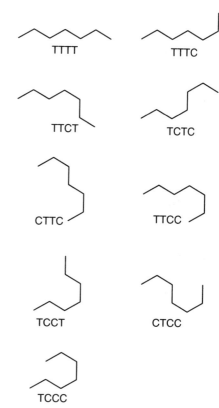

Figure 14. All conformers of a chain of length six (seven carbon atoms) superimposed on a graphite lattice.

Table 11. The ϕ_n Values for Different Conformers of a Chain of Length Six

Conformer	ϕ_n
T T T T	0.7783
T T T C	0.7555
T T C T	0.7476
T C T C	0.7340
C T T C	0.7297
T T C C	0.7079
T C C T	0.6971
C T C C	0.6965
T C C C	0.6558

those with a single *cis* fragment. These are followed by conformers with two *cis* fragments and last come the conformers with three *cis* portions. This agrees with our expectations and supports the interpretation of ϕ_n as an index of molecular folding. It would be difficult merely by inspection to choose between TTTC and TTCT the one that is more folded. Similarly it would be even more difficult to decide between TCTC, CTTC, TTCC, and TCCT about which conformer is more folded.

Folding as discussed here is not just a property of structures embedded in a plane (such as were the forms of long chains superimposed on the graphite lattice considered here). The approach is general and extends to curves in 3D space.[223]

6.7.3. Folding of Mathematical Curves

Could we think of a structure (mathematical object) that is even more folded than all-*cis* chains superimposed on the graphite lattice? Surely, one would be the all-*cisoid* chain superimposed on a square lattice. Even more so would be the all-*cisoid* chain superimposed on a triangular lattice. Finally, the most folded would be the all-*cis* chain superimposed on itself. It seems plausible to assume that $\phi = 0$ is the limit that represents the most folded mathematical object and we already have established that $\phi = 1$ is the limit for the unfolded structure.

In order to get more experience with the newly proposed index ϕ_n we will consider the leading eigenvalue λ of D/D matrices for several well-defined mathematical curves. We should emphasize that this approach is neither restricted to curves (chains) embedded on regular lattices, nor restricted to lattices on a plane. However, the examples that we will consider correspond to mathematical curves embedded on the simple square lattice associated with the Cartesian coordinates system in the plane, or a trigonal lattice. The selected curves show visibly distinct spatial properties. Some of the curves considered apparently are more and more folded as they grow. They illustrate the self-similarity that characterizes fractals.[224,225] A small portion of such curve has the appearance of the same curve in an earlier stage of the evolution. For illustration, we selected the Koch curve, the Hilbert curve, the Sierpinski curve and a portion of another Sierpinski curve, and the Dragon curve. These are compared to a spiral, a double spiral, and a worm-curve.

The Koch curve, which when closed forms a "snowflake," is the earliest illustration of a fractal in the literature.[226] The Peano curve,[227] the Hilbert curve,[228] and the Sierpinski curve[229] eventually, as their length increases, fill in the space of the square formed by their periphery. Of the curves considered, only another Sierpinski curve is a closed curve. Since we are interested in comparing chains, we cut the Sierpinski closed curve in half to obtain a chain. Hence, the label "Sierp" was used to remind us that this is but a half of the original curve. The Dragon curve[230,231] is of relatively new date, introduced by the physicist W. G. Harter and advertised by M. Gardner in his books and his column in *Scientific American*. In Figure 15 we illustrate the Koch curve, the Hilbert curve, the "Sierp" curve, and the Dragon curve.

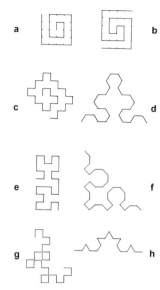

Figure 15. Illustration of spiral (a), double spiral (b), worm-curve (c), Sierpinski arrow (d), Hilbert curve (e), modified Sierpinski arrow (f), Dragon curve (g), and Koch curve (h).

The calculated λ and ϕ_n for the selected curves are sumarized in Table 12. The index n in ϕ_n indicates the length of the chain (or the size of the D/D matrix). We see that as n increases, ϕ_n decreases, as expected. A comparison of different curves again gives plausible results. Among the fractals considered the Koch curve appears the least

Table 12.   The Folding Index ϕ_n for Selected Mathematical Curves

Spiral	Double spiral	Worm path
ϕ_5 = 0.6376	ϕ_{10} = 0.5623	ϕ_{10} = 0.5663
ϕ_{10} = 0.6030	ϕ_{20} = 0.4802	ϕ_{20} = 0.4930
ϕ_{15} = 0.5543	ϕ_{30} = 0.4356	ϕ_{30} = 0.4407
ϕ_{20} = 0.5254		ϕ_{35} = 0.4240
ϕ_{25} = 0.4971		ϕ_{40} = 0.4025
ϕ_{30} = 0.4764		

Sierpinski arrow	Hilbert curve	Sierp
ϕ_{10} = 0.6376	ϕ_{16} = 0.4949	ϕ_8 = 0.6585
ϕ_{18} = 0.5878	ϕ_{32} = 0.4109	ϕ_{16} = 0.5704
ϕ_{27} = 0.5037	ϕ_{48} = 0.3464	ϕ_{24} = 0.4985
	ϕ_{64} = 0.3026	ϕ_{32} = 0.4592

Dragon curve	Koch's curve
ϕ_5 = 0.6216	ϕ_5 = 0.6672
ϕ_{10} = 0.5822	ϕ_{10} = 0.6424
ϕ_{15} = 0.5005	ϕ_{15} = 0.6023
ϕ_{20} = 0.4616	ϕ_{20} = 0.5857
ϕ_{25} = 0.4165	ϕ_{25} = 0.5600
ϕ_{30} = 0.3836	ϕ_{30} = 0.5404

folded. The Hilbert curve is apparently more folded than the Sierp curve, consistent with the fact that the ratio of the distance between the end points and the length of the curve having the same number of points is smaller for the Hilbert curve. The Dragon curve seems to be the most curved or the most folded. This agrees with the observation that the Dragon curve bends at every point, which was not the case with the Hilbert curve.

6.7.4. Similarity Based on D/D Matrices

The D/D matrices can be used to find relative similarity among different conformers. Often it is of interest to find the most similar conformers, since such structures can have similar properties. Similarity of conformers is of particular interest in the docking approach for matching small molecules to receptor cavities in proteins. Most molecules of interest in structure–activity studies can adopt several conformations that are connected by relatively low energy barriers. Sometimes it is of interest to search for the most different conformations.[232] Here we will consider the most similar and the least similar conformers of chains having six carbon atoms superimposed on a graphite lattice, already shown in Figure 14.

The D/D matrices for the "extreme" conformers, TTTT and TCCC, are given in Table 13. From a given D/D matrix we can obtain the path numbers, which are listed in Table 14 for all nine conformers of Figure 14. The paths of length one, p_1, are obtained by summing the elements of the D/D matrix corresponding to adjacent atoms. The paths of length two, p_2, are obtained by adding all elements in the D/D matrix corresponding to carbon atoms separated by two bonds. The paths p_k are obtained in

Table 13. The D/D Matrix for the TTTT and TCCC Conformer of the Chain of Seven Carbon Atoms Embedded on Graphite Lattice

			TTTT			
0	1.000	0.866	0.882	0.866	0.872	0.866
1.000	0	1.000	0.866	0.882	0.866	0.872
0.866	1.000	0	1.000	0.866	0.882	0.866
0.882	0.866	1.000	0	1.000	0.866	0.882
0.866	0.882	0.866	1.000	0	1.000	0.866
0.872	0.866	0.882	0.866	1.000	0	1.000
0.866	0.872	0.866	0.882	0.866	1.000	0
			TCCC			
0	1.000	0.866	0.882	0.750	0.529	0.289
1.000	0	1.000	0.866	0.667	0.433	0.200
0.866	1.000	0	1.000	0.866	0.667	0.433
0.882	0.866	1.000	0	1.000	0.866	0.667
0.750	0.667	0.866	1.000	0	1.000	0.866
0.529	0.433	0.667	0.866	1.000	0	1.000
0.289	0.200	0.433	0.667	0.866	1.0000	0

Table 14. Path Numbers for All Conformers of a Chain
Having Seven Carbon Atoms Embedded on Graphite Lattice

	T T T T	T T T C	T T C T
p_1	6.0000	6.0000	6.0000
p_2	4.3301	4.3301	4.3301
p_3	3.5277	3.3124	3.3124
p_4	2.5981	2.4821	2.3660
p_5	1.7436	1.5929	1.5211
p_6	0.8660	0.7638	0.7676
	T T C C	T C T C	T C C T
p_1	6.0000	6.0000	6.0000
p_2	4.3301	4.3301	4.3301
p_3	3.0972	3.0972	3.0972
p_4	2.0490	2.2500	1.9320
p_5	1.2502	1.5211	1.0583
p_6	0.5000	0.7638	0.5775
	T C C C	C T C C	C T T C
p_1	6.0000	6.0000	6.0000
p_2	4.3301	4.3301	4.3301
p_3	2.8819	2.8819	3.0972
p_4	1.6160	1.9330	2.3660
p_5	0.7292	1.2503	1.4422
p_6	0.2887	0.5774	0.5774

an analogous manner, by adding the elements in the D/D matrix corresponding to carbon atoms separated by k bonds. The numerical values for the similarity/dissimilarity among the nine conformers of the chain of length six are shown in Table 15, based on Euclidean metrics.

It is not surprising that the largest entry in Table 15, pointing to the least similar pair of conformers, belongs to the pair TTTT, TCCC. Other dissimilar pairs include:

Table 15. Similarity/Dissimilarity Matrix among the Nine Configurations of a Chain Having Seven Carbon Atoms Embedded on Graphite Lattice

	T T T T	T T T C	T T C T	T T C C	T C T C	T C C T	T C C C	C T C C	C T T C
T T T T	0	0.305	0.399	0.930	0.605	1.087	1.656	1.059	0.643
T T T C		0	0.136	0.649	0.325	0.817	1.381	0.799	0.342
T T C T			0	0.540	0.245	0.696	1.267	0.696	0.298
T T C C				0	0.428	0.237	0.742	0.256	0.379
T C T C					0	0.591	1.141	0.505	0.233
T C C T						0	0.582	0.288	0.579
T C C C							0	0.675	1.096
C T C C								0	0.520
C T T C									0

TTTC, TCCC (1.381); TTCT, TCCC (1.267); TCTC, TCCC (1.141); and so on. Visual inspection of Figure 14 conformers confirms that the obtained results are plausible.

It would be difficult by inspection of Figure 14 to decide which pair of conformers are the most similar. For example, one may think that TTTT and TTTC, or TTCT and TCTC may be the most similar, but the computation shows the most similar to be the pair TTTC, TTCT, which are at the "distance" 0.136. Next come the pair: TCTC, CTTC (0.233); TTCC, TCCT (0.237); TTCT, TCTC (0.245); TTCC, CTCC (0.256); and so on. The pair TTTT, TTTC, with the "distance" measure of 0.305, is clearly not among the most similar conformers, illustrating the difficulty of estimating similarity by visual perception.

6.8. INVARIANTS DERIVED FROM THE MOLECULAR GEOMETRY

If one knows the relative separations between atoms in a molecule, one can represent a molecule by its 3D distance matrix. The i, j elements of the geometric distance matrix are given by the Euclidean distance $d_{ij} = \sqrt{[(x_i - x_j)^2 + (y_i - y_j)^2 + (z_i - z_j)^2]}$. Here x, y, z are the Cartesian coordinates of the positions of individual atoms. The geometry-dependent distance matrix allows one to construct invariants that are sensitive to the molecular geometry.[233–244] We will refer to these matrices as G-matrices. For *cisoid* and *transoid* 1,3-butadiene we obtain the following geometry-dependent matrices assuming all CC bonds of unit length and all angles 120°:

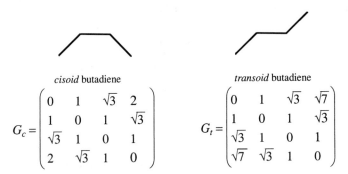

$$G_c = \begin{pmatrix} 0 & 1 & \sqrt{3} & 2 \\ 1 & 0 & 1 & \sqrt{3} \\ \sqrt{3} & 1 & 0 & 1 \\ 2 & \sqrt{3} & 1 & 0 \end{pmatrix} \qquad G_t = \begin{pmatrix} 0 & 1 & \sqrt{3} & \sqrt{7} \\ 1 & 0 & 1 & \sqrt{3} \\ \sqrt{3} & 1 & 0 & 1 \\ \sqrt{7} & \sqrt{3} & 1 & 0 \end{pmatrix}$$

From G-matrices we can extract the usual algebraic invariants, such as the eigenvalues and the characteristic polynomial. One can enumerate the paths of different lengths of G by using the entries corresponding to a pair of atoms (i, j) as the weight of the corresponding path. Using the matrix row sums we can also construct the generalized connectivity index $\chi\chi$.[233,234] The procedure is analogous to the construction of Balaban's J index from the row sums of the graph distance matrix.[245–247]

In Table 16 we give the generalized connectivity indices $(\chi\chi)$ for the six conformers of 1,3,5-hexatriene (chain of six atoms superimposed on the graphite lattice) based

Table 16.   The Molecular Connectivity Indices $\chi\chi$ for the Six Conformers of Hexatriene

	$^1\chi\chi$	$^2\chi\chi$	$^3\chi\chi$	ID
T T T	2.9634	2.2872	1.3405	13.2140
T T C	2.9717	2.3044	1.3561	13.2691
T C T	2.9682	2.3051	1.3530	13.2573
T C C	2.9832	2.3345	1.3810	13.3544
C T C	2.9744	2.3127	1.3631	13.2927
C C C	3.0000	2.3538	1.4049	13.4400

on matrix G. We see that $^1\chi\chi$ parallels the apparent folding of the conformers, the "linear" all-*transoid* conformer having the smallest $^1\chi\chi$ and the all-*cisoid* conformer having the largest $^1\chi\chi$. In Figure 16 we show the plot of the index $^1\chi\chi$ against the first molecular moments for hexatriene conformers.

As another illustration of use of structural invariants derived from the geometric matrices we will consider the chair and the boat conformers of cyclohexane. In Table 17 we give the geometry distance matrix for the chair structure. The geometry matrix for the boat conformer differs only in the entries corresponding to the carbon atoms at the "bow" and the "stern" positions of the "boat." The molecular path numbers for the chair and the boat conformers are:

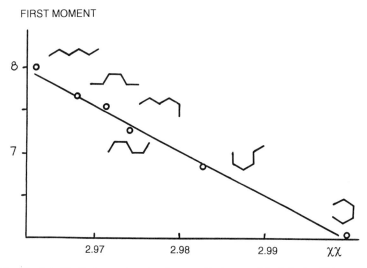

Figure 16.   The plot of the first moment against the generalized connectivity index for different conformers of the chain of length five.

Table 17.   The Geometry-Dependent Distance Matrix for
the Chair Conformation of Cyclohexane

$$\begin{pmatrix} 0 & 1 & 1.6330 & 1.9149 & 1.6330 & 1 \\ 1 & 0 & 1 & 1.6330 & 1.9149 & 1.6330 \\ 1.6330 & 1 & 0 & 1 & 1.6330 & 1.9149 \\ 1.9149 & 1.6330 & 1 & 0 & 1 & 1.6330 \\ 1.6330 & 1.9149 & 1.6330 & 1 & 0 & 1 \\ 1 & 1.6330 & 1.9149 & 1.6330 & 1 & 0 \end{pmatrix}$$

| chair | 3.0000 | 2.3600 | 1.4097 | 0.5669 | 0.1154 |
| boat | 2.9997 | 2.3639 | 1.4113 | 0.5675 | 0.1153 |

The difference between the two sets of path numbers is not large, but neither are the two structures very different: five out of six atoms in both molecules have the same coordinates. The corresponding ID (identification) numbers (given by the sum of all entries in the path sequence) are: 13.5421 and 13.4569.

6.9. MOLECULAR PROFILES

One can derive additional structural invariants from the row sums of the geometry matrix. Each row offers information about a single atom, hence allow one to extract novel *atomic* descriptors. Nonequivalent atoms in a structure will have as a rule different row sums. To obtain a *molecular* invariant we sum all of the row sums, or alternatively all of the entries in the matrix (or only entries above the diagonal, since the matrix is symmetrical). The derived number corresponds to the Wiener number of the graph theoretical distance matrix. The average atomic row sum is obtained by dividing the total sum by the number of rows. On the other hand, the first eigenvalue of the matrix G is bounded from below by the smallest and from the above by the largest row sum. In structures in which all of the rows have the same value (i.e., all of the atoms are equivalent, as in the case of the chair conformer of cyclohexane) the row sum will yield exactly the first eigenvalue. When the rows do not differ dramatically, the first eigenvalue will correlate with the average row sum of the matrix. We can now understand why the first eigenvalue of D/D matrices is a good measure of the molecular folding. In highly folded structures the row sums will be relatively small, because there will be many small entries for the interatomic distances while the graph theoretical distance continues increasing as the size of the structure increases. Consequently, in comparison with the row sums for a less folded structure the ϕ index for the former will be relatively small.

Individual row sums of a matrix do not represent an invariant, as mentioned, but the average row sum assigns a *single* number to a molecule that is a structural invariant. This number, except for normalization, can be viewed as the generalized "Wiener number." For molecular graphs, the Wiener number has been found to be a very useful molecular descriptor. It stands to reason to expect that its counterpart, the generalized Wiener number, or the average row sum derived from the G matrix, may similarly be found useful. We will refer to the average row sum of the geometry matrix as ^1D. Label D reminds us that this invariant was derived from interatomic distances.

Can we derive additional structurally related invariants? A single invariant can hardly be expected to suit diverse applications. A way to get additional invariants is to consider powers of the interatomic separations.[248–250] Let us write the G matrix:

$$G = \begin{pmatrix} g_{1,1} & g_{1,2} & g_{1,3} & g_{1,4} & \cdots & g_{1,n} \\ g_{2,1} & g_{2,2} & g_{2,3} & g_{2,4} & \cdots & g_{1,n} \\ g_{3,1} & g_{3,2} & g_{3,3} & g_{3,4} & \cdots & g_{1,n} \\ \cdots & \cdots & \cdots & \cdots & & \cdots \\ g_{n,1} & g_{n,2} & g_{n,3} & g_{n,4} & \cdots & g_{1,n} \end{pmatrix}$$

Then we define the elements of the matrix ^kG to be given by the powers $(g_{i,j})^k$. Hence:

$$^kG = \begin{pmatrix} g_{1,1}^k & g_{1,2}^k & g_{1,3}^k & g_{1,4}^k & \cdots & g_{1,n}^k \\ g_{2,1}^k & g_{2,2}^k & g_{2,3}^k & g_{2,4}^k & \cdots & g_{1,n}^k \\ g_{3,1}^k & g_{3,2}^k & g_{3,3}^k & g_{3,4}^k & \cdots & g_{1,n}^k \\ \cdots & \cdots & \cdots & \cdots & \cdots & \cdots \\ g_{n,1}^k & g_{n,2}^k & g_{n,3}^k & g_{n,4}^k & \cdots & g_{1,n}^k \end{pmatrix}$$

Observe that elements of ^kG are not obtained from the product of matrix G with itself using the standard matrix multiplication rules, thus $^kG \neq (G)^k$.

From the ^kG matrix we can extract invariant ^kD as the average row sum of ^kG. Using different k values we obtain a family of structurally related invariants:

$$^1D, {}^2D, {}^3D, {}^4D, {}^5D, {}^6D, {}^7D, {}^8D, \ldots$$

One can easily verify that in the case of *cisoid* and *transoid* butadiene we obtain from the previously shown matrices G_c and G_t the following sequences of the average row sum ^kD, respectively:

	cisoid butadiene	*transoid* butadiene
^1D	4.23205	4.55492
^2D	6.50000	8.00000
^3D	10.69615	13.35821
^4D	18.50000	35.00000
^5D	33.08846	81.90936
^6D	60.50000	200.00000
	. . .	. . .

As is to be expected, the ^kD values for the *cisoid* isomer are smaller than the corresponding values for the *transoid* isomer. The corresponding matrices differ only in the element (1,4) and (4,1). We see that the distance between the end carbon atoms is smaller in the *cisoid* case than in the *transoid* case. By raising the elements to higher powers, this difference is only to become more pronounced and eventually dominate the ^kD numbers for large k.

Clearly we can in this way generate many descriptors. Observe, however, that the magnitude of the successive members in the sequences ^kD increase as k increases. This is not a desirable aspect of a typical "power expansion." A straightforward way to curb this unlimited growth of the members of a sequence is to introduce suitable normalization. To obtain absolute convergence of the sequence of ^kD (i.e., convergence regardless of the size of the structure considered) one can select $1/k!$ as the normalization. This choice parallels the well-known expansion of the exponential function e^x. The coefficient $1/k!$ also appears in the Taylor expansion. In the case of *cisoid* and *transoid* butadiene the $1/k!$ normalized sequence of average row sums are

cisoid	transoid
4.23205	4.55492
3.25000	4.00000
1.78269	2.22637
0.77083	1.45833
0.27574	0.68258
0.08403	0.27778
. . .	. . .

Table 18. The Molecular Profiles for Nine-Membered Puckered Rings[a]

	B	C	B B	C C	B C	C B	T B	T C
^1D	22.14	22.41	21.37	23.45	23.78	22.53	21.36	22.54
^2D	33.55	34.46	30.71	38.15	38.69	34.47	31.50	34.82
^3D	36.23	37.73	31.00	44.60	44.82	37.22	33.51	38.53
^4D	30.76	32.43	24.34	41.21	40.78	21.62	28.38	33.75
^5D	21.63	23.02	15.70	31.66	30.71	12.67	20.11	34.62
^6D	13.01	13.94	8.61	20.88	19.78	6.45	12.28	15.41
^7D	6.34	7.36	4.11	12.08	11.15	2.90	6.60	8.46
^8D	3.20	3.45	1.74	6.23	5.59	1.17	3.17	4.14
^9D	1.35	1.46	0.66	2.91	2.53	0.43	1.37	1.82
^{10}D	0.52	0.56	0.23	1.24	1.04	0.14	0.54	0.73
^{11}D	0.18	0.20	0.07	0.48	0.39	0.04	0.20	0.27
^{12}D	0.06	0.06	0.02	0.18	0.14	0.01	0.07	0.09
^{13}D	0.02	0.02	0.01	0.06	0.05		0.02	0.03
^{14}D	0.01	0.01		0.02	0.01		0.01	0.01
^{15}D				0.01				

[a]C = chair, B = boat, CC = chair–chair, BB = boat–boat, CB = chair–boat, BC = boat–chair, TB = twist–boat, TC = twist–chair.

Table 19. The Molecular Profiles for *cata*-Condensed
Benzenoid Hydrocarbons Illustrated in Figure 18

	1	2	3	4	5
^1D	51.67	49.28	48.41	45.65	45.51
^2D	100.50	88.83	85.50	73.00	72.00
^3D	154.09	124.09	117.44	87.81	84.97
^4D	197.38	143.71	134.60	86.00	81.00
^5D	217.12	142.59	132.95	71.39	64.99
^6D	208.98	123.79	115.45	51.51	45.05
^7D	178.59	95.51	89.37	32.89	27.50
^8D	137.12	66.30	62.36	18.83	14.99
^9D	95.51	41.81	39.57	9.78	7.38
^{10}D	60.85	24.15	23.00	4.64	3.32
^{11}D	35.71	12.87	12.33	2.03	1.37
^{12}D	19.42	6.36	6.13	0.82	0.52
^{13}D	9.83	2.93	2.84	0.31	0.06
^{14}D	4.66	1.26	1.23	0.11	0.02
^{15}D	2.07	0.51	0.50	0.04	0.01
^{16}D	0.87	0.20	0.19	0.01	
^{17}D	0.35	0.07	0.07		
^{18}D	0.13	0.02	0.02		

We will refer to the normalized sequences of average interatomic distances and average higher powers of interatomic distances as *molecular profiles* and will continue to use the symbol ^kD for them.

We illustrated the construction of molecular profiles on 1,3-butadiene, a rather tiny molecule. The procedure is, however, general. It applies not only to arbitrary chemical structures in 3D space but also to an arbitrary object in 3D space that has been discretized (i.e., represented by a set of discrete points). In Table 18 we give molecular profiles for eight conformations of a nine-membered puckered C_9 ring illustrated in Figure 17. In Figure 18 we illustrated the molecular profiles (listed in Table 19) for *cata*-condensed PAH having four benzenoid rings and the same perimeter ($P = 18$).

The profiles of different benzenoids show considerable differences. The profiles can be used to discuss the degree of the relative similarity among benzenoids. In Table

Table 20. The Similarity/Dissimilarity Matrix for the
Five Benzenoids of Figure 18

	1	2	3	4	5
1	0	183	201	333	346
2		0	19	151	164
3			0	24	65
4				0	41
5					0

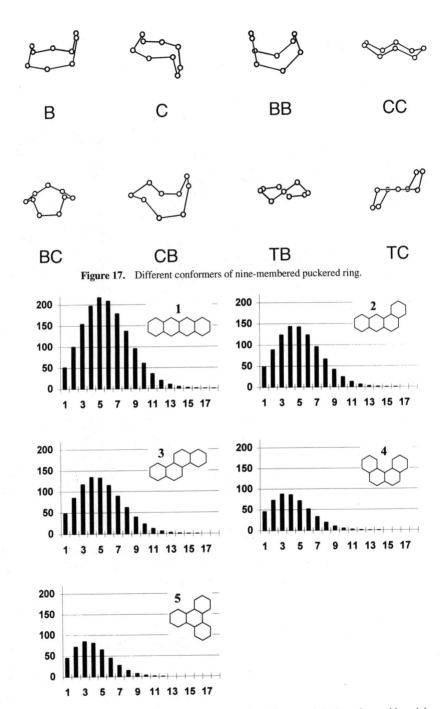

Figure 17. Different conformers of nine-membered puckered ring.

Figure 18. The molecular profiles for *cata*-condensed polycyclic aromatic hydrocarbons with periphery
$P = 18$.

20 we show the similarity/dissimilarity matrix for the five benzenoids of Figure 18. The small entries in Table 20 indicate similar structures while the large entries point to dissimilar structures. In case of benzenoids with perimeter $P = 18$, one finds that the largest entry in the similarity/dissimilarity table belongs to the pair tetracene (**1**)–triphenylene (**5**) signifying the least similar structures. As the most similar benzenoids with perimeter $P = 18$ one finds:

benzanthracene (**2**)–chrysene (**3**)
chrysene (**3**)–benzphenanthrene (**4**)
benzphenanthrene (**4**)–triphenylene (**5**)

6.9.1. Bond Profiles

If one is interested in the characterization of molecular local features, one should confine the matrix ^kD only to atoms that define the local environment of interest. When only atoms belonging to a local environment are included in the construction of the matrices, one obtains the sequence

$$L_1, L_2, L_3, L_4, L_5, L_6, L_7, L_8, \ldots$$

Here $L_k = \Sigma R_{k,i}/n$ is the average row sum of the matrix ^kD. If only atoms at the molecular periphery are included in the construction of the matrices, one obtains the sequence

$$S_1, S_2, S_3, S_4, S_5, S_6, S_7, S_8, \ldots$$

that will be discussed in the next section.

Moreover, when constructing the geometry-dependent matrix D_1 one is not restricted to consider only atoms. One can introduce any number of geometric points along chemical bond and in this way obtain sequences

$$B_1, B_2, B_3, B_4, B_5, B_6, B_7, B_8, \ldots$$

$$b_1, b_2, b_3, b_4, b_5, b_6, b_7, b_8, \ldots$$

that characterize the bonds of a molecule.[251] Here $B_k = \Sigma R_{k,i}/n$ is the average row sum and $b_k = \Sigma R_{k,i}/n^2$ is the average element of the matrix ^kD. The sequence b_k differs only in normalization. It is of interest when one compares models using a different number of points to represent a structure. Bond profiles constitute an important generalization of atomic molecular profiles since they allow one to arrive at a characterization of molecular connectivity. The geometric matrix does not explicitly record the information on the connectivity, which has to be inferred from the interatomic separations given.

Recently we elaborated the model by considering the connectivity in two polyhedra, cuboctahedron and twist cuboctahedron.[252] In Table 21 we give the profiles for two closely related cuboctahedra (shown in Figure 19). We enlarged the number of

Table 21. The Molecular Profiles for Two Closely
Related Cuboctahedra

	Cuboctahedron	Twist form
^1D	43.62453	43.61378
^2D	30.00000	30.00000
^3D	14.64826	14.65142
^4D	5.59722	5.59722
^5D	1.76451	1.76298
^6D	0.47465	0.47348
^7D	0.11156	0.11100
^8D	0.02330	0.02311
^9D	0.00438	0.00433
^{10}D	0.00075	0.00074

"ghost" sites along each edge of the two polyhedra gradually, each time reducing the separations between the points by half. As the number of the points that represent each polyhedron increased from $n = 12$ to several thousands, a convergence of the "per site" bond profiles b_i could be observed. In Figure 20 we show cuboctahedron as represented by 7, 15, and 31 points inserted into each edge. The computations were performed by inserting up to 127 points along each edge. As seen, relatively small numbers of points inserted along each edge offer a fair representation of molecular bonding and may suffice for many practical considerations.

6.9.2. Contour Profiles

The illustration of cuboctahedron immediately suggests an even broader generalization of the molecular profile approach to 3D molecular structure. Instead of confining the "ghost" points along the bonds in a molecule, the edges of embedded molecular graph, or the edges of a polyhedron, we can "distribute" the points inside the molecular volume or alternatively on the molecular surface. If the number of points distributed is large enough, we can expect a fair characterization of a molecule. Hence, the sequences

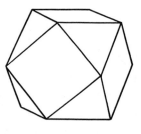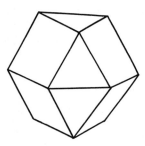

Figure 19. Cuboctahedron and twist cuboctahedron.

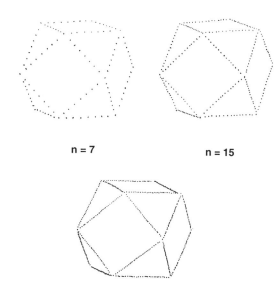

n = 7   n = 15

n = 31

Figure 20. Cuboctahedron as represented by insertion of 7, 15, and 31 equidistant points along each edge.

$$C_1, C_2, C_3, C_4, C_5, C_6, C_7, C_8, \dots$$

$$c_1, c_2, c_3, c_4, c_5, c_6, c_7, c_8, \dots$$

where $C_i = S\,R_{k,i}/n$ and $c_i = S\,R_{k,i}/n^2$ will lead to characterization of the molecule as defined by its contour. In the case of planar models, molecular contours will be closed curves and in the case of a 3D molecular model the contours will represent molecular surfaces. As discussed by Mezey,[253-255] if such contours are based on electron density, one can consider different threshold for the density. One then obtains contours of different form even for a single molecule. These contours can be connected or disconnected. All such cases can be treated by our approach equally well. We will illustrate the approach on the water molecule.[256] We have selected van der Waals radii of hydrogens and oxygen to define the molecular contours. First we will consider H_2O as a 2D model and then as a 3D model.

In Figure 21 we illustrate the structure of a water molecule as a projection on a plane. We have taken as the van der Waals radius of hydrogen 1.2 Å and as the van der Waals radius of oxygen 1.4 Å. The covalent O–H bond is taken to be 0.95 Å, and the H–O–H angle is taken as 105°. These are all rather approximate values as quoted by Watson,[257] but they are good enough to serve to illustrate the approach. The "jump" from the initial molecular profile $d_1, d_2, d_3, d_4, d_5, d_6, d_7, d_8$, to the contour profile $c_1, c_2, c_3, c_4, c_5, c_6, c_7, c_8, \dots$ corresponds to a "jump" from the discrete model to the continuum.

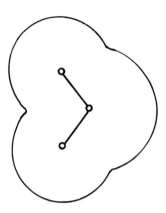

Figure 21. van der Waals contours for a planar model of H_2O.

In the cuboctahedron case, we were able to introduce the large number of the sites to represent each edge between the vertices of the polyhedron using the simple arithmetic mean to generate coordinates of new sites. In contrast, here such an approach is not possible since we want the new points to be "everywhere" inside the molecule, not only along the bonds. To arrive at approximately uniformly distributed points in the interior of the van der Waals contour of the molecule we select the coordinates of the points at random and then check that indeed the point is inside the molecular interior. In Figure 22 we illustrate distributions of 1000 and 5000 random points that represent a planar model of the H_2O molecule (i.e., the projection of H_2O on a plane).

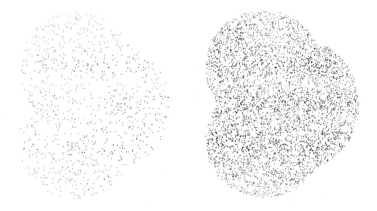

Figure 22. Representation of a planar model of H_2O by 1000 and 5000 random points, respectively.

The first question to consider is to find how many points need be taken in order to reach an apparent convergence for the contour sequences. As is known, the Monte Carlo calculations, of which the above is an illustration, converge but rather slowly. In order to increase the accuracy of the results by a factor of 10, one has to increase the number of points by a factor of 100. We found for the above case that after 2000, points fluctuations between successive calculations decrease dramatically.

The conditions for inclusion/exclusion of a randomly selected point are relatively simple. Each atom is represented by a circle:

oxygen $\qquad x_2 + y_2 = R_O^2$

hydrogen 1 $\qquad (x + a)^2 + (y + b)^2 = R_H^2$

hydrogen 2 $\qquad (x - a)^2 + (y - b)^2 = R_H^2$

where $-a$, $-b$ and $+a$, $+b$ are the coordinates of the center of two hydrogens relative to the oxygen that is placed in the center of the Cartesian coordinate system. The three equations lead to the constraints, shown as inequalities:

$$x^2 + y^2 \leq R_O^2$$

$$(x + a)^2 + (y + b)^2 \leq R_H^2$$

$$(x - a)^2 + (y - b)^2 \leq R_H^2$$

If the first inequality is satisfied, the random point is included in the list of new sites; if not, the second inequality is tested. Again, if the second inequality is satisfied, the point qualifies to be on the list; otherwise the third inequality is tested. If the third inequality is satisfied, the point is included; otherwise it is discarded and new two random numbers are tested to see if they will produce acceptable x, y random coordinates. Even though the three circles overlap, the sequential use of the conditions gives to the whole interior region the same weight.

When we consider H_2O as a 3D model, all that need be changed are the conditions for inclusion/exclusion of random points. Instead of overlapping circles we now have overlapping spheres, i.e., the conditions become:

$$x^2 + y^2 + z^2 \leq R_O^2$$

$$(x + a)^2 + (y + b)^2 + z^2 \leq R_H^2$$

$$(x - a)^2 + (y - b)^2 + z^2 \leq R_H^2$$

The two hydrogen atoms and oxygen are in the x, y coordinate plane. In Table 22 we collected for the leading entry of the contour profile the calculated averages in order to illustrate the degree of oscillations that reflect use of random numbers. As we see after use of 12,500 random points, the resulting c_1 profile fluctuates around the fourth

Table 22.   Variations in the Leading Entry of the Contour Profile for
H$_2$O as the Number of Random Points Increases

Number of points	c_1	Number of points	c_1
500	0.365020	7,000	0.359090
1000	0.360752	7,500	0.359712
1500	0.359130	8,000	0.359283
2000	0.358998	8,500	0.359194
2500	0.358495	9,000	0.359205
3000	0.359606	9,500	0.359247
3500	0.359646	10,000	0.359758
4000	0.359939	10,500	0.359155
4500	0.360070	11,000	0.358665
5000	0.359460	11,500	0.359039
5500	0.359982	12,000	0.358981
6000	0.358818	12,500	0.359006
6500	0.358502		

decimal place. Similar behavior is also found for other (higher) terms of the contour profile. The molecular contour profiles for the two models of H$_2$O are summarized below:

	2D		3D	
c_1	0.399227		0.359006	
c_2	0.980663	10^{-1}	0.796870	10^{-1}
c_3	0.181868	10^{-1}	0.134405	10^{-1}
c_4	0.275134	10^{-2}	0.186093	10^{-2}
c_5	0.353948	10^{-3}	0.220335	10^{-3}
c_6	0.397559	10^{-4}	0.228898	10^{-4}
c_7	0.397164	10^{-5}	0.212413	10^{-5}
c_8	0.357791	10^{-6}	0.178432	10^{-6}

First we see that the magnitudes for the profile amplitudes are small. This is not surprising, as water is a small molecule. We notice also that the amplitudes in the contour profile for the 3D model are smaller than the corresponding amplitudes in the 2D model. This is not so surprising. One can view the 3D model to consist of many 2D ones by slicing it horizontally. The cross sections of the spheres so obtained are then circles of smaller radius, hence, each cross section will result in smaller amplitudes for the corresponding contour profile. Apparently the contributions arising from the points that are in different layers of the "sliced" 3D model cannot reverse the dominant role of the "horizontal" contributions. This is in part because only the most distant points make larger contribution, and the number of such points (close to the molecular surface) is relatively small (compared to the number of points in the molecular interior).

6.10. SHAPE PROFILES

Shape is a very basic and generally well-understood concept. Yet its quantitative characterization, and even a qualitative characterization, appears elusive. In contrast to the idea of molecular structure that can be defined in terms of the components selected to emphasize the content of the structure, the shape has yet to be similarly defined. While we can differentiate shapes well and discriminate even between the smaller variations of shapes, hitherto no pragmatic definition of shape has emerged. It is therefore not surprising that not much progress has been reported concerning the representation and the characterization of shapes in general, including molecular shapes in particular.

A good starting point in trying to better understand shape is to list molecular properties that critically depend on shape. In this way we may be able to narrow down the task of quantitative representation of molecular shapes. By examining closely selected properties attributed to the shape of an object, we may hope to characterize the shape, at least partially. Such an approach corresponds to solving an inverse problem that, as is known, is not always possible. The classical example of an inverse problem is the question posed by Kac[258]: "Can you hear the shape of a drum?" The question asks if one can determine the *shape* of a drum by knowing all of the sounds (i.e., the frequencies and the amplitudes of the sounds) that it produces. In this case the answer is: No. That is, drums of different shape can produce precisely the same sound.

When the same problem is discretized and formulated in graph theoretical version, then again the answer is: No. Here graph eigenvalues correspond to the frequencies of the sound.[259] The fact that there are isospectral graphs signifies that the inverse problem is not unique. Isospectral graphs are graphs having the same set of eigenvalues. Several pairs of isospectral graphs are illustrated in Figure 23. The literature on isospectral graphs is abundant. For references see the papers cited in the select list of publications.[260–265]

Molecular properties, whose variations can be attributed to variations in molecular shape, include: odor, taste, optical dichroism (the "octant rule"), chirality, and drug–receptor interaction. In 1920 Ružička forwarded a theory that the character of an odoriferous substance is determined by its molecular shape, while the variations of this character depend on the osmophoric groups in a molecule.[12] This presented but one illustration of Emil Fisher's "lock and key" model for interaction of drugs and enzymes.[13]

The properties listed above illustrate interactions between *two* partners, and are not manifested by a single *isolated* object. Thus, in structure–activity studies, which include odor and taste, the shape of a small molecule has to match the cavity in the large molecule (receptor of a protein). In the case of optical dichroism the "other" component is circularly polarized light (electromagnetic radiation). Finally, chirality manifests itself only in the presence of a medium that can differentiate the enantiomers.

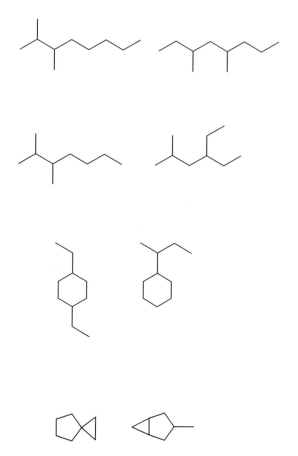

Figure 23.   A selection of isospectral graphs.

What is to be expected from a quantitative measure of shape? From the illustrations mentioned above, it would seem appropriate to have some index, or a couple of indices, that are sensitive to variations in the shapes. Hopefully then apparently similar shapes will have similar magnitudes for such indices, while widely different shapes will have sufficiently different characterization. With such indices one could quantitatively characterize the degree of matching between donor and receptor, or differentiate smaller variations among similar shapes.

A number of simple (and necessarily crude) measures of molecular shape have been outlined in the literature. One such measure was outlined in discussions of the chromatographic data for PAH.[266,267] These planar hydrocarbons have been "boxed" into the smallest possible rectangles (Figure 24). The ratio L/B, of the length (L) to the base (B) of the rectangle, was then taken as an index of the shape. Indeed, L/B is a

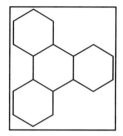 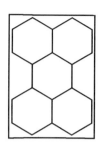

Figure 24. Illustration of "boxing" benzenoid hydrocarbon into a rectangle of length L and width B.

shape index, since the quotient is independent of the size. Moreover, for rectangles this may indeed be the simplest and at the same time a very true shape index. Geometrically speaking, however, benzenoids are not rectangles but are formed by fused hexagons. What is more important, the model of rectangles assumes that shapes of polycyclic aromatic compounds can be well characterized by *two* parameters (L and B). It is not difficult to find benzenoids of different shape that have the same L and B. The L/B index has some computational limitations. Computationally this approach associates with different molecules different degrees of accuracy. The ratio L/B describes better the shape of some benzenoids, while for others it does not offer equally good approximation.

We will show that the interatomic distances and their powers offer a better characterization of molecular shapes. The shape of an object is defined by its exterior, surface, or contour in the case of 2D objects. Hence, when characterizing molecular shape one should consider only the interatomic separations between atoms at the molecular periphery, excluding atoms in the interior part of the structure. In the case of *cata*-condensed benzenoids, all carbon atoms are on the periphery, hence the shape profile and the molecular profile of such compounds are the same. In the case of *peri*-condensed benzenoids, only carbon atoms at the periphery will be taken into account.

If we extend the similarity/dissimilarity analysis to all benzenoids having $P = 18$ (including contours of *peri*-condensed systems **6–12**), using shape profiles for their characterization we find as the least similar pairs:

1–5	tetracene–triphenylene
1–4	tetracene–benzo[c]phenanthrene
1–7	tetracene–benzo[d]pyrene
1–8	tetracene–perylene
1–10	tetracene–benzo[c,d,e]pyrene
1–11	tetracene–trinagulene
5–9	triphenylene–anthanthene

The *cata*-condensed benzenoids **1–5** having $P = 18$ have been shown in Figure 18 while the *peri*-condensed benzenoids **6–12** having $P = 18$ are shown in Figure 25. The

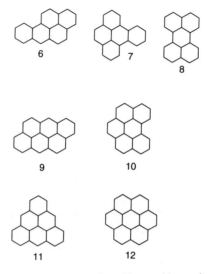

Figure 25.   The *peri*-condensed benzenoids examined.

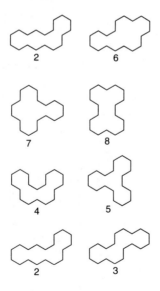

Figure 26.   The most similar benzenoids having perimeter $P = 18$.

observed dissimilarities are not surprising. Tetracene (**1**) appears the least similar to most benzenoids because of the dominant role of the longer interatomic separations.

Figure 26 illustrates the most similar pairs of benzenoids having perimeter $P = 18$. The similarity of benzo[c]phenanthrene (**4**)–triphenylene (**5**) and benzanthracene (**2**)–chrysene (**3**) was mentioned earlier; benzanthracene (**2**)–benzo[*a*]pyrene (**6**) are another such pair the similarity of which is apparent from visual inspection. However, the high similarity of benzo[*d*]pyrene (**7**)–perylene (**8**) is somewhat surprising. Indeed, the pair do not show visual similarity, but apparently have a similar distribution of interatomic separations, the descriptor that is critical in calculating their profiles. Hence, the most similar structures need not necessarily show apparent similarity. This is not the failure of the approach or the failure of the characterization of molecules but is a consequence of difficulties in separating different criteria when comparing structures. Our visual comparison is more likely to capture details of the contours of the molecule rather than an intrinsic property such as the interatomic separations.

6.10.1. Chromatographic Retention Indices

We will consider a set of $n = 16$ benzenoid hydrocarbons for which an analysis of the gas chromatographic retention indices I_N was reported. Among these there are nine *cata*-condensed benzenoid hydrocarbons (illustrated in Figure 27) and seven *peri*-condensed benzenoids (of Figure 28) for which chromatographic data were analyzed.[266,267] In *cata*-condensed benzenoids all carbon atoms are on the molecular periphery, hence the shape profile and the molecular profile (based on all carbon atoms) are the same (Table 23).

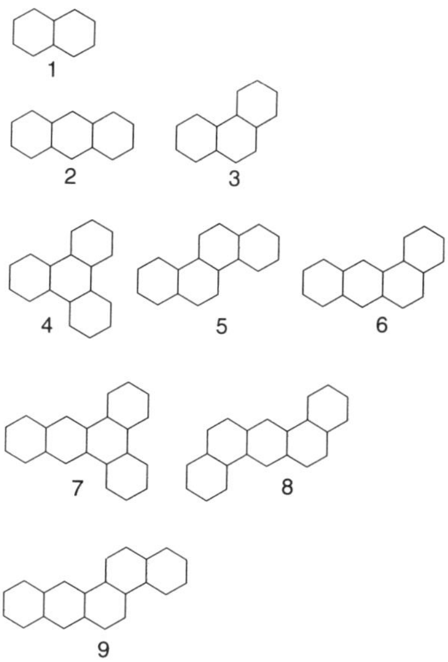

Figure 27. The structures of *cata*-condensed polycyclic aromatic hydrocarbons studied: **1**, naphthalene; **2**, anthracene; **3**, phenanthrene; **4**, triphenylene; **5**, chrysene; **6**, benzanthracene; **7**, dibenz(*a*,*c*)anthracene; **8**, dibenz(*a*,*h*)anthracene; **9**, benzo(*b*)chrysene.

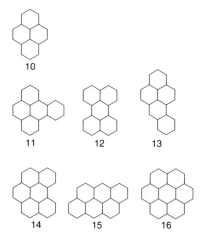

Figure 28. The structures of *peri*-condensed polycyclic aromatic hydrocarbons studied: **10**, pyrene; **11**, benzo(*d*)pyrene; **12**, perylene; **13**, benzo(*a*)pyrene; **14**, benzo(*c,d,e*)pyrene; **15**, anthranthrene; **16**, coronene.

We want to discuss the differences of the characterization of benzenoid hydrocarbons by the molecular profiles and by the shape profiles. Let us consider *peri*-condensed benzenoids more closely. In Table 24 are listed for the selected *peri*-condensed benzenoids (illustrated in Figure 28) the numerical values for the molecular

Table 23. The Molecular Profiles ^kD for Selected *cata*-Condensed Benzenoids
(Shown in Figure 27)

	1	2	3	4	5	6	7	8	9
^1D	17.85	31.19	32.54	45.51	48.41	49.28	65.59	69.89	70.52
^2D	20.50	44.64	50.00	72.00	85.50	88.83	124.95	148.00	151.95
^3D	17.59	48.49	59.49	84.97	117.44	124.09	180.99	245.50	258.30
^4D	12.31	43.29	58.81	81.00	134.60	143.71	214.50	339.47	367.87
^5D	7.31	33.00	49.85	64.99	132.95	142.59	216.10	403.45	451.82
^6D	3.78	21.98	36.95	45.05	115.45	123.79	189.65	420.51	487.67
^7D	1.73	13.01	24.30	27.50	89.37	95.51	147.55	390.19	469.11
^8D	0.70	6.93	14.34	14.99	62.36	66.30	103.13	326.17	406.71
^9D	0.26	3.35	7.67	7.38	39.57	41.81	65.43	248.00	320.71
^{10}D	0.08	1.48	3.74	3.32	23.00	24.15	38.43	172.88	231.79
^{11}D		0.61	1.68	1.37	12.33	12.87	20.34	111.24	154.56
^{12}D		0.23	0.70	0.52	6.13	6.36	10.10	66.45	95.62
^{13}D		0.08	0.27	0.19	2.84	2.93	4.67	37.03	55.15
^{14}D		0.03	0.10	0.06	1.23	1.26	2.02	19.34	29.79
^{15}D				0.02	0.50	0.51	0.82	9.49	15.12
^{16}D				0.01	0.20	0.20	0.32	4.40	7.24
^{17}D					0.07	0.07	0.11	1.93	3.28
^{18}D					0.02	0.02	0.04	0.80	1.41
^{19}D							0.01	0.32	0.58
^{20}D								0.12	0.22
^{21}D								0.04	0.08
^{22}D								0.02	0.03

Table 24. The Molecular Profiles ^kD for Selected *peri*-Condensed Benzenoids (Shown in Figure 28)

	10	11	12	13	14	15	16
^1D	36.48	52.20	52.48	54.51	59.32	61.18	66.77
^2D	52.00	84.65	86.00	95.45	98.77	107.50	114.00
^3D	55.41	102.44	105.70	128.62	122.51	143.83	144.54
^4D	48.08	100.35	105.62	143.95	122.87	158.69	147.75
^5D	35.48	82.99	89.39	138.82	103.93	150.24	127.06
^6D	22.87	59.50	65.75	117.89	76.15	125.06	94.45
^7D	13.13	37.68	42.79	89.48	49.25	93.04	61.86
^8D	6.80	21.38	24.97	61.38	28.51	62.64	36.20
^9D	3.21	10.99	13.20	38.39	14.94	38.50	19.15
^{10}D	1.40	5.16	6.38	22.04	7.15	21.77	9.24
^{11}D	0.56	2.23	2.84	11.70	3.15	11.40	4.10
^{12}D	0.21	0.90	1.17	5.76	1.28	5.55	1.68
^{13}D	0.07	0.33	0.45	2.65	0.49	2.53	0.64
^{14}D	0.02	0.12	0.16	1.14	0.17	1.08	0.23
^{15}D	0.01	0.04	0.05	0.46	0.06	0.43	0.08
^{16}D		0.01	0.02	0.18	0.02	0.17	0.02
^{17}D			0.01	0.06	0.01	0.06	0.01
^{18}D				0.02		0.02	
^{19}D				0.01		0.01	

profiles k_D. Similarly, in Table 25 the corresponding shape profiles k_S for the same *peri*-condensed benzenoids are given. Clearly the molecular profiles k_D and the shape profiles k_S are different since in the latter the atoms in the molecular interior do not make contributions. We assumed all C–C distances to be equal to one.

It is instructive to compare the shape profiles of Table 26. The smallest profiles are found for pyrene (**10**), benzo(*d*)pyrene (**11**), and perylene (**12**), and the largest profiles appear to be those of benz(*a*)pyrene (**13**) and anthanthrene (**15**). Why is this the case? Why, for example, isn't the profile of coronene (**16**) the largest, when this molecule has the largest number of fused rings? The answer is not difficult to find. The profile ^kD does not directly depend on the molecular size, but on interatomic separations. In benzo(*d*)pyrene and perylene the largest C·····C distance is only $\sqrt{28}$, while for anthanthrene it is $\sqrt{43}$. In coronene the largest C·····C distance is again only $\sqrt{28}$. However, while in benzo(*d*)pyrene there are fewer such distances, in coronene there are many, which results in the larger magnitudes for the components of the profile of coronene. Clearly as k increases, the larger interatomic separations assume the dominant role in defining the corresponding ^kD values in the profile since successively raising numbers to higher power favors the larger numbers.

The simple regression equation using the connectivity index χ as the molecular descriptor for the smaller benzenoids is given in Table 26. To improve the regression, Radecki, Lamparczyk, and Kaliszan[266,267] considered as an additional descriptor the ratio L/B of the length L and the width B of the rectangle within which the molecule

Table 25.   The Shape Profiles ^kS for Selected *peri*-Condensed Benzenoids
(Shown in Figure 28)

	10	11	12	13	14	15	16
^1S	33.93	49.13	49.24	51.29	52.66	54.29	56.11
^2S	51.50	83.33	84.00	93.83	95.33	103.50	108.00
^3S	57.53	104.35	106.51	130.65	126.20	147.40	150.50
^4S	51.58	104.78	109.00	149.71	132.78	170.26	165.00
^5S	38.90	88.19	93.97	146.85	116.29	166.71	149.29
^6S	25.44	64.02	70.10	126.28	87.37	142.27	115.10
^7S	14.74	40.91	46.12	96.75	57.54	107.83	77.37
^8S	7.68	23.37	27.13	66.84	33.75	73.59	46.14
^9S	3.64	12.07	14.43	42.02	17.85	45.71	24.74
^{10}S	1.59	5.69	7.01	24.23	8.60	26.05	12.05
^{11}S	0.64	2.47	3.13	12.89	3.80	13.71	5.38
^{12}S	0.24	0.99	1.29	6.37	1.56	6.71	2.22
^{13}S	0.08	0.37	0.50	2.93	0.59	3.06	0.85
^{14}S	0.03	0.13	0.18	1.26	0.21	1.31	0.30
^{15}S		0.04	0.06	0.51	0.07	0.53	0.10
^{16}S		0.01	0.02	0.20	0.02	0.20	0.03
^{17}S			0.01	0.07	0.01	0.07	0.01
^{18}S				0.02		0.03	
^{19}S				0.01			

Table 26.   The Regression Equations for the Chromatographic
Retention Data for Smaller Benzenoids

(1) $I_N = 609.68\ \chi - 1345.14$

$S = 123.0, R = 0.9947, F = 1301$

(2) $I_N = 611.933\ \chi + 275.99\ L/B - 1736.85$

$S = 98.0, R = 0.9969,$ and $F = 1030$

(3) $I_N = 74.01\ ^1D + 598.39$

$S = 251.0, R = 0.9776, F = 302$

(4) $I_N = 133.61\ ^1D - 25.68\ ^2D - 145.71$

$S = 125.1, R = 0.9949, F = 629$

(5) $I_N = 235.57\ ^1D - 116.01\ ^2D + 27.36\ ^3D - 604.26$

$S = 120.0, R = 0.9957, F = 456$

(6) $I_N = 74.01\ ^1\Omega + 1598.40$

(7) $I_N = 74.01\ ^1\Omega - 25.68\ ^2\Omega + 1598.40$

(8) $I_N = 74.01\ ^1\Omega - 25.68\ ^2\Omega + 27.36\ ^3\Omega + 1598.40$

can be embedded. Then one obtains equation (2) of Table 26. Since equation (1) is already very good, it is not easy to improve on it. The improvement of the regression (1) is marginal. Although the standard error S dropped by 20%, the Fisher ratio F, which is an independent measure of the quality of the regression, has somewhat decreased. A decrease in F indicates a decrease in the significance of the regression. Other researchers continued to use the L/B ratio as a descriptor in chromatographic studies of planar benzenoid compounds.[268] The L/B variable appears to supplement the connectivity index χ in the correlation of the chromatographic data of benzenoids. However, the interpretation that the chromatographic data are shape dependent, as we will see, needs a clarification.[269]

Let us test the new distance-dependent descriptors to see how well they can describe the chromatographic data of benzenoids. The use of ^1D in a simple regression gives equation (3) of Table 26. Despite the fact that the correlation coefficient R is high, the above equation is inferior when compared to equation (1). When ^1D and ^2D are used together, we obtain equation (4) of Table 26. The regression equation (4) is comparable in quality to the regression equation (1) based on the connectivity index but is associated with a smaller Fisher ratio F. Use of three descriptors finally gives equation (5) of Table 26.

If, instead of ^1D, ^2D, and ^3D, we employ the corresponding orthogonalized descriptors, we obtain the following regression equations (6)–(8) of Table 26. Observe the constancy of the coefficients appearing in equations (6)–(8) in contrast to the corresponding regression equations (3)–(5), which did not use orthogonal descriptors.

In conclusion, we have to admit that the geometry-dependent descriptors ^3D do not lead to as good a correlation as that based on χ and L/B. The reason is that ^kD are *global* molecular descriptors, while χ is a *local*, bond additive descriptor, which apparently better reflects the structure–chromatographic retention relationship. However, the index ^kD can help us to clarify the question:

6.10.2. Are Chromatographic Retention Indices Shape Dependent?

We would like to critically examine the interpretation of the regression equation (2) that it supports a notion that the chromatographic retention index for benzenoids is shape dependent. As one can see from equation (1), most of the correlation between the structure and the chromatographic retention index as the property is described by the connectivity index that is bond additive and size dependent. The conclusion therefore ought to be that the chromatographic retention values are mostly bond additive and size dependent. Because of this pronounced bond-additivity, clearly we can understand why ^kD indices that are global are inferior in this particular application.

Are the chromatographic retention indices not shape dependent? It appears that the retention indices of benzenoids should not be viewed as a shape-dependent property, despite the fact that the regressions of the chromatographic indices were improved by the use of L/B. We can justify this claim by comparing the regressions using molecular descriptors that characterize better molecular shape than L/B. If the molecular property is shape dependent, then use of shape profiles ^1S, ^2S, ^3S ... should give a better regression than use of the corresponding molecular profiles ^1D, ^2D, ^3D .... In the former only the carbon atoms on the molecular periphery contribute in the construction of the average interatomic distances and the average powers of the distances. For *cata*-condensed benzenoids, both ^kD and ^kS are identical, but for *peri*-condensed benzenoids there is a difference between the two sets of descriptors. When we use ^kS descriptors in the regressions of the chromatographic retention indices we obtain:

(9) $I_N = 74.00\ ^1S$ $+ 766.67$

(10) $I_N = 142.70\ ^1S\ -27.68\ ^2S$ $- 139.03$

(11) $I_N = -271.72\ ^1S\ -309.67\ ^2S - 95.72\ ^3S$ $+ 2073.38$

The corresponding values for S, R, and F are, respectively:

$$
\begin{array}{lll}
S = 470.1 & R = 0.9189 & F = 76 \\
S = 462.84 & R = 0.9274 & F = 40 \\
S = 211.66 & R = 0.9864 & F = 144
\end{array}
$$

A comparison with the corresponding values of equations (3)–(5) shows that the descriptors ^1D, ^2D, ^3D, in which the atoms in the interior as well as those at the periphery contribute, are superior in describing chromatographic indices of benzenoids. The superiority of ^kD over ^kS descriptors becomes even more pronounced if we limit the analysis only to *peri*-condensed benzenoids for which the descriptors ^kD and ^kS differ. In Table 27 we compare the regressions using the shape descriptors ^kS and the "full" descriptors ^kD.

We have to conclude that the chromatographic retention index for planar benzenoids is inherently not a shape-dependent property, i.e., it is not a property that will be determined solely by the atoms on the molecular periphery. One may wonder: Why

Table 27.   Comparison of the Regression Analysis Using the Shape Profile Descriptors ^kS and Molecular Profile Descriptors ^kD

^1S	^2S	^3S	Const.	R	S	F
62.66			1658.81	0.8106	483.9	11
189.06	−46.30		−481.10	0.8374	497.7	6
−214.90	284.20	−94.02	1404.92	0.9946	105.5	121

^1D	^2D	^3D	Const.	R	S	F
69.69			976.80	0.9454	269.2	51
143.59	−29.14		−341.62	0.9928	108.5	172
−377.71	419.84	−134.85	2707.6	0.9979	65.3	319

is *L/B* relatively good if shape is *not* the structural feature here of much importance? A close look at *L/B* shows that while it is a good shape descriptor (particularly for rectangles), it reflects indirectly to some degree the presence of interior carbon atoms. For example, when *L* and *B* are approximately equal (as in triphenylene and coronene), the interior carbons contribute to the "round" shape of such benzenoids. Hence, there will be continuing need for shape descriptors, including possibly *L/B*. However, we expect that in such instances one will obtain better results by using the more realistic shape descriptors ^1S, ^2S, ^3S.

6.11. PERIPHERY CODES

In a series of articles Mezey and Arteca studied topological properties of molecular surfaces in order to quantify molecular shape.[270–274] They characterized the shape by considering curvatures of portions of the molecular surface that have distinct topological properties. We will focus attention on a characterization of the molecular shape by binary molecular codes. Instead of considering the general problem of molecular shapes, we will consider a simpler task, namely, the characterization of the shapes of planar benzenoids. It will be revealed that the approach applies to the characterization of the shape of an arbitrary closed planar curve. A satisfactory code for the periphery of a simple benzenoid should be linear, have structural origin, be simple, have similar lengths for objects of similar size, be unique, and should allow reconstruction.

In Figure 29 we have illustrated the simplest shapes on a hexagonal map, those given by the periphery of benzene, naphthalene, anthracene, phenanthrene, phenalene, and pyrene. A simple way to represent the periphery of such objects parallels the description of a route on a dodecahedron outlined in the book by Rouse Ball[275] and

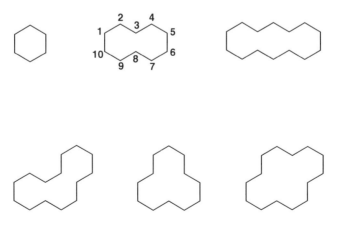

Figure 29.   The simplest shapes on the hexagonal grid.

LLRLRRRLLL...

Figure 30. Representation of a (cyclic) path on a trivalent graph.

illustrated in Figure 30. As one moves over the graph of dodecahedron at each point there is a choice of left (L) and right (R) path. Hence, a sequence of L and R symbols will fully describe the history of movements over cubic graphs. The graphite lattice is trivalent, hence figures embedded on a graphite lattice can be similarly described. We will use this approach to describe the peripheries of benzenoid shapes. Since closed curves have "inside" and "outside" regions, instead of L and R we may use I (for inside) and O (for outside), or simply the binary labels 1 and 0, respectively. The binary codes are not new in chemistry. For example, Balaban used binary codes to generate distinct annulenes.[276] Periphery codes for square-cell configurations (animals) using complex plane representation were also considered.[277]

Our approach applied to benzene gives a unique binary code: 1 1 1 1 1 1, since all six carbon atoms and all six C–C bonds are equivalent. Because in naphthalene there are *three* nonequivalent carbon atoms and *two* nonequivalent directions of circling around, there are several possible binary codes. The form of the code depends on the selection of the starting atom. For a clockwise circling we obtain the following codes:

Start	Code
1	1 1 0 1 1 1 1 0 1 1
2	1 0 1 1 1 1 0 1 1 1
3	0 1 1 1 1 0 1 1 1 1
4	1 1 1 1 0 1 1 1 1 0
5	1 1 1 0 1 1 1 1 0 1 etc.

The five codes listed are in fact the only possible codes for naphthalene. The code reflects the presence of the center of symmetry in naphthalene and other symmetries of the naphthalene. From the code derived by starting at atom 1 we see that the molecule has a plane of reflection, because the first part of the code is the same as the last part of the code. The code can be written simply as $(1\ 1\ 0\ 1\ 1)^2$ with the repeating section also having symmetry.

We have to make canonical rules in order to select one of these five codes as unique. In the case of naphthalene the code initiated by atom 3 corresponds to the

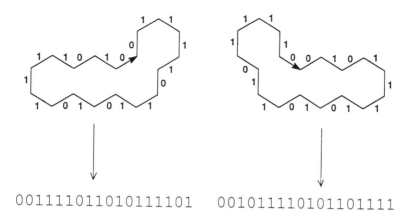

00111101101011101   0010111101011011111

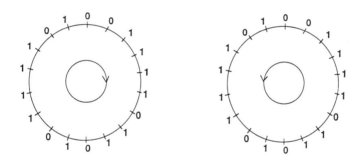

Figure 31. The periphery codes C* and *C for benzanthracene.

smallest binary number possible while the code derived by starting with atom 4 corresponds to the *largest* binary number. Either of the two is unique and can be used as the canonical code. We will use the smallest binary number as the canonical form for the periphery code.

Benzenoids having a larger periphery P can assume different shape forms. In the case $P = 14$ (i.e., the periphery has 14 carbon–carbon bonds) there are three different shapes that correspond to the peripheries of anthracene, phenanthrene, and pyrene. The minimal binary codes for the three $P = 14$ shapes are:

Phenanthrene	0 0 1 1 1 1 0 1 1 0 1 1 1 1
Anthracene	0 1 0 1 1 1 1 0 1 0 1 1 1 1
Pyrene	0 1 1 0 1 1 1 0 1 1 0 1 1 1

Figure 32. Chiral and achiral benzenoids with the perimeter $P = 18$.

By shifting the origin of the code suitably the same codes can be written as:

Phenanthrene	(0 1 1 1 1 0 1) (1 0 1 1 1 1 0)
Anthracene	(1 1 0 1 0 1 1) (1 1 0 1 0 1 1)
Pyrene	(1 0 1 1 1 0 1) (1 0 1 1 1 0 1)

In this form the codes reflect the symmetry of the underlying shapes. The codes for the shapes of anthracene and pyrene, which possess a higher symmetry, not only have the two brackets identical, but within each of the brackets the code is symmetrical. In the case of phenanthrene, which is of lower symmetry, the part of the code within the brackets lacks symmetry.

Chirality in chemistry is associated with 3D molecular forms. It is a consequence of the asymmetrical substitution of tetrahedral carbons. However, at the beginning of this century Kelvin had defined chirality for objects in n-dimensional spaces. Chiral

Table 28. The Smallest Binary Codes for the Benzenoids Having Periphery $P = 18$

1	Tetracene	0 1 0 1 0 1 1 1 1 0 1 0 1 0 1 1 1 1
2	Benzanthracene	0 0 1 0 1 1 1 1 0 1 0 1 1 0 1 1 1 1
3	Chrysene	0 0 1 1 0 1 1 1 1 0 0 1 1 0 1 1 1 1
4	Benzphenanthrene	0 0 0 1 1 1 1 0 1 1 0 1 1 0 1 1 1 1
5	Triphenylene	0 0 1 1 1 1 0 0 1 1 1 1 0 0 1 1 1 1
6	Benzo[a]pyrene	0 0 1 1 0 1 1 1 0 1 1 0 1 0 1 1 1 1
7	Benzo[d]pyrene	0 0 1 1 1 0 1 1 0 1 1 1 0 0 1 1 1 1
8	Perylene	0 0 1 1 1 0 1 1 1 0 0 1 1 1 0 1 1 1
9	Anthanthene	0 1 0 1 1 1 0 1 1 0 1 0 1 1 1 0 1 1
10	Benz[c,d,e]pyrene	0 0 1 1 1 0 1 1 0 1 1 0 1 1 0 1 1 1
11	Triangulene	0 1 0 1 1 1 0 1 0 1 1 1 0 1 0 1 1 1
12	Coronene	0 1 1 0 1 1 0 1 1 0 1 1 0 1 1 0 1 1

Chiral pairs

1	Benzanthracene	0 0 1 1 1 1 0 1 1 0 1 0 1 1 1 1 0 1
3	Chrysene	0 0 1 1 1 1 0 1 1 0 0 1 1 1 1 0 1 1
6	Benzo[a]pyrene	0 0 1 1 1 1 0 1 0 1 1 0 1 1 1 0 1 1
9	Anthanthene	0 1 0 1 1 0 1 1 1 0 1 0 1 1 0 1 1 1

objects are those that cannot be superimposed with their mirror image.[278] This includes chirality for objects in a 2D space that recently Buda and Mislow discussed for the case of asymmetrical triangles in a plane.[279] Hence, we can distinguish the two forms of benzanthracene, although it is arbitrary which one we call left and which we call right (Figure 31). That is, the two forms of benzanthracene, which are the mirror image of each other, cannot be brought into complete overlap by sliding the molecules within the plane in which they are embedded. It should be noticed that in 2D space the presence of a center of symmetry does not preclude chirality, while in 3D it does.

In Table 28 we give the smallest binary codes for benzenoids having periphery $P = 18$ (illustrated in Figure 32). Four of the twelve benzenoid forms are chiral. These chiral shapes either have no symmetry (except the symmetry of the plane in which they are embedded), or have only a center of symmetry. The former is the case with benzanthracene and benzo(a)pyrene, the latter is the case with chrysene and anthanthrene. One enantiomer of these 2D chiral shapes (molecules) is listed in the upper part of Table 28, where the structures have been ordered according to the smallest binary code, while the other enantiomer is shown in the lower part of Table 28.

Each form of the chiral shapes has a different code. Since the corresponding codes do not show symmetry (beyond the center of symmetry and the plane of the molecule that are here irrelevant), one can deduce from the binary code whether a shape is chiral or not.[280] If a code when read from left to right and then from right to left, gives a different result, then the shape is chiral. If a code is palindromic, i.e., when read from left to right and then from right to left, gives the same answer, the structure is achiral. If codes of enantiomers are read from right to left (instead of from left to right), they would give the codes of their enantiomers (Figure 31).

Figure 33. The 19 achiral shapes of hexagonal grid having perimeter $P = 22$.

6.11.1. Similarity Derived from Periphery Codes

Different characterizations will reflect different aspects of the similarity among objects. We will consider here how the similarity is reflected when benzenoids are characterized by their binary periphery codes. Consider the 19 achiral shapes of Figure 33 having periphery $P = 22$ labeled as A–S. Their binary codes are listed in Table 29. We will use the Hamming distance as the *index of similarity/dissimilarity*. A Hamming

Table 29.   The Binary Codes for Achiral Benzenoids
Having Perimeter $P = 22$

Molecule	Periphery code
A	0 0 0 0 1 1 1 1 0 1 1 0 1 1 0 1 1 0 1 1 1 1
B	0 0 0 1 1 1 0 1 1 1 0 1 0 1 0 1 1 1 0 1 1 1
C	0 0 1 0 0 1 1 1 1 0 1 1 0 1 0 1 1 0 1 1 1 1
D	0 0 1 0 1 1 1 1 0 1 0 0 1 1 1 0 1 1 0 1 1 1
E	0 0 1 0 1 1 1 1 0 1 0 0 1 1 1 1 0 0 1 1 1 1
F	0 0 1 0 1 1 1 1 0 1 0 1 1 0 1 0 1 1 1 1 0 1
G	0 0 1 1 0 0 1 1 1 1 0 1 0 1 1 0 1 0 1 1 1 1
H	0 0 1 1 0 0 1 1 1 1 0 1 1 0 0 1 1 0 1 1 1 1
I	0 0 1 1 0 1 1 0 1 1 0 1 1 0 1 1 0 0 1 1 1 1
J	0 0 1 1 0 1 1 1 0 1 0 1 1 0 1 0 1 1 1 0 1 1
K	0 0 1 1 0 1 1 1 0 0 1 1 1 0 1 1 0 0 1 1 1 1
L	0 0 1 1 0 1 1 1 0 1 1 0 0 1 1 0 1 1 1 0 1 1
M	0 0 1 1 1 0 1 0 1 1 0 1 1 0 1 1 0 1 0 1 1 1
N	0 0 1 1 1 0 0 1 1 1 0 1 1 0 1 0 1 1 0 1 1 1
O	0 0 1 1 1 0 0 1 1 1 1 0 0 1 1 1 0 0 1 1 1 1
P	0 0 1 1 1 0 1 0 1 1 1 0 0 1 1 1 0 1 0 1 1 1
Q	0 1 0 1 0 1 0 1 1 1 1 0 1 0 1 0 1 0 1 1 1 1
R	0 1 0 1 1 0 1 0 1 1 1 0 1 0 1 1 0 1 0 1 1 1
S	0 1 0 1 1 0 1 1 0 1 1 0 1 0 1 0 1 1 0 1 1 1

distance[281] is the standard parameter that counts the coincidences between binary sequences (of equal length). The larger the distance between two codes, the less similar are the codes and the less similar are the corresponding shapes. Since the periphery codes are cyclic, the Hamming distances shown in Table 30 are obtained by shifting the codes one against the other in order to produce the maximal overlap and the minimal distance.

Table 30 shows relatively small variations in the magnitude of the similarity indices. The most similar shapes are associated with the Hamming distance of 4, while for the most dissimilar shapes the Hamming distance is 10. Disappointingly, however, we see that there are 28 pairs of the "most similar" shapes, namely:

(A, E), (A, I), (A, M), (A, S), (C, N), (C, S), (D, E), (D, F), (F, G), (F, J),
(G, I), (G, R), (H, I), (H, J), (H, N), (I, K), (I, M), (J, L), (J, R), (M, N),
(M, P), (M, R), (M, S), (O, P), (P, Q), (P, R), (Q, R), (R, S)

Thus, apparently the binary codes do not well discriminate among most similar benzenoid forms. In contrast, among the same benzenoids we find fewer "least similar" shapes. The least similar benzenoids—(D, R), (F, S), (H, S), (I, L), (L, N)—appear rather distinct, in agreement with our common perceptions of dissimilarity.

In order to arrive at a more selective index of similarity we have extended the comparison of codes by considering longer sequences of digits in the binary codes that are compared. This would be tedious to do by hand, but by using a computer program one can easily derive a revised similarity/dissimilarity table.[282] The new table (Table

Table 30. The Hamming Distance for the Periphery Codes of Table 28

	A	B	C	D	E	F	G	H	I	J	K	L	M	N	O	P	Q	R	S
A	0	8	6	6	4	8	6	8	4	8	8	8	4	8	8	8	8	6	4
B		0	8	8	8	8	8	6	8	8	6	8	6	6	6	8	8	8	8
C			0	8	6	8	6	6	8	6	6	8	6	4	8	8	8	6	4
D				0	4	4	6	8	8	6	6	6	8	6	6	8	8	10	8
E					0	6	8	8	6	8	6	8	8	6	8	8	8	8	8
F						0	4	8	8	4	8	8	8	6	8	8	8	8	10
G							0	4	6	6	8	6	6	6	6	6	8	4	6
H								0	4	4	6	8	6	4	8	8	8	8	10
I									0	6	4	10	4	6	8	8	8	8	8
J										0	6	4	6	6	8	6	6	4	8
K											0	8	8	6	8	8	8	8	8
L												0	8	10	8	8	8	8	8
M													0	4	8	4	6	4	4
N														0	8	8	8	8	8
O															0	4	8	8	8
P																0	4	4	8
Q																	0	4	8
R																		0	4
S																			0

Table 31. The Generalized Hamming Distance for the Cyclic Codes of Table 28 Using Increasing Portions of the Code for Comparison

	A	B	C	D	E	F	G	H	I	J	K	L	M	N	O	P	Q	R	S
A	0	12	10	10	8	12	12	12	8	12	13	13	8	12	13	13	13	11	7
B		0	11	12	13	13	12	10	12	12	11	12	10	9	11	12	13	13	13
C			0	13	10	12	10	10	13	11	10	14	10	8	13	12	12	11	8
D				0	5	8	12	12	10	10	10	10	13	11	14	13	12	14	12
E					0	10	12	12	10	13	9	13	12	12	11	11	10	14	12
F						0	5	10	12	6	12	11	11	11	12	11	10	9	13
G							0	5	10	9	12	11	10	10	10	10	11	6	10
H								0	8	6	10	9	10	8	12	13	12	11	15
I									0	9	5	14	6	10	10	11	12	12	10
J										0	9	5	9	9	13	9	10	5	11
K											0	12	10	9	11	11	13	14	14
L												0	14	14	12	12	12	10	12
M													0	6	11	5	10	6	5
N														0	10	11	12	12	10
O															0	6	12	12	14
P																0	8	6	10
Q																	0	8	12
R																		0	6
S																			0

31) shows a greater discrimination among the shapes. Of the initial 28 pairs of the "most similar" shapes when the comparison was based on matching of single entries of two codes (Table 30), now only the following 8 pairs remain the most similar (at distance five):

$$(D, E), (F, G), (G, H), (I, K), (J, L), (J, R), (M, P), (M, S)$$

Similarly, instead of the five "least similar" cases of Table 30, now we find only a single such case (the pair H, S).

6.12. CHIRALITY

We can now consider a quantitative approach to chirality. Chirality can be defined for objects of n-dimensional space.[278] When $n = 2$ we have the case of chirality of objects embedded in a plane. For example, benzanthracene (Figure 33) when embedded in a plane is chiral, since it cannot be brought into coincidence with its mirror image if we allow only sliding a figure *within* the plane. Transformation of one of the benzanthracene "2D enantiomers" into the other requires one to take the molecule *out of the plane* in which it is embedded. Hence, an object that is chiral in a lower dimensional space will not remain chiral if placed into a space of a higher dimension.

Early papers on chirality can be traced to Kitaigorodski[283] who was interested in correlation between chirality and chemical properties. Kauzmann and collaborators,[284] Ugi,[285] and others[280–289] were interested in chirality functions in their study of attaching ligands to an achiral molecular frame. A quantitative approach to chirality, i.e., attempt to assign the "degree of chirality" to chiral systems, was rather unexplored until recently. Background information on chirality measures can be found in a recent paper by Zabrodsky and Avnir,[286] who themselves proposed a novel measure of chirality. There are several alternative approaches to estimate "the degree of chirality." Recently an attempt for a unified approach to quantitative chirality has been initiated.[287] There are no disputes as to whether a system is chiral or not, but deciding which pair of chiral objects is "more chiral" and which is "less chiral" is far from obvious. We will outline here briefly an approach based on binary periphery codes. We will consider smaller *cata*-condensed benzenoids, but the approach holds for planar benzenoids, and by extension to arbitrary planar figures.

There are two kinds of chirality measures[288]:

1. Measures of the first kind: those that gauge the extent to which a chiroid object differs from an achiral object
2. Measures of the second kind: those that gauge the extent to which two enantiomers differ

The definition of chirality:

DEFINITION: An object X^* is chiral if and only if it is not superimposable on its mirror image $^*X$.

In cases when a 2D object is represented by the binary code, the above definition "translates" into:

DEFINITION: An object is chiral if and only if its binary code is not a palindrome, i.e., C^* is not equal to its mirror image $^*C$.

Here we will consider the chirality measure of the second kind for planar benzenoids. Our measure of chirality will be based on the degree of similarity/dissimilarity among the pair of enantiomers. We will consider all chiral benzenoids with the periphery $P = 22$, i.e., a total of 28 pairs which are shown in Figure 34. Clearly by inspection of Figure 34 it is hard to guess which pair or pairs of benzenoids are the most chiral. It is even difficult to speculate which pair or pairs of benzenoids would be the least chiral. We will therefore use Hamming distances between the codes of the two enantiomers to find the most dissimilar (hence, most chiral) and the most similar (i.e., the least chiral).

We have already outlined the procedure for a quantitative measure of the degree of similarity or dissimilarity on achiral structures. Here we will limit the comparison among the codes only to those belonging to enantiomers. First we will compare the codes (by sliding one code against the other, which are viewed as cyclic) and look for coincidences of a single digit in two codes. The Hamming distance is given by the count of these coincidences of a single digit between two such codes. In the first column of Table 32 we listed the resulting similarities. Again, just as in the case of similarity/dissimilarity among achiral benzenoids the discrimination between different pairs is very limited. Most of the enantiomers have a Hamming distance of four. Four pairs (A, C, Q, and U) have a Hamming distance of six, signifying more chiral systems.

If we increase the length of the portion of the code to be used for comparison, we increase the resolution power of the approach. The remaining columns of Table 32 show the similarity/dissimilarity values obtained when longer portions of the codes were used in determining the coincidences. For example, using two consecutive digits of the code we would have four possibilities:

1st code	11	10	01	00
2nd code	11	10	01	00
Distance	1	1	1	1

Any other combination of the pair of digits, like 11/10, makes no contribution to the generalized Hamming distance using *two* digits as the basis for comparison of codes.

With two consecutive digits as the basis for the comparison of the codes we doubled the number of distinct chiral classes. The most chiral pair is now ($^*C$, C^*) at the distance 11. Table 32 gives the history of increased differentiation among chiral enantiomers as the length of the "window" of comparison is gradually increased. Entries in Table 32 can be used to rank chiral benzenoids.[282] In the case of the

Figure 34. Chiral benzenoids with the perimeter $P = 22$.

Table 32. The Similarity between the Enantiomers of the Benzenoids of Figure 34

	1	2	3	4	5	6	7	. . .	Sum
A	6	10	12	14	16	17	18	. . .	285
B	4	8	9	10	11	12	13	. . .	229
C	6	11	14	16	18	19	20	. . .	301
D	4	6	8	10	12	14	16	. . .	253
E	4	8	10	11	12	13	14	. . .	242
F	4	8	12	13	14	15	16	. . .	264
G	4	6	8	10	12	14	15	. . .	246
H	4	8	10	11	12	13	14	. . .	242
I	4	8	10	11	12	13	24	. . .	242
J	4	8	10	11	12	13	24	. . .	242
K	4	8	10	11	12	13	14	. . .	242
L	4	8	12	16	18	20	22	. . .	298
M	4	6	8	10	12	14	15	. . .	246
N	4	8	10	12	14	16	18	. . .	274
O	4	6	8	10	12	14	15	. . .	246
P	4	6	8	10	12	14	16	. . .	262
Q	6	10	12	14	15	16	17	. . .	278
R	4	6	8	10	12	14	16	. . .	261
S	4	8	11	12	13	14	15	. . .	254
T	4	6	8	10	12	14	16	. . .	253
U	6	10	12	14	15	16	17	. . .	278
V	4	8	10	12	14	16	18	. . .	274
W	4	8	12	16	17	18	19	. . .	289
X	4	8	12	14	16	18	20	. . .	290
Y	4	6	8	10	12	14	15	. . .	246
Z	4	8	10	12	14	16	18	. . .	278
AA	4	8	10	12	13	14	15	. . .	253
BB	4	6	8	10	12	14	16	. . .	240

benzenoid considered here one finds that maximal differentiation among chiral systems is obtained when using a portion of the code of length six. As we increase the length of the "window" of the comparison the measure of similarity/dissimilarity increases, and eventually reaches the maximal value 22, which is the length of the code. Some chiral pairs reach this maximal value sooner, some later. As reported,[282] by the time we compare the codes using the portion of length 16, for all benzenoids of Figure 34 we have reached the maximal value for dissimilarity.

Which pair of enantiomers of Figure 34 is most chiral? How will we decide, when the relative magnitudes for the similarity/dissimilarity depend on the length of the "window" used for the comparison? One way to obtain a single index for the measure of chirality is to add all of the entries in each row and view the total as the measure of chirality. The last column of Table 32 lists the total up to the "windows" of length 16. If we are to extend the table to the maximal "paths" of length 22, the relative values would not change since the additional values are all constant. From the last column of Table 32 we can read as the most chiral:

C (301), L(298), X(290), W(289), A(285), Q, U, and Z (all 278), . . .

The above is not the only possible or even plausible index of chirality that can be extracted for Table 32. Rather than exploring alternatives, let us emphasize the significance of the approach outlined. The main result of our scheme is to assign to chiral structures a *sequence*, not a single index. Thus, the two most chiral benzenoids C and L are represented by the respective sequences

$$\text{C} \quad 6, 11, 14, 16, 18, 19, 20, \ldots$$

$$\text{L} \quad 4, \; 8, 12, 16, 18, 20, 22, \ldots$$

If we want to answer the question which of the two benzenoids is more chiral, we have to decide which of the two sequences is "bigger." There would be no difficulties in establishing which sequence is "bigger" if the entries in one sequence are always bigger than the entries in the second sequence. But such (obvious) cases are few. As seen in the above case, at the beginning sequence C dominates sequence L, but later the relative magnitudes of the corresponding entries in both sequences reverse. Hence, it is not clear which of the sequences should be viewed as dominant. Comparison of sequences was considered in the mathematical literature at the beginning of this century by the Scottish mathematician Muirhead.[290–292] According to Muirhead, instead of looking at the original sequences, one should consider the corresponding sequences of partial sums and base the judgment of dominance on the latter. Hence, if we are to compare the two sequences

$$\text{A} \quad a_1, a_2, a_3, a_4, \ldots$$

$$\text{B} \quad b_1, b_2, b_3, b_4, \ldots$$

we should compare the sequences of the partial sums:

$$\text{A}' \quad a_1, a_1 + a_2, a_1 + a_2 + a_3, a_1 + a_2 + a_3 + a_4, \ldots.$$

$$\text{B}' \quad b_1, b_1 + b_2, b_1 + b_2 + b_3, b_1 + b_2 + b_3 + b_4, \ldots.$$

Then if the following holds:

$$a_1 \geq b_1$$

$$a_1 + a_2 \geq b_1 + b_2$$

$$a_1 + a_2 + a_3 \geq b_1 + b_2 + b_3$$

$$a_1 + a_2 + a_3 + a_4 \geq b_1 + b_2 + b_3 + b_4$$

$$\cdots$$

$$a_1 + a_2 + a_3 + a_4 + \cdots + a_n \geq b_1 + b_2 + b_3 + b_4 + \cdots + b_n$$

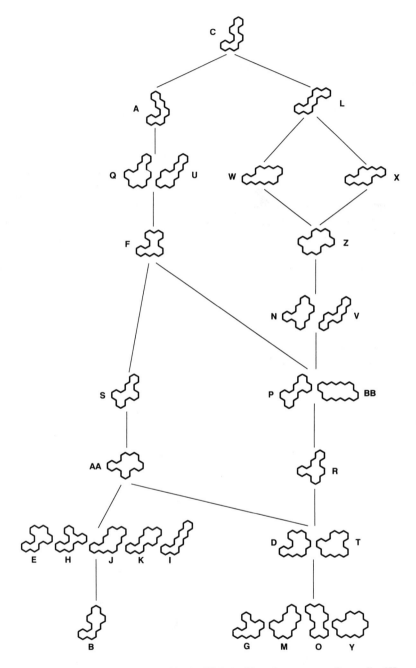

Figure 35. Partial order of chiral benzenoids ($P = 22$) derived from the sequences of generalized Hamming distance for cyclic binary codes.

we say that sequence A dominates sequence B. If the above inequalities are not satisfied, we conclude that the objects A and B (represented by the corresponding sequences) are not comparable, i.e., neither A dominates B nor B dominates A.

For our case of the sequences for C and L we obtain the following partial sums:

$$C \quad 6, 17, 31, 47, 65, 84, 114, \ldots$$

$$L \quad 4, 12, 24, 40, 58, 78, 100, \ldots$$

Because the trend shown above continues, we can conclude that the C sequence dominates L, or alternatively that C is more chiral than L. Observe that the conclusion is based on ordering of structures, not on the numerical value of an index. When this analysis is extended to all 28 chiral compounds of Figure 34, we obtain the partial ordering shown in Figure 35.

From Figure 35 we conclude that indeed C is the most chiral benzenoid. Equally we see that A is more chiral than Q and U and F and all structures that follow. On the other hand, L is more chiral than both W and X, which dominate Z and the structures that follow. However, we cannot answer which, A or L, is more chiral. These two benzenoids are located in different branches of the partial order diagram, and hence are not comparable. Yet even though A and L are not comparable, they both are dominated by C and they both dominate P, BB, R, and other benzenoids that follow.

A number of questions remain open, including extension of the present approach to objects in 3D space. Are the binary shape codes sufficiently general, i.e., can they be extended to 3D structures? We cannot answer these questions since they have not yet been considered. We will, however, outline briefly an approach to the 3D case. A generalization from planar contours to surfaces in a 3D space is possible even if not unique. For example, we first derive for a 3D surface a set of 2D contours. Such are, for example, the electrostatic maps that have found use in QSAR in the well-known CoMFA method.[293] Once we have the contours, each contour can be encoded by the binary code as illustrated in this chapter. A list of such contours, with specifications of their relative positions, will lead to a 3D code for molecular shapes. Comparisons of 3D shapes will now be more involved.

Such considerations may provoke new directions yet to be investigated, but at this initial stage one should not forget the anecdote of Michael Faraday, who at the end of his lecture about the discovery of electromagnetic induction was asked by someone in the audience: "Sir, for what good is your new discovery?" Faraday answered: "For what good is a new born baby?"

REFERENCES

1. L. Pauling, *The Nature of the Chemical Bond*, Cornell University Press, Ithaca, New York (1939).
2. W. Heitler and F. London, *Z. Phys. 44*, 455 (1927).
3. P. A. M. Dirac, *Proc. R. Soc. (London) Ser. A 179*, 714 (1929).
4. S. F. Boys, *Proc. R. Soc. (London), Ser. A 200*, 542 (1950); *201*, 125 (1950).

5. M. Randić, *J. Math. Chem. 4*, 157 (1990).

6. M. Randić, *J. Chem. Educ. 69*, 713 (1992).

7. M. Randić and N. Trinajstić, *New J. Chem. 18*, 179 (1994).

8. M. Randić and C. L. Wilkins, *Int. J. Quantum Chem. 18*, 1005 (1980).

9. M. Randić and C. L. Wilkins, *Chem. Phys. Lett. 63*, 332 (1979).

10. M. Randić and C. L. Wilkins, *J. Chem. Phys. 83*, 1525 (1979).

11. V. Prelog and L. Ružička, *Helv. Chim. Acta 27*, 66 (1944).

12. L. Ružička, *Chem. Z. 44*, 93, 129 (1920).

13. E. Fischer, *Ber. Dtsch. Chem. Ges. 27*, 2985 (1884).

14. C. Hansch, *Acc. Chem. Res. 2*, 232 (1969).

15. C. Hansch and T. Fujita, *J. Am. Chem. Soc. 86*, 1616 (1964).

16. P. G. Mezey, *J. Chem. Inf. Comput. Sci. 32*, 650 (1992).

17. N. Trinajstić, M. Randić, D. J. Klein, D. Babić, and Z. Mihalić, *Croat. Chem. Acta 68*, 241 (1995).

18. D. Babić, A. T. Balaban, and D. J. Klein, *J. Chem. Inf. Comput. Sci. 35*, 515 (1995).

19. A. T. Balaban, D. Babić, and D. J. Klein, *J. Chem. Educ. 72*, 693 (1995).

20. A. Baeyer, *Ber. Dtsch. Chem. Ges. 33*, 3771 (1900).

21. W. J. Wiswesser, *A Line-Formula Chemical Notation*, Thomas Y. Crowell, New York (1954).

22. H. L. Morgan, *J. Chem. Doc. 5*, 107 (1965).

23. J. Lederberg, *Proc. Natl. Acad. Sci. USA 53*, 134 (1965).

24. A. T. Balaban, *Tetrahedron 25*, 2949 (1969).

25. R. C. Read, in: *Graph Theory and Computing* (R. C. Read, ed.), pp. 153–182, Academic Press, New York (1972).

26. J. E. Dubois, in: *Chemical Applications of Graph Theory*, (A. T. Balaban, ed.), pp. 333–370, Academic Press, New York (1976).

27. N. Lozac'h, A. L. Goodson, and W. H. Powell, *Angew. Chem. Int. Ed. Engl. 18*, 887 (1979).

28. J. V. Knop, W. R. Muller, Z. Jeričević, and N. Trinajstić, *J. Chem. Inf. Comput. Sci. 21*, 91 (1981).

29. W. C. Herndon, in: *Chemical Applications of Topology and Graph Theory* (R. B. King, ed.), Elsevier, Amsterdam (1983).

30. A. L. Goodson, *Croat. Chem. Acta 56*, 315 (1983).

31. R. C. Read, *J. Chem. Inf. Comput. Sci. 23*, 135 (1983).

32. M. Randić, *J. Chem. Inf. Comput. Sci. 26*, 136 (1986).

33. A. L. Goodson and N. Lozac'h, *Croat. Chem. Acta 59*, 547 (1986).

34. W. C. Herndon and A. J. Bruce, *J. Math. Chem. 2*, 155 (1988).

35. D. Weininger, *J. Chem. Inf. Comput. Sci. 28*, 31 (1988).

36. W. R. Muller, K. Szymanski, J. V. Knop, and N. Trinajstić, *J. Chem. Inf. Comput. Sci. 34*, 960 (1994).

37. P. Hansen, B. Jaumard, C. Lebatteux, and M. Zheng, *J. Chem. Inf. Comput. Sci. 34*, 782 (1994).

38. A. L. Goodson, C. L. Gladys, and D. E. Worst, *J. Chem. Inf. Comput. Sci. 35*, 969 (1995).

39. W.-D. Ihlenfeldt and J. Gasteiger, *J. Chem. Inf. Comput. Sci. 35*, 663 (1995).

40. I. Strokov, *J. Chem. Inf. Comput. Sci. 35*, 939 (1995).

41. J. L. Schultz and E. S. Wilks, *J. Chem. Inf. Comput. Sci. 36*, 786 (1996).

42. R. C. Read and D. G. Corneil, *J. Graph Theory 1*, 339 (1977).

43. G. Gatty, *J. Graph Theory 3*, 95 (1979).

44. M. Razinger, K. Balasubramanian, and M. E. Munk, *J. Chem. Inf. Comput. Sci. 33*, 197 (1993).

45. S. Bohanec and M. Perdih, *J. Chem. Inf. Comput. Sci. 33*, 719 (1993).

46. I. P. Bangov, *J. Chem. Inf. Comput. Sci. 32*, 123 (1992).

47. G. Ruecker and C. Ruecker, *J. Chem. Inf. Comput. Sci. 31*, 123 (1991).

48. L. H. Hall and L. B. Kier, *Quant. Struct. Act. Relat. 9*, 115 (1990).

49. W. C. Herndon and J. E. Leonard, *Inorg. Chem. 22*, 554 (1983).

50. M. Randić, G. M. Brissey, and C. L. Wilkins, *J. Chem. Inf. Comput. Sci. 21*, 52 (1981).

51. E. V. Konstantinova, *J. Chem. Inf. Comput. Sci. 36*, 54 (1996).

52. M. Randić, *Acta Chim. Slov.* (in press).

53. R. S. Cahn, C. Ingold, and V. Prelog, *Angew. Chem. Int. Ed. Engl. 5*, 383 (1966).
54. V. Prelog and G. Helmchen, *Angew. Chem. Int. Ed. Engl. 21*, 567 (1982).
55. M. Razinger and M. Perdih, *J. Chem. Inf. Comput. Sci. 34*, 290 (1994).
56. P. Mata, A. M. Lobo, C. Marshall, and A. P. Johnson, *Tetrahedron Asymmetry 4*, 657 (1993).
57. M. Perdih and M. Razinger, *Tetrahedron Asymmetry 5*, 835 (1994).
58. M. Randić, Distance-based canonical labels, *Croat. Chem. Acta* (submitted).
59. N. Trinajstić, *Chemical Graph Theory*, CRC Press, Boca Raton, Florida (1992).
60. A. T. Balaban, in: *Chemical Applications of Graph Theory* (A. T. Balaban, ed.), Academic Press, New York (1976).
61. A. Graovac, I. Gutman, N. Trinajstić, and T. Živković, *Theor. Chim. Acta 26*, 67 (1972).
62. I. Gutman and N. Trinajstić, *Top. Curr. Chem. 42*, 49 (1973).
63. A. Graovac, I. Gutman, and N. Trinajstić, *Topological Approach to the Chemistry of Conjugated Molecules*, Springer-Verlag, Berlin (1977).
64. I. Gutman and O. E. Polansky, *Mathematical Concepts in Organic Chemistry*, Springer-Verlag, Berlin (1986).
65. I. Gutman, M. Milun, and N. Trinajstić, *MATCH 1*, 171 (1975).
66. I. Gutman, M. Milun, and N. Trinajstić, *J. Am. Chem. Soc. 99*, 1692 (1987).
67. J.-I. Aihara, *J. Am. Chem. Soc. 98*, 2750 (1986).
68. M. Randić, *Chem. Phys. Lett. 38*, 68 (1976).
69. M. Randić, *Tetrahedron 33*, 1905 (1977).
70. M. Randić, *Pure Appl. Chem. 52*, 1587 (1980).
71. M. Randić, V. Solomon, S. C. Grossman, D. J. Klein, and N. Trinajstić, *Int. J. Quantum Chem. 32*, 35 (1987).
72. M. Randić, L. L. Henderson, R. Stout, and N. Trinajstić, *Int. J. Quantum Chem. Quant. Chem. Symp. 22*, 127 (1988).
73. D. J. Klein and N. Trinajstić, *Pure Appl. Chem. 61*, 2107 (1989).
74. D. J. Klein, *Top. Curr. Chem. 153*, 57 (1990).
75. N. Trinajstić and D. Plavšić, *Croat. Chem. Acta 62*, 711 (1989).
76. S. Nikolić, N. Trinajstić, and D. J. Klein, *Comput. Chem. 14*, 313 (1990).
77. D. J. Klein and N. Trinajstić, eds., *Valence Bond Theory and Chemical Structure*, Elsevier, Amsterdam (1990).
78. W. T. Simpson, *J. Am. Chem. Soc. 75*, 593 (1953).
79. S. P. McGlynn, L. G. Vanquickenborne, M. Kinoshita, and D. G. Carroll, *Introduction to Applied Quantum Chemistry*, Holt, Reinhart & Winston, New York (1972).
80. W. C. Herndon, *J. Am. Chem. Soc. 95*, 2404 (1973).
81. W. C. Herndon, *Thermochim. Acta 8*, 225 (1974).
82. W. C. Herndon and M. L. Ellzey, Jr., *J. Am. Chem. Soc. 96*, 663 (1974).
83. W. C. Herndon, *Isr. J. Chem. 20*, 270 (1980).
84. M. Randić, *J. Am. Chem. Soc. 99*, 444 (1977).
85. M. Randić and N. Trinajstić, *J. Am. Chem. Soc. 106*, 4428 (1984).
86. R. Breslow, *Acc. Chem. Res. 6*, 393 (1973).
87. R. Breslow, W. Horspool, H. Sugiyama, and W. Vitale, *J. Am. Chem. Soc. 88*, 3677 (1966).
88. K. Hafner, R. Fleischer, and K. Fritz, *Angew. Chem. 77*, 42 (1965).
89. F. T. Smith, *J. Chem. Phys. 34*, 793 (1961).
90. A. T. Balaban and F. Harary, *Tetrahedron 24*, 2505 (1968).
91. A. T. Balaban, *Tetrahedron 25*, 2949 (1969).
92. O. E. Polansky and D. H. Rouvray, *MATCH 2*, 63 (1976).
93. J. V. Knop, W. R. Müller, K. Szymanski, and N. Trinajstić, *Computer Generation of Certain Classes of Molecules*, SKTH, Zagreb (1985).
94. H. Joela, *Theor. Chim. Acta 39*, 241 (1975).
95. S. El-Basil, *Int. J. Quantum Chem. 21*, 771 (1982).

96. S. El-Basil, *Int. J. Quantum Chem. 21*, 779 (1982).
97. S. El-Basil, *Int. J. Quantum Chem. 21*, 793 (1982).
98. S. El-Basil, *MATCH 13*, 209 (1982).
99. S. El-Basil, *Top. Curr. Chem. 153*, 274 (1990).
100. I. Gutman, *Theor. Chim. Acta 45*, 309 (1977).
101. S. El-Basil, *J. Chem. Soc. Faraday Trans. 2 82*, 299 (1986).
102. S. El-Basil and M. Randić, *Adv. Quantum Chem. 24*, 239 (1992).
103. E. Clar, *The Aromatic Sextet*, John Wiley & Sons, New York (1972).
104. J. W. Armitt and R. Robinson, *J. Chem. Soc. 1922*, 827 (1922).
105. J. W. Armitt and R. Robinson, *J. Chem. Soc. 1925*, 1604 (1925).
106. M. Randić and T. Pisanski, *Rep. Mol. Theory 1*, 107 (1990).
107. M. Randić, D. Plavšić, and N. Trinajstić, *Struct. Chem. 2*, 543 (1991).
108. M. Randić, *J. Mol. Struct. (Theochem) 229*, 139 (1991).
109. M. Randić *Croat. Chem. Acta* (submitted).
110. M. Randić *Polycyclic Aromatic Compounds* (in press).
111. M. Randić, *MATCH* (submitted).
112. M. Randić, D. J. Klein, S. El-Basil, and P. Calkins, *Croat. Chem. Acta* (in press).
113. M. Randić, *Int. J. Quantum Chem.* (in press).
114. W. C. Herndon, M. L. Ellzey, Jr., and K. S. Raghuveer, *J. Am. Chem. Soc. 100*, 2645 (1978).
115. S. H. Bertz, *J. Chem. Soc. Chem. Commun. 1981*, 818 (1981).
116. M. Randić, *Int. J. Quantum Chem. Quantum Chem. Symp. 7*, 187 (1980).
117. A. T. Balaban, *Rev. Roum. Chim. 21*, 1049 (1976).
118. E. V. Konstatinova and V. A. Skorobogatov, *J. Chem. Inf. Comput. Sci. 35*, 472 (1995).
119. C. Liang and Y. Jiang, *J. Theor. Biol. 158*, 231 (1992).
120. C. Liang and K. Mislow, *J. Math. Chem. 15*, 1 (1994).
121. L. V. Quintas and P. J. Slater, *MATCH 12*, 75 (1981).
122. P. J. Slater, *J. Graph Theory 6*, 89 (1982).
123. A. T. Balaban and L. V. Quintas, *MATCH 14*, 213 (1983).
124. M. Razinger, J. R. Chretian, and J. E. Dubois, *J. Chem. Inf. Comput. Sci. 25*, 23 (1985).
125. K. Szymanski, W. R. Müller, J. Knop, and N. Trinajstić, *J. Chem. Inf. Comput. Sci. 25*, 413 (1985).
126 K. Szymanski, W. R. Müller, J. Knop, and N. Trinajstić, *Croat. Chem. Acta 59*, 719 (1986).
127. D. A. Chalcraft, *J. Graph Theory, 14*, 314 (1990).
128. O. Ivanciuc and A. T. Balaban, *J. Math. Chem. 11*, 155 (1992).
129. H. Hosoya, U. Nagashima, and S. Hyugaji, *J. Chem. Inf. Comput. Sci. 34*, 428 (1994).
130. E. Estrada, *J. Chem. Inf. Comput. Sci. 35*, 1022 (1995).
131. M. Randić, *J. Am. Chem. Soc. 97*, 6609 (1975).
132. L. B. Kier, W. J. Murrey, M. Randić, and L. H. Hall, *J. Pharm. Sci. 65*, 1226 (1976).
133. H. Wiener, *J. Am. Chem. Soc. 69*, 17 (1947).
134. M. Randić, *Chem. Phys. Lett. 211*, 478 (1993).
135. M. Randić, X. Guo, T. Oxley, and H. Krishnapriyan, *J. Chem. Inf. Comput. Sci. 33*, 709 (1993).
136. M. Randić, X. Guo, T. Oxley, H. Krishnapriyan, and L. Naylor, *J. Chem. Inf. Comput. Sci. 34*, 361 (1994).
137. D. J. Klein, Z. Mihalić, D. Plavšić, and N. Trinajstić, *J. Chem. Inf. Comput. Sci. 32*, 304 (1992).
138. D. J. Klein, I. Lukovits, and I. Gutman, *J. Chem. Inf. Comput. Sci. 35*, 50 (1995).
139. H. Hosoya, *Bull. Chem. Soc. Jpn. 44*, 2332 (1971).
140. A. Hermann and P. Zinn, *J. Chem. Inf. Comput. Sci. 35*, 551 (1995).
141. M. Randić, D. A. Morales, and O. Araujo, *J. Math. Chem.* (in press).
142. M. Randić, D. A. Morales, and O. Araujo, *J. Math. Chem.* (in press).
143. M. Randić, *Croat. Chem. Acta 67*, 415 (1994).
144. M. Randić, (to be published).
145. M. Randić, *J. Chem. Inf. Comput. Sci. 32*, 57 (1992).

146. M. Randić, *J. Chem. Inf. Comput. Sci. 32*, 686 (1992).
147. M. Randić, *Theor. Chim. Acta 92*, 97 (1995).
148. S. S. Tratch, M. I. Stankevich, and N. S. Zefirov, *J. Comput. Chem. 11*, 899 (1990).
149. L. H. Hall, in: *Computational Chemical Graph Theory* (D. H. Rouvray, ed.), Nova Sci. Publ., Commack, New York (1990).
150. D. J. Klein and M. Randić, *J. Math. Chem. 12*, 81 (1993).
151. M. Randić, A. F. Kleiner, and L. M. DeAlba, *J. Chem. Inf. Comput. Sci. 34*, 277 (1994).
152. A. A. Balandin, *Usp. Khim. 9*, 390 (1940).
153. M. Randić and N. Trinajstić, *Croat. Chem. Acta 67*, 1 (1994).
154. M. Randić, *Croat. Chem. Acta 66*, 289 (1993).
155. M. Randić and P. G. Seybold, *SAR and QSAR 1*, 77 (1993).
156. D. E. Needham, I. C. Wei, and P. G. Seybold, *J. Am. Chem. Soc. 110*, 4186 (1988).
157. S. C. Basak and G. D. Grunwald, *J. Chem. Inf. Comput. Sci. 35*, 366 (1995).
158. S. C. Basak, V. R. Magnuson, G. J. Niemi, and R. R. Regal, *Discrete Appl. Math. 19*, 17 (1988).
159. A. R. Katrizky and E. V. Gordeeva, *J. Chem. Inf. Comput. Sci. 33*, 835 (1993).
160. D. A. Morales and O. Araujo, Preprint (1991).
161. A. Finizio, F. Sicbaldi, and M. Vighi, *SAR and QSAR 3*, 71 (1995).
162. R. W. Taft, in: *Steric Effects in Organic Chemistry* (M. S. Newman, ed.), pp. 556–675, John Wiley & Sons, New York (1956).
163. M. Randić, *J. Chem. Inf. Comput. Sci. 31*, 311 (1991).
164. M. Randić, *New J. Chem. 15*, 517 (1991).
165. M. Randić, *Croat. Chem. Acta 64*, 43 (1991).
166. M. Randić, *J. Comput. Chem. 14*, 363 (1993).
167. M. Randić, *Int. J. Quantum Chem. Quantum Biol. Symp. 21*, 215 (1994).
168. H. Hotelling, *J. Educ. Psychol. 24*, 417, 489 (1933).
169. L. Pogliani, *Curr. Top. Pept. Protein Res. 1*, 119 (1994).
170. L. Pogliani, *J. Phys. Chem. 98*, 1494 (1994).
171. L. Pogliani, *J. Phys. Chem. 99*, 925 (1995).
172. L. Pogliani, *Amino Acids 6*, 41 (1994).
173. L. Pogliani, *J. Chem. Inf. Comput. Sci.* (in press).
174. D. Amić, D. Davidović-Amić, and N. Trinajstić, *J. Chem. Inf. Comput. Sci. 35*, 136 (1995).
175. N. H. Nie, C. H. Hull, J. G. Jenkins, K. Steinbrenner, and D. H. Bent, *SPSS: Statistical Package for Social Sciences*, McGraw–Hill, New York (1975).
176. M. Šoškic, D. Plavšić, and N. Trinajstić, *J. Chem. Inf. Comput. Sci. 36*, 146 (1996).
177. M. Randić, in: *Concepts and Applications of Molecular Similarity* (M. A. Johnson and G. Maggiora, eds.), pp. 77–145, John Wiley & Sons, New York (1990).
178. M. Randić, in: *Mathematical Methods in Contemporary Chemistry* (S. I. Kuchanov, ed.), Chapter 1, pp. 1–100, Gordon & Breach, New York (1996).
179. M. A. Johnson and G. Maggiora, eds., *Concepts and Applications of Molecular Similarity*, John Wiley & Sons, New York (1990).
180. P. D. Walker, G. M. Maggiora, M. A. Johnson, J. D. Petke, and P. G. Mezey, *J. Chem. Inf. Comput. Sci. 35*, 568 (1995).
181. L. B. Kier and H. L. Hall, *Molecular Connectivity in Chemistry and Drug Research*, Academic Press, New York (1976).
182. L. B. Kier and H. L. Hall, *Molecular Connectivity in Structure–Activity Analysis*, John Wiley & Sons, New York (1986).
183. L. B. Kier and L. H. Hall, *J. Pharm. Sci. 70*, 583 (1981).
184. L. H. Hall and L. B. Kier, *Eur. J. Med. Chem. 16*, 399 (1981).
185. L. B. Kier and H. L. Hall, *J. Pharm. Sci. 65*, 1806 (1976).
186. J. R. Platt, *J. Chem. Phys. 15*, 419 (1947).
187. A. T. Balaban, I. Motoc, D. Bonchev, and O. Mekenyan, *Top. Curr. Chem. 114*, 21 (1983).

188. D. Bonchev, *Information Theoretic Indices for Characterization of Chemical Structures*, Research Studies Press, Chichester (1983).
189. M. Randić, *J. Chem. Inf. Comput. Sci.* (in press).
190. M. Randić, *MATCH 7*, 3 (1989).
191. M. Randić, *J. Magn. Res. 39*, 431 (1980).
192. M. Randić and N. Trinajstić, *MATCH 13*, 271 (1982).
193. M. Randić and P. C. Jurs, *Quant. Struc.-Act. Relat. 8*, 39 (1989).
194. M. Randić, *J. Math. Chem. 4*, 157 (1990).
195. M. Randić, *Int. J. Quantum Chem. 23*, 1707 (1983).
196. M. Randić, *Acta Pharm.* (to be submitted).
197. B. Lucić, S. Nikolić, N. Trinajstić, and D. Juretić, *J. Chem. Inf. Comput. Sci. 35*, 532 (1995).
198. H. F. Hameka, *J. Chem. Phys. 34*, 1966 (1961).
199. M. Randić, *Chem. Phys. Lett. 53*, 602 (1978).
200. T. G. Schmalz, D. J. Klein, and B. L. Sandleback, *J. Chem. Inf. Comput. Sci. 32*, 54 (1992).
201. D. J. Klein, *Int. J. Quantum Chem. Quantum Chem. Symp. 20*, 153 (1986).
202. R. D. Poshusta and M. C. McHughes, *J. Math. Chem. 3*, 193 (1989).
203. M. C. McHughes and R. D. Poshusta, *J. Math. Chem. 4*, 227 (1990).
204. V. Kvasnička and J. Pospichal, *J. Chem. Inf. Comput. Sci. 35*, 121 (1994).
205. M. Randić, *J. Comput. Chem. 12*, 970 (1991).
206. M. Randić and J. Cz. Dobrowolski, *J. Math. Chem.* (submitted).
207. M. Randić and S. C. Basak, work in progress.
208. M. Randić, *Chemometrics Intell. Lab. Syst. 10*, 213 (1991).
209. J. C. Slater, *Phys. Rev. 36*, 57 (1930).
210. W. J. Hehre, R. F. Stewart, and J. A. Pople, *J. Chem. Phys. 51*, 2657 (1969).
211. Y. Sakai and T. Anno, *J. Chem. Phys. 60*, 625 (1974).
212. M. Barisz, G. Jashari, R. S. Lall, and N. Trinajstić, in: *Chemical Applications of Topology and Graph Theory* (R. B. King, ed.), p. 222, Elsevier, Amsterdam (1983).
213. O. Mekenyan, D. Bonchev, A. Sabljić, and N. Trinajstić, *Acta Pharm. Jugosl. 37*, 75 (1987).
214. O. Ivanciuc and A. T. Balaban, *MATCH 30*, 117 (1994).
215. E. J. Kupchik, *Quant. Struct. Act. Relat. 8*, 98 (1989).
216. A. T. Balaban, *MATCH 21*, 115 (1986).
217. L. Lovasz and J. Pelikan, *Period. Math. Hung. 3*, 175 (1973).
218. A. T. Balaban, D. Ciubotariu, and M. Medeleanu, *J. Chem. Inf. Comput. Sci. 31*, 597 (1991).
219. F. Harary, *Graph Theory*, Addison–Wesley, Reading, Massachusetts (1969).
220. D. Cvetković, M. Doob, and H. Sachs, *Spectra of Graphs—Theory and Application*, 2nd ed., Academic Press, New York (1982).
221. D. Cvetković and P. Rowlinson, *Linear and Multilinear Algebra 28*, 3 (1990).
222. R. H. Martin, *Angew. Chem. Int. Ed. Engl. 13*, 649 (1974).
223. M. Randić and S. Bobst, work in progress.
224. B. B. Mandelbrot, *The Fractal Geometry of Nature*, Freeman, San Francisco (1982).
225. H.-O. Peitgen, H. Jurgens, and D. Saupe, *Chaos and Fractals*, Springer-Verlag, Berlin (1992).
226. H. von Koch, *Ark. Mat. 1*, 681 (1904).
227. G. Peano, *Math. Ann. 36*, 157 (1890).
228. D. Hilbert, *Math. Ann. 38*, 459 (1891).
229. W. Sierpinski, *C. R. Acad. Paris 160*, 302 (1915).
230. The Dragon curve was discovered by J. E. Heighway and analyzed by J. E. Heighway, W. G. Harter, and B. A. Banks; see M. Gardner, *Mathematical Magic Show*, Chapter 15, Vintage Books, New York (1978).
231. D. E. Knuth and C. Davis, *J. Recreational Math. 3*, 66, 133 (1970).
232. A. R. Leach, *J. Chem. Inf. Comput. Sci. 34*, 661 (1994).
233. M. Randić, *Stud. Phys. Theor. Chem. 54*, 101 (1988).

234. M. Randić, *Int. J. Quantum Chem. Quantum Biol. Symp. 15*, 201 (1988).
235. M. Randić, B. Jerman-Blažić, and N. Trinajstić, *Comput. Chem. 14*, 237 (1990).
236. S. Nikolić, N. Trinajstić, Z. Mihalić, and S. Carter, *Chem. Phys. Lett. 179*, 289 (1989).
237. B. Bogdanov, S. Nikolić, and N. Trinajstić, *J. Math. Chem. 5*, 305 (1990).
238. K. Balasubramanian, *Chem. Phys. Lett. 169*, 224 (1990).
239. Z. Mihalić and N. Trinajstić, *J. Mol. Struct. (Theochem.) 232*, 65 (1991).
240. J. Ivanov, St. Karabunarliev, and O. Mekenyan, *J. Chem. Inf. Comput. Sci. 34*, 234 (1994).
241. L. Pogliani, *J. Chem. Inf. Comput. Sci. 34*, 801 (1994).
242. M. V. Diudea, D. Horvath, and A. Graovac, *J. Chem. Inf. Comput. Sci. 35*, 129 (1995).
243. K. Balasubramanian, *Chem. Phys. Lett. 235*, 580 (1995).
244. E. Estrada, *J. Chem. Inf. Comput. Sci. 35*, 714 (1995).
245. A. T. Balaban, *Chem. Phys. Lett. 89*, 399 (1982).
246. A. T. Balaban and P. Filip, *MATCH 16*, 163 (1984).
247. A. T. Balaban, *Pure Appl. Chem. 55*, 199 (1983).
248. M. Randić and M. Razinger, *J. Chem. Inf. Comput. Sci. 35*, 140 (1995).
249. M. Randić, *J. Chem. Inf. Comput. Sci. 35*, 373 (1995).
250. M. Randić and M. Razinger, *J. Chem. Inf. Comput. Sci. 35*, 594 (1995).
251. M. Randić, *J. Math. Chem. 19*, 375 (1996).
252. M. Randić and G. Krilov, *Int. J. Quantum Chem. Quant. Biol. Symp.* (in press).
253. F. Harary and P. G. Mezey, *J. Math. Chem. 2*, 377 (1988).
254. X. Luo, G. A. Arteca, and P. G. Mezey, *Int. J. Quantum Chem. 42*, 459 (1992).
255. P. D. Walker and P. G. Mezey, *J. Am. Chem. Soc. 115*, 12423 (1993).
256. M. Randić and G. Krilov, *Chem. Phys. Lett.* (submitted).
257. J. D. Watson, *Molecular Biology of the Gene* (2nd ed.), Benjamin, Menlo Park, California (1970).
258. M. Kac, *Am. Math. Mon. 73*, 1 (1966).
259. G. A. Baker, Jr., *J. Math. Phys. 7*, 2238 (1966).
260. M. E. Fisher, *J. Comb. Theory 1*, 105 (1966).
261. A. T. Balaban and F. Harary, *J. Chem. Doc. 11*, 258 (1971).
262. M. Randić, M. Barysz, J. Nowakowski, S. Nikolić, and N. Trinajstić, *J. Mol. Struct. (Theochem.) 185*, 95 (1989).
263. J. V. Knop, W. R. Müller, K. Szymanski, N. Trinajstić, A. F. Kleiner, and M. Randić, *J. Math. Phys. 27*, 2601 (1986).
264. M. Randić and A. F. Kleiner, *Ann. N.Y. Acad. Sci. 555*, 320 (1989).
265. Y. Jiang, *Sci. Sin. Ser. B 27*, 236 (1984).
266. A. Radecki, H. Lamparczyk, and R. Kaliszan, *Chromatographia 12*, 597 (1979).
267. R. Kaliszan, H. Lamparczyk, and A. Radecki, *Biochem. Pharmacol. 28*, 123 (1979).
268. A. Robbat, N. P. Corso, P. J. Doherty, and D. Marshall, *Anal. Chem. 58*, 2072 (1986).
269. M. Randić, *New J. Chem. 19*, 781 (1995).
270. P. G. Mezey, *Shape in Chemistry: Introduction to Molecular Shape and Topology*, VCH Publishers, New York (1993).
271. P. G. Mezey, in: *Reviews in Computational Chemistry* (K. B. Lipkowitz and D. B. Boyd, eds.), Chapter 7, VCH Publishers, New York (1990).
272. G. Arteca and P. G. Mezey, *J. Comput. Chem. 9*, 554 (1988).
273. G. Arteca and P. G. Mezey, *Int. J. Quantum Chem. 34*, 517 (1988).
274. P. G. Mezey, *J. Chem. Inf. Comput. Sci. 34*, 244 (1994).
275. W. W. Rouse Ball, *Mathematical Recreations and Essays*, 5th printing, p. 262, Macmillan Co., New York (1967).
276. A. T. Balaban, *Tetrahedron 27*, 6115 (1971).
277. P. D. Walker and P. G. Mezey, *Int. J. Quantum Chem. 43*, 375 (1992).
278. Lord Kelvin, *Baltimore Lectures on Molecular Dynamics and the Wave Theory of Light*, pp. 439, 619, C. J. Clay and Sons, London (1904).

279. A. B. Buda and K. Mislow, *J. Mol. Struct. (Theochem.) 232*, 1 (1991).

280. M. Randić and P. G. Mezey, *J. Chem. Inf. Comput. Sci.* (in press).

281. R. W. Hamming, *Bell Syst. Tech. J. 29*, 147 (1950).

282. M. Randić and M. Razinger, *J. Chem. Inf. Comput. Sci. 36*, 429 (1996).

283. A. Kitaigorodskii, *Organic Chemical Crystallography*, p. 230, Consultants Bureau, New York (1961).

284. W. Kauzmann, F. B. Clough, and I. Tobias, *Tetrahedron 13*, 57 (1961).

285. Z. Ugi, *Z. Naturforsch. 20B*, 405 (1965).

286. H. Zabrodsky and D. Avnir, *J. Am. Chem. Soc. 117*, 462 (1995).

287. G. Gilat, *J. Math. Chem. 15*, 197 (1994).

288. N. Weinberg and K. Mislow, *J. Math. Chem. 17*, 35 (1995).

289. A. B. Buda and K. Mislow, *Elem. Math. 46*, 65 (1991).

290. R. F. Muirhead, *Proc. Edinburgh Math. Soc. 21*, 144 (1903).

291. E. Ruch, *Acc. Chem. Res. 5*, 49 (1972).

292. G. H. Hardy, J. E. Littlewood, and G. Polya, *Inequalities*, Cambridge University Press, London (1934).

293. R. D. Cramer III, D. E. Patterson, and J. D. Bunce, *J. Am. Chem. Soc. 110*, 5959 (1988).

7

Chemical Graph Theory of Fullerenes

PATRICK W. FOWLER

7.1. INTRODUCTION

The dramatic story of the discovery of C_{60} and of a way of manufacturing this new molecule in gram quantities is now well known to professionals and amateurs of science alike.[1-6] C_{60} and its companion C_{70} have become the prototypes of a whole new class of molecules—the fullerenes—and the chemistry, physics, and materials science literature since 1985 and especially since 1990 has been flooded with reports of measurements, calculations, and speculations on their properties. For the theoretical chemist, one of the prime motivations for studying the fullerenes is that they show a clear link between geometrical/topological and electronic structure. This link will be reviewed here.

In order to see the question in context, it will be necessary to address a number of problems in fullerene systematics, namely: What is a fullerene? How many structures are hypothetically possible, and how may they be constructed? What spectra are to be expected of fullerenes? What is the connection between π-electronic properties and overall stability? The present chapter deals with some of these questions in relation to the central problem of geometric and electronic structure. A more detailed account that includes discussion of isomerization, growth, and formation may be found elsewhere.[7]

PATRICK W. FOWLER • Department of Chemistry, University of Exeter, Exeter EX4 4QD, England.

From Chemical Topology to Three-Dimensional Geometry, edited by Balaban. Plenum Press, New York, 1997

7.2. THE ISOMER PROBLEM

The isomer problem, simply stated, is to find how many distinct fullerenes could be assembled from n carbon atoms.

7.2.1. What Is a Fullerene?

The conventional definition, and the one to be used in the present chapter, is that a fullerene C_n is a carbon cage where the atoms define the vertices of a trivalent polyhedron with 12 pentagonal and $n/2 - 10$ hexagonal faces. Fullerene polyhedra of this type are realizable for $n = 20$ and for all even $n \geq 24$, i.e., $n = 20 + 2k \, (k \neq 1)$ where k counts the hexagonal faces. All experimentally characterized fullerenes comply with this definition, and in fact all so far found have in addition isolated pentagons; isolated-pentagon fullerenes are possible for $n = 60$ and for all even $n \geq 70$.

Almost any trivalent polyhedron, spherical, toroidal, or of higher genus, is at least conceivably a candidate for an sp^2 carbon framework if reasonable bond lengths and angles can be achieved, and so there has naturally been some discussion in the literature of extended definitions of fullerenes (e.g., Refs. 8–17). Exotic topologies are dealt with elsewhere in the present volume in the chapters by Kirby and Klein, but even within the class of pseudospherical polyhedra other face sizes are possible. The Euler formula for an n-vertex trivalent spherical polyhedron is

$$v + f = e + 2$$

with $v = n$, $e = 3n/2$, and therefore $f = (n/2) + 2$, giving

$$\sum_r (6 - r)f_r = 3f_3 + 2f_4 + f_5 - f_7 - 2f_8 - \cdots = 12$$

The conventional fullerene recipe of $f_5 = 12$ and $f_6 = n/2 - 10$ has advantages for carbon as it gives the nearest spherical analogue of the hexagonal sheet, with many rings having the size of 6 favored by π-electronic considerations and 12 pentagonal defects supplying the necessary curvature. There are suggestions that inclusion of some "square" rings in sub-C_{70} cages may be energetically favored as they could alleviate the strain of pentagon crowding,[18] and observations of kink and taper morphologies in some nanotubes appear to imply the presence of some heptagonal rings in these very large systems,[19] so that eventually it may be necessary to broaden the definition from 5-6 to 4-5-6-7 trivalent polyhedra. In the absence of firm experimental evidence to the contrary, it is assumed here that treatment of this wider class can be left for the future.

7.2.2. How Many Fullerenes Are There?

Within the simple definition of the previous section there are in general many isomeric cages for a given number of carbon atoms. It has been argued[20] that the number of fullerene isomers of C_n should increase asymptotically as n^9; explicit consideration of the known counts[7] suggests an even faster rise in the chemically important range of $20 \leq n \leq 100$. Most of these isomers will be highly reactive or thermodynamically unstable, or both, but it is important as a first step in a systematic treatment to establish the universe of possibilities from which observed fullerenes are drawn.

Several methods of enumeration have been developed for the fullerene problem,[21-26] and there is now general agreement on the isomer counts for $n \leq 150$, which is more than adequate for present chemical purposes. Two classes of method can be envisaged: one that takes a "vertical slice" of fullerenes of some particular type or symmetry[21] and one that takes a "horizontal slice" of all fullerenes at a given n.[22] Both have advantages, and one method of each type will be summarized here.

Fullerenes may belong to only a restricted set of symmetry point groups. Any symmetry axis or plane of a fullerene cage must pass through the cage in two or more special points: atom positions, bond centers, or face centers. The maximal site symmetry is C_{3v} for an atom in a trivalent cage, C_{2v} for a bond center, and C_{5v} or C_{6v} for a (pentagonal or hexagonal) face center. Consideration of the overall point groups compatible with these site groups, and more detailed arguments about the consequences of high-order axes,[27] show that a fullerene may have one of 28 maximal (or "topological") symmetries, i.e., I_h, I, T_h, T_d, T, D_{6h}, D_{6d}, D_6, D_{5h}, D_{5d}, D_5, D_{3h}, D_{3d}, D_3, D_{2h}, D_{2d}, D_2, S_6, S_4, C_{3h}, C_{3v}, C_3, C_{2h}, C_{2v}, C_2, C_s, C_i, C_1. (The maximal symmetry is the largest group of symmetry operations compatible with 3D realization of the fullerene molecular graphs as a polyhedron[28]; in specific electronic states or with particular chemical additions or substitutions the cage may adopt a lower physical point group.)

A method for counting the fullerenes within each symmetry can be based on a 60-year-old mathematical construction that has proved useful for classifying polyhedra, viruses, and geodesic domes. The original work by Goldberg,[29] as rediscovered by Caspar and Klug in the context of viruses,[30] implicit in the geodesic dome constructions of Buckminster Fuller,[31] and reviewed by Coxeter,[32] applies to structures of icosahedral (I_h or I) symmetry.

In the Coxeter construction, a net of 20 congruent equilateral master triangles is superimposed on the tessellation of the plane by smaller equilateral triangles (Figure 1). The net can be lifted off the plane and folded to give a spherical deltahedron that is the dual of an icosahedral fullerene: every small triangle of the tessellation corresponds to an atom of the final fullerene, and every point where six small triangles meet corresponds to a hexagon center. The 12 pentagons of the fullerene arise from the defects in the triangulation introduced where five master triangles of the net meet.

For icosahedral symmetry, the whole net is completely specified by one lattice vector, i.e, by an ordered pair of integers (a, b).[32] The atom count of the fullerene is proportional to the area of the planar net, and so it is easily shown that at least one icosahedral fullerene C_n exists for

$$n = 20(a^2 + ab + b^2) \qquad (I_+)$$

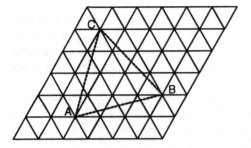

Figure 1. The Goldberg/Coxeter construction of an icosahedral triangulation of the sphere. Twenty copies of the large equilateral triangle fit together to form the net of a master icosahedron, which yields an icosahedral fullerene on taking the dual. The Coxeter parameters in this particular example are $a = 3$ and $b = 1$, i.e., to reach B from A, take 3 steps along the horizontal to the right, turn left through 60°, and take 1 step.

with $a \geq b$, $a > 0$, $b \geq 0$.

The code (a, b) gives information about the overall symmetry: fullerene cages with $b = a$ or $b = 0$ have the untwisted I_h symmetry, other combinations $a > b > 0$ have the chiral I symmetry and exist as enantiomeric pairs.

Numbers of the form $(a^2 + ab + b^2)$ are either divisible by 3 or exceed by one a multiple of 3, i.e,

$$(a^2 + ab + b^2) = 3m$$

or

$$(a^2 + ab + b^2) = 3m + 1$$

In the second series the cage has 20 equivalent atoms at the vertices of a dodecahedron, i.e., at the centers of the faces of the master icosahedron. In the first, all atoms fall into sets of 60 (I) or 60 and 120 (I_h) but with no set of atoms in the 20 special positions.

The Goldberg numbers characterize a set of fullerenes of increasing size in which all 12 pentagons are equivalent and (beyond C_{20}) isolated from one another. They have particular electronic properties as we will see later. Isomerism is possible, in that distinct (a, b) pairs may yield the same value of n. The smallest such case is $n = 980$ with an isomeric trio of $(a, b) = (7, 0)(I_h)$ and the chiral pair $(5, 3)(I)$; the smallest case with three distinct integer pairs is the set $(23, 4)$, $(21, 7)$, and $(17, 12)$ which all give $n = 12,740$.

The construction can be extended to other symmetries by adapting the shape of the net.[21,33] Cylindrical fullerenes with a fivefold rotational axis have nets consisting of 10 equilateral and 10 scalene master triangles arranged in five parallel strips, specified by a set of four integers; the net of a sixfold cylinder can be produced from the same code if an extra parallel strip is added. The cylindrical fullerenes C_n have

$$n = 10[k^2 + kl + l^2 + (il - jk)] \qquad (D_{5+})$$

$$n = 12[k^2 + kl + l^2 + (il - jk)] \qquad (D_{6+})$$

where D_{5+} is an abbreviation for I_h, I, D_{5d}, D_{5h}, D_5 and D_{6+} for D_{6d}, D_{6h}, D_6, and the integers obey appropriate restrictions.[33]

Tetrahedral ($T_+ = T_h$, T_d, and T) fullerenes can be constructed from a net based on the general twisted, truncated tetrahedron that has two sets of 4 equilateral and 12 scalene master triangles. Again, only four integers, with appropriate restrictions, are needed to code the structure. The vertex counts of the T_+ C_n fullerenes are[21]

$$n = 4[(i^2 + ij + j^2) + (k^2 + kl + l^2) + 3(il - jk)] \qquad (T_+)$$

All four equations have been used to count and construct fullerenes within the respective symmetry classes, and the tetrahedral construction in particular was central to the identification of a counterexample to the ring-spiral conjecture.[34] The Goldberg/Coxeter construction is demonstrably complete, though as the symmetry is lowered the number of parameters needed to specify the net increases and the removal of redundant codes becomes more difficult: Nets for several further groups have been devised[21] and the coding of threefold cylinders has been worked out in Exeter. As most fullerenes have no symmetry at all,[27] the method does not offer a practical solution for "horizontal" enumeration of all fullerenes at given n.

A "horizontal" method that has been used with a good deal of success is based on the conjecture that any fullerenes can be unwrapped or peeled as a continuous edge-sharing spiral strip of all $n/2 + 2$ faces.[22] The 12 positions of pentagons in this strip give a 1D code from which the 3D structure can be reconstructed, and if lexicographic ordering is applied to reduce the redundancy, we have a compact and readily computed representation of the fullerene. As the spiral encodes the adjacency matrix A of the fullerene, many relevant properties including cage symmetry, model Cartesian coordinates, spectroscopic signatures, and qualitative π-electronic structure can be recovered from it.[7,28] The spiral algorithm is also easily specialized to yield only isolated-pentagon cages, and can be extended if desired to count 4-5-6 and 4-5-6-7 analogues of the fullerenes.

Table 1 shows the numbers of distinct (nonenantiomorphic) fullerenes up to C_{100} and isolated-pentagon fullerenes to C_{140} found by the spiral algorithm. Tabulations of symmetry, electronic and spectroscopic properties, indicators of steric strain and illustrations of geometric structures for all fullerenes C_n $20 \leq n \leq 50$ and all isolated-pentagon fullerenes C_n $60 \leq n \leq 100$ are given in An Atlas of Fullerenes.[7] A listing of the program for generation of spiral codes is also given in that reference. It should be noted that in those cases where a complete set of calculations has been performed, all distinct fullerene isomers correspond to minima on the potential energy hypersurface, giving a posteriori justification for the consideration of the whole set in theoretical investigations.

Many problems in electronic structure, cage isomerization, and fullerene growth become accessible when a complete set of isomers is available for the testing of ideas. A nomenclature based on the spiral code forms part of the IUPAC proposals on systematic naming of the fullerenes; when labels based on atom count and symmetry

Table 1.   Enumeration of C_n Fullerene Isomers in the Range $n = 20$ to 100
(first three columns), and Isolated-Pentagon Isomers in the
Range $n = 70$ to 140 (last three columns)[a]

n	Isomers		n	IPR Isomers	
20	1	(1)	70	1	(1)
24	1	(1)	72	1	(1)
26	1	(1)	74	1	(1)
28	2	(3)	76	2	(3)
30	3	(3)	78	5	(6)
32	6	(10)	80	7	(9)
34	6	(9)	82	9	(12)
36	15	(23)	84	24	(34)
38	17	(30)	86	19	(33)
40	40	(66)	88	35	(56)
42	45	(80)	90	46	(78)
44	89	(162)	92	86	(161)
46	116	(209)	94	134	(252)
48	199	(374)	96	187	(349)
50	271	(507)	98	259	(483)
52	437	(835)	100	450	(862)
54	580	(1,113)	102	616	(1,179)
56	924	(1,778)	104	823	(1,606)
58	1,205	(2,344)	106	1,233	(2,401)
60	1,812	(3,532)	108	1,799	(3,502)
62	2,385	(4,670)	110	2,355	(4,645)
64	3,465	(6,796)	112	3,342	(6,568)
66	4,478	(8,825)	114	4,468	(8,820)
68	6,332	(12,501)	116	6,063	(11,997)
70	8,149	(16,091)	118	8,148	(16,132)
72	11,190	(22,142)	120	10,774	(21,326)
74	14,246	(28,232)	122	13,977	(27,763)
76	19,151	(38,016)	124	18,769	(37,313)
78	24,109	(47,868)	126	23,589	(46,907)
80	31,924	(63,416)	128	30,683	(61,069)
82	39,718	(79,023)	130	39,393	(78,476)
84	51,592	(102,684)	132	49,878	(99,343)
86	63,761	(126,973)	134	62,372	(124,282)
88	81,738	(162,793)	136	79,362	(158,258)
90	99,918	(199,128)	138	98,541	(196,532)
92	126,409	(252,082)	140	121,354	(242,126)
94	153,493	(306,061)			
96	191,839	(382,627)			
98	231,017	(461,020)			
100	285,914	(570,603)			

[a]For each entry the count is given first treating enantiomeric pairs as a single isomer, then (in parentheses) treating each enantiomer as distinct.

alone run out, as is bound to happen for large cages, the 12 pentagon positions in the spiral code can be used for unique identification of any spiralable fullerene.

The spiral proposal was always a conjecture in the mathematical sense: no proof was offered for its completeness in general, though it was shown that all fullerenes with high-order (C_5 or C_6) rotational axes have spirals.[27] After extensive attempts to prove or break the conjecture, a series of counterexamples was eventually found,[34] all having tetrahedral symmetry and all having 380 atoms or more. Further work, not yet published, has identified some dozens of fullerenes of lower symmetry without spirals, but none below the 380-atom threshold. From the point of view of accuracy of enumeration, these exceptions to the spiral conjecture, though interesting, are not important: the total count for $n = 380$ is expected to be many billions and is not obtainable with present methods and computers; a shortfall of 1 is therefore not too worrying from a practical or chemical point of view. The exceptional cages themselves, at least initially, have highly strained fused-triple arrangements of pentagons and so are unlikely to be synthesized in any case. Perhaps the most heartening news for users of the spiral algorithm is that the new superfast and mathematically complete enumeration technique based on Petrie paths and "pent-hex" puzzles that was recently developed by Brinkmann and Dress[26] confirms the spiral counts at least as far as $n = 170$. However obtained, the sets of isomers for $n \leq 100$ and beyond are now useful test-beds for hypotheses about stability and structure.

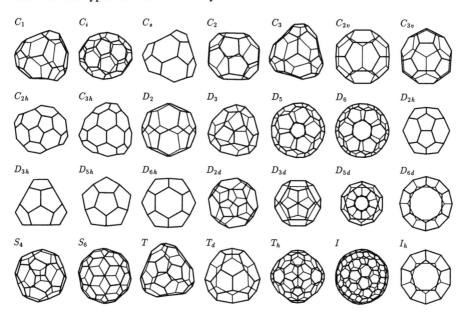

Figure 2. Representative fullerene (the smallest) for each of the 28 possible maximal point groups. The examples shown are C_n with: $n = 20$ (I_h), $n = 24$ (D_{6d}), $n = 26$ (D_{3h}), $n = 28$ (D_2, T_d), $n = 30$ (D_{5h}, C_{2v}), $n = 32$ (C_2, D_{3d}, D_3), $n = 34$ (C_s, C_{3v}), $n = 36$ (C_1, D_{2d}, D_{6h}), $n = 40$ (D_{5d}, C_3, D_{2h}), $n = 44$ (T, S_4), $n = 48$ (C_{2h}), $n = 56$ (C_i), $n = 60$ (D_5), $n = 62$ (C_{3h}), $n = 68$ (S_6), $n = 72$ (D_6), $n = 92$ (T_h), $n = 140$ (I).

Figure 2 shows the smallest fullerene of each of the 28 attainable maximal point symmetry groups. Symmetry counts[7] confirm the intuitively expected result that high symmetry is the exception rather than the rule, and that the overwhelming majority of fullerenes will have little or no symmetry. In this respect the early members C_{60} and C_{70} of the experimental series may be misleading guides to what is to be expected as their higher homologues are synthesized and separated. High symmetry is not necessarily a concomitant of stability.

7.3. FULLERENE NMR SPECTRA

The main tool used in the characterization of the higher fullerenes has been ^{13}C NMR spectroscopy.[35-40] In natural abundance, only 1 in 100 carbon nuclei is ^{13}C, and in ideal circumstances the ^{13}C NMR spectrum of a fullerene will be very simple. It will consist of a number of peaks, one for each set of equivalent atomic sites, with intensities proportional to the number of sites in each set.

The site symmetries available to a ^{13}C nucleus in a fullerene are limited: the nucleus may lie on a threefold axis (proper or improper) or on no axis, it may lie on zero, one, or three mirror planes. The site symmetry is therefore described by one of the groups C_{3v}, C_3, C_s, or C_1. The number of equivalent nuclei in a set is the ratio $|G|/|H|$ of the orders of the overall point group of the molecule and the site group: ^{13}C NMR peak heights within a single spectrum are inversely proportional to the orders of the respective site groups. A symmetry analysis of the vertex sets of a fullerene can therefore predict the numbers and relative intensities, though not the positions, of peaks in the ^{13}C NMR spectrum of a given hypothetical isomer.

The overall symmetry and assignment of nuclei to orbits can be done in several ways. Indirectly, the topological coordinates[28] obtained from the eigenvectors of the adjacency matrix can be used to construct a 3D geometric structure that is then checked for symmetry elements. Directly, the order of the point group follows from the number of redundant copies of each distinct face spiral.[7,34] NMR signatures are tabulated in Ref. 7 for all isomers with $20 \leq n \leq 50$ and all isolated-pentagon isomers with $60 \leq n \leq 100$.

Symmetry can impose powerful limitations on the form of the ^{13}C NMR spectrum. Clearly, if only four site groups are available, the idealized spectrum may have no more than four different peak heights; in fact, since the site groups C_{3v} and C_3 are mutually exclusive (no point group having two distinct sets of C_3 axes), the spectrum may contain at most *three* different peak heights. Any spectrum with a larger number of distinct peak heights is an indication of a mixture. The 28 fullerene point groups give NMR patterns as follows:

C_1 sites only: C_1, C_i, C_2, S_4, D_2, D_5, D_6 all heights equal
C_1 and C_s sites: C_s, C_{2h}, C_{2v}, D_{2h}, D_{5h}, D_{6h}, D_{2d}, D_{5d}, $D_{6d} \leq 2$ heights (2:1)
C_1 and C_3 sites: C_3, S_6, D_3, T, $I \leq 2$ heights (3:1)

C_1, C_s, and C_3 sites: C_{3h}, $T_h \leq 3$ heights (3:2:1)

C_1, C_s, and C_{3v} sites: C_{3v}, D_{3h}, D_{3d}, T_d, $I_h \leq 3$ heights (6:2:1)

It is important to bear in mind here that the physical symmetry of the fullerene molecule may be lower than the maximal symmetry of its graph. First- or second-order Jahn-Teller distortion of cages that would have open-shell or pseudoclosed electronic configurations in the maximal point group will reduce the symmetry and increase the number of NMR peaks. In reading the tabulations or the general rules given above, it will then be necessary to use the physical rather than the maximal groups.

Symmetry alone gives no information on ^{13}C NMR chemical shifts, but as experience accumulates[35–40] it may be possible to make useful estimates of shifts from the topological or geometrical characteristics of the carbon sites. A correlation[41] of calculated chemical shift with local bond angles through the POAV (π orbital axis vector[42]) parameter is a useful step in this direction, suggesting that progress can be made using any theoretical method that yields reasonable geometries.

Equal peak counts are often predicted for distinct fullerene isomers, but even so the requirement of compatibility can give a powerful reduction of the possibilities to be investigated. C_{76}, for example, has 19,151 fullerene isomers of which 50 give rise to the observed count of 19 peaks in maximal symmetry and a further 5 could conceivably do so under an appropriate reduction in symmetry. When taken together with the energetic consideration that only one of the 50 has isolated pentagons, counting peaks is sufficient to identify the experimental isomer as *the* isolated-pentagon D_2 C_{76}.[43] This coupling of symmetry and energetic arguments was also decisive in the identification of the experimental C_{84} product as a 2 : 1 mixture of D_2 and D_{2d} isolated-pentagon isomers[39] (22 and 23 in the list of 24 isolated-pentagon C_{84} fullerenes[28]). The same kind of approach is now being used for fullerene derivatives where frequently the NMR evidence is compatible with a number of isomers that differ substantially in computed stability so that the experimental isomer can be identified with a degree of certainty from a combination of NMR and theoretical data (e.g., Ref. 44).

Although this section has concentrated on the NMR spectrum, the topological and symmetry available encapsulated in the face spiral or the adjacency matrix can equally well be used to construct IR and Raman signatures for a set of fullerene isomers.[7] One practical problem with their use for comparison with experiment is that it seems to be a general feature of fullerenes and their compounds that many of the allowed IR and Raman bands have low intensity, so that the *number* of allowed peaks is not of itself such a valuable piece of information as in the case of the NMR spectrum.

7.4. ELECTRONIC STRUCTURE

This section deals with the overall qualitative features of fullerene electronic structure rather than with highly accurate calculations on specific cages. Some *ab initio* calculations are reviewed in Refs. 45 and 46.

7.4.1. The Model

The "obvious" qualitative model of electronic structure of fullerenes is based on simple Hückel theory. The edges of the polyhedral cage are taken to be spanned by a set of two-electron σ bonds, leaving to each atom one electron in an approximately radial π orbital to contribute to a surface π system. Appealing as this model is at first sight, it involves a number of assumptions that require some justification.

The separation of σ and π systems in a planar hydrocarbon is forced by symmetry, since σ orbitals are symmetric and π orbitals antisymmetric with respect to the molecular plane; on the spherical surface the separation of radial π orbitals from the radial and tangential σ orbitals is no longer forced but still holds approximately for energetic reasons. Explicit calculations on C_{60}, for example, show the molecular orbitals at or near the HOMO level to have predominantly π character,[47] and Hückel calculations in general give a good guide to the HOMO–LUMO gaps for fullerenes.

A second problem with the application of Hückel theory to general fullerenes is that a single α (β) parameter may not be appropriate for all atoms (bonds): the environments of atoms may vary from near-sp^3 (at the junction of three pentagons, for example) to near-sp^2 (at a hexagon junction in a graphitic "face" of some large fullerene, for example). Some dispersion in the parameters must therefore be expected. It is possible to account for this by using parameterized semiempirical methods, but to some extent this defeats the object of treating the whole class of molecules by a qualitative method from which global conclusions may be drawn.

For a first survey of the electronic structures of the fullerenes, we take the position that all variation in Hückel parameters and all implicit or explicit treatment of σ electrons is to be relegated to the "steric" factor and that the crude version of Hückel theory is taken to describe the pure π effects. The advantages of this approach for searching for formal rules of fullerene electronic structure will be made apparent in the following sections. It is also clear that the steric factor defined as above must be taken into account if realistic energies and energetic trends are to be obtained. Indeed, the steric factor seems to *dominate* relative stabilities of both lower and higher fullerenes, as will also become apparent later.

7.4.2. Types of Electronic Structure

A fullerene C_n has a nominal total of n π electrons and in the simplest Hückel model, the manifold of π orbital energies is obtained by diagonalizing the adjacency matrix ($A_{ij} = 1$ for atoms linked along an edge of the polyhedron, $A_{ij} = 0$ otherwise). As a trivalent, nonalternant polyhedron, a fullerene has an eigenvalue spectrum $-3 < \lambda_i \leq +3$ ($i = 1, \ldots, n$), leading to orbital energies $\alpha + \lambda_i \beta$ where the single Coulomb parameter α provides an origin and the resonance parameter β a scale for the topologically determined spectrum. The π-electron configuration is obtained by applying the Aufbau principle and Hund's rule, i.e., by filling the orbitals in order of increasing energy (decreasing λ_i) and maintaining maximal multiplicity during the process of filling any degenerate set.

A number of types of configuration can be envisaged: (1) open ($\lambda_{n/2} = \lambda_{n/2+1}$) where the HOMO energy level is only part filled; (2) pseudoclosed[21] ($\lambda_{n/2} > \lambda_{n/2+1} > 0$) where the LUMO level is nominally bonding and ready to accept more electrons; (3) properly closed ($\lambda_{n/2} > 0$, $\lambda_{n/2+1} \leq 0$) where the HOMO is bonding and the LUMO non- or antibonding; (4) metaclosed[48] ($\lambda_{n/2} \leq 0$, $\lambda_{n/2} \neq \lambda_{n/2+1}$) where even the HOMO is non- or antibonding.

Open-shell cages are expected to be kinetically unstable (reactive) and/or susceptible to first-order Jahn–Teller distortion if the maximal symmetry of the framework is high. Pseudoclosed cages have unused bonding capacity and, other things being equal, a small HOMO–LUMO gap that would render them liable to second-order Jahn–Teller distortion. Metaclosed cages would have electrons in antibonding orbitals and would therefore be expected to be metastable to distortion to some other topology; in fact, no strictly metaclosed fullerene isomers have been identified so far, though some large tetrahedral cages ($n > 600$) have a (part-filled) antibonding HOMO. The major class of interest from the point of view of π stability is therefore class 3, the properly closed shells, as these should have significant HOMO–LUMO gaps and should be chemically and thermodynamically stable if π-electronic effects are dominant.

A survey of the Hückel spectra of the lower fullerenes rapidly reveals that properly closed π shells are very much the exception rather than the rule; most fullerene isomers have pseudoclosed π configurations. The rare occurrences of properly closed shells can be almost entirely described by two magic number rules that are to the fullerenes what the Hückel $4n + 2$ is to monocyclic systems and the Wade $n + 1$ rule is to boranes. The two are the *leapfrog* and the *carbon cylinder* rules.

7.4.3. The Leapfrog Rule and Electron Deficiency

The possibility of a general rule connecting electronic and geometric structure for fullerenes first emerged from Hückel calculations on the icosahedral series described earlier, i.e., where C_n with $n = (i^2 + ij + j^2)$. The electronic configurations of the icosahedral fullerenes follow a remarkably simple rule[49]:

a. For an I or I_h fullerene with $60k$ atoms a properly closed shell with a filled fivefold degenerate HOMO is found for the neutral molecule.

b. For an I or I_h fullerene with $60k + 20$ atoms an open shell with two electrons in a fourfold degenerate HOMO is found for the neutral molecule.

Thus, I_h C_{60} is the first of an infinite magic-number series of neutrals and I_h C_{20}^{2+} the first of a series with closed-shell dications but open-shell neutrals. Inspection of the two series shows a general geometrical relationship in that any open-shell icosahedral fullerene C_n with $n = 60k + 20 = 20(3k + 1)$ can be converted into a larger icosahedral fullerene C_{3n} with $3n = 60(3k + 1)$ by a specific transformation, and all icosahedral closed-shell neutrals are so produced. The conversion operation is the so-called

 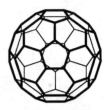

Figure 3. Action of the leapfrog operation on a dodecahedron: the 20-vertex trivalent dodecahedron is converted to a 32-vertex deltahedron by omnicapping, and then to the 60-vertex truncated icosahedron by taking the dual.

leapfrog transformation,[50] which derives its name from the fact that it jumps from one fullerene to another over an intermediate deltahedron, and its significance from the fact that it always yields a fullerene with a properly closed shell, no matter what the topology or symmetry of the starting fullerene.

As applied to a fullerene polyhedron, the leapfrog transformation can be realized as a simple two-stage procedure. The parent is first capped on every face and then the dual is taken, yielding a fullerene polyhedron of the same (maximal) symmetry but with three times as many vertices. For example, the leapfrog of I_h C_{20} is I_h C_{60}: the omnicapping of the dodecahedron leads to a 32-vertex deltahedron with twelve 5-coordinate and twenty 6-coordinate vertices, the dual of which is the C_{60} polyhedron with 12 pentagonal and 20 hexagonal faces (Figure 3).

An alternative route to the same result is to take the dual *first* (in this case giving the dual of the dodecahedral parent, i.e., the icosahedron), and then truncate the result by shaving off a small pyramid from every vertex (in this case producing, of course, the truncated icosahedron).[51]

After the operation has been carried out by whichever route, the final polyhedron has (1) a face of size r for every face of that size in the parent, but rotated through π/r from its original orientation; (2) an extra face of size $2d$ centered on each vertex of degree d of the parent; (3) an edge crossing every edge of the original at right angles, the set of such "perpendicular" edges exhausting the vertices of the leapfrog (Figure 4). A polyhedron with v, e, f vertices, edges, and faces has a leapfrog with $2e$, $3e$, and $v + f$ vertices, edges, and faces, respectively, from which it can be seen that the leapfrog is always trivalent, irrespective of the starting polyhedron.

In the case of a fullerene parent with n vertices, $n/2 - 10$ hexagonal, and 12 pentagonal faces, the leapfrog is itself a fullerene with $3n$ vertices and n extra hexagonal faces centered on the old vertex sites. Of the $9n/2$ edges of the leapfrog, $3n/2$ cross those of the parent at right angles.

The leapfrog operation has a field of action much wider than the set of fullerene polyhedra; it can be applied to other polyhedra, trivalent and nontrivalent, and in general to any map (graph embedded in a surface), and can be repeated an indefinite number of times to generate ever larger polyhedra.[48] Trivalent polyhedra may be classified by their position in such series of repeated leapfroggings: every trivalent

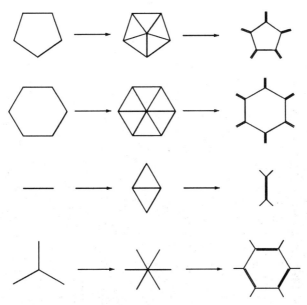

Figure 4. Transformations of structural components (faces of two types, vertices and edges) of a fullerene under the leapfrog operation. The double bonds indicate those edges of the leapfrog that derive from the parent, and give a consistent Fries Kekulé pattern with the maximal number of "benzenoid" rings and therefore the maximal stability on a localized model of π-electronic structure.

polyhedron can be regarded as an nth leapfrog $L^n P$ of a polyhedral parent P, where n (which may be zero) is the *leapfrog order* of the polyhedron.

The chemically significant fact about the leapfrog fullerenes is that, considered as neutral carbon frameworks C_{3n}, each has exactly $3n/2$ bonding and $3n/2$ antibonding π orbitals. As fullerenes exist for all $n = 20 + 2k$ $(k \neq 1)$, the leapfrog rule is thus[50]:

> For $n = 60 + 6k$ $(k \neq 1)$, a properly closed-shell fullerene C_n exists for each fullerene isomer of $C_{n/3}$, with a predictable symmetry, shape, and electronic configuration.

The claim about symmetry and shape follows from the geometric nature of the leapfrog construction as discussed earlier. The claim about the electronic configuration will now be justified. There are several discussions in the literature of the closed-shell aspect of the rule, ranging from pictorial MO-theoretic[50] and symmetry arguments[52] to a formal graph-theoretic proof.[53] The latter shows that a leapfrog C_n of *any* trivalent $C_{n/3}$ parent has a strictly antibonding LUMO

$$\varepsilon_{n/2+1} = \alpha + \lambda_{n/2+1}\beta \qquad \lambda_{n/2+1} < 0$$

and a bonding or nonbonding HOMO

$$\varepsilon_{n/2} = \alpha + \lambda_{n/2}\beta \qquad \lambda_{n/2} \geq 0$$

so that *no* leapfrog of a trivalent parent has an open or pseudoclosed shell. When *all* faces of the parent polyhedron have multiples of 3 edges, the bound on the HOMO cannot be improved; in cases such as the 4-6 triangle + hexagon trivalent polyhedra of Goldberg,[29] the HOMO is actually nonbonding and the cage falls at the limit of the metaclosed classification. When at least one face of the parent does *not* have a multiple of 3 edges, the bound is replaced by a strict inequality, the HOMO is bonding and separated by a gap from an antibonding LUMO, and hence the leapfrog has a properly closed shell. In particular, therefore, a leapfrog fullerene will always have a properly closed shell.

A more chemical rationalization of the π stability of a leapfrog fullerene is based on the special one-in-three of its edges that derive from edges of the parent. If every such edge of the leapfrog C_n is decorated with a formal double bond and all others with formal single bonds, a perfect "Fries" Kekulé structure is produced: it has the maximum possible number ($n/3$) of "benzenoid" hexagons and a formal single bond on every pentagon edge. The class of fullerenes for which a Fries structure is possible and the class of leapfrogs are one and the same.[54] Leapfrog fullerenes are thus those that are maximally stable on the Taylor aromaticity criterion.[55]

In a localized picture of the bonding, the occupied π molecular orbitals of the fullerene derive from local bonds along the double-bond parent-derived edges. Interaction of these local bonding orbitals may be expected to change their energies, but not their number or symmetries, and so the symmetry spanned by the occupied π orbitals of the leapfrog will be equal to the permutation representation of the edges of the parent

$$\Gamma_{occ}(L, C_n^0) = \Gamma_{\sigma}(e, P)$$

We have thus a model for the π-electronic configuration of the neutral leapfrog fullerene in terms of properties of the parent.

A further mathematical property of leapfrog fullerenes can be used to shed light on the characteristic electron-deficient chemistry of fullerenes in general. The definition needed is that of a Clar polyhedron.[48] A trivalent polyhedron is *Clar* if it can be colored in the following way: all faces are either black or white, with no two black faces touching and every vertex of the polyhedron belonging to exactly one black face. A C_{60} football or a cube is Clar, a tetrahedron or a dodecahedral C_{20} is not. The concept of the Clar fullerene gives precise meaning to the term *carbon football*: a Clar fullerene can be colored as a conventional black-and-white football. The black faces always include the 12 pentagons. The classes of Clar fullerenes and leapfrog fullerenes are again one and the same.

As Figure 5 shows, the Fries and Clar structures of a fullerene are complementary. Every "empty" face of the Fries structure (i.e., every one bordered by single bonds alone) can be colored white. Just as the Fries structure supplies a localized model for

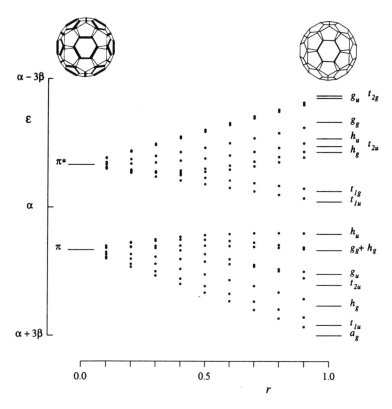

Figure 5. The Fries model of the electronic structure of C_{60}. The Hückel eigenvalues are plotted as a function of r, the relative size of the resonance integral for the "single" bonds to that of the formal double bonds of the Fries structure. The limiting case on the RHS is the neutral molecule with all bonds of equal strength, showing that all bonding levels of C_{60} derive from local bonding orbitals associated with the formal double bonds. The degeneracy of the levels is indicated by the symmetry label at the RHS.

the electronic configuration of the neutral leapfrog fullerene C_n, the Clar structure gives a localized configuration for the dodeca-anion C_n^{12-}. If a sextet of electrons is associated with every black face of the Clar structure, the cage carries an excess charge of 12 electrons and the corresponding configuration of the anion has occupied orbitals spanning the σ and π representations of the black faces, i.e., of the faces of the parent. A symmetry extension of the Euler theorem for trivalent polyhedra relates this to the permutation representation of the edges of the parent by[56]

$$\Gamma_{occ}(L, C_n^{12-}) = \Gamma_\sigma(e, P) + \Gamma_T + \Gamma_R$$

Combination of the two pictures shows the parentage of the bonding orbitals of the leapfrog fullerene *and* of the six lowest-lying antibonding MOs. Every leapfrog fullerene is expected to have six low-lying acceptor orbitals symmetry-matched to the

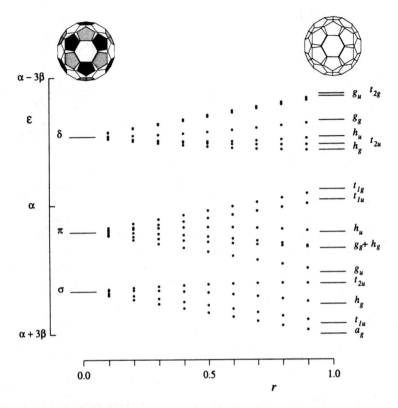

Figure 6. The Clar model of the electronic structure of C_{60}^{12-}. The Hückel eigenvalues are plotted as a function of r, the relative size of the resonance integral for the bonds outside the Clar rings to those inside. The limiting case on the RHS shows that all bonding levels of C_{60} plus the six low-lying orbitals of T_{1u} and T_{1g} symmetry derive from local bonding σ and π orbitals of the pentagons. The degeneracy of the levels is indicated by the symmetry label at the RHS.

translation and rotation vectors in the molecular point group. As leapfrog fullerenes are expected, other things being equal, to show the largest HOMO–LUMO gaps, this is a clear indication from qualitative MO theory that all fullerenes are expected to be electron deficient to some extent. As some or all of the low-lying levels move down to more bonding energies, the properly closed shells of the leapfrogs give way to pseudoclosed shells in which the lowest empty levels are formally bonding.

As a specific illustration, Figures 5 and 6 show the parentage of the occupied orbitals of C_{60}. On the LHS of the Fries correlation diagram (Figure 5) are the 30 π and 30 π^* orbitals of the localized double bonds along hexagon–hexagon edges of C_{60}. On interaction these spread out to give bonding and antibonding bands, but preserve the number and symmetry of components in each half of the spectrum. On the LHS of the Clar correlation diagram (Figure 6) are the 12 σ bonding, 24 π bonding, and 24

δ antibonding orbitals of the 12 Clar pentagons; on interaction the σ band remains bonding and the π band gives 18 cluster bonding orbitals plus the threefold degenerate LUMO and LUMO+1 sets which have symmetries T_{1u} and T_{1g} in I_h as expected from their identification with the translations and rotations in this group.

Leapfrog fullerenes are necessarily rather rare. By construction, they have isolated pentagons. C_{60} is both the smallest isolated-pentagon fullerene and the smallest leapfrog. The gap in the fullerene series at C_{22} leads to a gap at C_{66} in the leapfrogs; henceforward the counts for leapfrogs follow the isomer numbers for clusters of one-third their size, i.e, C_{60} (1), C_{72} (1), C_{78} (1), C_{84} (2), C_{90} (3), C_{96} (6), ... Thus, only about 1 in 10^5 of the first one and a half million fullerenes are leapfrogs.

7.4.4. The Carbon Cylinder Rule

As will not have escaped the reader, the leapfrog rule does not provide an explanation for the stable π configuration of C_{70}. The C_{70} cage can be constructed in imagination by splitting the C_{60} ball into equal hemispheres, twisting them relative to one another, and inserting a pentaphenylene belt of five hexagons. C_{70} is thus the smallest example of a cylindrical fullerene with C_{60} caps. It has a closed shell of doubly occupied bonding orbitals lying below a nonbonding LUMO. Exploration of the series of cylinders with 5- or 6-fold axes (D_{5+} and D_{6+} symmetries) reveals that this pattern is repeated whenever three more layers are added to the tubular section.[33] The carbon cylinder rule is therefore:

Properly closed shells are found for cylinders C_n with either hemi-C_{60} or hemi-C_{72} caps at the nuclearities

$$n = 10(7 + 3p), p = 0, 1, 2, \ldots \quad (D_{5+})$$

$$n = 12(7 + 3p), p = 0, 1, 2, \ldots \quad (D_{6+})$$

i.e., at C_{70}, C_{100}, C_{130}, . . . and C_{84}, C_{120}, C_{156}, ...

By construction, all such carbon cylinders have isolated pentagons. The periodicity in electronic configuration can be rationalized by considering the nodal pattern of the LUMO and relating it to solutions of the particle-on-a-cylinder Schrödinger equation. The LUMO of the closed-shell cylinders forms part of a sextet of low-lying acceptor orbitals,[56] and so the arguments that were developed more fully in the preceding section for leapfrogs also apply to this series.

Closed-shell fullerenes following the carbon cylinder rule are again rare (just 3 for $n \leq 100$) and again an experimentally characterized fullerene is the parent of an infinite series of properly closed shells. For $n < 112$ the leapfrog and cylinder rules together exhaust the list of properly closed shells. They are not complete for all n, however, as the next section points out.

Table 2. Enumeration of C_n Fullerene Isomers in the Range $n \leq 140$ with "Sporadic" Properly Closed Shells outside the Leapfrog and Carbon-Cylinder Magic-Number Rules

n	Isomers	n	Isomers	n	Isomers
112	1	122	1	132	4
114	0	124	3	134	7
116	1	126	0	136	9
118	0	128	3	138	4
120	1	130	3	140	12

7.4.5. Sporadic Closed Shells

In the formal division of electronic configurations into types, the dichotomy between properly closed and pseudoclosed systems is absolute. A LUMO energy of $\alpha - 10^{-6}\beta$ would qualify a cage as properly closed but an energy of $\alpha + 10^{-6}\beta$ would lead to a pseudoclosed classification, even though such small shifts in energy could have little consequence for the properties. It turns out that, as the nuclearity increases, properly closed shells of this marginal type start to appear. An explicit search of isolated-pentagon fullerenes up to C_{140} (Ref. 7) reveals 49 in the range $112 \leq n \leq 140$, as listed in Table 2. The series starts at 112 atoms, skips some n values, but then has members for every fullerene nuclearity above 128. All have tiny antibonding LUMO energies, and generally low symmetry. They are not extremal in resonance energy or band gap (e.g., the unique properly closed C_{116} has only the 15th largest gap of the 6063 isolated-pentagon isomers with this number of atoms), and this encourages the view that they are numerical "accidents," outside the neat regularities of the leapfrog and cylinder rules. From this perspective the classification as "sporadic" closed shells seems appropriate. It will be interesting to see if any more subtle pattern connects these apparent exceptions to the main rules, but the question is of more mathematical than chemical or physical interest and will not be pursued here.

7.4.6. Adequacy of the Electronic Model?

Within the limitations of simple Hückel theory, two remarkably simple rules give a coherent account of π stability for all currently chemically interesting fullerene sizes. C_{60} and C_{70} are awarded special status by the rules, and both rules are compatible with the observed association of stability and pentagon isolation. However, it is also clear from experiment that a π-only theory is insufficient to explain the known isomer patterns of higher fullerenes. In addition to C_{60} and C_{70}, fullerenes such as C_{76}, C_{78}, and C_{84} have been isolated and characterized by NMR; the first of these has a nuclearity that lies outside both magic-number rules, the second and third have been shown to be produced as mixtures of three and two isomers, respectively, with nuclearities that fit the leapfrog rule but with nonleapfrog structures. Electronic effects *are* important, for example, in the isomer preference of C_{76}, but steric factors are equally or more important for large n.

7.5. STABILITY AND STRUCTURE

As we have seen, the electronic properties of the fullerenes are largely codified by simple geometric/topological rules, and in any particular case a qualitative description of the π-electronic structure is available from a trivial calculation once the adjacency matrix is given. If a similarly systematic picture of overall stability is to emerge, it is necessary to find rules of thumb that give estimates of steric strain from the molecular graph alone. Of course, given a guess for the geometry of a particular isomer, it is always possible to carry out a semiempirical quantum-mechanical or molecular-mechanics optimization, and hence derive a total energy. Calculations of this type are feasible for small to moderately large fullerene systems, and the agreement between methods, at least for the most stable isomers, is usually good. Wholesale use of any such method will eventually be overwhelmed by the sheer number of isomeric possibilities, and in any case would not meet the requirement for a transparent rule based on the molecular graph itself. However, such calculations can be used to give clues to possible rules. Several rules have emerged.

7.5.1. The Minimum Pentagon-Adjacency Rule

Fullerene isomers are distinguished by the disposition of the necessary 12 pentagonal faces on the otherwise hexagonal framework. It was noted in the earliest discussions of fullerene stability[57,58] that separation of the pentagons one from another should lead to higher π-electronic stability and lower steric strain: fusion of two pentagons creates an antiaromatic 8-circuit in the cage and also a site of sharp curvature with local bond angles that are unfavorable for sp^2 carbons. Arguments such as these were used to establish the isolated-pentagon rule (IPR). All experimentally characterized fullerenes have isolated pentagons, and all calculations show that whenever isolation of pentagons can be achieved, the most stable isomer has isolated pentagons.

Below $n = 60$ and for $62 \leq n \leq 68$, the strict IPR has nothing to say about relative stabilities of fullerene isomers, since isolation of pentagons is impossible in these ranges. However, explicit calculations by a number of research groups using quite different methods[59-62] show that an individual energetic penalty of 70–150 kJ mol^{-1} can be associated with each pentagon adjacency, and that, roughly speaking, the fewer pentagon–pentagon bonds present in a cage, the better is its stability. A rule can be deduced[7]:

The most stable isomer of a particular fullerene will be an isomer with the smallest possible number N_P of pentagon–pentagon fusions.

The most extensive test of this idea is a study using the QCFF/PI semiempirical method to estimate the relative stabilities of all 1812 distinct fullerene isomers of C_{60}.[63,64] As Figure 7 shows, the total energy is monotonic in N_P. The best isomer at each value of N_P is more stable than any with a higher value, though there is some overlap in energy

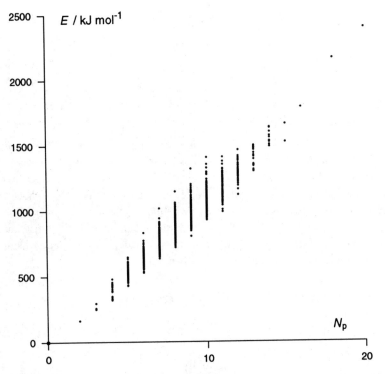

Figure 7. Variation of the relative energies of C_{60} isomers with N_P, the number of pentagon adjacencies. The energy E is calculated using the QCFF/PI method (Ref. 63) and N_P varies from zero (the isolated-pentagon I_h isomer of C_{60}) to 20 (the cylinder capped by two hemi-dodecahedra).

for successive values of this parameter. In particular, the unique isolated-pentagon I_h isomer of C_{60} is lower in energy by ~ 160 kJ mol^{-1} than its nearest rival which has $N_P = 2$. Similar trends are found for C_{40},[59] where the lowest achievable value of N_P is 10 and the two isomers with ten adjacencies lie ~ 120 and ~ 85 kJ mol^{-1} below the best isomer with $N_P = 11$. The most stable fullerene isomer at every value of $n \leq 70$ found in the simulated-annealing survey of Zhang *et al.*[62] has the minimum achievable N_P at that value of n. There is therefore ample evidence for a minimum pentagon-adjacency rule ($n \leq 70$) that shades into a pentagon isolation rule ($n \geq 70$).

A more specific form of the minimal adjacency rule has been suggested.[7] To each fullerene isomer it is possible to attach a pentagon neighbor index signature $\{p_k\}$ where p_k ($k = 0, \ldots, 5$) is the number of pentagons with k pentagonal neighbors. Trivially,

$$\sum_k p_k = 12$$

and

$$\frac{1}{2} \sum_k k p_k = N_P$$

The calculations in Ref. 62 highlight certain isomers within the minimal-N_P class as most stable. A rule compatible with these results is as follows:

The most stable isomer of the class with minimal N_P is one that minimizes the pentagon neighbor index p_k of maximal k.

For example, C_{46} has seven isomers with $N_P = 8$. Of these, one has pentagon neighbor signature $\{2, 4, 6, 0, 0, 0\}$, three have signatures $\{1, 6, 5, 0, 0, 0\}$, and three have signatures $\{0, 8, 4, 0, 0, 0\}$.[7] The most stable isomer according to the calculations of Zhang et al.[62] is one of the last three. The results of their studies for all cages with $n \leq 70$ are consistent with the above statement of the rule. The signature for an isolated-pentagon fullerene is $\{12, 0, 0, 0, 0, 0\}$ and so this refinement of the minimal adjacency rule again reduces to the IPR.

As we have seen, the IPR emerges as the limiting case of minimization of adjacencies. The criterion of isolation of pentagons ($n = 60$, $n \geq 70$) is a useful filter for fullerenes of moderate size; for $n \leq 100$ the fraction of distinct fullerenes that obey the IPR is a slowly increasing function of n which is still less than 2 in 1000 by the top of the range. This still leaves some 450 isomers of C_{100}, for example, to be distinguished by more detailed rules, but is a vast improvement on the total of 285,918 that would have to be considered without invoking the rule.[7]

7.5.2. Beyond the Isolated-Pentagon Rule

The IPR cuts down the number of isomers that must be considered, but still leaves a great deal of scope for significant variations in stability. Although the calculated range of energies of general fullerene isomers is large (e.g., ~ 2400 kJ mol^{-1} for C_{60}),[63] the IPR isomers span a much reduced range (e.g., ~ 255 kJ mol^{-1} for C_{84}),[65] and so any further "second-order" stability rules will be correspondingly more delicate than the IPR.

One suggestion, made by Raghavachari[66] on the basis of semiempirical MNDO calculations on C_{84}, is to use a hexagon neighbor index signature $\{h_k\}$ where h_k is the number of hexagonal faces that have exactly k hexagonal neighbors ($k = 0, \ldots, 6$ for general fullerenes, and $k = 3, \ldots, 6$ for IPR fullerenes). The idea is that once the preference of pentagonal rings for isolation has been met, attention should be shifted to the hexagons. In an ideal cage all hexagon environments should be the same, i.e., $\{h_k\}$ should have a single nonzero entry $h_{k^*} = (n/2 - 10)$ for some ideal $k = k^*$. Failing this uniformity, $\{h_k\}$ should show as little dispersion as possible. In the particular case that generated it,[66] this proposal cuts the field of 24 IPR C_{84} isomers down to three with signature $\{0, 0, 0, 0, 24, 4, 0\}$, two of which together form the experimental product[39] and have the lowest MNDO energies.

It turns out that a *perfect* hexagon neighbor signature is attained only at one of three specific limiting nuclearities, but between these cluster sizes optimal signatures in which no more than two h_k are nonzero can be defined from purely graph-theoretical considerations. A natural extension to the Raghavachari model for C_{84} is therefore the hexagon neighbor rule (HNR)[7,67]:

The most stable IPR isomers are expected to be those with the optimal hexagon neighbor index signatures defined by

$$(h_3, h_4, h_5, h_6) = \begin{cases} (80 - n, 3n/2 - 90, 0, 0) & \text{if } 60 \leq n \leq 80 \\ (0, 70 - n/2, n - 80, 0) & \text{if } 80 \leq n \leq 140 \\ (0, 0, 60, n/2 - 70) & \text{if } 140 \leq n \end{cases}$$

which reduce to

$$(h_3, h_4, h_5, h_6) = \begin{cases} (20, 0, 0, 0) & \text{if } n = 60 \\ (0, 30, 0, 0) & \text{if } n = 80 \\ (0, 0, 60, 0) & \text{if } n = 140 \end{cases}$$

The HNR applies a further filter to the set of isolated-pentagon isomers. It selects at most 44 optimal isomers of any one fullerene in the range $70 \leq n \leq 140$ (Table 3). It has not been proved by exhaustive calculation that these do in fact always include the most stable isomer in all cases, but the HNR does give a qualitative justification[7,67] of the empirical observation that leapfrog and carbon-cylinder isomers beyond C_{70} are *not* those found separated from the Krätschmer–Huffman soot.

Recall that a leapfrog fullerene is derived from its parent, $C_{n/3}$, by a construction in which each hexagon of the parent becomes surrounded by new hexagons in the leapfrog. This implies a *minimum* of $n/6 - 10$ hexagons in the leapfrog having the maximum hexagon neighbor index of $k = 6$. All leapfrogs in the range $60 < n < 140$ therefore have $h_6 > 0$ and *cannot* achieve the optimal signature. More detailed considerations show that all leapfrog fullerenes in the range $60 \leq n \leq 180$ have either

Table 3. Enumeration of C_n Fullerene Isomers in the Range $n = 70$ to 140 with Optimum Hexagon Neighbor Index Signatures

n	Isomers	n	Isomers	n	Isomers	n	Isomers
70	1	88	1	106	33	124	1
72	0	90	6	108	34	126	0
74	1	92	28	110	21	128	0
76	1	94	15	112	19	130	1
78	1	96	17	114	6	132	0
80	2	98	42	116	15	134	0
82	1	100	38	118	3	136	0
84	3	102	38	120	4	138	0
86	1	104	44	122	3	140	1

$h_3 > 0$ or $h_4 > 0$ or both and so all are nonoptimal on this criterion. Similar reasoning shows that carbon cylinders with $n = 70 + 30k$ and $k > 0$ have signatures $\{0, 0, 0, 10, 10, 10, n/2 - 40\}$ and those with $n = 84 + 36k$ and $k > 0$ have signatures $\{0, 0, 0, 0, 24, 12, n/2 - 46\}$, both differing markedly from the optimal. The conclusion is clear: for IPR fullerenes beyond C_{60} the steric and π-electronic effects in stability are generally acting in opposite directions. This contrasts strongly with the situation for lower fullerenes where the two effects cooperate to push the pentagons apart.

This competition of steric and electronic effects rationalizes the known experimental isomer distributions for C_{76}, C_{78}, and C_{84}. The conclusion from comparisons of total energies, Hückel properties, and hexagon-neighbor signatures[7,67] is that, while π-electronic effects are dominant for C_{76}, the balance is more even for C_{78}, and for C_{84} the steric effects have taken over the major role. The indications are that the relative importance of the π-electronic stability will continue to decline as n increases.

7.5.3. Beyond the Hexagon-Neighbor Rule

Both isolated-pentagon and hexagon-neighbor rules can be used to filter out high-energy isomers. Both imply that the presence of certain structural motifs will confer stability or instability on the fullerene cage. Attempts have been made to refine the rules by finding other significant motifs, but are hindered by the fact that many motifs are strongly correlated and so large samples of structures are needed if "good statistics" are to be obtained. The large-scale study of all C_{60} isomers[63] mentioned earlier shows, for example, that aggregation of hexagons, as measured by the number of fused hexagon triples or by the second moment of the hexagon-neighbor distribution, $H = \Sigma_k k^2 h_k$, is destabilizing. From a π-electronic point of view this appears surprising, but it is easily understood in terms of cage curvature: hexagon clusters are destabilizing because they enforce local planarity and hence imply high curvature and pentagon crowding in another part of the fullerene surface. Hexagon–pentagon–hexagon triples on the other hand allow the cage to distribute steric strain and hence have an overall stabilizing effect. The fact that calculated energy gains and losses per motif are smaller by a factor of 5 than the penalty per pentagon adjacency emphasizes the difficulty of finding simple but quantitative rules beyond the IPR. It will be interesting to see how the conclusions reached for general isomers of C_{60} transfer to the higher fullerenes.

7.6. PERSPECTIVE

By comparison with the situation when the first observation of C_{60} was made, there is now a much clearer theoretical understanding of the chemistry and physics of carbon. C_{60} is now known to be just one of many molecular allotropes of carbon, which can be counted, constructed as molecular graphs and in 3D space, and assigned to a limited number of point groups with characteristic NMR signatures. A general if not formally complete picture of the link between their electronic structures and molecular topology has been constructed, and several powerful rules of thumb are available to filter out more than 99% of the less stable isomers. Discrimination among the remaining short list of more-or-less favorable isomers is not yet fully codified, and relies on explicit calculation to supply the fine details of the energy order. Such calculations are readily available.

Although not discussed in the present chapter, a systematic picture of isomerization and family relationships among the fullerenes can be built up once the isomers have been catalogued.[7] Isomers fall into "Stone–Wales families"[68] within which interconversion by this hypothetical mechanism[69] would be possible. On the still-open questions of growth and formation, it has recently been shown[70] that all of the more stable isomers of the lower fullerenes ($n \leq 60$) can be linked in a topological realization of the fullerene-road model[71] of C_{60} formation that uses only C_2 insertion and the Stone–Wales isomerization process. The requirement is now to attach an energy dimension to these purely graph-theoretical schemes.[61,65]

Finally, over the next few years it is likely that a similar systematic approach will pay dividends in the study of the chemical derivatives of the fullerenes, where the fullerene problems of isomer counting, electronic and steric effects all reappear in another form. A bibliography on fullerene research has recently been published.[72]

REFERENCES

1. H. W. Kroto, J. R. Heath, S. C. O'Brien, R. F. Curl, and R. E. Smalley, *Nature 318*, 162 (1985).
2. W. Krätschmer, L. D. Lamb, K. Fostiropoulos, and D. R. Huffman, *Nature 347*, 354 (1990).
3. H. W. Kroto, *Angew. Chem. Int. Engl. 31*, 111 (1992).
4. R. E. Smalley, *The Sciences* p. 22 (1991).
5. J. Baggott, *Perfect Symmetry*, Oxford University Press, London (1994).
6. H. Aldersey-Williams, *The Most Beautiful Molecule*, Aurum Press, London (1994).
7. P. W. Fowler and D. E. Manolopoulos, *An Atlas of Fullerenes*, Oxford University Press, London (1995).
8. L. A. Chernozatonskii, *Phys. Lett. A 170*, 37 (1992).
9. S. Itoh, S. Ihara, and J. Kitakami, *Phys. Rev. B 47*, 1703 (1993).
10. S. Itoh, S. Ihara, and J. Kitakami, *Phys. Rev. B 47*, 12908 (1993).
11. S. Itoh and S. Ihara, *Phys. Rev. B 48*, 8323 (1993).
12. J. C. Greer, S. Itoh, and S. Ihara, *Chem. Phys. Lett. 222*, 621 (1994).
13. E. C. Kirby, R. B. Mallion, and P. Pollak, *J. Chem. Soc. Faraday Trans. 89*, 1945 (1993).
14. E. C. Kirby, *Croat. Chem. Acta 66*, 13 (1993).
15. E. C. Kirby, *Fullerene Sci. Technol. 2*, 395 (1993).
16. H. Sachs, *J. Chem. Inf. Comput. Sci. 34*, 432 (1994).
17. D. J. Klein, *J. Chem. Inf. Comput. Sci. 34*, 453 (1994).
18. Y.-D. Gao and W. C. Herndon, *J. Am. Chem. Soc. 115*, 8459 (1993).
19. S. Iijima, T. Ichihashi, and Y. Ando, *Nature 356*, 776 (1992).
20. C.-H. Sah, *Fullerene Sci. Technol. 2*, 445 (1994).
21. P. W. Fowler, J. E. Cremona, and J. I. Steer, *Theor. Chim. Acta 73*, 1 (1988).
22. D. E. Manolopoulos, J. C. May, and S. E. Down, *Chem. Phys. Lett. 181*, 105 (1991).
23. D. Babić, D. J. Klein, and C.-H. Sah, *Chem. Phys. Lett. 211*, 235 (1993).
24. D. Babić and N. Trinajstić, *Comput. Chem. 17*, 271 (1993).
25. X. Liu, D. J. Klein, T. G. Schmalz, and W. A. Seitz, *J. Comput. Chem. 12*, 1252 (1991).

26. G. Brinkmann and A. W. M. Dress, PentHex Puzzles: A reliable and efficient top-down approach to fullerene structure enumeration, *Proc. Nat. Acad. Sc.*, New York (in press).

27. P. W. Fowler, D. E. Manolopoulos, D. B. Redmond, and R. P. Ryan, *Chem. Phys. Lett. 202*, 371 (1993).

28. D. E. Manolopoulos and P. W. Fowler, *J. Chem. Phys. 96*, 7603 (1992).

29. M. Goldberg, *Tohoku Math. J. 43*, 104 (1937).

30. D. L. D. Caspar and A. Klug, *Cold Spring Harbor Symp. Quant. Biol. 27*, 1 (1962).

31. R. B. Fuller, *Ideas and Integrities*, Prentice–Hall, Englewood Cliffs, New Jersey (1963).

32. H. S. M. Coxeter, in: *A Spectrum of Mathematics* (J. C. Butcher, ed.), p. 98, Oxford University Press/Auckland University Press, London/Auckland (1971).

33. P. W. Fowler, *J. Chem. Soc. Faraday Trans. 86*, 2073 (1990).

34. D. E. Manolopoulos and P. W. Fowler, *Chem. Phys. Lett. 204*, 1 (1993).

35. R. Taylor, J. P. Hare, A. K. Abdul-Sada, and H. W. Kroto, *J. Chem. Soc. Chem. Commun. 1990*, 1423 (1990).

36. R. Ettl, I. Chao, F. Diederich, and R. L. Whetten, *Nature 353*, 149 (1991).

37. F. Diederich, R. L. Whetten, C. Thilgen, R. Ettl, I. Chao, and M. M. Alvarez, *Science 254*, 1768 (1991).

38. K. Kikuchi, N. Nakahara, T. Wakabayashi, S. Suzuki, H. Shiromaru, Y. Miyake, K. Saito, I. Ikemoto, M. Kainosho, and Y. Achiba, *Nature 357*, 142 (1992).

39. D. E. Manolopoulos, P. W. Fowler, R. Taylor, H. W. Kroto, and D. R. M. Walton, *J. Chem. Soc. Faraday Trans. 88*, 3117 (1992).

40. F. Diederich and R. L. Whetten, *Acc. Chem. Res. 25* 119 (1992).

41. U. Schneider, S. Richard, M. M. Kappes, and R. Ahlrichs, *Chem. Phys. Lett. 210*, 165 (1993).

42. R. C. Haddon, *Acc. Chem. Res. 25*, 127 (1992).

43. S. J. Austin, P. W. Fowler, G. Orlandi, D. E. Manolopoulos, and F. Zerbetto, *Chem. Phys. Lett. 226*, 219 (1994).

44. S. J. Austin, P. W. Fowler, J. P. B. Sandall, P. R. Birkett, A. G. Avent, A. D. Darwish, H. W. Kroto, R. Taylor, and D. R. M. Walton, *J. Chem. Soc. Perkin Trans. 2*, p. 1027 (1995).

45. G. E. Scuseria, in: *Buckminsterfullerenes* (W. E. Billups and M. A. Ciufolini, eds.), VCH Publishers, New York (1993).

46. J. Cioslowski, *Electronic Structure Calculations on Fullerenes and Their Derivatives*, Oxford University Press, London (1995).

47. P. W. Fowler, P. Lazzeretti, and R. Zanasi, *Chem. Phys. Lett. 165*, 79 (1990).

48. P. W. Fowler and T. Pisanski, *J. Chem. Soc. Faraday Trans. 90*, 2865 (1994).

49. P. W. Fowler, *Chem. Phys. Lett. 131*, 444 (1986).

50. P. W. Fowler and J. I. Steer, *J. Chem. Soc. Chem. Commun. 1987*, 1403 (1987).

51. R. L. Johnston, *J. Chem. Soc. Faraday Trans. 87*, 3353 (1991).

52. P. W. Fowler and D. B. Redmond, *Theor. Chim. Acta 83*, 367 (1992).

53. D. E. Manolopoulos, D. R. Woodall, and P. W. Fowler, *J. Chem. Soc. Faraday Trans. 88*, 2427 (1992).

54. P. W. Fowler, *J. Chem. Soc. Perkin Trans. 2*, p. 145 (1992).

55. R. Taylor, *J. Chem. Soc. Perkin Trans. 2*, p. 3 (1992).

56. P. W. Fowler and A. Ceulemans, *J. Phys. Chem. 99*, 508 (1995).

57. H. W. Kroto, *Nature 329*, 529 (1987).

58. T. G. Schmalz, W. A. Seitz, D. J. Klein, and G. E. Hite, *J. Am. Chem. Soc. 110*, 1113 (1988).

59. P. W. Fowler, D. E. Manolopoulos, G. Orlandi, and F. Zerbetto, *J. Chem. Soc. Faraday Trans. 91*, 1421 (1995).

60. J.-Y. Yi and J. Bernholc, *J. Chem. Phys. 96*, 8634 (1992).

61. R. L. Murry, D. L. Strout, G. K. Odom, and G. E. Scuseria, *Nature 366*, 665 (1993).

62. B. L. Zhang, C. Z. Wang, K. M. Ho, C. H. Xu, and C. T. Chan, *J. Chem. Phys. 97*, 5007 (1992).

63. S. J. Austin, P. W. Fowler, D. E. Manolopoulos, G. Orlandi, and F. Zerbetto, *J. Phys. Chem. 99*, 8076 (1995).

64. S. J. Austin, P. W. Fowler, D. E. Manolopoulos, and F. Zerbetto, *Chem. Phys. Lett. 235*, 146 (1995).

65. P. W. Fowler and F. Zerbetto, *Chem. Phys. Lett. 243*, 36 (1995).

66. K. Raghavachari, *Chem. Phys. Lett. 190*, 397 (1992).
67. P. W. Fowler, S. J. Austin, and D. E. Manolopoulos, in: *Chemistry and Physics of the Fullerenes* (K. Prassides, ed.), Kluwer Academic Press, Dordrecht (1994).
68. P. W. Fowler, D. E. Manolopoulos, and R. P. Ryan, *Carbon 30*, 1235 (1992).
69. A. J. Stone and D. J. Wales, *Chem. Phys. Lett. 128*, 501 (1986).
70. D. E. Manolopoulos and P. W. Fowler, in: *The Chemical Physics of the Fullerenes 10 (and 5) years Later* (W. Andreoni, ed.), Kluwer Academic Press, Dordrecht (1995).
71. J. R. Heath, in: *Fullerenes: Synthesis, Properties and Chemistry of Large Carbon Clusters* (G.S. Hammond and V. J. Kuck, eds.), *ACS Symp. Ser. 481*, 1 (1991).
72. A. Braun, A. Schubert, H. Maczella, and L. Vasvári, *Fullerene Research 1985–1993—A Computer-Generated Cross-Indexed Bibliography of the Journal Literature*, World Scientific Publ., Singapore (1995).

8

Recent Work on Toroidal and Other Exotic Fullerene Structures

E. C. KIRBY

8.1. INTRODUCTION

The carbon molecule buckminsterfullerene, C_{60}, is an object of great beauty and structural simplicity, and its recognition and synthesis represents a splendid achievement of late 20th century science.[1,2] Surely few chemists who read the reports were not moved and thrilled by that single sharp peak in the molecule's ^{13}C magnetic resonance spectrum.[3,4]

Much work* was, and still is, being done on the properties of this structure, on its possible isomerism, and on the properties of many other similar structures of varying size. In parallel, there arose the idea—it is difficult to know exactly when—that a sphere might not be the only hollow topological shape that a carbon cluster might assume, and interest in such possibilities has grown, steadily rather than explosively.[†]

---

*For an introduction and a number of important references, see, for example, Kroto's recent and delightful essay.[5] For a comprehensive recent report see also Ref. 6.

†Certainly one of the earliest papers containing any serious discussion on this topic at the molecular level came from the Galveston Group in 1988.[7] After this, according to the author's own bibliography, at least 1, 2, 7, 9, and 10 papers about toroidal structures were published in the years 1991–1995, respectively.

E. C. KIRBY • Resource Use Institute, 14 Lower Oakfield, Pitlochry, Perthshire PH16 5DS, Scotland.

From Chemical Topology to Three-Dimensional Geometry, edited by Balaban. Plenum Press, New York, 1997

Critics of such "exotic fullerene" work can still, correctly, point out that there is as yet no evidence for the existence of toroidal forms of carbon, nor any very obvious synthetic method to try. However, much the same was said about buckminsterfullerene before 1990; the fact that carbon tubules exist, and that more than one calculation method has shown that certain toroidal structures might have reasonable stability, does encourage some optimism.

8.2. SOME GENERAL PRINCIPLES AND DEFINITIONS

Common conventions of graph theory are followed here (see, e.g., Refs. 8–10), but a rigid distinction is not always maintained between chemical and mathematical contexts. So the term *graph* is used interchangeably in both its strict sense as a mathematical object and as a shorthand term for its realization as an actual (usually carbon) molecule. Some equivalent pairs of terms are treated as being synonymous; atom ≡ vertex, bond ≡ edge, valency ≡ degree, and so on.

As is natural in a young and developing subject, there is some fluidity in the usage of new terms. Consistent with the central importance of buckminsterfullerene and all of the studies its discovery inspired, a *fullerene* is usually taken to mean a spherically embedded trivalent cluster made up of 12 pentagons with two or more hexagons.[11] However, with interest expanding to embrace topological possibilities other than the sphere, there is now the need for a term that encapsulates what the whole group of structures have in common, namely, a boundless and essentially 2D network that curves on itself through the third dimension of space. The word *cage* could serve, but it suffers the disadvantage of having a well-known usual meaning in common English.

We therefore continue to use the unqualified term *fullerene* in this now conventional sense of a 3-connected trivalent graph embedded on a sphere, but do not state requirements as to genus and numbers of rings. In this way the meaning can be extended in an obvious way without undue strain to accommodate such objects as "toroidal" fullerenes and "Klein-bottle" fullerenes. [The only warning note to sound here is that these different topological formulations are usually *but not necessarily* unique. For example, taking the meaning outlined here, the cube (of genus zero) can be drawn as a "spherical" fullerene with six square faces *or* as a "toroidal" fullerene with four hexagonal faces.] It is possible *in principle* that there might one day be discovered a fullerene of genus-zero that could exist in either of two metastable forms: a polyhedron or a torus.

For more specific cases where the surface network has a simple uniform pattern of rings, we specify the surface type and refer, for example, to a *toroidal polyhex* (also called a *toroidal hexagonal system* by John and Walther[12]) or a *toroidal azulenoid*, and so on. Others[e.g., 13–15, 40–42] prefer to speak more generally of *torusenes* or *toroidal forms* of carbon or of graphitic carbon or graphite.

By a *polyhex* is meant a structure comprising an assembly of hexagons such that any two hexagons are either disjoint or have at least one common edge. (Note that this is a looser definition than that used for finite planar polyhexes,[10,16] where the hexagons are regular, and can have only one edge in common with another.)

The *connectivity* of a graph is the smallest number of vertices whose removal disconnects the graph. There is an analogous version applying to edges. For polyhedra, including fullerenes, a value of three is considered an essential feature—in order to have something of an "intuitively cagelike" character in 3D space. It is natural to extend this consideration to tori, but it seems somewhat inadequate, for it is possible to formulate a closed network covering a torus, that is 3-connected, but a cut through the torus tube will sever fewer than three edges. In this area the rules (about what should or should not be considered a toroidal fullerene) are fluid, and it may become useful to apply greater rigor as the subject develops. For this discussion we require only that the graph must be 3-connected, and that it can be (but need not be) embedded on a toroidal or other surface. In the case of a torus we implicitly require that it must be embedded so that at least one circuit embraces each of the circumferences (a and b in Figure 1), and the structure is therefore, in imagination at least, "anchored" to the required shape.

For a fuller understanding of the *genus of a surface*, reference should be made to mathematical works (a good introductory discussion is given by Devlin for example,[17] but loosely speaking genus refers to the number of holes a surface has. So an infinite plane, or the surface of any "solid" object like a polyhedron or sphere is of genus-zero, whereas a torus (one hole—see Figure 1) has a genus of one, and a double torus (Figure 2) a genus of two. A planar or zero-genus surface can be distinguished from others by the fact that *any* circle on it, wherever it is placed, can be contracted to a point without cutting or leaving the surface.

The *genus of a graph*, on the other hand, is the genus of the lowest genus surface on which the graph can be drawn without crossings (i.e., embedded). There are some efficient practical algorithms available for testing planarity, i.e., for testing whether or not $g = 0$, based on systematic techniques for constructing planar drawings. See, for example, Refs. 18 and 19 for useful reviews. In general, however, determining the genus of an arbitrary graph is not an easy matter.

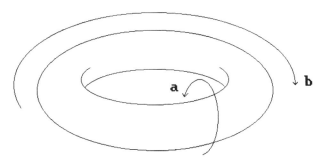

Figure 1. A torus.

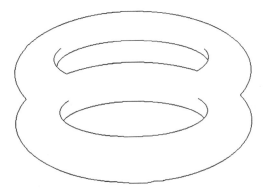

Figure 2.   A double torus.

Another important classification of closed surfaces is into ones that are *orientable* and ones that are *nonorientable*. The former have two sides, while the latter has only one. The difference may be exemplified by considering a rectangular strip of paper. If the ends are glued together without twisting, the result is a cylinder, having two sides. Given one twist, however, a Möbius strip forms, and this has one continuous (nonorientable) surface (Figure 3). A nonorientable surface can be characterized by its *cross-cap* number, but consideration of this is beyond relevance to the present discussion, for here we go no further than to distinguish a Klein bottle (whose surface is nonorientable) from a torus, as described later.

Some other terms are defined as they occur in the text.

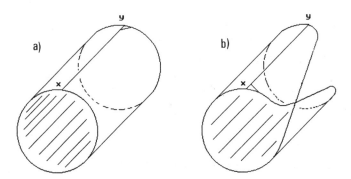

Figure 3.   A cylinder (a), and a Möbius strip, (b), made by cutting the cylinder at $x - y$, twisting, and rejoining.

8.3. TYPES OF ISOMERISM

Here we are concerned mainly with fullerene isomers that differ both in their connection pattern (and so usually their eigenvalues) and/or the kind of closed surface in which they are embedded. The simplest kind, with topologically spherical, zero-genus embedding, embraces C_{60} as a polyhedron, with all of its possible ring arrangements, including that of buckminsterfullerene itself, and analogous structures that vary in the number of carbon atoms. Cylinders, or tubules, have less symmetry, but are of essentially the same type. A very interesting special kind of cylinder studied recently (see later) is itself the thread of a helix.

Beyond this are higher-genus surfaces that can be embedded: first the familiar torus, the closed "doughnut-shaped" tube, and the less familiar Klein bottle. The latter is more complicated to deal with because it has a self-intersecting surface in 3D space. Unlike a sphere, either of these objects can be completely tessellated with hexagons, or in other words, a complete covering made from a single-ring-sized graphite sheet. On the other hand, the inevitable geometric distortion in comparison with a planar structure or a sphere is no less pronounced, and in fact is aggravated—for the curvature of any torus tube is inevitably tighter than that of a sphere with the same number of atoms or vertices.

A significant body of work has now been done on toroidal possibilities, with, so far, only brief consideration of double tori and of Klein bottles. There are many roots of possible isomerism among toroidal structures. They start with all of the same variations of surface ring sizes and ring disposition patterns as is seen for simple fullerenes (but within different constraints on face sizes and numbers). Given these, a range of possible nonisomorphic (connectional) isomers can be created by introducing a varying twist up to $\pm180°$ on the torus tube (imagined as twisting a cylinder in varying

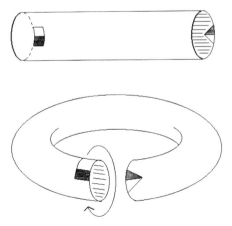

Figure 4. The formation of "twisted" toroidal fullerene isomers.

degrees around its axis before gluing the ends together: Figure 4). This is the kind that has been considered in most detail thus far. If the extent of twist goes beyond 180°, new stereoisomers will continue to be generated, but they will be isomorphic with earlier ones.

If a torus is imagined as made from a planar structure by gluing pairs of opposite sides, there will always be two stereoisomers (degenerate in the case of a square) depending on which gluing is done first. Another choice available is to fold the planar

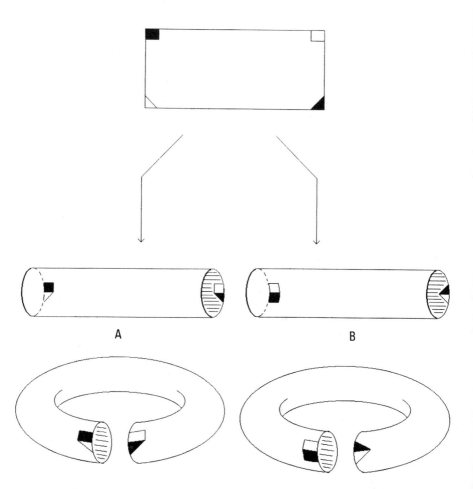

Figure 5. Two stereoisomeric (but isomorphic) toroidal fullerenes formed by gluing pairs of sides in different order: in (a) the rectangle is folded down into the plane of the paper and glued top to bottom; in (b) the rectangle is rotated 90° anticlockwise first. Both cylinders are then bent outwards from the paper before joining to form a torus. The two stereoisomers can be interchanged by a formal topological deformation process, but not by any process that is physically plausible. [This illustration is an adaptation of one used by Klein (Ref. 20a). Copyright 1994 American Chemical Society.]

structure the "other way" before gluing, thus reversing the inside and outside surfaces, but in other respects the resulting torus is not new (Figure 5). To exhaust all possibilities of twisting isomers it is necessary to consider doubly twisted and glued planar structures, i.e., making a cyclic permutation between *each* pair of glued sides simultaneously. A good way of picturing this is to say that if the single twists are done on a rectangle, then a *notched* rectangle is used for the double twisting (see Section 8.5.1). An algorithm for enumerating and generating nonisomorphic toroidal polyhexes is about to be submitted for publication.[23c]

There are also a number of other fairly obvious stereoisomeric possibilities (i.e., ones transparent to graph theory) that have as yet received no attention, e.g., multiple toroidal rings (or Klein bottles) with a catenane type of interlocking; knots tied in the torus tube; concentric multilayered toroidal tubes.

There are thus *many* theoretically possible isomers of many possible types, especially when increasingly large sizes are considered, but many also have a low priority for investigation because of apparent implausibility within a chemical context. An excellent review of carbon isomerism, which goes into some of the topological aspects in greater depth than here, has been written by Klein and Liu[21a]; another is by Ceulemans and Fowler.[21b]

8.4. SOME USEFUL FUNDAMENTAL RELATIONSHIPS

For a graph with n vertices, m edges, and f faces, with g being the genus, and v_i the degree of any vertex i, the generalized Euler relationship gives

$$(1) \qquad n - m + f = 2 - 2g$$

The general handshaking lemma gives

$$(2) \qquad \sum_{i=1}^{n} v_i = 2m$$

and so, for a cubic graph

$$(3) \qquad m = 3n/2$$

It follows that in general,

$$(4) \qquad f = n/2 + 2 - 2g$$

and in particular

$$(5) \qquad f_{planar} = n/2 + 2$$

$$(6) \qquad f_{toroidal} = n/2$$

(7) $$f_{\text{doubly toroidal}} = n/2 - 2$$

If we stipulate that every vertex should be shared by three faces, then

(8) $$\sum (\text{face size}) = 3n$$

Every face is a circuit in the graph. It therefore follows that as a bounding condition there must be enough circuits that are sufficiently small for there to be a set of circuits where $\sum$ (circuit size) $\leq 3n$. This can sometimes be a useful practical check on feasible structures. A similar, but less useful, upper bound can be written.

8.5. CONCEPTUAL CONSTRUCTION

In this section we consider various methods for arriving at a connection table or adjacency diagram for higher-genus fullerenes with given characteristics. The process is qualified as "conceptual" in order to emphasize that we are not yet ready to talk about practical synthetic chemistry!

8.5.1. Folding and Gluing

This method has great appeal because of its analogy with the practical way, discussed earlier, in which one might make a real torus-shaped object out of a sheet of semistiff material. In its simplest form a rectangle is folded and two opposite sides glued to make a cylinder. If this is bent round, and the (now circular) ends joined, the result is a toroidal tube. In this chemical context we can imagine a toroidal cage being made in the same way, by starting with a planar rectangular polycyclic structure, and joining and fusing each pair of opposite sides (Figure 6).

In this way the adjacency of such a toroidal network can conveniently be represented as a planar rectangular polycyclic structure with opposite sides marked as being identical, i.e., in a labeling, the set of vertices along one side is the same, and in the same sequence, as those of the opposite side (or of a cyclic permutation of them). For toroidal polyhexes this is equivalent to the diagrammatic approach discussed elsewhere. Klein[20] uses the term *parity* in discussing which identifications are possible.

Although the simplest shape for this purpose *is* the rectangle, the technique is by no means confined to this, and it can be any shape where suitable pairs of "sides" can be identified and joined, and nonrectangular shapes are often necessary for generating some of the possible isomers. Some fairly obvious stipulations[22] are required of the perimeter of this diagram:

1. Every vertex on the perimeter must be identified with at least one other that is also on the perimeter. On the other hand, an edge may or may not be so identified with another edge.

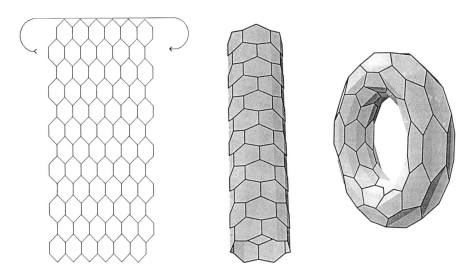

Figure 6. "Folding and gluing" a polyhex sheet, first to a cylinder, and then to a toroidal polyhex. [The cylinder and the torus are reproduced by kind permission of Dr. Peter John, Technische Universität Ilmenau, from drawings published by John and Walther (Ref. 12).]

2. A corollary of this is that every vertex label on the perimeter appears at least twice. Two of them (at the "corners") appear three times.

3. Identified pairs of vertices cannot both be of degree-3, i.e., a vertex of degree-3 must be matched with one of degree-2 (otherwise the single vertex that they represent will be of more than degree-3).

4. A pair of identified degree-2 vertices are incident to two unique edges, and are associated with a new ring that is present on the torus but not in the precursor.

5. Traveling around the perimeter in one (arbitrary) direction, the vertices must be divisible into four consecutive sets *A, B, C,* and *D* where *A & C* and *B & D* contain equal numbers (*A* and *B* may or may not be the same size). The vertices of *A* must match [in the sense of (1)–(3) above] with some cyclic permutation of the vertices of *C read backwards,* and similarly there must be a matching of *B* and *D*. The need for one of the pair to be read backwards arises from the fact that a consecutive numbering of the perimeter "turns back" toward the starting position from the furthest point (which must be between *B* and *C*).

6. The requirements can also be given an appropriate geometric interpretation. In this section the term *rectangle* has been used as elaborated by Kirby *et al.*[23a] in a rather loose sense, but with an obvious meaning, for *subgraphs* of the infinite honeycomb lattice; polyhexes whose strict rectangularity is exhibited only by the perimeter of their inner duals. Some workers[20] prefer to consider

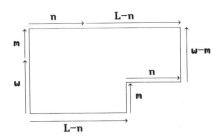

Figure 7. The "notched rectangle" used by Klein (Ref. 20a) for constructing a "doubly twisted" poly-hex torus. The boundaries enclose a polyhex section cut (through edges) from a planar hexagon lattice, and the labels show how sections of its perimeter are paired for identification of same-edges on the torus.

For a horizontally aligned polyhex, L represents the number of hexagons cut by a horizontal line that cuts but does not coincide with any edges, and w is twice the number of hexagons cut by a vertical line through horizontal edges. (Reprinted from Ref. 20a with permission. Copyright 1994 American Chemical Society.)

a geometrically accurate excision of the honeycomb lattice, in which edges are cleaved at their centers rather than adjacent to vertices, meaning that it is always dangling edges, rather than vertices, that are identified. The results, of course, are the same.

8.5.1.1. Toward a Characterization of the Planar Polyhex Precursor of a Toroidal Polyhex

Following the preceding section, when a planar polycyclic precursor of a torus is a polyhex, there are further properties it must have, although only limited work has been done on this characterization.[22]

1. A polyhex precursor with h rings and n_i internal vertices, when all perimeter edges are fused by folding, generates a toroidal polyhex with $2h + 2 - n_i$ fewer vertices. (This is because a planar polyhex has[24] $4h + 2 - n_i$, and a toroidal polyhex $2h$ vertices.)

2. The matching rules for identified pairs of sides (Section 8.5.1) can be formulated in geometric terms. Thus (using the terminology of Gutman and Cyvin,[24] (shown in Figure 8) we can pair

   a. A fissure with a hexagon of type L_1, L_2, L_3, A_2, P_2, P_3, or P_4
   b. A bay with an L_1, L_3, A_2, or P_2 hexagon
   c. A cove with an L_1 or P_2 hexagon
   d. A fjord with an L_1 hexagon

   and there are other possibilities when creation of a new hexagon is allowed. With this number of combinations, it has to be said that no significant simplification is achieved by this geometric approach.

3. In general there appears to be nothing unique about polyhex precursors or their derived tori. A given polyhex can be folded in different ways to give different toroidal polyhexes, and a given toroidal polyhex can be derived from more than one planar polyhex.

It is apparent that neither these specific points, nor the more general ones of the preceding section afford any easy answer to the question: Which polyhex(es) can be

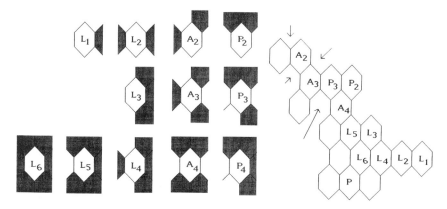

Figure 8. The 12 possible hexagon modes in a benzenoid system, and four perimeter concavities [Gutman and Cyvin (Ref. 24); reproduced by permission of Springer-Verlag GmbH & Co. KG].

folded and glued, and how, in order to make a given toroidal polyhex? Each case must be examined carefully and individually.

8.5.2. Stitching

Here, this term, a shorthand one for the process of making a garment by stitching together pieces of cloth already cut to the right size and shape, is used as a metaphor for constructing a fullerene. It is a more complex version of the simple "folding and gluing" already described, which includes also connection by the process of adding

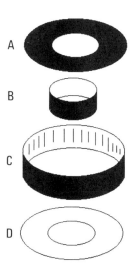

Figure 9. An exploded view of sections for making a torus by a method of Itoh and Ihara (Ref. 25). Each one is a polyhex; (a) and (d) are flat rings of a coronoid shape, while (b) and (c) are cylindrical. The dimensions and shapes are chosen so that pentagons and heptagons (see Figure 10) are formed at suitable positions.

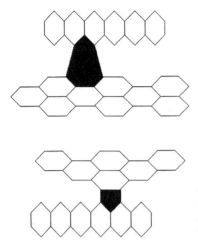

Figure 10. An example of pentagon and heptagon formation, arising from the connecting of polyhex sheets that are oriented in different planes, during the construction of toroidal fullerenes (see Figure 9). (From Ref. 25.)

new edges. Several, rather than only one suitable section and shape of lattice may be joined. Itoh and Ihara,[25] for example, make up a torus by connecting together two annular and two cylindrical nets of hexagons, with the pentagons and heptagons necessary to alleviate strain being formed at chosen positions (like holes in a garment) in conjunction with hexagons of changed orientation (see Figures 9 and 10).

Others have used the technique of joining together several preformed cylinders, end to end, to form a torus.[26,27]

8.5.3. Tiling

The mathematical term *tiling* derives in an obvious sense from the physical process of completely covering a surface with tiles—flat objects that all correspond to one or more particular shapes. In this chemical context, a 2D network can be regarded as a "tiling" of some imaginary surface with the rings, which, in the case of closed objects with a boundless surface such as fullerenes, are also "faces." The process therefore provides another approach to the general problem of enumeration; of determining how many isomers there are for a given size, distribution of face sizes, and topology.

There are a number of textbooks[28–30] on the subject which give the mathematical background. Unfortunately, its seems (1) that there are few really general theorems without symmetry constraints and (2) that some otherwise interesting results are irrelevant to this study because the number of face sides meeting at a point is variable, whereas here this number must always be three, since we are considering the construction of cubic (3-valent) graphs.

Avron and Berger[31,32] recently applied tiling rules[30] in an interesting way to construct molecules that approximate the family of tori corresponding to the surface of revolution with Cartesian representation

$$R\{(1 - \cos\theta/\eta)\cos\phi,(1 - \cos\theta/\eta)\sin\phi,\sin\theta/\eta\}$$

Here θ and ϕ are the toroidal coordinates. R (> 0) and η (> 1) represent size and shape, respectively, with large η corresponding to thin tori (the authors give a more precise formulation[32]). They tile, not the original network (with atoms at the vertices), but rather its *dual* where vertices are at the centers of the original faces. By this means, because the original graph is 3-valent, a network containing pentagons, hexagons, and heptagons is reduced to one having only triangles. The precise shape of the tiles is needed for classification purposes (right-angled isosceles triangles are used). The atoms are then allowed to relax from the positions predicted by this tiling and they reach equilibrium according to some interatomic potential.[33] A *generation* is a class of tiles, all of the same size, and an approximate relationship between η (shape) and the number of generations m can be expressed as

$$(\eta + 1)/(\eta - 1) \approx 2^{[m/2]}$$

where $[m/2]$ is the integral part of $m/2$, and the fatter the torus, the more generations are needed.

These tilings may be characterized (Table 1) in terms of four numbers: m (the number of generations, m-generation tori being those that can be built by matching together tiles of m different sizes); g_1 and g_m (the intuitive meaning of g being the length of a longitudinal circumference); and z (roughly the latitudinal length of a unit cell).

A program that evaluates the Hückel spectra of up to two generation tori has been written in MATHEMATICA to run on PCs with a 386 or better processor.[33]

8.5.4. Modification of an Already Existing Lattice

In this method, an existing toroidal lattice of known structure, usually a polyhex, is modified by insertion or excision of vertices and edges. Thus, Borštnik and

Table 1. A Selection of Variously Sized Toroidal Fullerenes with Five-, Six-, and Seven-Membered Rings, and Their HOMO–LUMO Values (β units) Compared with C_{60}, Buckminsterfullerene (*)[a,b]

m	2	4	2	5	2	2	3	
g_1	2	2	2	2	4	6	4	
g_m	2	1	2	2	4	6	2	
z	2	1	3	1	4	6	6	
n	120	120	180	240	480	1080	1200	60*
HOMO	0.43	0.04	0.45	0.19	0.16	0.08	0.07	0.62
LUMO	−0.27	0.04	−0.18	0.01	−0.22	−0.02	−0.02	−0.14
Inner radius (Å)	4.05	2.02	6.87	2.09	8.33	12.57	14.15	
Outer radius (Å)	7.29	6.02	10.14	8.49	14.21	21.17	23.42	
Height (Å)	2.93	4.57	3.37	6.89	5.68	8.00	8.13	7.38

[a]Taken from the work of Berger and Avron (Refs. 31 and 32).

[b]These were generated according to tiling rules; m, g_1, g_m, and z being four parameters of an algorithm that generates and characterizes the structure (m is the number of generations; g_1, g_m, and z are specified edge-counts: Refs. 31 and 32 should be consulted for details).

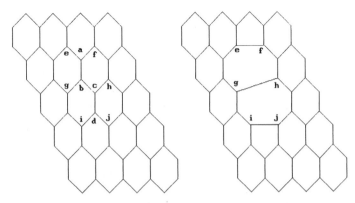

Figure 11. A connectional technique for narrowing, and introducing five- and seven-membered ring pairs to the graphitic tube of a torus in order to alleviate curvature strain. (From Ref. 26.)

Lukman[26] excise four vertices (a–d), resulting in two "back-to-back" azulene skeletons (two pentagons and two heptagons) being formed at a (narrowed) section of the tube (Figure 11). An alternative construction mode, rearranging two edges within a double naphthalene unit, results in a different orientation of the azulene units without loss of vertices (Figure 12),[22] and the technique can be used for several other kinds of face formation.

8.6. ISOMORPHISM

In favorable and comparatively simple cases, such as the all-hexagon-faced toroidal polyhexes, and other graphs made by applying tiling rules, their identities are apparent from the methods of construction. Where this is not so, however, it can often be just as difficult with this, as with any other group of structures to decide whether two of them are isomorphic.

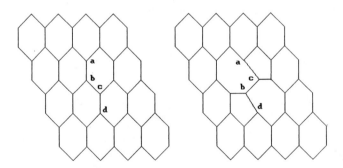

Figure 12. Another face-changing connectional rearrangement that creates five- and seven-membered rings. (This results in a different orientation and no loss of vertices.)

The best method currently available for identifying a particular graph appears to be the combination of its set of eigenvalues and its $T(G)$ matrix, following work by Liu and Klein.[34] The $T(G)$ matrix is a graph invariant; an ordered "eigenvalue-eigenprojector" table with columns labeled by distinct eigenvalues, and rows by vertices. The combination of the eigenvalue spectrum with this matrix appears to be a very powerful tool (two structures are considered nonisomorphic if either their eigenspectra or their $T(G)$ matrices differ). It is known that its discriminatory power is not absolute, but, nevertheless, Liu and Klein, after extensively testing it, wrote "we know of no reasonably 'molecular' graph G where $T(G)$ fails to be faithful."

Other, much cruder preliminary checks that can be helpful include a count of progressively increasing size of circuits and a ring-size adjacency matrix (see Section 8.9.1.2).

8.7. ENUMERATION

Various approaches to the enumeration of connectional isomers have been discussed: folding and gluing of a planar lattice section in every possible way,[20,35–37] the application of tiling rules,[31,32] the application of hexagon matrix rules,[23a,c] and others.[12,38] There appears to have been only one report[36] giving explicit enumeration

Table 2. An Enumeration of Toroidal Polyhexes with up to 30 Hexagons (60 Vertices)[a]

3	2	1a	13	2	1	18	4	1	22	2	1	25	3	1b	28	6	1
4	2	1	13	3	1	18	5	1a	22	3	1	25	4	1b	28	7	1
5	2	1b	13	4	1	18	6	1	22	4	1	25	5	1	28	8	1
6	2	1a	14	2	1c	18	9	1	22	5	1	25	10	1	28	13	1cd
6	3	1	14	3	1	6	0	3a	22	6	1	5	0	5b	28	14	1
7	2	1c	14	4	1cd	19	2	1	22	11	1	26	2	1	14	4	2
7	3	1	14	7	1	19	3	1	23	2	1	26	3	1	14	6	2
8	2	1	15	2	1ab	19	4	1	23	3	1	26	4	1	29	2	1
8	3	1	15	3	1b	19	8	1	23	4	1	26	5	1	29	3	1
8	4	1	15	4	1b	20	2	1b	23	5	1	26	7	1	29	4	1
9	2	1a	15	5	1a	20	3	1b	24	2	1a	26	8	1	29	5	1
9	3	1	15	6	1	20	4	1b	24	3	1	26	13	1	29	9	1
3	0	3a	16	2	1	20	5	1	24	4	1	27	2	1a	30	2	1ab
10	2	1b	16	3	1	20	6	1	24	5	1a	27	3	1	30	3	1b
10	3	1b	16	4	1	20	9	1b	24	6	1	27	4	1	30	4	1
10	5	1	16	7	1	20	10	1	24	7	1	27	5	1a	30	5	1a
11	2	1	16	8	1	10	4	2b	24	8	1a	27	6	1	30	6	1
11	3	1	8	4	2	21	2	1ac	24	9	1	27	9	1	30	7	1b
12	2	1a	4	0	4	21	3	1	24	10	1	9	3	3a	30	8	1ab
12	3	1	17	2	1	21	4	1c	24	11	1a	9	6	3a	30	9	1b
12	4	1	17	3	1	21	5	1a	24	12	1	28	2	1c	30	10	1
12	5	1a	17	4	1	21	7	1	12	4	2a	28	3	1	30	11	1a
12	6	1	18	2	1a	21	8	1a	12	6	2	28	4	1cd	30	12	1b
6	4	2a	18	3	1	21	9	1c	25	2	1b	28	5	1	30	15	1
															15	5	2

[a] Some fully arenoid structures are marked a–d: a = fully benzenoid; b = fully naphthalenoid; c = fully anthracenoid; d = fully phenanthrenoid. (Note that in Ref. 36 there is a misprint, where 30-3-1 was starred as fully benzenoid instead of 30-2-1, and 14-6-2 was omitted.) The notation is explained in Section 8.9.1.1 (see also Fig. 16).

results. A listing of distinct toroidal polyhexes from C_{12} (6 hexagons) to C_{60} (30 hexagons), 142 in all, is shown in Table 2. A computer-operated algorithm capable of generating many nonisomorphic toroidal polyhexes will shortly become available.[23c] However it be done, reliably complete enumeration is quite a complicated task, especially for systems of mixed ring size.

8.8. THE DRAWING OF TOROIDAL FULLERENES

Understanding of these structures can be greatly helped by constructing models or making drawings with 3D perspective, but this itself is by no means a trivial task. It is relatively simple to draw a solid torus, either by simply drawing suitably matched ellipses (consider Figure 1), or from the equation of a solid of revolution, using any one of several available mathematical plotting programs. Unless the chemist happens also to be an artist, however, tessellating such a drawing with faces by freehand drawing is difficult, partly because of the need for accurate perspective, and partly because, even on the "real" 3D object, ring sizes and shapes must inevitably be distorted by the curvature.

MATHEMATICA is one computer software package that does allow some reasonable drawings to be made (Figure 6), and there is some scope for experimentation by manipulation,[12,39] but, in general, a more promising approach seems to be use of the *NiceGraph* suite of programs[40–42] (see Figure 13). In Ref. 40 the authors recommend techniques for a quick determination of molecular geometries, especially when many

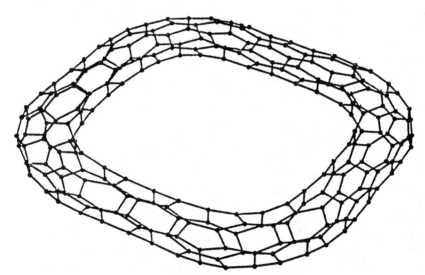

Figure 13. "NiceGraph" drawing of a toroidal azulenoid cage with ten pentagons (reprinted from Ref. 40). NiceGraph appears to have a significant and important ability, namely, in favorable cases it can recognize genus.

isomers of a given molecule have to be searched. They investigated and developed algorithms based on treating a graph as a mechanical system of balls and springs for which there exists a geometrical configuration of minimum energy. The Fruchterman and Reingold algorithm[43] appears to be the best. They also tested the algorithm based on adjacency matrix eigenvectors.[44–46] This is a highly intriguing and at first sight somewhat mysterious algorithm. If the eigenvalues of an adjacency matrix are ordered, and three consecutive sets of the associated eigenvectors are taken, then if each triad of elements (say x_{2i}, x_{3i}, x_{4i} for the ith elements of eigenvectors 2, 3, and 4) is regarded as the coordinates of vertex$_i$ in 3D space, a "rather decent 3D drawing" of the graph is obtained. The efficacy of this method was recently confirmed by the author while working on toroidal and double-toroidal polyhexic structures.[42c]

Some drawings of striking elegance have been produced by Japanese workers.[14,25] They have used CHEM3D for drawing molecules and MACDRAWII for brushup (on a MacIntosh).[47]

8.9. SOME PARTICULAR CLASSES OF STRUCTURE

8.9.1. Toroidal Fullerenes

8.9.1.1. A Three-Integer Code for Toroidal Polyhexes

The closed but boundless network of hexagons embedded as an all-hexagon toroidal polyhex can be represented by an infinite planar lattice, on which the set of hexagons forms a parallelogram which repeats itself endlessly in two dimensions (i.e., it is doubly periodic). Figure 14 shows an example of a torus with nine hexagons (A–I).

A parallelogram that identifies the set of hexagons in this way, also defines the structure to the point of isomorphism. If the center of one arbitrarily selected hexagon is selected both as the corner of a parallelogram and as a Cartesian origin (Figure 15 shows such a scheme applied to Figure 14), then vectors that define the adjacent corners, also define the parallelogram, and hence the lattice. A four-element square matrix

$$\begin{bmatrix} a & b \\ c & d \end{bmatrix}$$

can be used. Here (a,c) and (b,d) give coordinates of the two points (measured in units of one hexagon width). However, it is always possible to choose one of the axes to coincide with a side of the chosen parallelogram. By this means, c is always given a value of zero, and is thus redundant, so that a, b, d may be read as a three-integer-array code. This defines the structure, but is not unique; to achieve this, a rule for choosing between possible alternative arrays must be stated. The rule chosen[23a] is that d and b, in that order, should have minimum values, and the notation adopted was to write "TPH(a-b-d)" as the standardized symbol for a particular toroidal polyhex, or simply

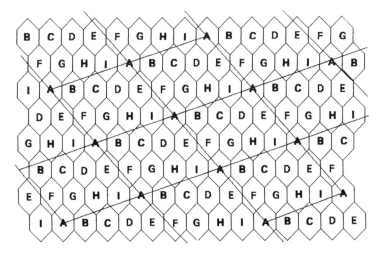

Figure 14. The infinitely-repeating-pattern representation of a toroidal polyhex on a planar lattice. This example has 9 hexagons (labeled A–I) and 18 vertices.

the unprefixed string (*a-b-d*) for any valid (but not necessarily unique) representation. Figure 16 shows an example of this.

8.9.1.2. Properties of the 2 × 2 Toroidal Polyhex Matrix

It was explained above that the three-element code used to characterize a toroidal polyhex is derived from a convenient special case of the matrix

$$\begin{bmatrix} a & b \\ c & d \end{bmatrix}$$

where $c = 0$, obtained through a careful choice of axes.

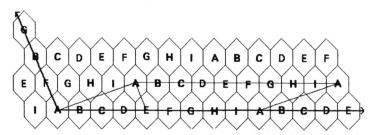

Figure 15. A Cartesian system applied to Figure 14 in a way that simplifies representation of the toroidal surface. The unit of length is the distance between the centers of two hexagons that share a common edge. The choices of a 120° (rather than 60°) angle between axes, and of a left-to-right vector as positive are adopted as standard conventions. In this example the coordinate positions of repeating "A" hexagons at $(a,c) = (9,0)$, and $(b,d) = (4,1)$, abbreviated to "(9-4-1)" suffice to define the parallelogram, and hence the toroidal surface.

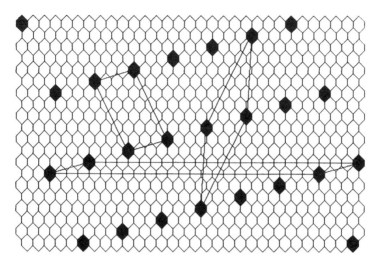

Figure 16. A 24-hexagon toroidal polyhex. Filled hexagons all represent the same hexagon on the torus. These are where this arbitrarily marked hexagon occurs in this endlessly repeating representation of the toroidal surface (only one such hexagon needs to be identified). Parallelograms are drawn to show different ways in which the set of 24 hexagons may be defined. By rotating and/or taking a mirror image where necessary, and applying the method described in the text, this gives the equivalent and equally valid integer codes 24-4-1, 24-21-1, 8-4-3, 8-7-3, and 6-1-4 and 6-3-4. Minimizing the third and then the second element yields TPH(24-4-1) as the unique descriptor.

A useful general property is that the number of hexagons is equal to the absolute value of $ad - bc$.

Rules for manipulation of the matrix elements have been given[23a] which enable isomorphic equivalences of matrices to be detected, as an alternative to the graphical approach of the preceding section.

Rule 1:

$$\begin{bmatrix} a & b \\ c & d \end{bmatrix} \equiv \begin{bmatrix} a & b + ka \\ c & d + kc \end{bmatrix}$$

where k is an integer, plus or minus, i.e., any integral multiple of the first column may be added to or subtracted from the second one, whereupon any polyhex matrix can be reduced to the form

$$\begin{bmatrix} a & b \\ 0 & d \end{bmatrix}$$

with $a > b$, $d > 0$, $b \geq 0$; equivalent to choosing one side of the parallelogram as one of the axes, as discussed above. It can be shown that up to six equivalent reduced matrices can exist. Two further rules can be applied to this reduced matrix:

Rule 2:

$$\begin{bmatrix} a & b \\ 0 & d \end{bmatrix} \equiv \begin{bmatrix} a & d-b \\ c & d \end{bmatrix}$$

Rule 3:

$$\begin{bmatrix} a & b \\ 0 & d \end{bmatrix} \equiv \begin{bmatrix} ad/h & xd \\ 0 & h \end{bmatrix}$$

Here h is the highest common factor of a and b, and x is the multiplicative inverse of b/h modulo a/h (or, put another way, x is an integer such that $xb - ya = h$ for some integer y). Rules 2 and 3 are self-inverse, i.e., a second application restores the original matrix.

An additional, fourth, rule was introduced to cater for a special case that arises from the automorphism of the C_4 graph, when a toroidal polyhex consists of an even number of squares through which the torus is threaded, that is, they have the code (2-b-d) with $b = 0$ or 1, and d is even. Such a case is the pair of toroidal polyhexes (2-0-4) and (2-1-4) in Figure 17. These are distinct, but are isomorphic. When the number of squares is odd, this difference does not arise. Its origin is the automorphism peculiar to the C_4 graph that leaves alternate vertices fixed while permuting the others. We therefore write

Rule 4:

$$\begin{bmatrix} 2 & 0 \\ 0 & d \end{bmatrix} \equiv \begin{bmatrix} 2 & 1 \\ 0 & d \end{bmatrix}$$

This rule, illustrated by lattice representations (Figure 17), shows the care that is needed in interpretation, for in this particular case, the diagrammatic approach alone does not bring this equivalence to light; (2-0-4) and (2-1-4) are *not* superimposable, although it can be shown that they are isomorphic graphs.

Finally, a fifth, and obvious, rule needs to be stated in order to construct a rule set that is useful for comprehensive enumeration.[23c]

Rule 5:

$$\begin{bmatrix} a & 0 \\ 0 & d \end{bmatrix} \equiv \begin{bmatrix} d & 0 \\ 0 & a \end{bmatrix}$$

8.9.2. Klein Bottles

The possibility of a carbon cage network that can be regarded as a topological embedding in the surface of a Klein bottle (Figure 18) is an obvious extension from the consideration of tori. In 3D space it has a "self-intersecting" surface, which it is necessary to define, and, the essential surface being a network, it is possible for it to intersect with itself with or without a physical join (Figure 19). The ring-interlocked, catenane type of linking is attractively simple, but, at least for carbon structures, such

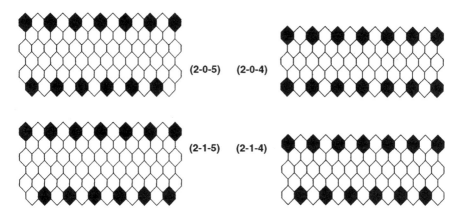

Figure 17. Two pairs of (2-*b*-*d*) toroidal polyhex patterns (filled ones track the repeated occurrence of a marker hexagon). If these are carefully examined it can be seen that the patterns for (2-0-5) and (2-1-5), where *d* is odd, are superimposable (after reversing one of the images) and are therefore identical. When *d* is even, as in (2-0-4) and (2-1-4), they are not the same. They *are*, however, isomorphic (see text).

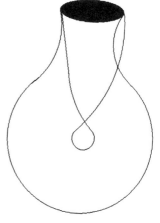

Figure 18. A Klein bottle drawn to show its relationship to the torus. It has one surface (there is no "inside and outside") which intersects with itself (Figure 19). Whereas a torus can be formed by joining the two ends of a cylinder "end-face to end-face," a Klein bottle is the result of joining it "end-face behind end-face." It can also be formed by joining together the perimeters of two Möbius bands.

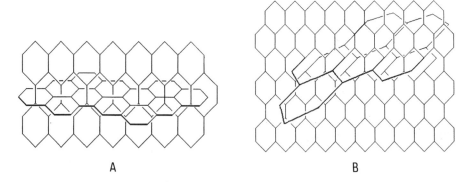

Figure 19. Two possible ways in which two networks could intersect for embedding in the surface of a Klein bottle: (a) constrained by interlocking, but with no connectional join, and (b) connected. Mode (a) is unlikely to be feasible for elemental carbon networks although it might be possible for larger polyene systems, while if (b) is adopted, the system has to be one of mixed ring sizes (and has additional geometric strain).

an arrangement is likely to be chemically plausible only for large rings. Alternatively, a physically linked structure is possible, but this necessarily has "larger than normal" rings around the intersection so that a uniform tessellation is not possible. Klein[20] and Kirby[48] have discussed the formation of Klein bottle structures by folding and gluing; Klein and Zhu[35] extended their treatment of Kekulé structure enumeration for toroidal polyhexes to include Klein bottles. There is tentative evidence for there being some consistent differences between toroidal and Klein-bottle Hückel numerical properties (see Section 8.9.1.2).[48] In principle the techniques used for encoding toroidal polyhexes should be applicable, but a detailed formulation of rules is more complicated, and results are incomplete.[22] (Thomassen[49] has discussed tessellations of both tori and Klein bottles with hexagons.)

8.9.3. Double Tori

Formally, a torus may be described as a sphere with one handle, so that a double torus is one with two handles; in practical terms, an imaginary object such as shown in Figure 2 covered with a closed network. Very little work seems to have been done so far. John and Walther[12] mention them, and Kirby[48] described a stitching process for making one or two examples by joining the ends of a cylinder (see Figure 20) to the surface of a toroidal polyhex. The resultant objects were "near polyhexes," with a polyhex covering of the surface everywhere except around the join, and in that paper[48] it was raised as an open question as to whether a double torus *can* be tessellated with hexagons alone. In fact, the impossibility of this can be seen as an obvious consequence of the generalized Euler relationship, for the number of faces must be $n/2 - 2$ (Section

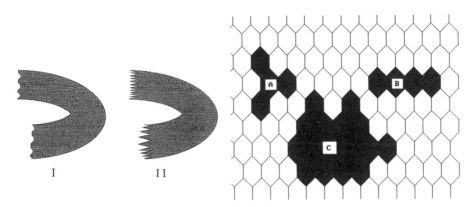

Figure 20. Cylindrical polyhexes which, by bridging two holes in the surface of a toroidal polyhex, can form a double torus by inserting or gluing edges. A hole of type A can be joined by gluing a cylinder of type I, and B by connecting with edges to one of type II, whereas most holes (e.g., C) are more complicated. (Note that none of the results have a purely polyhex surface.)

8.4). Since the sum of face sizes for a cubic graph must add up to $3n$, it is not possible to achieve the required equality with hexagons, for

$$(9) \qquad 6f = 6(n/2 - 2) \neq 3n$$

and the single torus or Klein bottle are the *only* objects that can sustain a closed graphite sheet.

The doubly toroidal C_{60} isomer mentioned[48] had 24 hexagonal faces and 4 of 9 vertices each, totaling 28 faces, and conforming with equation (7), Section 8.4. A quite different kind of double torus was recently suggested by Klein and Liu.[21a] This looks rather like a cotton reel and consists of two separate tori connected by a cylinder (Figure 21). It involves the merging of one lattice sheet onto the surface of another (not passing

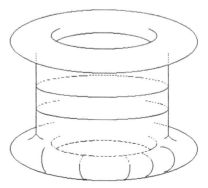

Figure 21. Another kind of "double torus" in the sense of being two separate tori connected with a single (cylindrical) lattice sheet. (From Ref. 21a.)

through it, as for a Klein bottle). Again, if it is made from a 3-valent lattice, then nonhexagons occur around the seam. [The nature of this kind of join is also discussed in Ref. 50, where joined all-hexagon networks are achieved by introducing a row of 4-valent ("diamond") carbon atoms where three network surfaces meet.]

8.9.4. Helical Tubes

In topological terms these are more closely related to simple spherical and tubular cage forms, but, because of the geometric similarity of a helix to a torus, it is of interest to note that the possibility of helically coiled tubular forms of carbon lattice (with five-, six-, and seven-membered rings) has been studied by Ihara et al.[51] They calculate that various such forms should be energetically and thermodynamically stable, and point to the mechanical stability of systems that are seen both in nature (e.g., DNA, beanstalks) and among man-made mechanical structures such as spiral springs. The authors also speculate briefly about variant forms like the supercoiled structure, nested helical forms, and the double and triple helices seen in biological systems—and whether these could be constructed as carbon cages.

The paper mentioned[51] contains some beautifully clear drawings of helically coiled forms; the rather "scale-like" network faces enhancing the serpentiform appearance.

8.10. MATTERS RELATING TO THE ENERGY AND STABILITY OF TOROIDAL FULLERENES

8.10.1. Adjacency Matrices and Their Eigenvalues for Toroidal Polyhexes

Three cases have been distinguished,[23a,c] according to the numerical properties of a, b, d and the nature of the cyclic pattern of hexagons on the torus.

Structures with up to 7200 vertices (3600 hexagons) can be dealt with under Cases I and II. Beyond this, certain structures come under Case III only, which is more complicated to deal with. Specifically, Case I can be used if and only if b is divisible by the highest common factor of a and d. It covers all polyhexes with fewer than 1800 vertices and 900 hexagons (and many, although not all, larger ones). It can be proved that at least one of the matrices equivalent to $(a\text{-}b\text{-}d)$ has this required property unless ad is divisible by the squares of three different primes, but often it can even then, i.e., this prime number condition is necessary but not sufficient for exclusion from consideration under this Case.

The smallest number that is divisible by the squares of three distinct prime numbers is 900 (equivalent to $2^2 \times 3^2 \times 5^2$), and TPH(450-5-2) is an example of a toroidal polyhex that must be brought under a higher Case.

The eigenvalues for Case I are given by

(10) $\lambda^2 = 3 + 2\{\cos 2\pi u/a + \cos 2\pi(ur/a - v/d) + \cos 2\pi[u(r-1)/a - v/d]\}$

or

(11) $\lambda^2 = 1 + 8 \cos \pi u/a \cos \pi(ur/a - v/d) \cos \pi[u(r-1)/a - v/d]$

$(0 \le u < a$ and $0 \le v < d$ in both cases).

Case II requires that there must be no integer greater than one that divides all of a, b, d; the case is therefore a useful supplement or alternative to Case I and, together with Case I, allows all structures with up to 7200 vertices (3600 hexagons) to be dealt with.

The eigenvalues are given by

(12) $\lambda^2 = 3 + 2\{\cos 2\pi kp/a + \cos 2\pi k(rdp - 1)/ad + \cos 2\pi k[(r-1)dp - 1]/ad\}$

or

(13) $\lambda^2 = 1 + 8 \cos \pi kp/a \cos \pi k(rdp - 1)/ad \cos \pi k[(r-1)dp - 1]/ad$

$(0 \le k < ad$ in both cases).

Case III caters, in addition, for toroidal polyhexes that do not meet either of the previous conditions. They cannot have fewer than 7200 vertices [or 3600 hexagons, since a toroidal polyhex with $2h$ vertices has h hexagons—see equation (6), Section 8.4], and this Case has not been subjected to an explicit general treatment. The essential features are discussed, with an example, in an appendix of Ref. 23a. Algorithms for compiling adjacency matrices and their eigenvectors are given in the same paper; it is a relatively simple matter to write down the eigenvectors, and an adjacency matrix, when a systematic labeling system is adopted. A more generalized mathematical treatment was recently discussed by Haigh.[23b]

8.10.2. Aspects of Hückel Spectrum Values

There are structures either with or without odd-sized rings that have nonbonding orbitals (zero eigenvalues). Although few specific figures have been recorded, Kirby et al.[23] reported on 13 toroidal polyhex isomers of C_{60}, and found E_π values of 1.43–1.60 per atom, with HOMO–LUMO gaps varying from 0 to 2, thus straddling the value for buckminsterfullerene C_{60}, and with several being rather similar (all energy figures being in beta units).

Itoh and Ihara[25] have given some figures for eight geometrically optimized toroidal C_{240} isomers containing five-, six-, and seven-membered rings. HOMO–LUMO gaps ranged from 0 to 0.50, and delocalization energies were near 0.565, between the values for C_{60} (0.553) and graphite (0.576).

Examination of a small and arbitrary sample of isomer pairs within the size range C_{16} to C_{84} suggested that Klein bottle structures have the same HOMO–LUMO gap and slightly lower E_π values than their corresponding toroidal analogues.[48] More recent work by the author confirmed HOMO–LUMO constancy, but found the generalization about E_π false. The number of results, however, remains small.

8.10.3. Kekulé Structures of Toroidal Polyhexes

In general, the thinner the tube for a given torus size, the higher is its Kekulé count. Workers in Galveston[7] obtained a Kekulé count of 41,297 for a C_{60} toroidal polyhex with a fairly thick tube (from the description given it appears to be what could now be called 6-3-5 or TPH(15-5-2); see Section 8.9.1), and a very high figure[7,23a] of 1,860,500 for the thinner-tubed TPH(30-2-1). As a general trend this is somewhat counterintuitive, insofar as K may be taken as a measure of chemical stability. Care should be taken, however, in generalizing from this particular example, for although it can be embedded in closed fashion upon a torus, it is actually a graph-theoretically planar structure, and so is not necessarily typical. It is, in any case, apt to be misleading to read too much into simple K values of structures that can have destabilizing circuits (in this particular case 4-circuits are present).

Klein and Zhu[35] have given a technique for evaluating K which notes that each Kekulé structure on a toroidal network corresponds to a *self-avoiding walk system* (SAWS) around the torus. Using this correspondence, together with the "transfer-matrix" method (which keeps count of the number of ways in which one local state may follow another across a basic cell), explicit formulas are developed that enable a table to be written, giving numbers of Kekulé structures as functions of L for a given twisted torus $T_m(L,2h)$ (Sections 8.3 and 8.5) in terms of h, m, and R. (L is the length, and $2h$ the width of the corresponding planar polyhex that would result from skinning the torus; m denotes the degree of twisting, and R is the number of SAWS.)

Sachs and John[38,52] devised another method in which K is evaluated as a determinant of a size not more than half the number of vertices, and in some special cases simple formulas can be derived; for example, to a toroidal polyhex that can be represented as a folded parallelogram with a side having three hexagons, the following result can be applied:

$$K = 2^{q+1} + 3^q + s$$

Here q = (number of hexagons)/3; s = 2 if q is odd or 3 if q is even. Examination of the C_{60} toroidal polyhexes shows that one, TPH(30-10-1), conforms to this type. Here $q = 30/3 = 10$ (even), so that

$$K_{TPH(30-10-1)} = 2^{11} + 3^{10} + 3 = 61,100$$

The absolute value of the tail coefficient of the characteristic polynomial[53] (which is known to give K for benzenoids[10]), however, is only 3069 (Table 3), and this apparent discrepancy is another reminder that these 3D structures differ in important ways from simple planar benzenoids. All conjugated circuits in the latter are stabilizing,[10] but for tori, the closure of the lattice around the tube allows the presence of destabilizing circuits in some cases.

Table 3.   The Square Root of the Tail
Coefficient of the Characteristic
Polynomial for the 30 Hexagon,
60 Vertex, Toroidal Polyhexes[a]

Code	$\sqrt{a_n}$
30-2-1	0
30-3-1	95,139
30-4-1	25,344
30-5-1	0
30-6-1	21,483
30-7-1	21,483
30-8-1	0
30-9-1	17,019
30-10-1	3,069
30-11-1	0
30-12-1	16,128
30-15-1	3,2769
15-5-2	26,829

[a]Actually calculated as the product of nonnegative
eigenvalues.

The apparent (even if rather deceptive) simplicity of these structures gives the feeling that simple algorithms requiring negligible computation, such as the Gordon–Davison method for unbranched *cata*-condensed benzenoids[54] or its reformulated "numeral-in-hexagon" version[24] ought to be available for these structures (work cited in Refs. 24 and 10 extends the essential technique to *some peri*-cyclic benzenoids), but up to now, none appear to have been published.

8.10.4. Fully Arenoids

Following recognition of the significance of the aromatic sextet,[55] the concept of fully benzenoid hydrocarbons was introduced[56] by Clar and Zander in 1958 and has been discussed in various places (with some slight differences in the terminology used).[e.g.,24,57–64] Fully benzenoid hydrocarbons are particularly stable benzenoid structures which can be represented by a set of benzenoid rings (i.e., six-membered rings that are capable of cyclic conjugation within the ring) that are connected to each other by single bonds, and which account for all of the carbon atoms. In equivalent graph-theoretical terms, such a structure is a subgraph of the infinite hexagon lattice that has a 2-factor comprising a set of disjoint hexagons.

As Aihara and Tamaribuchi recently pointed out,[64] graphite can be classified as an infinite fully benzenoid. Some boundless toroidal polyhex surfaces also fall into this category, and examples are listed in Table 2. It was concluded that for a toroidal polyhex encoded (Section 8.9.1.1) as *a-b-d*, it will be fully benzenoid if and only if both *a* and (*b* + *d*) are multiples of 3.[36] Note: this is *also* a necessary and sufficient condition for the appearance of zero eigenvalues in these structures.[23a]

Somewhat later, this concept was generalized in an obvious manner from fully benzenoid structures to ones that are fully arenoid.[65–68] This, however, has brought to light a further distinction that can be made. Knop *et al.*,[69] analyzing the concept of fully benzenoid hydrocarbons, reminded readers that the Clar structures of such molecules do not account for all of the Kekulé structures (the bonds connecting the sextets are not essentially single) and concluded that "the only truly fully-benzenoid hydrocarbon is benzene." A similar case can be argued for higher fully arenoids. In Figure 22, for example, following this argument, (a) is regarded as fully naphthalenoid, but not (b). Although this narrower definition seems to pinpoint a subclass of structures that is significant, others[70] have argued for retaining the wider definition, which includes fully benzenoid hydrocarbons and their obvious fully arenoid analogues together with benzene and those structures considered by Knop *et al.*[69] to be special cases. This wider definition is used here to consider toroidal polyhexes. [Indeed there appears to be no possibility of a fully arenoid toroidal polyhex conforming to the stricter criterion, because any given hexagon is (topologically) indistinguishable from any other.[22b]] Only a little work has been done[22,36,62] but, as well as noting that various directly comparable fully arenoid structures are possible, it was pointed out that there also exist new possibilities. For example, finite planar benzenoids cannot be both fully benzenoid *and* fully naphthalenoid, but certain cylindrical and toroidal species can. Figure 23 shows an example of a toroidal form of C_{60} that is both fully benzenoid and fully naphthalenoid. Other fully arenoid types would be of great interest[22]; e.g., tori that are completely or partially fully azulenoid. A suitable network of five and seven numbered rings can be called "azulenoid,"[37] but to be fully azulenoid the structure must also have hexagonal rings.[22b]

Closely related to the concept of being fully arenoid is that of being *s-circuit 2-factorable*, which was discussed by Kirby.[36] The process of 2-factoring is well known in graph theory, being the identification of a subgraph in which every vertex

A   B

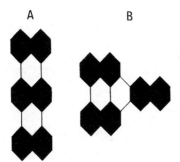

Figure 22. Two fully naphthalenoid structures. [But, following Knop *et al.* (Ref. 69) one would regard only (a) as being "truly" fully arenoid, because in (b), Kekulé structures can be drawn that involve the "empty" hexagons.]

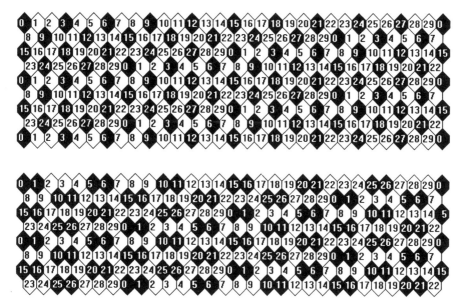

Figure 23. A C_{60} toroidal polyhex, in fact TPH(30-8-1) (Section 8.9.1.1), that can be viewed as either fully benzenoid, with the vertices accounted for by ten hexagons (in this case 0, 3, 6, 9, 12, 15, 18, 21, 24, and 27), or as fully naphthalenoid, with six hexagon-pairs (in this case 0-1, 5-6, 10-11, 15-16, 20-21, and 25-26). This concurrence does not occur among finite planar polyhexes. Note also that (again unlike the planar series) the "full"-hexagon set is not unique. This example could equally well be described as fully benzenoid with the ten-hexagon set 1, 4, 7, 10, 13, 16, 19, 22, 25, 28.

is of degree-2; in other words a set of rings that will account for every vertex. For fully *cata*-condensed benzenoid structures (no internal vertices) the results are usually the same, but clearly for fully pericondensed benzenoid structures (these do have internal vertices) they will not be. There is also an additional subtlety: a full set of 2-factorings will, in many cases, include a set of identical rings through which the torus tube is "threaded." These rings, unlike the face circuits "on the surface," and those in finite planar benzenoids,[24] do *not* necessarily correspond to $4n + 2$ circuits, and may therefore be potentially destabilizing.

8.10.5. *Ab Initio* Calculations

Toroidal forms of carbon were predicted to be stable on the basis of molecular dynamics simulations using a Stillinger–Weber-type potential.[13–15,51] To test these predictions, the stability of a C_{120} torus[71] was compared with that of C_{60} using *ab initio* self-consistent field calculations and ionization potentials determined by Koopman's theorem (Table 4). The C_{120} structure investigated had D_{5d} symmetry, and appears to

Table 4.   A Computed Geometry for a C_{120} Torus with Five-, Six-, and Seven-Membered
Rings by Ihara, Itoh, and Kitakami[a,b]

Atom No.	Coordinates (in Å) of each atom relative to the torus center		
	x	y	z
1	2.294	0.000	0.353
2	2.510	0.000	1.933
3	2.209	2.583	2.554
4	3.139	1.303	2.554
5	4.709	1.401	2.357
6	2.787	4.045	2.357
7	5.437	0.000	1.973
8	4.258	4.099	1.727
9	5.214	2.783	1.727
10	6.392	0.000	0.689
11	5.884	2.693	0.277
12	4.379	4.764	0.277

[a]Ref. 13.

[b]Relaxed positions of inequivalent atoms are shown. Those of the rest of the structure can be deduced from its D_{5d} symmetry. There are 10 pentagons, 10 heptagons, and 40 hexagons.

have 60 faces consisting of ten 5-rings, ten 7-rings, and forty 6-rings. Bond lengths straddle the values found for C_{60} with, overall, a tendency for them to be relatively stretched (because of the curvature). The electronic structure is comparable, and within errors of calculation, the binding energies for this torus and for C_{60} may be taken as approximately equal. The HOMO–LUMO gaps—0.28 (toroidal C_{120}) and 0.27 (buckminsterfullerene)—also are quite close.

8.10.6. Molecular Mechanics

Molecular mechanics calculations have been carried out on several toroidal structures, including both a toroidal polyhex, and certain mixed systems containing 5-, 6-, and 7-ring sizes.[26] Not surprisingly, suitable mixed systems are more favorable with regard to mechanical strain, and the energy of formation becomes comparable to values for medium-size fullerenes.

It is the bending energy required to form a torus from a planar rectangular lattice that is considered. The energy of formation of polyhex tori from tubule segments of various lengths was studied. The energy per carbon atom varies approximately as the inverse square of n (the number of atoms). This is explained by the fact that curvature is inversely proportional to n, and bending energy is proportional to the square of the curvature. Taking C_{60} as the standard to aim at, the results show that a pure toroidal *polyhex* should have more than 300 carbon atoms, but that once 5- and 7-membered rings are introduced, the threshold drops to about half that size, i.e., C_{150} might be stable in this toroidal form.

In Refs. 40–42, the geometries from this approach are compared with those generated by NiceGraph programs (Section 8.8).

8.10.7. The Effects on a Finite Lattice of Having No Periphery

The stability of a finite planar benzenoid is known to depend on the shape of its periphery. As any polyhex series increases in size, its π-electronic properties should approach those of graphite, but, since in the toroidal structures there is no periphery to start with, these might be expected to converge more quickly, and Hosoya has confirmed that this is indeed the case.[72]

8.11. CONCLUDING REMARKS

At present, the justification for this work in terms of practical chemistry is distinctly limited. The case rests, essentially, on asking the question: We now know about C_{60}, buckminsterfullerene, so, if a sphere, why not a torus, and so on?

To my knowledge, no electromagnetic radiation spectrum of interstellar space or of combustion processes has been shown to have peaks suggestive of toroidal structures. Nor has anyone suggested any obvious possible mechanism for their formation (although neither is it by any means certain yet how even spheres are formed!). On the other hand, there are encouraging signs, and in 1993 Kirby *et al.* wrote of toroidal polyhexes that

> whilst they do not have the same near-perfect symmetry of buckminsterfullerene, [they] do have their own considerable aesthetic appeal and, once formed, we see no good reason why they, too, should not be stable, despite the important reservations we have expressed concerning their inherent geometric strain, and the practical problems of formation. Numerous interesting and possibly useful properties could be predicted for carbon* structures of this class.[23a]

In the same year, independently approaching the matter from a different standpoint, Ihara *et al.* wrote:

> We have proposed toroidal forms of the carbon structure. . . . The toroidal carbon forms are predicted to be energetically and thermally stable. We believe that the structures will someday be discovered or synthesized.[13]

So, are these reported ideas, of carbon perhaps existing in strange new shapes and topologies the beginnings of a new chapter in fullerene research that will continue to develop from its origin—the discovery of buckminsterfullerene—or will they come to be seen as merely a footnote and a dead-end? Who knows? Meanwhile and in either case though, the ideas are beautiful, and can help scientists to refine their concepts while groping their way toward what is or is not reality.

---

*King[73] recently reminded me that chemists should not become too narrowly focused on carbon as the only element of interest. If the field is widened, this can only increase the probability of intriguing new structures being realized somewhere.

ACKNOWLEDGMENTS

It is with pleasure that I thank A. T. Balaban (Bucharest), J. Berger (Haifa), B. Borštnik (Ljubljana), G. Brinkmann (Bielefeld), S. Cyvin (Trondheim), P. W. Fowler (Exeter), A. Graovac (Zagreb), C. W. Haigh (Swansea), H. Hosoya (Tokyo), S. Ihara (Tokyo), Y. Jiang (Nanking), P. John (Ilmenau), M. Kaufman (Ljubljana), R. B. King (Athens, Georgia), D. J. Klein (Galveston), R. B. Mallion (Canterbury), T. Pisanski (Ljubljana), B. Plestenjak (Ljubljana), P. Pollak (Canterbury), H. Sachs (Ilmenau) S. S. Tratch (Moscow), and D. Veljan (Zagreb) for points of information, comments, or discussions.

REFERENCES

1. E. Osawa, *Kagaku (Kyoto)* [in Japanese] *25*, 854–863 (1970).
2. H. W. Kroto, J. R. Heath, S. C. O'Brien, R. F. Curl, and R. E. Smalley, *Nature 318*, 162–163 (1985).
3. R. Taylor, J. P. Hare, A. K. Abdul-Sada, and H. W. Kroto, *J. Chem. Soc. Chem. Commun.* 1423–1425 (1990).
4. R. D. Johnson, G. Meijer, and D. S. Bethune, C_{60} has icosahedral symmetry, *J. Am. Chem. Soc. 112*, 8983–8984 (1990).
5. H. Kroto, The first predictions in the buckminsterfullerene crystal ball, *Fullerene Sci. Technol. 2*, 333–342 (1994).
6. K. M. Kadish and R. S. Ruoff, eds., *Recent Advances in the Chemistry and Physics of Fullerenes and Related Materials*, The Electrochemical Society, Pennington, New Jersey (1994).
7. T. G. Schmalz, W. A. Seitz, D. J. Klein, and G. E. Hite, Elemental carbon cages, *J. Am. Chem. Soc. 110*, 1113–1127 (1988).
8. F. Harary, *Graph Theory*, Addison–Wesley, Reading, Massachusetts (1969).
9. R. J. Wilson, *Introduction to Graph Theory*, Oliver & Boyd, Edinburgh (1972).
10. N. Trinajstić, *Chemical Graph Theory*, 2nd ed. (D. J. Klein and M. Randić, eds.), CRC Press, Boca Raton, Florida (1992).
11. J. Castells, Some comments on fullerene terminology, nomenclature, and aromaticity, *Fullerene Sci. Technol. 2*, 367–379 (1994).
12. P. E. John and B. Walther, Sketch about possible structures and numbers of perfect matchings of toroidal hexagonal systems, Technische Universität Ilmenau, Fakultät für Mathematik und Naturwissenschaften (1994).
13. S. Ihara, S. Itoh, and J. Kitakami, Toroidal forms of graphitic carbon, *Phys. Rev. B 47*, 12908–12911 (1993).
14. S. Itoh and S. Ihara, Toroidal forms of graphitic carbon. II. Elongated tori, *Phys. Rev. B 48*, 8323–8328 (1993).
15. S. Itoh and S. Ihara, Toroidal form of carbon C_{360}, *Phys. Rev. B 47*, 1703–1704 (1993).
16. A. T. Balaban, ed., *Chemical Applications of Graph Theory*, Academic Press, New York (1976).
17. K. Devlin, *Mathematics: The New Golden Age*, Penguin Books, London (1988).
18. R. C. Read, *Graph Theory and its Applications*, (B. Harris, ed.), pp. 51–78, Academic Press, New York (1970).
19. R. C. Read, "Algorithms in Graph Theory," in *Applications of Graph Theory* (R. J. Wilson and L. W. Beineke, eds.), pp. 381–417, Academic Press, New York (1979).
20. (a) D. J. Klein, Elemental benzenoids, *J. Chem. Inf. Comput. Sci. 34*, 453–459 (1994); (b) R. Tosić and S. J. Cyvin, Hypothetic Structure for the C_{60} Cluster: A toroid, *Zb. Rad. Priv. Mat. Fak. Univ. Novom Sadu 22* (1992).

21. (a) D. J. Klein and X. Liu, Elemental carbon isomerism, *Int. J. Quantum Chem. 28*, 501–523 (1994); (b) A. Ceulemans and P. W. Fowler, Symmetry extensions of Euler's theorem for polyhedral, toroidal and benzenoid molecules, *J. Chem. Soc. Faraday Trans. 91*, 3089–3093 (1995).

22. (a) E. C. Kirby, unpublished work; (b) E. C. Kirby, Fully arenoid toroidal fullerenes, both benzenoid and non-benzenoid, *MATCH 33*, 147–156 (1996).

23. (a) E. C. Kirby, R. B. Mallion, and P. Pollak, Toroidal polyhexes, *J. Chem. Soc. Faraday Trans. 89*, 1945–1953 (1993); (b) C. W. Haigh, unpublished work and paper in preparation; (c) E. C. Kirby and P. Pollak, paper in preparation; (d) E. C. Kirby, R. B. Mallion, and P. Pollak, On the question of counting spanning trees in labeled, nonplanar graphs, *Molecular Physics 83*, 599–602 (1994).

24. I. Gutman and S. J. Cyvin, *Introduction to the Theory of Benzenoid Hydrocarbons*, Springer-Verlag, Berlin (1989).

25. S. Itoh and S. Ihara, Isomers of the toroidal forms of graphitic carbon, *Phys. Rev. B Condensed Matter 49*, 13970–13974 (1994).

26. B. Borštnik and D. Lukman, Molecular mechanics of toroidal carbon molecules, *Chem. Phys. Lett. 228*, 312–316 (1994).

27. B. I. Dunlap, *Phys. Rev. B 46*, 1933 (1992).

28. D. A. Klarner, *The Mathematical Gardener*, Wadsworth International, Belmont, California (1981).

29. E. A. Lord and C. B. Wilson, *The Mathematical Description of Shape and Form*, Ellis Horwood, Chichester (1984).

30. B. Grünbaum and G. C. Shepard, *Tiling and Patterns*, Freeman, San Francisco (1987).

31. J. E. Avron and J. Berger, Tiling rules for toroidal molecules, *Phys. Rev. A 51*, 1146–1149 (1995).

32. J. Berger and J. E. Avron, A classification scheme for toroidal molecules, *J. Chem. Soc. Faraday Trans. 91*, 4037–4045 (1995).

33. J. Berger and J. E. Avron, personal communications (February and April 1995).

34. X. Liu and D. J. Klein, The graph isomorphism problem, *J. Comput. Chem. 12*, 1243–1251 (1991).

35. D. J. Klein and H. Zhu, Resonance in elemental benzenoids, personal communication of a manuscript (August 1994).

36. E. C. Kirby, Cylindrical and toroidal polyhex structures, *Croat. Chem. Acta 66*, 13–26 (1993).

37. E. C. Kirby, On toroidal azulenoids and other shapes of fullerene cage, *Fullerene Sci. Technol. 2*, 395–404 (1994).

38. (a) H. Sachs, Graph theoretical means for calculating Kekulé and Hückel parameters in benzenoid and related systems, *J. Chem. Inf. Comput. Sci. 34*, 432–435 (1994). (b) H. Sachs, P. Hansen, and M. Zheng, Kekulé count in tubular hydrocarbons, personal communication of a monograph (1994).

39. P. E. John and B. Walther, Mögliche Strukturen und Linearfaktoranzahlen toroidaler hexagonaler Systeme, Technische Universität Ilmenau (Poster, Symposium für Theoretische Chemie 27/9–1/10/93).

40. T. Pisanski, B. Plestenjak, and A. Graovac, NiceGraph Program and its applications in chemistry, *Croat. Chem. Acta 68*, 283–292 (1995).

41. B. Plestenjak, T. Pisanski, and A. Graovac, Generating fullerenes at random, *J. Chem. Inf. Comput. Sci. 36*, 825–828 (1996).

42. (a) A. Fitnik, T. Pisanski, A. Graovac, D. Lukman, and B. Borštnik, unpublished work; (b) M. Kaufman, E. C. Kirby, T. Pisanski, and B. Plestenjak, unpublished work.

43. T. Fruchterman and E. Reingold, *Software Pract. Exper. 21*, 1129–1164 (1991).

44. D. E. Manolopoulos and P. W. Fowler, *J. Chem. Phys. 96*, 7603 (1992).

45. J. Shawe-Taylor and T. Pisanski, Characterizing graph drawing with eigenvectors, Technical Report CSD-TR-93-20, Royal Holloway, University of London, Department of Computer Science, Egham, Surrey TW20 0X, England (1995).

46. C. D. Godsil, *Algebraic Combinatorics*, Chapman & Hall, London (1993).

47. S. Ihara, personal communication (March 1995).

48. E. C. Kirby, Remarks upon recognising genus and possible shapes of chemical cages in the form of polyhedra, tori and Klein bottles, *Croat. Chem. Acta 68*, 269–282 (1995).

49. C. Thomassen, Tilings of the torus and the Klein bottle and vertex-transitive graphs on a fixed surface, *Trans. Am. Math. Soc. 323*, 605–635 (1991).

50. A. T. Balaban, D. J. Klein, and C. A. Folden, Diamond–graphite hybrids, *Chem. Phys. Lett. 217*, 266–270 (1994).

51. S. Ihara, S. Itoh, and J. Kitakami, Helically coiled cage forms of graphitic carbon, *Phys. Rev. B 48*, 5643–5647 (1993).

52. H. Sachs and P. John, unpublished work (1994).

53. D. Cvetković, M. Doob, and H. Sachs, *Spectra of Graphs—Theory and Application*, Deutscher Verlag der Wissenschaften and Academic Press, New York (1980).

54. M. Gordon and W. H. T. Davison, *J. Chem. Phys. 20*, 428 (1952).

55. J. W. Armit and R. Robinson, *J. Chem. Soc.* 1604 (1925).

56. E. Clar and M. Zander, *J. Chem. Soc.* 1861 (1958).

57. E. Clar, *Polycyclic Hydrocarbons*, Academic Press, New York (1964).

58. E. Clar, *The Aromatic Sextet*, John Wiley & Sons, New York (1972).

59. J. R. Dias, Total resonant sextet benzenoid hydrocarbon isomers and their molecular orbital and thermodynamic characteristics, *Thermochim. Acta 122*, 313–337 (1987).

60. J. R. Dias, Benzenoid series having a constant number of isomers. 3. Total resonant sextet benzenoids and their topological characteristics, *J. Chem. Inf. Comput. Sci. 31*, 89–96 (1991).

61. J. V. Knop, W. R. Müller, K. Szymanski, and N. Trinajstić, On the enumeration of 2-factors of polyhexes, *J. Comput. Chem. 7*, 547–564 (1986).

62. E. C. Kirby, Why can so few benzenoids be completely drawn with Clar's resonant sextets? An analysis using 'branching graphs' and a 'coiled-hexagon code,' *J. Chem. Soc. Faraday Trans. 86*, 447–452 (1990).

63. P. Fowler and T. Pisanski, Leapfrog transformations and polyhedra of Clar type, *J. Chem. Soc. Faraday Trans. 90*, 2865–2871 (1994).

64. J. Aihara and T. Tamaribuchi, Aromatic character of graphite intercalation compounds, *J. Chem. Soc. Faraday Trans. 90*, 3513–3516 (1994).

65. (a) S. Nikolić, M. Randić, D. J. Klein, and D. Plavšić, The conjugated-circuit model: Application to benzenoid hydrocarbons, *J. Mol. Struct. 198*, 223–237 (1989); (b) M. Randić, D. J. Klein, N. Zhu, N. Trinajstić, and T. Zivković, Aromatic properties of fully-benzenoid hydrocarbons, *FIZIKA A 3*, 61–75 (1994).

66. I. Gutman, J. Brunvoll, B. N. Cyvin, E. Brendsdal, and S. J. Cyvin, Essentially disconnected fully acenoids, and fully-arenoid hydrocarbons, *Chem. Phys. Lett. 219*, 355–359 (1994).

67. I. Gutman, S. J. Cyvin, V. Petrović, and A. Teodorović, Fully-naphthalenoid hydrocarbons and their conjugation modes, *Polycyclic Aromatic Compounds 4*, 183–189 (1994).

68. J. Brunvoll, B. N. Cyvin, S. J. Cyvin, E. C. Kirby, and I. Gutman, Fully-naphthalenoid hydrocarbons: Enumeration and classification, *Polycyclic Aromatic Compounds 4*, 219–229 (1995).

69. J. V. Knop, W. R. Müller, K. Szymanski, S. Nikolić, and N. Trinajstić, *MATCH 29*, 81 (1993).

70. I. Gutman and S. Cyvin, Fully-arenoid hydrocarbons, *MATCH 30*, 93–102 (1994).

71. J. C. Greer, S. Itoh, and S. Ihara, Ab initio geometry and stability of a C_{120} Torus, *Chem. Phys. Lett. 222*, 621–625 (1994).

72. (a) H. Hosoya and Y. Tsukano, 2-Dimensional torus benzenoids whose electronic state rapidly converges to graphite, Conference paper, ISNA-7, International Symposium on Novel Aromatic Compounds, Victoria, British Columbia (July 1992); (b) H. Hosoya, Cyclic fence graphs, Conference paper, *MATH/CHEM/COMP 93*, International Course and Conference on the Interfaces among Mathematics, Chemistry and Computer Science, Rovinj, Croatia, June 1993; (c) H. Hosoya, Y. Okuma, Y. Tsukano, and K. Nakada, Multilayerd cyclic fence graphs: Novel cubic graphs related to the graphite network, *J. Chem. Inf. Comput. Sci. 35*, 351–356 (1995).

73. R. B. King, personal communication (December 1994).

9

All-Conjugated Carbon Species

DOUGLAS J. KLEIN and HONGYAO ZHU

9.1. INTRODUCTION

9.1.1. History and Motivation

There has long been an interest in *conjugation* of bonds, at least intimated in Kekulé's classic idea[1,2] about the nature of benzene. It was rather early realized (e.g., by Erlenmeyer[3] but more completely by Armit and Robinson[4]) in cases with more than one classical bonding pattern that (because of the novel "emergent" properties) the consequent substances are not mixtures of molecules exhibiting different individual bonding patterns but rather that individual molecules are in some sense simultaneously all of the possible structures. With the advent of quantum mechanics this simultaneity of structures was realized by Pauling[5,6] to correspond to a superposition of different basis wave functions, each basis wave function corresponding to an individual chemical structure. The idea of superposition of more classically interpretable wave functions was perhaps first indicated in Heitler and London's paper[7] on H_2. But before too long Rumer[8] described a set of wave functions in clear correspondence with classical chemical-bonding structures and Pauling[5,6] immediately emphasized the relevance of the superposition idea, garbing it in a classical view under the title of "resonance theory" and finally publishing his masterwork[9] *The Nature of the Chemical Bond*. Amusingly the idea that even when there is but a single neighbor-bonded classical

DOUGLAS J. KLEIN and HONGYAO ZHU • Texas A&M University at Galveston, Galveston, Texas 77553-1675

From Chemical Topology to Three-Dimensional Geometry, edited by Balaban. Plenum Press, New York, 1997

structure there may still be minor (superposition) contributions from higher-energy (more reactive) structures is traceable back to Thiele's turn-of-the-century discussion[10] of the character of polyenes. The quantum-mechanical view has developed now so that high-accuracy *ab initio* computations are available (for smaller species) within the "resonance-theoretic" view, e.g., in Gerrat, Raimondi, and Cooper's work.[11,12] On the semiempirical (more widely applicable) side there too has been progress with the recognition[13] of a systematic *hierarchy of models* (earlier on developed by other workers). Of course the quantum-mechanical work was also developed in rather different manners, particularly via the molecular-orbital-SCF and molecular-orbital-CI methods, as emphasized in many standard quantum-chemistry texts.[14,15] But notably too the classical thinking did not end with the development of the quantum-mechanical ideas, perhaps most neatly illustrated by Clar's ideas concerning[16] "aromatic sextets," such as may be viewed to be related to the final models in the mentioned hierarchy of models. The rather vast literature on "aromaticity" concerns consequences of conjugation, reviews[17,18] and texts[19–21] for "bench" organic chemists often being presented in qualitative resonance-theoretic language.

Of all systems conjugated or not, a fundamental category must be those species composed from a single element, most notably carbon, which through its singular propensity for self-combination gives rise to the whole field of organic chemistry. Here, though, the (outwardly directed) tetravalence often results in an outwardly radiating multiplicative increase in structural mass which usually is only capped off by other elements with unidirectional valence (as hydrogen). For the case of elemental carbon, graphite has long been recognized to be fully conjugated, while the other long-recognized carbon allotrope (diamond) is completely unconjugated (or saturated). For some time there were theoretical speculations as to other possible stable structures, many of which would be fully conjugated: infinite networks,[22–30] long poly-yne chains,[31–35] or finite molecular structures.[36–40] Moreover, there were occa-

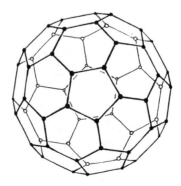

Figure 1.   The "uniquely elegant" truncated-icosahedron C_{60} structure.

sional reports of ill-characterized alternative allotropic forms of carbon,[41-49] presumably with much conjugation. But a change occurred after the 1985 suggestion of Kroto et al.[50] of the "uniquely elegant" truncated-icosahedron conjugated molecular structure of Figure 1 (corresponding to an unusually prominent C_{60} mass-spectral peak from their molecular-beam experiment). Theoretical speculations and computations proliferated—and all the more so a few years later after Krätschmer et al.[51] isolated macroscopic quantities of this novel molecular species. Interest has been further engendered by the preparation: of at least a dozen other fully conjugated cages (as indicated in several articles[52-57]); of Iijima's[58] polyhex conjugated carbon nanotubes; and of poly-yne[59] (to at least a length of 200 carbon atoms, with some "minor" heteroatomic groups at the chain ends).

9.1.2. Preview

In the present discussion the special yet fundamental category of fully conjugated elemental carbon species is considered from a fairly global point of view. Several points of structural inquiry naturally arise:

- What are the possible structures?
- How do the structures relate to σ-electron stability?
- How do the structures relate to π-electron stability?
- How do the structures relate to other properties?

(And of course there are questions of dynamics or process, as to synthetic routes, but they are not considered here.) Though the listed property question is certainly important, here priority is given to the first three questions since they relate to just what structures might actually be stable. That is, they relate to the classical areas of isomerism and stereochemistry. From the view of quantum chemistry, different isomeric structures are seen (and treated) as individual minima in a many-dimensional potential-energy hypersurface. But a formalized classical approach is in fact typically the view adopted in stereochemistry texts[60,61] and remains of crucial value to indicate possibilities and to provide a framework for thought and discussion. The present focus in the isomerism for a single element contrasts with the most complete developments concerning substitutional isomerism, for which there is elegant mathematical theory.[62,63] Some part of the current focus does overlap with an area of topological isomerism that has recently been developed[64-69] to deal with the special circumstance of knotting and linking in chains. In the area of focus here in addition to one-dimensional chains which can knot and link, we also have two-dimensional surfaces and three-dimensional networks with the possibility of further topological characteristics. Moreover, because the structures turn out to be strained to some extent, the question of the π-electron stability contribution is rather nontrivial so far as a global qualitative picture goes. Rather naturally one can anticipate that curvature measures for chains or surfaces following the overall structures will be of relevance. Thence a mixture of graph theory, topology, and differential geometry becomes of interest in describing these systems—the development being a natural extension of earlier work[70–72] from here in Galveston. Further, it is anticipated that the types of ideas developed here may be more widely applicable than to just the all-conjugated all-carbon species of present focus.

9.2. GRAPHICO-TOPOLOGICAL VIEW OF CHEMICAL STRUCTURE

9.2.1. General Three-Stage Description

At the first stage of description a molecular structure is naturally represented[73] as a *graph G*, specified in terms of vertices (corresponding to atoms) and of edges (corresponding to bonds). Traditionally a molecular graph is constrained to be *connected*. Since often the patterns of conjugation are not fully determined for a single molecular structure, only the σ-bonds are to be included in the graph at this first stage, graph theory potentially playing a key role in characterizing the manner of deployment of the remaining π-bonds. For our *all-conjugated* species a graph is to consist solely of vertices of degree either 2 or 3 (respectively corresponding to carbon atoms connected via either 2 or 1 π-bond). That is, classical ideas are invoked to avoid transient species or (poly)radicals. Of course, then too it is of importance to identify the chemically relevant graphs *G* which may be suitably supplemented with π-bonds (2 and 1 into each carbon of degree 2 or 3 in *G*), so that this is kept in mind. Now two graphs are *isomorphic* if there is a one-to-one correspondence between their vertices such that adjacency is preserved. Two molecular structures corresponding to nonisomorphic graphs with equal numbers of vertices represent *graphical isomers*, the idea of which is somewhat like that[60,61,74] of "valence isomers" except we have not paid attention to the placement of multiple bonds.

Next there is a second stage of description wherein some topological aspects are incorporated (which is often not so formally done in chemistry, though there is a well-developed hopefully useful mathematical theory[75–78]). Thence the graphs are to be embedded in Euclidean 3D space E_3 (or perhaps later on embedding in a 1D or 2D manifold may be of interest). Formally an *embedding* of a graph in a manifold *M* is a mapping that takes each vertex of *G* to a point in *M* and each edge of *G* to a (closed) section of a continuous curve in *M* such that: first, all of the vertex image points are distinct; second, all of the edge image curves are disjoint except possibly at their end points; and, third, a curve section corresponding to edge *e* begins and ends at points corresponding to the vertices that *e* connects in *G*. Indeed, it is standard practice to present pictorially a graph as an embedding. Chemically of course embeddings are quite reasonable things to think about since atoms and even bonds tend to exclude one another from the region of space that each occupies (even more so than single points and curves do). Now a key point is that different embeddings of one graph can be

Figure 2.   Three possible linking patterns for molecular cycles.

topologically distinct, such as is the case for knotted and unknotted cycles. Somewhat formally two topological structures are *homeomorphic* if there is a point-by-point one-to-one correspondence that preserves nearness between points. Thence if the topological structures are just the points PG of an embedding to which a graph G is mapped, then homeomorphism becomes equivalent to graph isomorphism. But attention may also be paid to the manifold into which the graph is embedded. Two embeddings are said to be *topologically equivalent* if there is a mapping between the two such that it is a homeomorphic mapping both for the PG and for the complement of PG in the manifold in which the embeddings are made. (This type of "topological equivalence" is known in topology[68,69,78] as *isotopic equivalence*, which as nomenclature seems not so suitable for use in chemistry where "isotopic" already has another meaning.) Topologically inequivalent embeddings of the same graph in E_3 correspond to *topological isomers*. Notably it now makes sense to relax the traditional condition that the molecular graph be connected, since two cycles may be *linked* (in more than one way) as in Figure 2. Recently knotting and linking in molecules have become of much general interest.[64–68]

At a third stage of description one might seek to take[71,72] into account some strain features of molecular structures associated with atomic sizes, bond lengths, and bond angles. In a rather simple yet elegant fashion this can be done in terms of an intermediate embedding, into 1D or 2D manifolds (or complexes) that are in turn to be embedded into E_3. For example, a sequence of degree-2 vertices would be embedded in a 1D manifold (or curve) which should be linear (when itself is embedded in E_3) to avoid angle strain. Or a network of degree-3 vertices as for graphite would be embedded in a 2D surface. More particularly a *digonally hybridized* atom representing a degree-2 vertex has three possible arrangements of the π-bonds as in Figure 3 and is well known to be least strained when the overall π-bond directions are opposite (180° apart). Thence such vertices and the adjacent parts of their adjoining bonds are here to be embedded in a section of a curve, here termed a *segment*. A *trigonally hybridized*

Figure 3.   The π-bonding arrangements for degree-2 vertex.

Figure 4. The π-bonding arrangements for degree-3 vertex.

atom representing a degree-3 vertex also has three possible arrangements of its π-bond as shown in Figure 4 and is well known to be least strained when the overall σ-bond directions are coplanar at 120° to one another. Thence such vertices and the adjacent parts of their adjoining bonds are viewed to be embedded in suitable 2D manifolds locally each equivalent to a disk and here termed a *patch*. But also the segments and patches are to be extended or joined together so as ultimately to encompass all of the vertices and bonds of a molecule: first, pairs of trigonally hybridized vertices with roughly parallel orientations of hybridization are such that their local regions are fused to give a local region that includes both vertices and that is equivalent to a disk; and second, other adjacent pairs of vertices are such that their local manifolds are fused together at a single point. The local structures for different cases of fusion (digonal + digonal; digonal + trigonal; trigonal + trigonal with parallel orientations; and trigonal + trigonal with perpendicular orientations) may be depicted as in Figure 5. In principle there may be ambiguity whether one classifies an adjacent pair of trigonal vertices as having parallel or perpendicular orientations but this is presently ignored. The subcase where a trigonal vertex has three adjacent trigonal vertices each oriented perpendicular to the central vertex, as in Figure 6 evidently is to be avoided, since the central vertex is then precluded from making a conventional π-bond with any one of its neighbors. Examples and development for some further precision in the ideas are found in the next section.

The foregoing three-stage description now set up offers a chemically plausible framework for the discussion of molecular structure going substantively beyond the simple graph-theoretic framework while stopping significantly short of a full-blown molecular geometry. The overall picture is of an *embedding complex* comprised of surfaces and curve segments on which a graph is embedded and which in turn is embedded in E_3. Hopefully this proposed framework exhibits a mathematical elegance allowing significant formal manipulation and prediction, e.g., as to the possible isomers and some of their characteristics. In particular it is to be argued that at least some estimate of geometric strain implicit in a structure can be made.

A   B   C   D

Figure 5. The local structures for different cases of fusion: (a) digonal + digonal; (b) digonal + trigonal; (c) trigonal + trigonal with parallel orientations; and (d) trigonal + trigonal with perpendicular orientations.

Figure 6.   Three adjacent trigonal vertices each oriented perpendicular to the central vertex.

9.2.2. Examples

Perhaps a few explicit examples are in order indicating the variety of possible graphical molecular structures as in Figure 7. In the first cases it is seen that several structures may lead to the same complex. Though the example species listed may look somewhat exotic, most have been seriously considered[79–85] as candidates for the C_{20} species that occur[86–89] in the high-temperature gases that arise in typical preparations of fullerenes. The understanding of just which species occur for different atom counts presumably should engender a significant step toward understanding the mechanism of formation of fullerenes,[90–96] and perhaps of carbon soots in general.[90] Here C_{20}

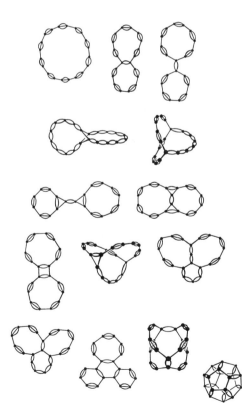

Figure 7.   A collection of all-conjugated C_{20} species, especially as associated to different "complexes."

bears special relevance in possibly being near a crossover from cyclic structures for fewer atoms to branched structures for more atoms.

A further example is the buckminsterfullerene molecule of Figure 1 as would have a complex identified to a spherical surface. Indeed for any of the fullerenes (C_{60}, C_{70}, C_{76}, C_{78}, C_{84}, and so on) the associated complex is homeomorphic to a sphere. The graphite (or honeycomb) lattice has a complex corresponding to the Euclidean plane E_2.

9.2.3. Elaboration

Evidently there is a range of conceivable all-conjugated carbon structures, and some more systematic investigation of this range of possibilities seems desirable. As a step toward this end it is convenient to make even more explicit the type of local surface into which a trigonal vertex is to be embedded and the manner of joining together of the patches. We imagine three possibilities depending on whether the vertices are to be linked to one, two, or three other trigonal vertices of parallel hybridal orientation as in Figure 8. Here the local patches are shaded and viewed to be polygonal in character with several boundary points corresponding to vertices of the polygon located at positions where the boundary line does not follow a smooth path. There is such a boundary point on any bond connecting the central graph vertex to a vertex that is not a trigonal one of parallel orientation. The boundary edges between the boundary points are smooth, being straight if they are designated to be across bonds joining a pair of parallelly oriented trigonal vertices, and otherwise being concave. Then for a full molecule the straight boundary edges of these local patches are all to be joined in pairs. The remaining convex edges are never paired but are such that the point where two convex edges meet is shared with another point either from a 1D curve or from another 2D patch. For a ring of trigonal sites their patches are to mutually meet in the center of the ring.

In constructing possible all-conjugated molecular structures one may instead of beginning with the graph begin with the different overall complexes into which the graph is to be embedded. The overall complex has within it segments that may be combined to form connected 1-manifolds of maximal size, each of which is here called a *max-segment* or *string*—such a string is to include a bond from a digonal to a trigonal site, and also a bond between two trigonal sites if they are perpendicularly hybridized. Further a complex has connected 2-manifolds of maximal size, each here called a *max-surface*. Maximality implies that each string can share a point only with a

Figure 8.   Three types of imagined patches at trigonal sites.

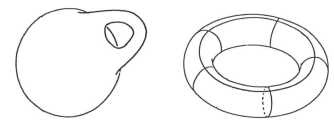

Figure 9. Two homeomorphic forms for a torus.

max-surface and each max-surface can share a point only with a string. Each string is either connected at both ends or else it is closed. For a closed string or a string both ends of which are connected to the same max-surface there is the possibility of knots and of linking to other strings.

A systematic listing of the topological possibilities for complexes in terms of max-components is conceivable. That is, the conceivable complexes could be identified in terms of the numbers and types of max-segments and max-surfaces. We let c denote a string that is a cycle, S_c denote a generic closed (i.e., edgeless) surface, and S_p denote a punctured surface (i.e., a surface obtained from a closed surface by cutting one or more disks from it). It is a fundamental result[76,78] of topology that such surfaces are of two kinds, namely, *orientable* and *nonorientable*, though it is only the orientable ones with an inside and an outside with which we presently concern ourselves. An orientable S then is topologically characterized up to homeomorphism by its *genus* $g(S)$, this being the number of "handles" on the surface; thus, $g(S) = 0$ for a sphere (or anything topologically equivalent to it), whereas $g(S) = 1$ for a torus (which also itself could conceivably be knotted in E_3). For example, two homeomorphic forms for a torus are depicted in Figure 9 (where the first form more clearly depicts a "handle"). A rather different type of "handle" occurs with what we term a 1-handled *basket* B_1 which is just an S_p with one puncture having a string connecting between two points on the boundary of the puncture. For example, three cases of 1-handled baskets are shown in Figure 10, each case being topologically inequivalent, though the last two

Figure 10. Three cases of 1-handled baskets.

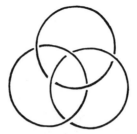

Figure 11. A higher-order linking for three cycles—the Borromean-ring structure.

are homeomorphic (these two differing just in terms of the "internal" linking of its handle). Also 2-handled baskets B_2 are possible. The beginnings of a systematic listing for the finite topological possibilities are indicated in Table 1. The possibility for higher-order linking for three cycles mentioned there is indicated via the "Borromean-ring" structure as in Figure 11 where no pair of the rings is linked. Clearly there are many possibilities, increasing with intricacies as ever more strings and max-surfaces are allowed. But beyond this there are generally many graphs that may be properly embedded on each of these complexes. Moreover, ultimately questions about extended systems are to be addressed—involving polymeric species, sheets, or crystals with one, two, or three dimensions of infinite extent. To reasonably pare the possibilities that arise, attention may be directed: to whether a graph allows π-bonding patterns, and to strain indicators for a graph embedding. But first (in the next section) some

Table 1. Possible Topological Complexes

No. of strings		No. of max-surfaces		
c	open	S_c	S_p	Description
1	0	0	0	c, knotted or not
0	0	1	0	S_c, sphere or S_c with $g(S_c) > 0$
2	0	0	0	two c linked
1	0	1	0	c inside S_c
0	1	0	1	1-handled basket B_1, possibly with "internal" linking
0	0	2	0	one S_c inside another, or two higher-genus S_c linked
3	0	0	0	three c linked (possibly at higher order)
2	0	1	0	two c (linked or not) inside S_c
1	1	0	1	1-handled basket linked with c
0	2	0	1	2-handled basket, or double-punctured S_p with handle at each puncture
1	0	2	0	c inside S_c inside another S_c, or c and S_c inside S_c
0	1	1	1	1-handled basket inside S_c, or 1-handled basket and S_c linked
0	1	0	2	two S_p connected by string
0	0	3	0	S_c inside S_c' inside S_c'', or S_c and S_c' (maybe linked) inside S_c'' or S_c inside S_c' linked to S_c'', or mutually interlinked S_c, S_c', S_c''

aspects of the complexes and the sorts of graphs that may be embedded on a given complex are considered.

9.3. INVARIANTS AND RELATIONS

There are a number of different fundamental invariants associated with our three-stage description. They relate not only to classical questions of stoichiometry, but also to questions of strain as mediated by some basic differential-geometric theorems involving curvatures.

9.3.1. Graphico-Topological Invariants

The simplest relations have to do with atom and bond counts. If $N_i(G) = N_i$ denotes the number of vertices of degree i in graph G and $E(G) = E$ denotes the number of edges, then

$$2 N_2 + 3 N_3 = 2 E$$

(for graphs such as ours with just degree-2 and -3 vertices). Of course, letting $N(G) = N$ be the total number of vertices (or atoms), one has

$$N_2 + N_3 = N$$

But there are some additional relations involving invariants of more fundamental graphico-topological content.

First, for a graph G a simple invariant is the difference $N(G) - E(G)$ between the vertex and edge counts for G. Clearly this (which is very nearly the cyclomatic number of G) is additive in terms of any disconnected components G might have. It takes the value 0 for a simple cycle and the value 1 for a double cycle such as naphthalene.

Second, a fundamental topological invariant for a geometric complex K is the *Euler characteristic* $\chi(K)$. This may be defined[97,98] to be a real-valued function invariant under homeomorphism such that

$$\chi(A) + \chi(B) = \chi(A \cup B) + \chi(A \cap B)$$

Moreover the zero of this function is chosen to coincide with the empty set ϕ, and the scale is chosen so that the value for a one-point set is 1. That is, $\chi(\phi) = 0$ and $\chi(\cdot) = 1$. As a result one sees immediately that for a complex with disconnected components the overall characteristic is simply the sum of those for the components. But also the defining relation determines the function values for connected complexes. For instance, to find the value for a line segment one might consider two line segments A from -1 to 0 (on the x-axis) and B from 0 to 1, whence $A \cup B$ is a line segment (from -1 to $+1$) while $A \cap B$ is a point (at 0); then since all line segments are homeomorphically equivalent, the defining relation reduces to

$$\chi(\text{segment}) + \chi(\text{segment}) = \chi(\text{segment}) + \chi(\text{point})$$

and the value for a line segment is determined (as 1). Similarly for a disk (homeomorphically equivalent to a square) one might consider two adjacent squares sharing a common edge, whence the defining relation leads to

$$\chi(\text{disk}) + \chi(\text{disk}) = \chi(\text{disk}) + \chi(\text{segment})$$

and the value for a disk is determined. Continuing in such a fashion one may begin a list of values:

$$\chi(\text{line segment}) = 1$$
$$\chi(\text{disk}) = 1$$
$$\chi(\text{cycle}) = 0$$
$$\chi(\text{sphere surface}) = 2$$
$$\chi(\text{torus surface}) = 0$$
$$\chi(\text{1-handled basket}) = 0$$
$$\chi(\text{double torus surface}) = -2$$
$$\chi(\text{cylinder, without ends}) = 0$$

A fundamental result of some interest is: if the complex K is a closed orientable surface S, then $\chi(S)$ is unique up to homeomorphism and is $2 - 2g(S)$.

Third, for the case of embedding a graph G in a complex K one can rather naturally identify what is roughly the Euler characteristic of what is left of the complex after subtracting the image of G. More explicitly we count the number of faces (or cells) discounting the concave-sided cells arising with the patch-construction of Section 9.2. That is, this invariant $\chi(K \backslash G)$ is essentially the number $R_3(G)$ of the all-trigonal-hybridized rings occurring in G. Thus, for the examples at the beginning of Section 9.2 $\chi(K \backslash G) = R_3(G) = 0$ while the value is 32 for buckminsterfullerene (of Figure 1).

The fundamental graphico-combinatorial result[76–78] we are currently after relates the topological, graphical, and embedding invariants of the preceding three paragraphs.

EULER–POINCARÉ THEOREM. *For a graph G (suitably) embedded in an allowed complex K*

$$N(G) - E(G) + R_3(G) = \chi(K)$$

As an illustration the second and third examples of Section 9.2.2 have $N = 20$, $E = 20$, and $R_3 = 2$ while the associated complex has $\chi(K) = 1$. For the case of buckminsterfullerene (of Figure 1) $N = 60$, $E = 90$, and $R_3 = 32$ (there being 12 pentagonal and 20 hexagonal faces) while the associated complex is a sphere with $\chi(S) = 2$.

Beyond the invariants so far discussed there are many others, even for the graphical first stage of description. Such is indeed a major interest in "chemical graph theory," e.g., as reviewed elsewhere.[73] In this context for the fullerenes there are some initial studies[100] indicating (among other things) that this class of structures provides a sensitive test for the ability of a graph invariant to distinguish among similar graphical structures. Here though we focus on some invariants relating rarer directly to the second and third stages of description.

9.3.2. Differential-Geometric Invariants and Strain

First one may consider the case of a cycle within a graph. If the cycle is small at least as far as the count of trigonally hybridized sites goes, then there is ring strain, an idea that indeed is of classical origin.[101–105] The total curvature around a cycle (measured as the net angle through which a walker along the curve turns) is 2π. The digonal sites favor no turn while each trigonal site favors a turn of $2\pi/6$, so that if there are fewer than six trigonal sites in a cycle there should be some strain; indeed a plausible strain measure for a cycle C is

$$s(C) = \begin{cases} 2\pi\{1 - N_3(C)/6\}, & N_3(C) < 6 \\ 0, & N_3(C) > 6 \end{cases}$$

The total strain for a graph G could then be taken as the sum of the cycle contributions.

One may view these ideas from a more differential-geometric view. There one looks first at *differential* (*linear*) *curvature*, definable as the rate at which a walker on a curve (in Euclidean space) changes direction. More commonly[106,107] this curvature k at a point p on a curve is defined (equivalently) as the inverse of the radius of a circle drawn tangent to the curve at p such that the circle "kisses" the curve in the sense that second derivatives match at p. The value of 2π comes from integrating k around a closed cycle. Also granted a cycle, a more refined strain measure would check to see if the curvature of a cycle being embedded into matches the locations of the trigonal sites. Though these differential-geometric reinterpretations of the relatively straightforward ring-strain ideas might seem overly mathematicized, it turns out that they offer a guide for the more challenging case of strain in surfaces, as of course should be anticipated to be crucial to understanding fullerenes and related species.

Now we address the case of a surface S that is presumed to be embedded without self-intersection in E_3. To characterize the shape of S consider the curvatures along different directions at a point in the surface to have minimum and maximum values k_1 and k_2 at that point. Then[106,107] the *Gaussian curvature* (at that point of S) is defined as

$$\kappa = \pm | k_1 k_2 |$$

where the sign is "+" if the k_1 and k_2 orientations of curvature are toward the same side of S, and "−" if the k_1 and k_2 orientations of curvature are toward opposite sides of S. Such curvature entails strain, regardless of sign, so that the *absolute Gaussian* (or *isotropic*) *curvature* $|\kappa|$ should[71,72] perhaps be of chemical importance. But also even if $k_1 = 0$ while $k_2 \neq 0$, there is still strain, so that the *absolute anisotropic curvature*

$$\lambda = (\,|\,k_1\,|\,-\,|\,k_2\,|\,)^2$$

should[71,72] also be of relevance. But yet further there is a "curvature mismatch" contribution to the overall curvature strain if the curvature implicit in the chemical structure does not match that of the surface S in which it is embedded. Thus, we imagine[72] a total curvature strain for a structure G embedded in a surface S taking the form

$$\mathcal{E}(G,S) \approx a\!\int_S |\,\kappa\,|\,dA + b\!\int_S \lambda\,dA + c\sum_\alpha^G \varepsilon(\alpha)$$

where a, b, c are suitable parameters, the integrals are over the area of the surface, and the last sum is over local *curvature mismatch* terms $\varepsilon(\alpha)$ associated with the different rings (or faces) α of the embedding. These $\varepsilon(\alpha)$ are developed in the next paragraph.

9.3.3. Curvature Mismatch

To obtain a curvature mismatch one would presumably need to compare differential-geometric curvature against some sort of graphical "combinatorial curvature." An indication of how to do this is obtained from the consideration of three (or more) planar faces intersecting at a point i in E_3 such that each face has two i-incident edges each shared with exactly one i-incident edge of one of the other faces. These could be polyhedron faces meeting at a vertex i of the polyhedron. Now as a measure of "nonplanarity" or "curvature" at a site i one might define an *angle defect* as the deviation from 2π of the sum of the face angles $\phi_{\alpha i}$ at the vertex i,

$$\phi_i = 2\pi - \sum_\alpha^i \phi_{\alpha i}$$

For example, for a cube the eight angle defects are each $\pi/2$, whereas for a regular tetrahedron each of the four angle defects is π. Indeed, if for a convex (geometric) polyhedron, one first rounds off ever so slightly the sharp corners and edges to obtain a smooth surface, and second integrates the Gaussian curvature over the local region around a single vertex, then the result is just the angle defect of that vertex (before rounding, all as is readily seen from the "Gauss map" form for integrated curvatures[106]). But such angle defects corresponding to an (integrated) geometric curvature can also be related to more graph-theoretic quantities (there being a natural correspondence of vertices and edges of faces to vertices and edges of graphs). Thus, for a graph G suppose that i labels a trigonal site that is in three rings of sizes n_1, n_2, and n_3. Then when each ring is embedded in its own plane in space these rings form polygons with average face angles of $\phi_{\alpha i} = \pi - 2\pi/n_\alpha$, $\alpha = 1, 2, 3$. Thence having in mind such embeddings, a *combinatorial curvature* for site i is defined as

$$\Delta_\alpha = 2\pi - \sum_\alpha^{\sim i} (\pi - 2\pi/n_\alpha)$$

The total combinatorial curvature is the sum over local contributions. As an alternative to identifying local combinatorial curvatures to trigonal sites, one might instead identify the curvature contribution of each member site in a ring (or face of a polyhedron to be)—since each trigonal site can potentially contribute to three rings, we define a combinatorial curvature at face α to be. Thence from the formula for Δ_i we take from the first term a portion $2\pi/3$ for each of the n_α sites in ring α, and from the face-angle sum we take just the portion $\pi - 2\pi/n_\alpha$ for each site in α, thereby obtaining a *combinatorial curvature* for face α

$$\Delta_\alpha = \sum_i^{\sim \alpha} [\frac{2\pi}{3} - (\pi - 2\pi/n_\alpha)] = \frac{\pi}{3}(6 - n_\alpha)$$

where the sum is over the trigonal sites in the ring. This result can also be viewed as the combinatorial curvature for site α of the dual graph (embedded in S).

It is of crucial importance to understand that if the combinatorial and differential-geometric curvatures do not match, then there is strain. One extreme illustration of this is in terms of *Schlegel diagrams* of convex polyhedra; such a diagram is obtained as a view of a transparent polyhedron's edges seen by a single eye located close to (and outside) one of the faces. For example, for buckminsterfullerene one has Figure 12. But if one takes this to be a geometric form the geometric curvature is 0 (except perhaps at the "outside face") while the combinatorial curvature of the previous paragraph is

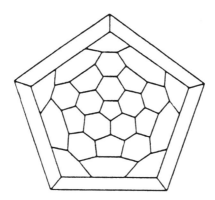

Figure 12. A Schlegel diagram for buckminsterfullerene.

spread out uniformly over all vertices, so that there is a rather dramatic mismatching of these two types of curvature—and this is reflected in the wide range of bond lengths manifested in the Schlegel diagram. On the other hand, if one takes a planar graphite lattice and stereographically projects it onto the distant pole of a sphere setting with its near pole on the graphitic plane, then one obtains a structure on the surface of the sphere that has "uniform" geometric curvature while the combinatorial curvature is 0 everywhere (except perhaps at the distant pole), so that there is again a dramatic mismatching of curvatures, though in the opposite sense (to the Schlegel-diagram example)—and again this is reflected in a wide range of bond lengths on the sphere. It is worth emphasizing that the classical[105] ring-strain ideas concern bond-angle mismatches, whereas the curvature-mismatch strain for surfaces involves the stronger force constants for bond-length dilatations. Thence even local matching is needed if strain is to be avoided. That is, if they do not match in some region containing ring α, then there arises a consequent strain at ring α involving bond dilatation in order to accommodate the graph-theoretic combinatorial curvature to the geometrically realized curvature. A strain at α due to the region of rings within a graphical distance n of α then seems to be (proportional to)

$$\varepsilon_n(\alpha) \equiv \left| \sum_{\beta}^{n,\alpha} \left(\Delta_\beta - \int_\beta \kappa dA \right) \right|$$

and an overall curvature mismatch for face α

$$\varepsilon(\alpha) \equiv \sum_n \varepsilon_n(\alpha)$$

This then goes into the overall strain formula at the end of Section 9.3.2.

9.4. CLOSED SURFACES

The case where the complex K is just a *closed* (i.e., edgeless) surface S is a rather special case, involving all sites of trigonal hybridization. Significantly this case includes the fullerenes. But rather interestingly there are powerful fundamental mathematical results, applicable to a considerably richer diversity of possible structures than the usual fullerenes.

9.4.1. General Theorematic Results

One of the most fundamental results[106,107] involves an area integral over all (Gaussian) curvatures on a surface S:

GAUSS–BONNET THEOREM. *For an orientable closed smooth surface S in E_3, the total Gaussian curvature integrated over S is a topological invariant*

$$\int_S \kappa dA = 2\pi\chi(S)$$

That is, for the surfaces S around the outside of a basketball, or of an (American) football, or of a discus, the total Gaussian curvature integrated over all of S has the same value 4π, whereas for the surfaces around a doughnut or around a coffee mug, the result is another characteristic value 0.

A special natural type of closed surface[77,108] embedded in E_3 is that comprised entirely of planar polygonal *faces*. Indeed, a formal study of such so-called "polyhedra" [especially more "regular" ones with $\chi(S) = 0$] dates back to the Greeks. But we wish to make no constraint on $\chi(S)$ and to emphasize this will here refer to these so unconstrained closed orientable surfaces with polygonal faces as *topohedra*. The special case of genus 0 (or Euler characteristic 2) will be called a *polyhedron*. A fundamental result for general topohedra is:

DESCARTES'S THEOREM. *If S is a topohedral surface in E_3, then*

$$\sum_i^S \phi_i = 2\pi\chi\ (S)$$

where the sum is over all angle defects ϕ_i of the vertices i of S.

In fact, this result may be viewed to be the same as that of the Gauss–Bonnet theorem if we recall that the angle defects are essentially the net (i.e., integrated) Gaussian curvatures associated with each vertex.

A topohedron S (or also the associated graph embedding) is also characterized by counts N, E, and F of its vertices, edges, and faces. These counts are related by:

EULER'S THEOREM. *If a cellular embedding of G with N vertices and E edges in a closed surface S gives F faces, then*

$$N - E + F = \chi(S)$$

This last theorem applies to rather general graphs though in the present context we are interested in *3-connected* graphs, these being those connected graphs that require the removal of at least 3 edges to separate it into two fragments. Further this theorem may be recast in a useful manner. Let the number of n-sided faces be denoted by F_n and keep in mind incidence relations between vertices, edges, and faces as in the first equations of Section 9.3. Then Euler's theorem from above reduces to:

EULER–EBERHARD SPECIALIZATION. *For trivalent graphs G cellularly embedded in a closed surface S,*

$$3F_3 + 2F_4 + F_5 + 0 - F_7 - 2F_8 - 3F_9 - 4F_{10} - \cdots = 6\chi(S)$$

This theorem has received widespread note,[109–111] but not so for an equivalent form:

COMBINATORIAL-CURVATURE SPECIALIZATION. *For trivalent graphs cellularly embedded in a closed surface S,*

$$\sum_i \Delta_i = 2\pi\chi(S) = \sum_\alpha \Delta_\alpha$$

where the Δ_i and Δ_α are combinatorial curvatures for vertex i and face α, respectively. Particularly with this last form in hand along with our identification of angle defects (as curvatures) we see an equivalence among the theorems of Gauss–Bonnet, of Descartes, and of Euler.

Next there are some inequalities that apply to the absolute curvatures appearing in the curvature-strain expression of Section 9.3.2. These allow often fairly tight bounds on the corresponding strains. First we have:

ABSOLUTE-ISOTROPIC-CURVATURE BOUND. *For an orientable closed smooth surface S,*

$$\int_S |\kappa|\, dA \ge 4\pi[1 + g(S)]$$

This in fact "readily" follows from the Gauss–Bonnet theorem, as indicated elsewhere.[72] General results for the integrated absolute anisotropic curvature seem to be more difficult to obtain. But there is the:

EXTENDED WILLMORE CONJECTURE. *For a closed smooth orientable surface S,*

$$\int_S \lambda\, dA \ge 8(\pi - 2)\pi g(S)$$

In fact, the usual Willmore conjecture[112,113] is for $g(S) = 1$ and strictly concerns the integral over $k_1^2 + k_2^2$, but the present conjecture implies the usual one.

In organic chemistry, hexagonal (benzenoid) rings play a special role, so that some sort of constraint to approach this circumstance might be of interest. In fact, rings of size 3 are quite strained, while 4-membered conjugated rings (and to a slightly lesser extent 8-membered conjugated rings) are unstable—this being the content of the standard $4n+2$ Hückel rule. Thence it seems plausible to define a cellular embedding of a trivalent G on S to be *benzenoid* if all faces are hexagonal and more generally to be *qua-benzenoid* if it has a minimal number [consistent with $\chi(S)$] of nonhexagonal rings any of which required to occur are taken as pentagonal or heptagonal. The preceding theorem readily leads to a characterization of these special embeddings:

EULERIC COROLLARY. *For trivalent qua-benzenoid graphs cellularly embedded in a closed surface S, all $F_n = 0$ for $n \ne 6$ except for*

$$F_5 = 12 \quad \text{and} \quad F_7 = 0 \qquad \text{for } \chi(S) = 2$$

$$F_5 = 6 \quad \text{and} \quad F_7 = 0 \qquad \text{for } \chi(S) = 1$$

$$F_5 = 0 \quad \text{and} \quad F_7 = -2\chi(S) \quad \text{for } \chi(S) \neq 0$$

Notably benzenoid graphs arise if and only if $\chi(S) = 0$, i.e., if S is a torus (or a suitable nonorientable surface discussed briefly in Section 9.4.4). The qua-benzenoids on spherically homeomorphic surfaces are usually termed *fullerenes*, for which many special results have been developed and will be discussed more explicitly before long.

Finally, which of the many embeddings of a graph G into a surface S might be "equivalent" is of chemical relevance, because such equivalence classes can each correspond to a chemical (stereo)isomer. Two embeddings ψ and φ are *combinatorially equivalent* if there is a one-to-one correspondence between vertices, edges, and faces of ψ and φ such that incidences between vertices, edges, and faces are preserved. That is, if i, e, and α are vertex, edge, and face images under ψ corresponding to i', e', and α' under φ, then

$$i \sim e \Leftrightarrow i' \sim e'$$

$$e \sim \alpha \Leftrightarrow e' \sim \alpha'$$

where "~" indicates "incident to."

9.4.2. Polyhedra

There are a number of strong results that apply for the special case of closed surfaces homeomorphic to a sphere. The graph embedding on such an S then gives in essence a polyhedron, as is asserted[108] by:

THE STEINITZ–WHITNEY THEOREM. *If S is homeomorphic to a sphere, then each planar 3-connected graph without a cut-vertex corresponds to a unique combinatorial equivalence class of polyhedra.*

Though a powerful classification result for topologically spherical S, this theorem does not seem to extend (simply) to $\chi(S) \neq 0$. Still for the $\chi(S) = 0$ case this theorem reduces the characterization of these equivalence classes to a standard graph-theoretic problem.

This theorem does not distinguish between chiral polyhedra (as is evident since the molecular graph G does not). But this is well handled if one instead pays attention to the embedding of G in the surface S (presently homeomorphic to a sphere), all as is done by the *topological equivalence* of Section 9.2.1. Then[114]:

WHITNEY–STEINITZ REFINEMENT. *If S is homeomorphic to a sphere, then each combinatorial equivalence class corresponds either to one or two topological equiva-*

lence classes of embeddings depending on whether mirror embeddings are topologically equivalent or distinct, respectively.

In the case of a single such achiral topological equivalence class there often (maybe always) is an achiral embedding within the class. If the condition $\chi(S) = 0$ were removed from the hypothesis in this theorem, the consequent statement would not generally be true, as is further indicated in Section 9.4.4.

There are further results that concern structural stability in terms of a classical view of the distribution of double bonds. If one can place double bonds onto the graph embedding with exactly one double bond incident at each site (or alternatively if there exists a spanning subgraph every vertex of which is univalent), then such a placement (or corresponding univalent subgraph) is termed a *Kekulé structure*. Standard classical chemical theory views such structures as a "necessity" (or nearly so) for a stable structure, so that the following is of relevance:

KEKULÉ-STRUCTURE THEOREM. *Every trivalent graph cellularly embeddable on a sphere admits at least three Kekulé structures.*

Amusingly, the proof[115] utilizes the famous[116,117] "four-color theorem"—indeed the proof follows readily from comments[118] over a century ago. The present result thence indicates resonance is always conceivable for such carbon cages. Thus, a graphic resonance-theoretic approach seems a possibility such as indeed we have pursued.[70,71,119–123]

Next slightly less standard but still classical chemical-bonding ideas[16,123–130] suggest that conjugated molecular structures are more chemically stable (and more "aromatic") if Kekulé structures are admitted with larger numbers of hexagonal rings containing three double bonds, in which case such rings are said to be *alternating* (at least in the associated Kekulé structures). Moreover, fundamental quantum-chemical rationale supports[126,131–133] this view. Thus, it is meaningful[123,129,130] to define a cellular trivalent graph embedding to be a Clar-sextet structure if there exists a Kekulé structure wherein every double bond is a member of two (i.e., the maximum number) so alternating hexagonal faces. On the other hand, Fowler *et al.*[134,135] have noted from Hückel calculations on fullerene cages that there seems to be a stable class of fullerenes, namely, the so-called *leapfrog* cages,[136,137] each such having a graph G constructable from another fullerene G_1 (of one-third as many vertices) via a twofold process: first, capping every face of the embedding G_1 on a sphere S to obtain G_2; then second, taking the dual of G_2 to obtain G. Here we relax the restriction that G_1 (and G_2) be fullerenes to the weaker condition of a trivalent polyhedral graph. The relevant theorem here is:

LEAPFROG-CLAR-SEXTET THEOREM. *Clar-sextet graphs are exactly the leapfrog graphs.*

That the fullerenic leapfrog cages are included in the Clar-sextet cages was first proved[138] though the converse also holds.[123] Thence this theorem provides evidence in addition to other evidence for "stable" π-electronic structures for leapfrog cages. Notably (as is seen from Fowler's leapfrog transformation, which multiplies the number of vertices in G_1 by 3 in forming G) the truncated icosahedron is the smallest Clar-sextet cage (and the next largest has $v = 72$ vertices as may be seen from the Grünbaum–Motzkin theorem below).

9.4.3. Fullerenes

For the (qua-benzenoid) polyhedral cages of trigonally hybridized sites with all rings of size 5 or 6 even more has been established. A first result[139] concerns the existence of such cages:

GRÜNBAUM–MOTZKIN THEOREM. *For every even vertex count $v \geq 24$ there exists at least one fullerene, and the sole smaller fullerene is the dodecahedron with $v = 20$.*

In fact, the proof of this is constructive, providing examples of such structures. A further focus notes that the Hückel $4n + 2$ rule is usually extended to a statement concerning cycles around pairs (or even triples) of fused rings—and in particular, the eight-membered cycle around any two fused pentagons should contribute toward instability. Thence[70,71,140] fullerenes with no abutting pair of pentagons are defined as *preferable* or *isolated-pentagon-rule* fullerenes. Parallel to the preceding theorem there is another result[115]:

PREFERABLE-FULLERENE THEOREM. *For every even vertex count $v \leq 70$ there exists at least one preferable fullerene, and the sole smaller preferable fullerene is the truncated icosahedron with $v = 60$.*

Again, the proof is constructive. Further, all fullerenes so far experimentally obtained in macroscopic quantities fall into the class of this theorem—and the two smallest such preferable fullerenes (at $v = 60$ and $v = 70$) corresponding to the two species most readily experimentally obtained.

Beyond the theorems of the preceding sections the generation of explicit cage graphs is desired—followed by a characterization of these cages. The case of fullerene graphs embeddable on a spheroidal graph has a simple (one-to-one) correspondence to combinatorial equivalence classes of embedding, but the generation of all of these graphs seems to offer a significant challenge, though the fullerene cages of icosahedral symmetry are complete,[120,134,141–148] as well as some aspects[71,135,149] for other higher-symmetry cases.

There is one comprehensive brute-force approach to generation of graph cages embeddable on a spheroidal surface, though the computational expense increases rapidly (perhaps exponentially) with vertex number N, though a previous presenta-

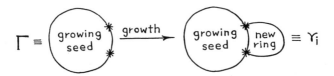

Figure 13. The manner of growing a seed in comprehensive fullerene generation.

tion[149] warrants improvement. In this approach Schlegel diagrams are grown, starting from an initial seed (say a pentagonal ring) and adding rings at different locations around the boundary of the current stage of the growing seed. At any (nonfinal) stage of an acceptable growing seed Γ there are a number (≥ 1) of pairs of currently divalent sites on the boundary such that no other divalent sites lie on "the" boundary between these two sites, as in Figure 13 (where the asterisks locate the two considered initially divalent sites). Between *every* such pair a new ring is grown (if possible with the given number of intervening already trivalent sites on the boundary) in two possible ways, as a pentagon or as a hexagon, each result as indicated in Figure 13. To minimize the length of the rapidly accumulating list L of growing seeds Γ, the seeds are monitored and deleted if they cannot yield a desired candidate cage. This is done keeping track of the numbers of different *types* of faces, edges, and vertices generated—where these types are respectively defined to be characterized by: the size of a face; the sizes of the two faces on either side of an edge; and the sizes of the three faces incident at a vertex. Thence the admissible (final) types are: 5 or 6 for faces; (5,5), (5,6), or (6,6) for edges; and (5,5,5), (5,5,6), (5,6,6), or (6,6,6) for vertices. We denote the (final) numbers of these various types of graph embedding pieces by a type label appended to every subscript to F, E, or N. Then defining p and q as $E_{(5,5)}$ and $N_{(5,5,5)}$ it may be shown that[140]:

> FULLERENE CAGE-TYPING THEOREM. *For a fullerene with $p \equiv e_{(5,5)}$, $q \equiv N_{(5,5,5)}$, and N vertices the numbers of faces, edges, and vertices of the various types are*

$$F_{(5)} = 12 \qquad E_{(5,5)} = p \qquad\qquad N_{(5,5,5)} = q$$
$$F_{(6)} = (N/2) - 10 \qquad E_{(5,6)} = 60 - 2p \qquad N_{(5,5,6)} = 2p - 3q$$
$$E_{(6,6)} = N - 60 + p \qquad N_{(5,6,6)} = 60 - 4p + 3q$$
$$N_{(6,6,6)} = N - 60 + 2p - q$$

Thence in growing a seed the current counts of these types of substructures are monitored, discarding a potential growing seed if the number of any one type exceeds the final desired values for target graphs with chosen fixed values for N, p, q. Thus, a list C of candidate cages results.

A final matter of concern is the elimination of redundant graphs from the cage list C. This is done by our own effective graph isomorphism testing algorithm,[150] though there apparently were algorithms already developed elsewhere.[151] Our scheme scales with time much the same (i.e., $\sim N^3$) as involved in diagonalizing a graph's adjacency (or Hückel) matrix. The algorithm as applied to two graphs G_1 and G_2 ends up returning one of three responses:

1. G_1 and G_2 not isomorphic
2. G_1 and G_2 isomorphic
3. G_1 and G_2 isomorphism undetermined

But notably for all (of many thousands) of the cages tested, only the (favorable) first or second answers have resulted. In fact, such a favorable response has been obtained for all "molecular" graphs we have tested, including those selected because of their presumed difficulty.

For example, we obtain the resultant (combinatorial equivalence classes of) 60-site fullerenes given in Table 2. Notably this result (particularly the number in the $q = 3$ column) corrects[152] an earlier table[121] where rather larger but incomplete initiating seeds were used in an attempt to limit the number of members of the list G (before elimination of duplicates). The currently reported generation assumes certainly correct very minimal seeds, so that all cages are generated. An individual listing of the cages is made by Fowler and Manolopoulos.[153] Another generation scheme[154] for graphs makes assumptions of "spirality" and of nonisospectrality—both assumptions

Table 2. Fullerene Isomers for $v = 60$ Sites

$p\backslash q$	0	1	2	3	4	5	6	7	8	9	10	q-sum
0	1											1
1	0											0
2	1	0										1
3	3	0	0									3
4	17	0	0									17
5	81	5	0	0								86
6	215	39	0	0	0							254
7	210	147	6	0	0							363
8	145	214	54	0	0	0						413
9	23	132	131	11	0	0	0					297
10	7	28	116	42	4	0	0					197
11	0	1	31	54	10	0	0	0				96
12	1	0	6	16	25	2	0	0	0			50
13	0	0	0	2	10	7	0	0	0			19
14	0	0	0	0	3	5	2	0	0	0		10
15	0	0	0	0	0	0	2	0	0	0	0	2
16		0	0	0	0	1	0	0	0	0		1
17			0	0	0	0	0	0	0			0
18				0	0	0	0	1	0	0		1
19					0	0	0	0	0	0		0
20							0	0	0	0	1	1

seemingly fairly reliable though there are[155] counterexamples to the first. For the second assumption the results from the application of our algorithm yield:

FULLERENE SPECTRALITY THEOREM. *Among the fullerene polyhedra there are no isospectral pairs of fewer than 72 vertices, and among the fullerenes with isolated pentagons there are no isospectral pairs of fewer than 98 vertices.*

There is another generation scheme[156,157] replacing the "spirality" assumption by a weaker one, though still an assumption. Also beyond the restriction to fullerenes (but still for degree-3 polyhedral graphs) it may be noted that there are[158] counterexamples to the nonisospectrality assumption.

As another point of interest one can inquire what might be the conditions that the absolute-anisotropic curvature and curvature mismatch terms both become their minimum value 0. Evidently the extended Willmore conjecture indicates that in order to achieve this we must have $g(S) = 0$, and we might further constrain attention to the fullerenes (i.e., trivalent-graph embeddings in a sphere such that all faces are pentagonal or hexagonal). Then it may be shown:

FULLERENE CURVATURE THEOREM. *The only fullerenes such that the curvature mismatch and the absolute-anisotropic curvature both vanish simultaneously are the dodecahedron and the truncated icosahedron.*

Of course the dodecahedron structure is unfavorable for other reasons: because it is open shell (at icosahedral symmetry) and because the absolute-isotropic curvature per site is rather high (i.e., 60/20 = 3 times as high). Thence the truncated-icosahedron structure is singled out as exceptional in several ways: first, by the preferable-fullerene theorem of this section; second, by the Clar-sextet leapfrog theorem (along with the comment shortly following the theorem) in Section 9.4.2; and third, by the present fullerene curvature theorem. Thus, there is a threefold justification of Kroto and colleagues' original suggestion[50] of the "uniquely elegant" truncated-icosahedron structure for C_{60}.

Aside from the formal theory there are several hundred papers now presenting quantum-chemical computations on buckminsterfullerene, other fullerenes, and their derivatives; also the experimental literature is comparably rich. There seems to be relatively little consideration of (nonfullerene) polyhedral cages containing other-size rings—but in this direction there is some[159,160] identification of other possible stable cages.

9.4.4. Toroidal Surfaces

Of the infinite hierarchy of finite closed surfaces beyond those homeomorphic to a sphere, a candidate[71,161–168] for the next most important case is that of the toroidal surface S with $\chi(S) = 0$. Typical topohedral embeddings on a torus may be obtained from a rectangular fragment cut from the graphite lattice, say as in Figure 14. First the

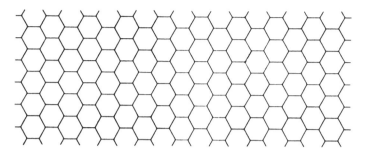

Figure 14. A rectangular fragment cut from the graphite lattice.

top and bottom edges are joined together and second the left and right edges are joined together to yield a *torus*. Interestingly, topologically inequivalent embeddings can result depending on the "order" in which the two joinings are done, at least so long as we also have S embedded in E_3. This is indicated in Figure 15. That is, having joined together the same bonds crossing the edges of the rectangle, the same graph results though we have different consequent stereoisomers. But there are yet more topocombinatorial possibilities[166] if, e.g., on the left in Figure 15 one end of the tube is twisted through a full rotation before making the final joining. (The cyclic arrow on the left in Figure 15 indicates the manner of this twist.) For higher genuses beyond the torus surely many more such possibilities arise. That is, nothing like the Steinitz–Whitney theorem holds for $\chi(S) = 0$.

But even with the torus further possibilities arise in connection with the embedding of S into E_3. Again with reference to the middle left side of Figure 15 we can

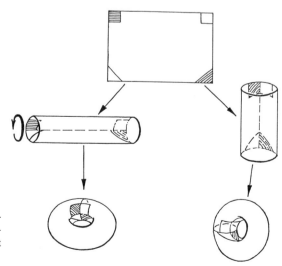

Figure 15. Two topologically in-equivalent embeddings for the construction of graphically equivalent toroidal networks.

think of the open cylinder as corresponding to the sides of a (somewhat thick) cord, but before forming it into a torus the cord could be knotted. As a consequence knot theory becomes of relevance. But in the present context this issue seems separable from the purely graph-theoretic aspects and generally entails additional (anisotropic) curvature strain.

For the toroidal case all of the benzenoid graphs (such as are possible via the Euleric corollary of Section 9.4.1) have been identified,[164–168] with a crucial part[166] of the completeness of the construction going back to some mathematical work.[169,170] The possible topological equivalence classes of embeddings are a further problem but even all of these may have been found.[168] The problem of Kekulé-structure counts was first done for a few cages,[71] then for a few classes of tori,[161–163] and is now completely solved.[167] Moreover, a general analytic treatment of the Hückel eigenspectrum has been made.[165,168]

Finally a few comments on curvature strain are in order. The torus has absolute Gaussian curvature of at least 8π, double that of a sphere (as seen from the Absolute-Isotropic-Curvature-Bound theorem of Section 9.4.1), while also the torus has aniso-tropic curvature presumably of at least $8\pi(\pi - 2)$ from the Willmore conjecture (also of Section 9.4.1). Further yet if the structure is benzenoid (i.e., constructed entirely from hexagonal rings), then there is curvature mismatch, which, however, we believe can be relieved in regions of negative Gaussian curvature by "crinkling" the embed-ding surface so that faces are not very planar. Presumably the curvature mismatch favors tori with narrow tubes, much like the tire of a bicycle. Notably there must always be some sort of crinkling (or nonplanarity) of the rings, since degree-3 vertices in a topohedral structure can only have positive angle defects, which then correspond to only positive Gaussian curvature—whereas a torus must have 4π worth of negative Gaussian curvature.

One idea[171–180] to relieve curvature mismatch strain is to introduce five- and seven-membered rings in regions of positive and negative Gaussian curvature. Nu-merous computations have been made[172–180] on these structures.

9.4.5. Other Closed Surfaces

There is of course a whole sequence of (oriented) surfaces of ever increasing genus, with ever more handles. Here the absolute-Gaussian and absolute-anisotropic curvatures would ever increase, roughly linearly with genus. Notably the requisite larger rings, say seven-membered, should tend to ameliorate the curvature mismatch. The limit of negatively infinite genus leads to the interesting extended negatively curved surface networks of Section 9.5.4.

Another mathematically amusing case involves the nonorientable surface S with $\chi(S) = 0$, this surface being the so-called *Klein bottle* (named after the renowned geometer Felix Klein, who devised the surface[181]). The condition $\chi(S) = 0$ admits benzenoid embeddings, and one can imagine the associated graphs made much as for the torus—for the Klein bottle the top and bottom sides of the strip are joined as before

Figure 16. From rectangular sheet to Klein bottle.

but when joining the left and right sides of the strip the top left is joined to the bottom right (while the bottom left joins to the top right). Indeed such nonorientable surfaces cannot be embedded in E_3 without self-intersections, as indicated in Figure 16. Because of this self-intersection the purely benzenoid species are not feasible. But if a decoration as in Section 9.6.4 is made, then there should be room for strings of bonds to interpenetrate through the centers of the hexagons. Thus, it might be that at least in principle chemistry offers a way around this intersection problem. Finally, there is[76,78] a whole sequence of nonorientable closed surfaces with integer $\chi(S) \leq 1$—the case with equality here corresponding to the "projective plane."

9.5. EXTENDED STRUCTURES

Here there are many possibilities, especially if systems with differently hybridized sites are considered. Only the purely hybridized systems are considered in the present section, but many examples with mixed hybridization may be generated from the pure hybridized systems via the poly-yne transformation of Section 9.6.5.

9.5.1. Pure-Digonal One-Dimensionally Extended Structures

Here the possibilities are particularly straightforward, involving just an infinite chain. Such species indeed were suggested[33,34] almost 20 years ago but only recently well characterized[59] (not quite in infinite chains, but in chains of about 200 atoms, with minor heteroatomic end-groups). Straight chains would be unstrained, in contrast to small cycles[31,32] (which presumably would be unstable below about 10 atoms). Small cycles (of less than about 20 atoms) are believed[182-186] to appear in the high-temperature gases occurring during the preparation of fullerenes, and the work of Diederich et al.[35] aims at the synthesis of macroquantities of such cycles.

Much as already noted in Section 9.2.2 for finite cycles one could imagine knotting or linking for the infinite chain case. These would, however, generally entail excess curvature strain.

9.5.2. Pure-Trigonal One-Dimensionally Extended Structures

Here there is a more substantive range of possibilities for long tubular polymers. From a chemical or physical viewpoint one anticipates that end-structures may be neglected. Mathematically one sees that the ends of the tube can be viewed to be capped so that $\chi(S) = 2$ while the vertex, edge, and ring counts all scale with the length of the polymer, so that the Euler characteristic is negligible, in the high-polymer limit. If cyclic boundary conditions are imposed, one in effect deals with a torus, for which $\chi(S) = 0$. The earlier Euler theorems of Section 9.4.1 still retain content perhaps as is most easily seen if the one-dimensionally extended structure has a simple (*reduced*) *unit cell*, such being a smallest section of the structure which through (an Abelian group of) symmetry transformations can be carried to every other part of the overall structure. Then one obtains a per-cell Euler's theorem:

$$n - e + f = 0$$

and

$$3f_3 + 2f_4 + f_5 = f_7 + 2f_8 + 3f_9 + 4f_{10} + \cdots$$

where n, e, f, and f_m are vertex counts, bond counts, ring counts, and m-ring counts all per cell. The symmetry is not really needed—the same relation applies if we let n (= 1), e, f, and f_m be simply the mean numbers of sites, edges, rings, and m-rings per site. Rings larger and smaller than size 6 evidently are to be (exactly) balanced, and the possibility of no nonhexagonal rings arises, whence the tubes are uniform. Translationally symmetric tubes with other than size-6 rings seem to have not yet been investigated. But they can be important in making local adjustments to combinatorial curvature to match to a corresponding geometric curvature to allow bends or changes in diameter of otherwise uniform tubes.

The case of an infinite tube with all rings of size 6 is of special interest since they might be termed *benzenoid*. Of special note is the experimental preparation[58,187-194] of such tubes, including multilayer tubes[195-202] and heteroatom-filled[203-207] tubes. Moreover, for this case the possible topological equivalence classes of graph embeddings for these benzenoid structures are[208] completely characterized. From a given hex-ring one can imagine walks radiating out in three directions passing in one side of a hexagon and out the opposite side, as indicated in Figure 17. After some number of steps along one pair of paths an intersection occurs (for otherwise one would end up with the graphite lattice). We let the first intersection be of t_+ and t_- steps along these two paths with $t_+ \geq t_- \geq 0$ (and the center-to-center distance between adjacent hexagons corresponding to a "step"). The pair t_+, t_- then completely characterizes the structure (since fusing on adjacent hexagons in an iterative fashion fills out the tube). One may note that the graph admits achiral structures only when $t_- \neq 0$ and $t_+ \neq t_-$.

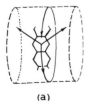

(a)

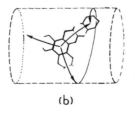

(b)

Figure 17. Indications of the manner of subsequent self-intersection of three polyacene chains radiating from a single hexagon.

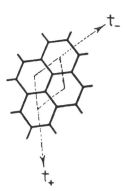

Figure 18. Typical reduced unit cell enclosed within dotted lines; also shown are the directions taken to make the t_+ and t_- counts of the text.

Further, such tubes can be mathematically generated[209-220]: first, by cutting a long strip of width $w = t_+ + t_-/2$ hexagons, much as for the toroidal case in Section 9.4.5 (but now with the strip extending toward ∞); then second, by joining top and bottom together, the bottom of a column of hexagons being joined to the top of a column of hexagons t_- positions away along the strip. (From consideration of this strip construction alone it is evident that one generates tubes—the three-path argument preceding guarantees that this construction exhausts all possibilities.) But also the three-path construction reveals a graph-theoretic group of symmetry transformations (also called graph automorphisms) with a reduced unit cell of two sites, as enclosed by the dotted lines in Figure 18. This then enables[208,215-220] an analytic solution to the Hückel problem. It is found that (with all equal resonance integrals) this leads to 0 band gaps whenever $t_+ - t_-$ is an integral multiple of 3. Standard band-theoretic ideas would then indicate a Peierls distortion opening up a gap, though quantum computations[215-220] for representative cases exhibit unusually small such splittings. Of course, as the radius of the tubes increases, the band gap approaches ever more closely the graphitic limit regardless of the difference $t_+ - t_-$. Also, it might be mentioned that all of these benzenoid tubes have many Kekulé structures (since the long strips in the above construction have[221] many, and this number can only be increased on joining the top and bottom of the strip). In fact, the exact analytic Kekulé-structure enumerations[168] for tori can be carried over to the infinite tubes.

There of course is the possibility of translationally nonsymmetric tubes. In this area the idea of inserting pentagons to begin a capping off followed by an insertion (of an equal number of heptagons to reinstate tube propagation) in effect changes the diameter of the tube, while placing a pentagon and heptagon on opposite sides introduces a bend in the tube, as has been suggested[222] and utilized in constructing[223] helically bent nanotubes.

9.5.3. Pure-Trigonal Graphically Planar Sheets

Graphite is of course the classical example here, though there are many other possibilities. The Euler relation of the preceding subsection continues to apply for 2D

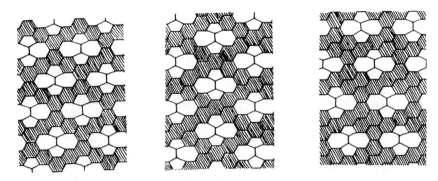

Figure 19. Examples of 2D carbon nets with discrete pyracylenic unit embedded in the hexagonal lattice.

sheets, and of course the translational case is simplest. As for fullerenic polyhedral cages we anticipate that five- and seven-membered rings should lead to the least energetic cost especially if the pentagons and heptagons abut while avoiding pentagon–pentagon (or heptagon–heptagon) abutments. Restriction to pentagons and heptagons then reveals that their numbers per unit cell exactly match. Examples of small regions of sheets based on unit cells containing one or more pentagon–heptagon pair are shown in Figure 19 and a somewhat comprehensive study has been made[224] (including "conjugated-circuit" computations).

Of course there are numerous possibilities without translational symmetry. Of special note in this area are those structures that are composed entirely of hexagons except for a local "defect," most simply a single nonhexagonal ring. Indeed, just such structures with a single anomalous ring have been studied.[225–228] Though such locally defected networks are planar from a purely graph-theoretic viewpoint, such an anomalous nonhexagonal ring must introduce combinatorial curvature, and thence too (to avoid curvature mismatch) geometric curvature is needed (to avoid strain). For a single anomalous ring of size less than 6 there then arises a cone, such as has in fact been experimentally observed.[229] For a single anomalous ring of size greater than 6 there would arise some sort of "pleated-skirt" structure. Beyond a single-ring defect there are possible other local defects say from two or more nonhexagonal rings within the local region; in any case the total combinatorial curvature is an important invariant which through the avoidance of curvature mismatch dictates the "conicalness" of the defect. With a combinatorial curvature of 2π the defect becomes a cap on an extended tube such as in Section 9.5.2.

9.5.4. Pure-Trigonal Uniformly Negatively Curved Networks

Here the possibility of extended structures with a size-extensive amount of negative combinatorial curvature is noted. That is, we introduce an Euler characteristic

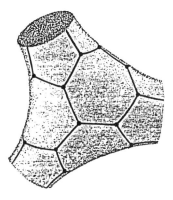

Figure 20. A building block containing negative combinatorial curvature graphs.

$x = x(K)$ per site and demand that it be strictly negative (even for the infinite graph limit). The Euler relation then devolves[230] to

$$\sum_i (6 - i) f_i = x$$

Of course the negative combinatorial curvature then implicates a corresponding geometric curvature. Indeed, there are such geometric surfaces, which are translationally symmetric, on which the corresponding negative combinatorial curvature graphs can be embedded, and for which solid-state quantum computations have been made.[231–234] An example showing a portion of such a structure is depicted in Figure 20. Apparently several of these should be quite stable if ever they could be prepared. There has been some work[235,236] embedding structures on negatively curved surfaces of the special type called *minimal*, these being those for which the two principal linear curvatures k_1 and k_2 add to 0 at every point on the surface—sometimes too these surfaces are required to be uniform in that k_1 and k_2 are constant everywhere. Another scheme by which to generate overall negatively curved structures is described in Section 9.6.6. All of the above-quoted references involve three-dimensionally extended networks, but with the construction of Section 9.6.2 two- or even one-dimensionally extended structures are readily imagined, though often with patches of positive curvature in local regions above and below the "plane" of the structure.

9.6. TRANSFORMATIONS AND DECORATIONS

There are a number of formal transformations that have been considered for generating one or more (new) structures from another already available, though some processes too have been viewed as experimental possibilities.

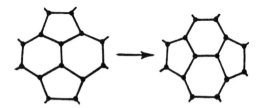

Figure 21.   The Stone–Wales pyracylene rearrangement.

9.6.1. Isomerizations and Stone–Wales Rearrangement

This transformation for all purely trigonal vertices may be viewed to "rotate" a single edge as shown in Figure 21 where this fragment is generally just a local portion of a whole connected to more sites around the outer vertices (where only two incident vertices are shown). Here the top and bottom rings increase in size by 1 while the left and right rings decrease in size by 1. Thus, if initially the top and bottom rings are pentagons while the left and right rings are hexagons, the transformation proposed by Stone and Wales[237] will change a fullerene into another fullerene—sometimes this then is also called the "pyracylene" rearrangement. This transformation (so restricted to the subset of fullerenes) has been viewed as a possibility for the actual mechanism of rearrangement in (very) hot fullerenes, though it seems[238,239] to be fairly high in energy. As a scheme for generating new fullerenes from an initial one it also is incomplete, e.g., starting from buckminsterfullerene only a portion of the possible 1812 C_{60} fullerenes are generated. Indeed, there is[240] a C_{60} fullerene rearrangement graph with 44 disconnected components.

There are other possible fairly local rearrangement processes conceivable, such as in Figure 22, which has been suggested by Curl[241,242] as a possibility to correspond to a physically realized mechanism. Moreover, it interconnects[243] all 1812 fullerenes of 60 atoms, though not so for 70-atom fullerenes. Still this is[244] much "better" than for the ordinary Stone–Wales rearrangement. There are[240] several other at least formally conceivably rearrangement processes.

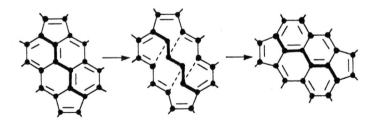

Figure 22.   A generalized pyracylene rearrangement.

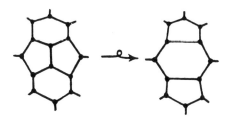

Figure 23. A simple local process for losing two atoms.

9.6.2. Edge Deletions

Again for the case of purely trigonal sets of vertices there is a simple local process whereby a single edge is deleted as in Figure 23. Here the top and bottom rings decrease in size by 1, while if the initial left and right rings are of sizes a and b, then the resulting new central ring is of size $a + b - 4$. The case with the initial top and bottom rings being hexagons and the initial left and right rings being pentagons has been suggested by Curl[242] as a possible mechanism corresponding to an atom-pair deletion process experimentally observed[245,246] in suitably laser-excited fullerene gas. One could also imagine[247] the inverse of this process as a way to build up larger cages from smaller ones.

9.6.3. Leapfrog Transformations

This process again for trigonal sites has been much discussed by Fowler and co-workers[134,136,137] and is typically viewed as applied to one fullerene graph G_1 to obtain another fullerene G. The process is usually viewed in terms of two steps:

a. Cap every face of the embedding G_1 on a sphere S to obtain G_2; then
b. Take the dual of G_2 to obtain G.

The local aspect of the process may be illustrated as in Figure 24. But also the same overall result can be obtained via an alternative two-step process[120]:

a′. Take the dual of G_1 to obtain G_1' (wherein the roles of vertices and faces are interchanged from G_1); then
b′. Truncate the vertices of G_1' to obtain G.

The local aspect of this process may be illustrated as in Figure 25. Starting with a fullerene one ends up with another fullerene with three times as many vertices, which also has the special characteristic of a Clar-sextet cage, as discussed in Section 9.4.2. Notably the leapfrog transformation may be applied to structures much more general[243,248–250] than fullerenes.

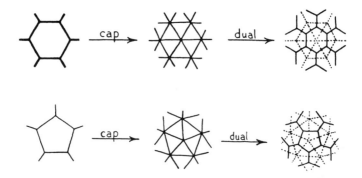

Figure 24. The cap–dual process for leapfrog transformation.

9.6.4. Transformation Algebra

The leapfrog transformation may be viewed as a representative member of a whole algebra of transformations. To see this let us denote the capping and dualing transformations of (a) and (b) above as **C** and **D**, respectively, while the vertex truncation process of (b′) above is denoted by **T**. Then the equivalence between the two formulations for the leapfrog transformation may be written as

$$\mathbf{D\,C = T\,D}$$

(where we take the convention that the operator on the right of a product is applied first). One might thus consider what these operators generate, perhaps augmented by other operators, such as **C′**, which just caps nontriangular faces, or **T′**, which truncates vertices in such a way that the new truncation faces just touch at the centers of the

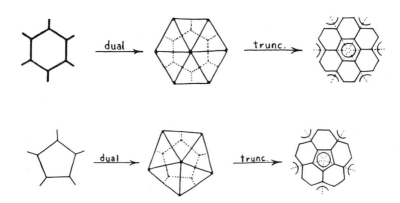

Figure 25. The dual–truncation process for leapfrog transformation.

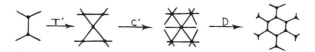

Figure 26. The transformation **DC′T′** takes one fullerene to another.

original edges. In general this generates some sort of *semigroup* (generally lacking proper inverses to yield a group). Of course the subsemigroup of elements each of which carries the subset of fullerenes into itself is of special interest, and notably this contains more than just powers of the leapfrog transformation. For example, the transformation **DC′T′** takes one fullerene G_1 to another G, as indicated in Figure 26. Here one may verify that the graph $G = \mathbf{DC'T}G_1$ has four times as many vertices as G_1 (whereas $\mathbf{TD}G_1$ has three times as many).

Sah[251] has described a yet more general way to extend the leapfrog transformation to a whole sequence of possibilities. These transformations may be represented within our algebra as $\mathbf{DT}_{(a,b)}\mathbf{D}$, where a and b are integers such that $a > 0$ and $a \geq b \geq 0$. The particularities of the "new" transformation are conveniently described in terms of the triangular lattice: $\mathbf{T}_{(a,b)}$ acts on the dual polyhedron to replace each triangular face by a larger equilateral triangular section of the triangular lattice, namely, that with side corresponding to a displacement of $a\alpha + b\beta$ with α and β unit vectors as indicated in Figure 27. Here $(a,b) = (1,1)$ yields the usual leapfrog transformation, while $(a,b) = (2,0)$ yields that of **DC′T′**. The overall semigroup could be augmented further by the local processes of Sections 9.6.1 and 9.6.2. Restriction of the application of these local processes to suitable positions could even augment the fullerene-preserving subsemigroup.

9.6.5. Poly-yne Replacements

There are a large number of possibilities for structures with different sites of different hybridization. Though a systematic comprehensive treatment beyond the foundations of Sections 9.2 and 9.3 does not yet seem to be established, there is[252] a neat way to generate many structures by a decoration of the purely trigonal structures.

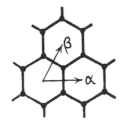

Figure 27. The triangular unit cell.

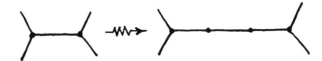

Figure 28.   A poly-yne decoration for a local single bond.

One simply imagines various (possibly all) bonds between trigonal sites to be replaced by chains of even numbers of digonally hybridized sites, e.g., as in Figure 28. For each Kekulé structure of the original structure there is a corresponding Kekulé structure in the decorated structure: a single bond in the original Kekulé structure corresponds to alternating single and triple bonds in the new Kekulé structure, while a double bond in the original Kekulé structure corresponds to a chain of all double bonds in the new Kekulé structure. With the proviso of the preservation of at least one Kekulé structure the decoration can be extended to include the possibility of replacements by some chains of odd numbers of digonally hybridized sites if such odd chains are only put in positions where double bonds occur in at least one original Kekulé structure.

The resulting structures from this decoration seem to have only been investigated in a few cases: for buckminsterfullerene[252] and for graphite.[253] Clearly the decoration is rather generally applicable. But moreover one might conceive of other replacements. For example, a single trigonal site could be replaced by a triangle of trigonal sites (thence being equivalent to a vertex truncation). Any piece cut out of a surface with a given number of dangling bonds might be replaced by another with the same number of dangling bonds.

9.6.6. Joining Pairs of Cycles

A classical transformation combining two structures is via fusion. For two cycles each with an adjacent pair of sites each combined with no more than two other nonhydrogenic atoms an *edge fusion* merges the two pairs to a single pair as in Figure 29. Thence this is applicable to the presently considered structures if one initially has two pairs of adjacent digonally hybridized sites. It is realized that the two cycles involved in fusion need not be simple cycles but could have other attachments (away from the fusion).

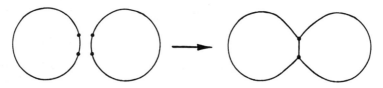

Figure 29.   The edge fusion for joining pairs of cycles.

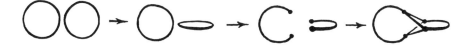

Figure 30. The scheme for a twist-coupling of two cycles.

Other types of fusion are possible, particularly: site-fusion where two digonally hybridized sites are merged; and ring-fusion where two like-sized rings are merged together. But this usually leads to tetrahedrally hybridized sites not considered within the focus of the present discussion.

Aside from fusion there is at least one more fundamental way to join two cycles. In a *twist-coupling* of two cycles, the two cycles are placed adjacent to one another (much as preparing for fusion, while not yet touching) but then twisting one 90° about their common axis, breaking one bond in each cycle, and forming new intercycle bonds, as shown in Figure 30. Evidently two digonal sites in each cycle are converted to intercycle-bonded trigonal sites, with perpendicular hybridization between cycles. This sort of coupling generalizes to couplings between two cages in a useful way, as indicated in the next subsection.

9.6.7. Face-to-Face Twist-Couplings

It is perhaps worthwhile to indicate a construction, which seems especially convenient to generate "negatively curved" structures. We begin with two fullerenic cages and consider a face-to-face *twist-coupling* as in Figure 31. That is, two equally sized rings from two cages are brought up parallel to one another, rotated so that the two rings together are at the end-positions of an antiprism, and finally the bonds

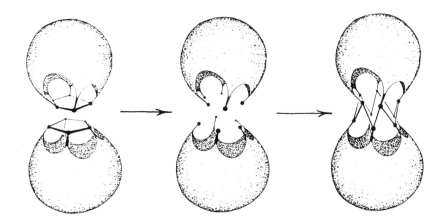

Figure 31. The scheme for a face-to-face twist-coupling of fullerenes.

internal to the rings are broken and then re-formed along the lateral edges interconnecting the two ends of the antiprism. During this twist-coupling each ring so twist-coupled disappears from the new surface while the adjacent rings each increase in size by one. This conserves the global combinatorial curvature (at 4π), but the adjacent rings in growing larger may introduce local negative combinatorial curvature. Of course twist-couplings may be repeated to interconnect several different cages together. Indeed, we may imagine fullerenic cages set down with centers at positions on a lattice, such that the adjacent cages are arranged with parallel faces to make twist-couplings, which then are carried out to obtain an extended negatively curved surface structure. The two-dimensionally extended structures alluded to at the end of Section 9.5.4 simply entail placing the fullerenic cages centered on the vertices of a 2D lattice. Perhaps too it should be noted that centering the fullerenic cages on the vertices of a giant polyhedron (possibly a fullerene) leads to the possibility of rather novel "superfullerenes."

Other sorts of hybridization-conserving combinings of cages are conceivable. For instance, one may: first, bring two cages up to one another with like faces parallel to one another; second, delete all of the atoms in the two parallel faces (leaving dangling bonds); and third, join the dangling bonds in pairs.

9.7. OVERVIEW

In the presentation here an outline toward an extension of current organic stereochemical theory has been attempted for an area of recently expanding interest. A view of the structural possibilities going beyond the simple graph-theoretic structure to embeddings of the graph in appropriate geometric complexes has been advocated. The consequent features not only of the molecular graph G but also of the surface S and of the embedding of G in S seem to lead far toward a characterization of the conceivable stereoisomers. Thence the presentation here may be viewed as a development of classical stereochemical theory, to be utilized by both experimentalists and quantum chemists. Despite the restriction to carbon and to conjugation, a rich variety of structural possibilities is revealed, though in fact emphasis has been placed on those structures correlating with either closed or extended surfaces. Even if one considers closed (i.e., edgeless) complexes there are many other possibilities, such as the two shown in Figure 32 where there would be a string of perpendicularly trigonally hybridized sites along the seam where three surfaces meet, the local structure appearing as in Figure 33. Table 1 is just a bare beginning of a listing for the embedding complexes. Also some other possibilities are derivable from the given ones vias the poly-yne replacement of Section 9.6.5.

Beyond the matter of structural possibilities, many questions remain unanswered, most prominently: first, as to what sorts of novel "emergent" properties might arise; and second, as to what means there might be to synthesize the various conceived possible structures. Of course, current quantum-chemical methods especially of a

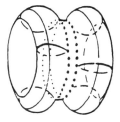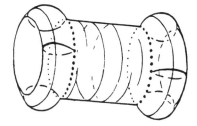

Figure 32. Two "complexes" leading to conceivable all-conjugated carbon networks.

semiempirical nature may help (and indeed to some extent already have helped) address this first question. But at least to the present authors the second question concerning rational synthesis seems quite opaque. Still the range of possibilities seems intriguing, and might encourage others to address some of the indicated questions.

Notably in the context of our advocated description the curvature-related ideas marshaled in Section 9.3 elaborate on an important stereochemical aspect, and these ideas lead in later sections to insight concerning strain and its effects. For example, the global shape of large (hypothetical) fullerene shapes is better understood, and the truncated-icosahedral structure of C_{60} is verified to be "uniquely elegant."

Still further developments and work seem likely—and hopefully useful (even at a stage preceding full-blown quantum-chemical computations). Hopefully too the ideas described here might be extended beyond the presently presumed limits of conjugation to include sp^3 tetrahedrally hybridized carbons, and yet further extended to include other possible elements.

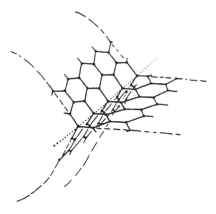

Figure 33. The local structure where three graphite-lattice sheets meet.

ACKNOWLEDGMENTS

Acknowledgment is made to numerous colleagues (especially here in Galveston) for helpful discussions and to the Welch Foundation of Houston, Texas for financial support of the research.

REFERENCES

1. A. Kekulé, *Bull. Acad. R. Belg. 19*, 557 (1865).
2. A. Kekulé, *Liebigs Ann. Chem. 137*, 129 (1866).
3. E. Erlenmeyer, *Liebigs Ann. Chem. 316*, 57 (1901).
4. J. W. Armit and R. Robinson, *J. Chem. Soc. 38*, 827 (1922).
5. L. Pauling, *J. Chem. Phys. 1*, 280 (1933).
6. L. Pauling and G. W. Wheland, *J. Chem. Phys. 1*, 362 (1933).
7. W. Heitler and F. W. London, *Z. Phys. 44*, 455 (1927).
8. G. Rumer, *Nachr. Ges. Wiss. Goettingen Math. Phys. Kl. 1932*, 337 (1932).
9. L. Pauling, *The Nature of the Chemical Bond*, Cornell University Press, Ithaca, New York (1939).
10. F. K. J. Thiele, *Liebigs Ann. Chem. 306*, 89, 125 (1899).
11. D. L. Cooper, J. Gerrat, and M. Raimondi, *Nature 323*, 699 (1986).
12. J. Gerrat, D. L. Cooper, and M. Raimondi, in: *Valence-Bond Theory and Chemical Structure* (D. J. Klein and N. Trinajstić, eds.), Elsevier, Amsterdam (1990).
13. D. J. Klein, *Top. Curr. Chem. 153*, 57 (1990).
14. F. Pilar, *Elementary Quantum Chemistry*, McGraw–Hill, New York (1968).
15. I. N. Levine, *Quantum Chemistry*, Allyn & Bacon, Boston (1974).
16. E. Clar, *The Aromatic Sextet*, John Wiley & Sons, New York (1972).
17. Articles in *Aromaticity, Pseudo-Aromaticity, Anti-Aromaticity* (E. D. Bergmann and B. Pullman, eds.), Israel Academy of Science, Jerusalem (1971).
18. B. Y. Simkin, V. I. Minkin, and M.N. Glukhovtsev, *Adv. Heterocycl. Chem. 56*, 303 (1993).
19. P. J. Garratt, *Aromaticity*, McGraw–Hill, New York (1971).
20. D. Lewis and D. Peters, *Facts and Theories of Aromaticity*, Macmillan & Co., London (1975).
21. D. Lloyd, *Non-benzenoid Conjugated Carbocyclic Compounds*, Elsevier, Amsterdam (1984).
22. A. T. Balaban, C. C. Rentia, and E. Ciupitu, *Rev. Roum. Chim. 13*, 231, 1233 (1968).
23. A. T. Balaban, *Comput. Math. Appl. 17*, 397 (1987).
24. R. Hoffmann, T. Hughbanks, M. Kertesz, and P. H. Bird, *J. Am. Chem. Soc. 105*, 7831 (1983).
25. V. Stankevich, M. V. Nikerov, and D. A. Bochvar, *Russ. Chem. Rev. 53*, 670 (1987).
26. K. M. Merz, Jr., R. Hoffmann, and A. T. Balaban, *J. Am. Chem. Soc. 109*, 6772 (1987).
27. R. H. Baughman, H. Eckerdt, and M. Kertesz, *J. Chem. Phys. 87*, 6687 (1987).
28. R. Biwas, R. M. Martin, R. J. Needs, and O. H. Nielsen, *Phys. Rev. B 30*, 3210 (1984).
29. R. L. Johnston and R. Hoffman, *J. Am. Chem. Soc. 111*, 810 (1989).
30. D. Boercker, *Phys. Rev. B 44*, 11592 (1991).
31. K. S. Pitzer and E. Clementi, *J. Am. Chem. Soc. 81*, 4477 (1959).
32. K. S. Pitzer, in: *Molecular Orbitals in Chemistry, Physics and Biology* (P. O. Löwdin and B. Pullman, eds.), pp. 281–291, Academic Press, New York (1964).
33. M. J. S. Dewar, G. P. Ford, and H. S. Rzepa, *Chem. Phys. Lett. 50*, 262 (1977).
34. M. Kertesz, J. Koller, and A. Azman, *J. Chem. Phys. 68*, 2779 (1978).
35. F. Diederich, Y. Rubin, O. L. Chapman, and N. S. Goroff, *Helv. Chim. Acta 77*, 1441 (1994).
36. D. E. H. Jones, *New Sci. 32*, 245 (1966).
37. Z. Yoshida and E. Osawa, *Aromatic Compounds*, Section 5.6.2, Kagakudojan, Kyoto (1971).
38. J. Castells and F. Serratosa, *J. Chem. Educ. 60*, 941 (1983).
39. D. A. Bochvar and E. G. Galipern, *Dokl. Akad. Nauk SSSR 209*, 610 (1973).

40. R. A. Davidson, *Theor. Chim. Acta 58*, 193 (1981).
41. A. E. Goresy and G. Donnay, *Science 161*, 363 (1968).
42. A. G. Whittaker and G. M. Wolten, *Science 178*, 54 (1972).
43. R. Hayatsu, R. G. Scott, M. H. Studier, R. S. Lewis, and E. Anders, *Science 209*, 1515 (1980).
44. A. G. Whittaker, E. J. Watts, R. S. Lewis, and E. Anders, *Science 209*, 1512 (1980).
45. A. G. Whittaker, *Science 200*, 763 (1978).
46. A. G. Whittaker, M. E. Neudorffer, and E. J. Watts, *Carbon 21*, 597 (1983).
47. J. Hoerni and J. Weigle, *Nature 164*, 1088 (1949).
48. L. Pauling, *Proc. Natl. Acad. Sci. USA 56*, 1646 (1966).
49. S. Ergun, *Carbon 6*, 141 (1968).
50. H. W. Kroto, J. R. Heath, S. C. O'Brien, R. F. Curl, and R. E. Smalley, *Nature 318*, 162 (1985).
51. W. R. Krätschmer, L. D. Lamb, K. Fostiropoulos, and D. R. Huffman, *Nature 347*, 354 (1990).
52. R. Taylor, J. P. Hare, A. K. Abdul-Sada, and H. W. Kroto, *J. Chem. Soc. Chem. Commun. 20*, 1423 (1990).
53. R. Ettl, I. Chao, F. Diederich, and R. L. Whetten, *Nature 353*, 149 (1991).
54. F. Diederich, R. L. Whetten, C. Thilgen, R. Ettl, I. Chao, and M. M. Alvarez, *Science 254*, 1768 (1991).
55. F. Diederich, R. Ettl, Y. Rubin, R. L. Whetten, R. Beck, M. Alvarez, S. Anz, D. Sensharma, F. Wudl, K. Khemani, and A. Koch, *Science 252*, 548 (1991).
56. D. Ben-Amotz, R. G. Cooks, L. Dejarme, J. C. Gunderson, S. H. Hoke II, B. Kahr, G. L. Payne, and J. M. Wood, *Chem. Phys. Lett. 183*, 149 (1991).
57. K. Kikuchi, N. Nakahara, T. Wakabayashi, M. Honda, H. Matsumiya, T. Monkawi, S. Suzuki, H. Shiromaru, K. Saito, K. Yamauchi, I. Ikemoto, and Y. Achiba, *Chem. Phys. Lett. 186*, 177 (1992).
58. S. Iijima, *Nature 354*, 56 (1991).
59. R. J. Lagow, J. J. Kampa, H.-C. Wei, S. L. Battle, J. W. Genge, D. A. Laude, C. J. Harper, R. Bau, R. C. Stevens, J. R. Haw, and E. Munson, *Science 267*, 363 (1994).
60. K. Mislow, *Introduction to Stereochemistry*, Benjamin, New York (1965).
61. E. Eliel and S. Wilen, *Stereochemistry of Organic Compounds*, John Wiley & Sons, New York (1994).
62. G. Polya and R. C. Read, *Combinatorial Enumeration of Groups, Graphs, and Chemical Compounds*, Springer-Verlag, Berlin (1987).
63. S. Fujita, *Symmetry and Combinatorial Enumeration in Chemistry*, Springer-Verlag, Berlin (1991).
64. H. L. Frisch and E. Wasserman, *J. Am. Chem. Soc. 83*, 3789 (1961).
65. V. I. Sokolov, *Russ. Chem. Rev. 42*, 452 (1973).
66. D. M. Walba, in: *Chemical Applications of Topology and Graph Theory* (R. B. King, ed.), pp. 17–32, Elsevier, Amsterdam (1983).
67. D. M. Walba, R. M. Richards, S. J. Sherwood, and R. C. Haltiwanger, *J. Am. Chem. Soc. 103*, 6213 (1981).
68. D. W. Sumners, *J. Math. Chem. 1*, 1 (1987).
69. Articles by D. W. Sumners, by D. M. Walba, by J. Simon, by E. Flapan, and by D. P. Jonish and K. C. Millett in: *Graph Theory and Topology in Chemistry* (R. B. King and D. H. Rouvray, eds.), Elsevier, Amsterdam (1987).
70. T. G. Schmalz, W. A. Seitz, D. J. Klein, and G. E. Hite, *Chem. Phys. Lett. 130*, 203 (1986).
71. T. G. Schmalz, W. A. Seitz, D. J. Klein, and G. E. Hite, *J. Am. Chem. Soc. 110*, 1113 (1988).
72. D. J. Klein and X. Liu, *Int. J. Quantum Chem. S28*, 501 (1994).
73. N. Trinajstić, *Chemical Graph Theory*, CRC Press, Boca Raton, Florida (1992).
74. A. T. Balaban, M. Banciu, and V. Ciorba, *Annulenes, Benzo-, Hetero-, Homo-Derivatives and Their Valence Isomers*, CRC Press, Boca Raton, Florida (1987).
75. D. Hilbert and S. Cohn-Vossen, *Geometry and the Imagination*, Chelsea, New York (1952).
76. P. J. Giblin, *Graphs, Surfaces and Homology*, Chapman & Hall, London (1981).
77. J. L. Gross and T. N. Tucker, *Topological Graph Theory*, John Wiley & Sons, New York (1987).
78. J. Stillwell, *Classical Topology and Combinatorial Group Theory*, Springer-Verlag, Berlin (1980).
79. V. Parasuk and J. Almlof, *Chem. Phys. Lett. 184*, 187 (1991).

80. M. Sawtarie, M. Menon, and K. R. Suvvaswamy, *Phys. Rev. B 49*, 7739 (1994).

81. P. R. Taylor, E. Bylaska, J. H. Weare, and R. Kawai, *Chem. Phys. Lett. 235*, 558 (1995).

82. D. A. Plattner and K. N. Houk, *J. Am. Chem. Soc. 117*, 4405 (1995).

83. X. Jing and J. R. Chelikowsky, *Phys. Rev. B 46*, 15503 (1992).

84. G. von Helden, M. T. Hsu, N. G. Gotts, P. R. Kemper, and M. T. Bowers, *Chem. Phys. Lett. 204*, 15 (1993).

85. V. A. Schweigert, A. L. Alexandrov, Y. N. Morokov, and V. M. Bedanov, *Chem. Phys. Lett. 235*, 221 (1995).

86. G. von Helden, N. G. Gotts, and M. T. Bowers, *Nature 363*, 60 (1993).

87. J. Hunter, J. Fye, and M. F. Jarrold, *J. Phys. Chem. 97*, 3460 (1993).

88. M. Ehbrecht, M. Raerber, F. Rohmund, V. V. Smirnov, O. Stelmakh, and F. Huisken, *Chem. Phys. Lett. 214*, 34 (1993).

89. H. Schwartz, *Angew. Chem. Int. Ed. Engl. 32*, 1412 (1993).

90. Q. L. Zhang, S. C. O'Brien, J. R. Heath, Y. Liu, R. F. Curl, H. W. Kroto, and R. E. Smalley, *J. Phys. Chem. 90*, 525 (1986).

91. D. J. Klein and T. G. Schmalz, in: *Quasicrystals, Networks, and Molecules of Fivefold Symmetry* (I. Hargittai, ed.), pp. 239–246, Springer-Verlag, Berlin (1990).

92. R. E. Haufler, Y. Chai, L. P. F. Chibante, J. Conceico, C. Jin, L.-S. Wang, S. Maryuma, and R. E. Smalley, *Mt. Res. Soc. Symp. Proc. 206*, 627 (1991).

93. J. R. Heath, in: *Fullerenes* (G. S. Hammond and V. J. Kuck, eds.), pp. 1–23, American Chemical Society, Washington, DC (1992).

94. T. Wakabayashi and Y. Achiba, *Chem. Phys. Lett. 190*, 465 (1992).

95. T. Belz, H. Werner, F.Semlin, U. Klengler, M. Wesemann, B. Tesche, E. Zeitler, A. Reller, and R. Schlögl, *Angew. Chem. Int. Ed. Engl. 33*, 1866 (1994).

96. T. Wakabayashi, H. Shiromaru, K. Kikuchi, and Y. Achiba, *Chem. Phys. Lett. 201*, 470 (1993).

97. H. Hadwiger, *J. Reine & Angew. Math. 194*, 101 (1955).

98. G.-C. Rota, in: *Studies in Pure Mathematics* (L. Mirsky, ed.), pp. 221–233, Academic Press, New York (1971).

99. Y. A. Shashkin, *The Euler Characteristic*, MIR Publ., Moscow (1984).

100. A. T. Balaban, X. Liu, D. J. Klein, D. Babic, T. G. Schmalz, W. A. Seitz, and M. Randić, *J. Chem. Inf. Comput. Sci. 35*, 396 (1995).

101. A. Baeyer, *Chem. Ber. 18*, 2269 (1885).

102. N. L. Allinger, M. T. Tribble, M. A. Millar, and D. W. Wertz, *J. Am. Chem. Soc. 93*, 1637 (1971).

103. E. M. Engler, J. D. Andose, and P. v. R. Schleyer, *J. Am. Chem. Soc. 95*, 8005 (1973).

104. R. Huisgen, *Angew. Chem. Int. Ed. Engl. 25*, 297 (1986).

105. Articles in *Molecular Structure and Energetics*, Vols. 1 and 2 (J. F. Liebman and A. Greenberg, eds.), Verlag-Chemie, New York (1986 & 1987).

106. D. J. Struik, *Classical Differential Geometry*, Dover Publications, New York (1988).

107. T. Y. Thomas, *Tensor Analysis and Differential Geometry*, Academic Press, New York (1965).

108. B. Grünbaum, *Convex Polytopes*, Chapter 13, Interscience, New York (1967).

109. D'A. Thompson, *Growth and Form*, 2nd ed., Cambridge University Press, London (1943).

110. C. S. Smith, *Metal Interfaces*, pp. 65–113, American Society of Metals, Cleveland, Ohio (1952).

111. A. F. Wells, *The Third Dimension in Chemistry*, Chapter 2, Oxford University Press, London (1956).

112. J. Langer and D. A. Singer, *Bull. London Math. Soc. 16*, 531 (1984).

113. R. Osserman, *Am. Math. Mon. 97*, 731 (1990).

114. H. Whitney, *Am. J. Math. 54*, 150 (1932).

115. D. J. Klein and X. Liu, *J. Math. Chem. 11*, 199 (1992).

116. K. I. Appel and W. Haken, *Ill. J. Math. 21*, 429 (1977).

117. K. I. Appel, W. Haken, and J. Koch, *Ill. J. Math. 21*, 491 (1977).

118. P. G. Tait, *Proc. R. Soc. Edinburgh 10*, 729 (1878–1880).

119. D. J. Klein, T. G. Schmalz, G. E. Hite, and W. A. Seitz, *J. Am. Chem. Soc. 108*, 1301 (1986).

120. D. J. Klein, W. A. Seitz, and T. G. Schmalz, *Nature 323*, 703 (1986).
121. X. Liu, D. J. Klein, W. A. Seitz, and T. G. Schmalz, *J. Comp. Chem. 12*, 1265 (1991).
122. X. Liu, T. G. Schmalz, and D. J. Klein, *Chem. Phys. Lett. 188*, 550 (1992).
123. X. Liu, D. J. Klein, and T. G. Schmalz, *Fullerene Sci. Technol. 2*, 405 (1994).
124. K. Fries, *Liebigs Ann. Chem. 454*, 121 (1927).
125. K. Fries, R. Walter, and K. Schilling, *Liebigs Ann. Chem. 516*, 248 (1935).
126. W. C. Herndon, *Thermochim. Acta 8*, 225 (1974).
127. M. Randić, *Tetrahedron 33*, 1905 (1977).
128. D. J. Klein, *J. Chem. Educ. 69*, 691 (1992).
129. M. Randić, S. Nikolic, and N. Trinajstić, *Croat. Chem. Acta 60*, 595 (1987).
130. R. Taylor, *J. Chem. Soc. Perkin Trans. 2*, 3 (1991).
131. W. T. Simpson, *J. Am. Chem. Soc. 73*, 593 (1953).
132. W. T. Simpson, *J. Am. Chem. Soc. 74*, 6285 (1954).
133. D. J. Klein and N. Trinajstić, *Pure Appl. Chem. 61*, 2107 (1989).
134. P. W. Fowler, *Chem. Phys. Lett. 131*, 444 (1986).
135. P. W. Fowler, J. E. Cremmona, and J. I. Steer, *Theor. Chim. Acta 73*, 1 (1988).
136. R. J. Johnston, *J. Chem. Soc. Faraday Trans. 87*, 3353 (1991).
137. D. R. Woodall and P. W. Fowler, *J. Chem. Soc. Faraday Trans. 88*, 2427 (1992).
138. P. W. Fowler, *J. Chem. Soc. Perkin Trans. 2*, 145 (1992).
139. B. Grünbaum and T. S. Motzkin, *Can. J. Math. 15*, 744 (1963).
140. H. W. Kroto, *Nature 329*, 529 (1987).
141. M. Goldberg, *Tohoku Math. J. 43*, 104 (1937).
142. J. S. Rutherford, *J. Math. Chem. 14*, 385 (1993).
143. A. C. Tang, Q. S. Li, C. W. Liu, and J. Li, *Chem. Phys. Lett. 201*, 465 (1993).
144. A. C. Tang, F. Q. Huang, Q. S. Li, and R. Z. Liu, *Chem. Phys. Lett. 227*, 579 (1994).
145. A. C. Tang and F. Q. Huang, *Phys. Rev. B 51*, 13830 (1995).
146. P. W. Fowler, *J. Chem. Soc. Faraday Trans. 86*, 2073 (1990).
147. Y.-N. Chiu and B.-C. Wang, *J. Mol. Struct. 283*, 13 (1993).
148. Y.-N. Chiu, P. Ganelin, X. Jiang, and B.-C. Wang, *J. Mol. Struct. 312*, 215 (1994).
149. X. Liu, D. J. Klein, T. G. Schmalz, and W. A. Seitz, *J. Comput. Chem. 12*, 1252 (1991).
150. X. Liu and D. J. Klein, *J. Comput. Chem. 12*, 1260 (1991).
151. B. Mackay, *Congressus Numerant. 30*, 45 (1981).
152. T. G. Schmalz, X. Liu, and D. J. Klein, *Chem. Phys. Lett. 192*, 331 (1992).
153. P. W. Fowler and D. E. Manolopolous, *An Atlas of Fullerenes*, Oxford University Press, London (1994).
154. D. E. Manolopolous, J. C. May, and S. E. Down, *Chem. Phys. Lett. 181*, 105 (1991).
155. D. E. Manolopolous and P. W. Fowler, *Chem. Phys. Lett. 204*, 1 (1993).
156. D. Babic and N. Trinajstić, *Comput. Chem. 17*, 271 (1993).
157. D. Babic, D. J. Klein, and C. H. Sah, *Chem. Phys. Lett. 211*, 235 (1993).
158. H. Hosoya, U. Nagashima, and S. Hyugaji, *J. Chem. Inf. Comput. Sci. 34*, 428 (1994).
159. Y.-D. Gao and W. C. Herndon, *J. Am. Chem. Soc. 115*, 8459 (1993).
160. D. Babic and N. Trinajstić, *Chem. Phys. Lett. 237*, 239 (1995).
161. R. Tosic and S. J. Cyvin, *Zb. Rad. Priv. Mat. Fak. Univ. Novom Sadu 22* (1992).
162. H. Sachs, *J. Chem. Inf. Comput. Sci. 34*, 432 (1994).
163. M. Randić, Y. Tsukano, and H. Hosoya, *Nat. Sci. Rep. Ochanumizu Univ. 45*, 101 (1994).
164. E. C. Kirby, *Croat. Chem. Acta 66*, 13 (1993).
165. E. C. Kirby, R. B. Mallion, and P. Pollak, *J. Chem. Soc. Faraday Trans. 89*, 1945 (1993).
166. D. J. Klein, *J. Chem. Inf. Comput. Sci. 34*, 453 (1994).
167. D. J. Klein and H.-Y. Zhu, *Disc. Appl. Math. 67*, 157 (1996).
168. J. Szucs, D. J. Klein, and H.-Y. Zhu, unpublished results.
169. A. Altshuler, *Disc. Math. 1*, 299 (1972).

170. A. Altshuler, *Disc. Math. 4*, 201 (1973).

171. B. I. Dunlap, *Phys. Rev. B 46*, 1933 (1992).

172. S. Itoh, S. Ihara, and J. Kitakami, *Phys. Rev. B 47*, 1703 (1993).

173. S. Ihara, S. Itoh, and J. Kitakami, *Phys. Rev. B 47*, 12908 (1993).

174. E. G. Galpern, I. V. Stankevich, A. L. Chistyakov, and L. A. Chernozatonskii, *Fullerene Sci. Technol. 2*, 1 (1994).

175. S. Itoh and S. Ihara, *Phys. Rev. B 48*, 8323 (1993).

176. S. Itoh and S. Ihara, *Phys. Rev. B 49*, 13970 (1994).

177. J. C. Greer, S. Itoh, and S. Ihara, *Chem. Phys. Lett. 222*, 621 (1994).

178. B. Borstnik and D. Lukman, *Chem. Phys. Lett. 228*, 312 (1994).

179. J. K. Johnson, B. N. Davison, M. R. Pederson, and J. Q. Broughton, *Phys. Rev. B 50*, 17575 (1994).

180. J. E. Avron and J. Berger, *Phys. Rev. A 51*, 1146 (1995).

181. F. Klein, *Theorie der Algebraischen Funktionen und Ihre Integrale*, Teubner, Leipzig (1882).

182. E. A. Rohlfing, D. M. Cox, and A. Kaldor, *J. Chem. Phys. 81*, 3322 (1984).

183. W. Weltner and R. VanZee, *Chem. Rev. 89*, 1713 (1989).

184. P. Gerhardt, S. Loffler, and K. H. Homann, *Chem. Phys. Lett. 137*, 306 (1987).

185. P. P. Radi, M. T. Hsu, J. Brodbelt-Lustig, M. Rincon, and M. T. Bowers, *J. Chem. Phys. 92*, 4817 (1990).

186. D. K. Bohme, *Chem. Rev. 92*, 1487 (1992).

187. M. Ge and K. Sattler, *Science 260*, 515 (1993).

188. M. Ge and K. Sattler, *Appl. Phys. Lett. 64*, 710 (1994).

189. D. S. Bethune, C. H. Klang, M. S. deVries, G. Gorman, R. Savoy, J. Vazquez, and R. Beyers, *Nature 363*, 605 (1993).

190. S. Iijima and T. Ichihashi, *Nature 363*, 603 (1993).

191. S. Wang and D. Zhou, *Chem. Phys. Lett. 225*, 165 (1994).

192. J. M. Lambert, P. M. Ajayan, P. Bernier, J. M. Planeix, V. Brotons, B. Coq, and J. Castaing, *Chem. Phys. Lett. 226*, 364 (1994).

193. D. T. Colbert, J. Zhang, S. M. McClure, P. Nikolaev, Z. Chen, J. H. Hafner, D. M. Owens, P. G. Kotula, C. B. Carter, J. H. Weaver, A. G. Rinzler, and R. E. Smalley, *Science 266*, 1218 (1994).

194. Y. Saito, M. Okuda, M. Tomita, and T. Hayashi, *Chem. Phys. Lett. 236*, 419 (1995).

195. S. Iijima, P. M. Ajayan, and T. Ichihashi, *Phys. Rev. Lett. 69*, 3100 (1992).

196. P. M. Ajayan, T. Ichihashi, and S. Iijima, *Chem. Phys. Lett. 202*, 384 (1993).

197. Y. Saito, T. Yoshikawa, M. Inayak, M. Tomita, and T. Hayashi, *Chem. Phys. Lett. 204*, 277 (1993).

198. N. Hatta and K. Murata, *Chem. Phys. Lett. 217*, 398 (1994).

199. S. Amelinckx, X. B. Zhang, D. Bernaerts, X. F. Zhang, V. Iranov, and J. B. Nagy, *Science 265*, 635 (1994).

200. L. A. Chernozatonskii, Z. J. Kosakovskaja, A. N. Kiselev, and N. A. Kiselev, *Chem. Phys. Lett. 228*, 94 (1994).

201. E. Dujardin, T. W. Ebbesen, H. Hiura, and K. Tanigaki, *Science 265*, 1850 (1994).

202. N. Kprinarov, M. Marinov, G. Pchelarov, M. Konstantinova, and R. Stefanov, *J. Phys. Chem. 99*, 2042 (1995).

203. P. M. Ajayan and S. Iijima, *Nature 361*, 333 (1993).

204. R. S. Ruoff, D. C. Lorents, B. Chan, R. Malhorta, and S. Subramoney, *Science 259*, 346 (1993).

205. P. M. Ajayan, T. W. Ebbesen, T. Schihashi, S. Iijima, K. Tanigaki, and H. Hiora, *Nature 362*, 522 (1993).

206. S. C. Tsang, P. J. F. Harris, and M. L. H. Green, *Nature 362*, 520 (1993).

207. S. C. Tsang, Y. K. Chen, P. J. F. Harris, and M. L. H. Green, *Nature 372*, 159 (1994).

208. D. J. Klein, W. A. Seitz, and T. G. Schmalz, *J. Phys. Chem. 97*, 1231 (1993).

209. D. H. Robertson, D. W. Brenner, and J. W. Mintmire, *Phys. Rev. B 45*, 12592 (1992).

210. J. W. Mintmire, D. H. Robertson, B. F. Dunlap, R. C. Mowrey, D. W. Brenner, and C. T. White, *Mater. Res. Symp. Proc. 247*, 339 (1992).

211. N. Hamada, S. Sawada, and A. Oshiyama, *Phys. Rev. Lett. 68*, 1579 (1992).
212. K. Tanaka, K. Okahara, M. Okada, and T. Yamabe, *Chem. Phys. Rev. Lett. 191*, 469 (1992).
213. Y.-D. Gao and W. C. Herndon, *Mol. Phys. 77*, 585 (1992).
214. R. Saito, M. Fujita, G. Dresselhaus, and M. S. Dresselhaus, *Phys. Rev. B 46*, 1804 (1992).
215. C. T. White, D. H. Robertson, and J. W. Mintmire, *Phys. Rev. B 47*, 5485 (1993).
216. B. J. Dunlap, *Phys. Rev. B 49*, 5643 (1994).
217. K. Okahara, K. Tanaka, H. Aoki, T. Sato, and T. Yamabe, *Chem. Phys. Lett. 219*, 462 (1994).
218. J. Aihara, *J. Phys. Chem. 98*, 9773 (1994).
219. J. Aihara, T. Yamabe, and H. Hosoya, *Synth. Met. 64*, 309 (1994).
220. R. A. Jishi, L. Ventataraman, M. S. Dresselhaus, and G. Dresselhaus, *Phys. Rev. B 51*, 11176 (1995).
221. D. J. Klein, G. E. Hite, W. A. Seitz, and T. G. Schmalz, *Theor. Chim. Acta 69*, 409 (1986).
222. S. Ihara, S. Itoh, and J. Kitakami, *Phys. Rev. B 48*, 5643 (1993).
223. K. Akagi, R. Tamura, M. Tsukade, S. Itoh, and S. Ihara, *Phys. Rev. Lett. 74*, 2307 (1995).
224. H.-Y. Zhu, A. T. Balaban, D. J. Klein, and T. P. Zivkovic, *J. Chem. Phys. 101*, 5281 (1994).
225. A. T. Balaban, D. J. Klein, and X. Liu, *Carbon 32*, 357 (1994).
226. J. R. Dias, *J. Chem. Inf. Comput. Sci. 22*, 139 (1982).
227. S. J. Cyvin, *J. Math. Chem. 9*, 389 (1992).
228. S. J. Cyvin, J. Brunvoll, and B. N. Cyvin, *Chem. Phys. Lett. 205*, 343 (1993).
229. M. Ge and K. Sattler, *Chem. Phys. Lett. 220*, 192 (1994).
230. B. Grünbaum and G. C. Shepard, *Tilings and Patterns*, Freeman, San Francisco (1986).
231. D. Vanderbilt and J. Tersoff, *Phys. Rev. Lett. 68*, 511 (1992).
232. T. Lenosky, X. Gonze, M. Teter, and V. Elser, *Nature 355*, 333 (1992).
233. R. Phillips, D. A. Drabold, T. Lenonsky, G. B. Adams, and O. F. Sankey, *Phys. Rev. B 46*, 1941 (1992).
234. S. J. Townsend, T. J. Lenonsky, D. A. Muller, C. S. Nichols, and V. Elser, *Phys. Rev. Lett. 69*, 921 (1992).
235. H. Terrones and A. L. Mackay, *Chem. Phys. Lett. 207*, 45 (1993).
236. H. Terrones and A. L. Mackay, *J. Math. Chem. 15*, 183 (1994).
237. A. J. Stone and D. J. Wales, *Chem. Phys. Lett. 128*, 501 (1986).
238. G. E. Scuseria, *Chem. Phys. Lett. 195*, 534 (1992).
239. R. L. Murray, D. L. Stout, G. K. Odom, and G. E. Scuseria, *Nature 366*, 665 (1993).
240. D. Babic and N. Trinajstić, *Comput. Chem. 17*, 271 (1993).
241. R. F. Curl, *Carbon 30*, 1149 (1992).
242. R. F. Curl, *Philos. Trans. R. Soc. 343*, 19 (1993).
243. D. Babic, S. Bassoli, M. Casartelli, F. Cataldo, A. Graovac, O. Ori, and B. York, *Mol. Simul. 14*, 395 (1995).
244. A. T. Balaban, T. G. Schmalz, H.-Y. Zhu, and D. J. Klein, *J. Mol. Struct. (Theochem.) 363*, 291 (1996).
245. S. C. O'Brien, J. R. Heath, R. F. Curl, and R. E. Smalley, *J. Chem. Phys. 88*, 220 (1988).
246. P. O. Radi, T. L. Bunn, P. R. Kemper, M. E. Molchan, and M. T. Bowers, *J. Chem. Phys. 88*, 2809 (1988).
247. C. J. Brabec, A. Maiti, and J. Bernholc, *Chem. Phys. Lett. 219*, 473 (1994).
248. J. R. Dias, *Chem. Phys. Lett. 204*, 486 (1993).
249. P. W. Fowler, S. J. Austin, O. J. Dunning, and J. R. Dias, *Chem. Phys. Lett. 224*, 123 (1994).
250. D. Babic, N. Trinajstić, and D. J. Klein, *Croat. Chem. Acta 67*, 37 (1994).
251. C. H. Sah, *Croat. Chem. Acta 66*, 1 (1993).
252. R. H. Baughman, D. S. Galvao, C. Cui, Y. Wang, and D. Tomanek, *Chem. Phys. Lett. 204*, 8 (1993).
253. R. H. Baughman, H. Eckhardt, and M. Kertesz, *J. Chem. Phys. 87*, 6687 (1987).

10

Applications of Topology and Graph Theory in Understanding Inorganic Molecules

R. BRUCE KING

10.1. INTRODUCTION

The diverse chemical applications of topology[1] and graph theory[2-5] have had a significant impact in a number of areas of inorganic chemistry including the structure and bonding in boron cage compounds, metal clusters, coordination compounds, and solid-state materials as well as polyhedral rearrangements. This chapter provides an overview of the applications of these areas of mathematics in understanding inorganic molecules. Further details of many aspects of the material presented in this chapter are given in a book recently published by the author[6] and the literature references cited therein.

Many aspects of applications of topology and graph theory in inorganic chemistry relate to the description of chemical structures by means of polyhedra. In this connection a polyhedron may be regarded as a set consisting of (zero-dimensional) points, namely, its *vertices*; (one-dimensional) lines connecting some of the vertices, namely, its *edges*; and (two-dimensional) surfaces formed by the edges, namely, its *faces*. Polyhedra can appear in the structures of inorganic molecules as *coordination*

R. BRUCE KING • Department of Chemistry, The University of Georgia, Athens, Georgia 30602-2556.

From Chemical Topology to Three-Dimensional Geometry, edited by Balaban. Plenum Press, New York, 1997

polyhedra in which the vertices represent ligands surrounding a central atom or as *cluster polyhedra* in which the vertices represent multivalent atoms and the edges represent bonding distances. *Deltahedra*, in which all faces are triangles, are a special type of polyhedra that often appear in inorganic structures. The topology of a polyhedron can be described by a graph, called the 1-skeleton of the polyhedron.[7] The vertices and edges of the 1-skeleton correspond to the vertices and edges, respectively, of the underlying polyhedron.

This chapter focuses on applications of graph theory and topology in inorganic chemistry as suggested by the polyhedra used to describe inorganic structures. Within this general outline the following topics are discussed:

1. Coordination polyhedra and their relationship to atomic orbitals
2. Cluster polyhedra as found in boranes, metal carbonyl clusters, coinage metal clusters, and post-transition element clusters
3. Polyhedra found in polyoxometalates

10.2. COORDINATION POLYHEDRA

10.2.1. Polyhedral Topology

Before considering coordination polyhedra it is first useful to consider the static topology of polyhedra in general. Of fundamental importance are relationships between possible numbers and types of vertices (v), edges (e), and faces (f) of polyhedra. In this connection the following elementary relationships are particularly significant[8]:

1. *Euler's relationship:*

(1)
$$v - e + f = 2$$

This arises from the properties of ordinary 3D space.

2. *Relationship between the edges and faces:*

(2)
$$\sum_{i=3}^{v-1} if_i = 2e$$

In equation (2) f_i is the number of faces with i edges (e.g., f_3 is the number of triangular faces, f_4 is the number of quadrilateral faces). This relationship arises from the fact that each edge of the polyhedron is shared by exactly two faces. Since no face can have fewer edges than the three of a triangle, the following inequality must hold in all cases:

(3)
$$3f \le 2e$$

3. *Relationship between the edges and vertices:*

(4)
$$\sum_{i=3}^{v-1} iv_i = 2e$$

In equation (4) v_i is the number of vertices of *degree i* (i.e., having i edges meeting at the vertex). This relationship arises from the fact that each edge of the polyhedron connects exactly two vertices. Since no vertex of a polyhedron can have a degree less than three, the following inequality must hold in all cases:

(5)
$$3v \leq 2e$$

4. *Totality of faces:*

(6)
$$\sum_{i=3}^{v-1} f_i = f$$

5. *Totality of vertices:*

(7)
$$\sum_{i=3}^{v-1} v_i = v$$

Equation (6) relates the f_i's to f and equation (7) relates the v_i's to v.

In generating actual polyhedra, the operations of capping and dualization are often important. *Capping* a polyhedron $\mathcal{P}_1$ consists of adding a new vertex above the center of one of its faces $\mathcal{F}_1$ followed by adding edges to connect the new vertex with each vertex of $\mathcal{F}_1$. This capping process gives a new polyhedron $\mathcal{P}_2$ having one more vertex than $\mathcal{P}_1$. If a triangular face is capped, the following relationships will be satisfied where the subscripts 1 and 2 refer to $\mathcal{P}_1$ and $\mathcal{P}_2$, respectively: $v_2 = v_1 + 1$; $e_2 = e_1 + 3$; $f_2 = f_1 + 2$. Such a capping of a triangular face is found in the capping of an octahedron to form a capped octahedron, i.e.,

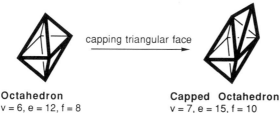

capping triangular face

Octahedron
v = 6, e = 12, f = 8

Capped Octahedron
v = 7, e = 15, f = 10

In general, if a face with f_k edges is capped, the following relationships will be satisfied: $v_2 = v_1 + 1$; $e_2 = e_1 + f_k$; $f_2 = f_1 + f_k - 1$. An example of such a capping process converts a square antiprism into a capped square antiprism, i.e.,

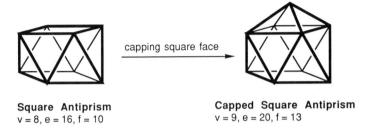

Square Antiprism
v = 8, e = 16, f = 10

Capped Square Antiprism
v = 9, e = 20, f = 13

A given polyhedron P can be converted into its dual P^* by locating the centers of the faces of P^* at the vertices of P and the vertices of P^* above the centers of the faces of P. Two vertices in the dual P^* are connected by an edge when the corresponding faces in P share an edge. An example of the process of dualization is the conversion of a trigonal bipyramid into a trigonal prism, i.e.,

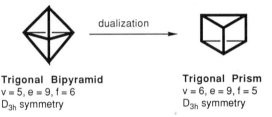

Trigonal Bipyramid
v = 5, e = 9, f = 6
D_{3h} symmetry

Trigonal Prism
v = 6, e = 9, f = 5
D_{3h} symmetry

The process of dualization has the following properties:

1. The numbers of vertices and edges in a pair of dual polyhedra P and P^* satisfy the relationships $v^* = f$, $e^* = e$, $f^* = v$, in which the starred variables refer to the dual polyhedron P^*. Thus, in the case of the trigonal bipyramid (P)/trigonal prism (P^*) dual pair depicted above, $v^* = f = 6$, $e^* = e = 9$, $f^* = v = 5$.
2. Dual polyhedra have the same symmetry elements and thus belong to the same symmetry point group. Thus, in the example above, both the trigonal bipyramid and the trigonal prism have the D_{3h} symmetry point group.
3. Dualization of the dual of the polyhedron leads to the original polyhedron
4. The degrees of the vertices of a polyhedron correspond to the number of edges in the corresponding face polygons in its dual.

The problem of the classification and enumeration of polyhedra is a complicated one. Thus, there appear to be no formulas, direct or recursive, for which the number of combinatorially distinct polyhedra having a given number of vertices, edges, faces, or any given combination of these elements can be calculated.[9,10] Duijvestijn and Federico have enumerated by computer the polyhedra having up to 22 edges according to the numbers of vertices, edges, and faces and their symmetry groups and present a summary of their methods, results, and literature references to previous work.[11] Their work shows that there are 1, 2, 7, 34, 257, 2606, and 32,300 topologically distinct polyhedra having 4, 5, 6, 7, 8, 9, and 10 faces or vertices, respectively. Tabulations are available for all 301 (= 1 + 2 + 7 + 34 + 257) topologically distinct polyhedra having 8 or fewer faces[12] or 8 or fewer vertices.[13] These two tabulations are essentially equivalent by the dualization relationship discussed above.

10.2.2. Atomic Orbitals and Their Quantum Numbers

In order to understand the coordination polyhedra encountered in inorganic chemistry, it is useful to understand the properties of the atomic orbitals on the central atom, which can hybridize to form the observed coordination polyhedra. In this connection atomic orbitals correspond to the one-particle wave functions Ψ, obtained as *spherical harmonics* by solution of the following second-order differential equation in which the potential energy V is spherically symmetric:

$$(8) \qquad \frac{\partial^2 \Psi}{\partial x^2} + \frac{\partial^2 \Psi}{\partial y^2} + \frac{\partial^2 \Psi}{\partial z^2} + \frac{8\pi^2 m}{h^2} (E - V)\, \Psi = \nabla^2 \Psi + \frac{8\pi^2 m}{h^2} (E - V)\, \Psi = 0$$

Thus, the spherical harmonics Ψ obtained by solving equation (8) are functions of either the three spatial coordinates x, y, and z or the corresponding *spherical polar coordinates* r, θ, and ϕ defined by the equations

(9a)
$$x = r \sin\theta \cos\phi$$

(9b)
$$y = r \sin\theta \sin\phi$$

(9c)
$$z = r \cos\theta$$

Furthermore, a set of linearly independent wave functions can be found such that Ψ can be factored into the following product:

$$(10) \qquad \Psi(r,\theta,\phi) = R(r){\cdot}\Theta(\theta){\cdot}\Phi(\phi)$$

in which the factors R, Θ, and Φ are functions solely of r, θ, and ϕ, respectively. Since the value of the radial component $R(r)$ of Ψ is completely independent of the angular coordinates θ and ϕ, it is independent of direction (i.e., *isotropic*) and therefore remains unaltered by any symmetry operations. For this reason all of the symmetry properties of a spherical harmonic Ψ, and thus of the corresponding wave function or atomic orbital, are contained in its angular component $\Theta(\theta){\cdot}\Phi(\phi)$. Furthermore, each of the three factors of Ψ [equation (10)] generates a quantum number. Thus, the factors $R(r)$, $\Theta(\theta)$, and $\Phi(\phi)$ generate the quantum numbers n, l, and m_l (or simply m), respectively. The *principal quantum number n*, derived from the radial component $R(r)$, relates to the distance from the center of the sphere (i.e., the nucleus in the case of atomic orbitals). The *azimuthal quantum number l*, derived from the factor $\Theta(\theta)$ in equation (10), relates to the number of nodes in the angular component $\Theta(\theta){\cdot}\Phi(\phi)$, where a *node* is a plane corresponding to a zero value of $\Theta(\theta)\cdot\Phi(\phi)$ or Ψ, i.e., where the sign of $\Theta(\theta)\cdot\Phi(\phi)$ changes from positive to negative. Atomic orbitals where $l = 0, 1, 2$, and 3 have 0, 1, 2, and 3 nodes, respectively, and are conventionally designated as s, p, d, and f orbitals, respectively. For a given value of the azimuthal quantum number l, the *magnetic quantum number m_l* or m, derived from the factor $\Phi(\phi)$ in equation (10), may take on all $2l + 1$ different values from $+l$ to $-l$. There are therefore necessarily $2l + 1$ distinct orthogonal orbitals for a given value of l corresponding, for example, to 1, 3, 5, and 7 distinct s, p, d, and f orbitals, respectively. The magnetic quantum number, m, relates to the distribution of the electron density of the atomic orbital relative to the z axis. Thus, if the nucleus is in the center of a sphere in which the z axis is the polar axis passing through the north and south poles, an atomic orbital with $m = 0$ has its electron density oriented toward the north and south poles of the sphere whereas an atomic orbital with the maximum possible value of $|m|$, i.e., $\pm l$, has its maximum electron density in the equator of the sphere.

Table 1 depicts the shapes of the s, p, and d orbitals and Table 2 depicts the shapes of two different sets of f orbitals.[14–16] In this connection the following points are of interest:

1. All atomic orbitals are *orthogonal*, i.e., the overlap between different atomic orbitals is zero.
2. The extent of the s orbital (with $l = 0$) is independent of direction, i.e., the s orbital is spherically symmetrical and its wave function Ψ_s is independent of θ and ϕ. In view of this and in view of the uniqueness of the ns orbital for a given value of n, there is no polynomial designation for the s orbital.

Table 1. Properties of s, p, and d Atomic Orbitals

Type	Nodes	Polynomial	Angular function	Appearance
s	0		independent of θ, ϕ	spherically symmetrical
p	1	x	$\sin\theta\cos\phi$	
p	1	y	$\sin\theta\sin\phi$	
p	1	z	$\cos\theta$	
d	2	xy	$\sin^2\theta\sin 2\phi$	
d	2	xz	$\sin\theta\cos\theta\cos\phi$	
d	2	yz	$\sin\theta\cos\theta\sin\phi$	
d	2	$x^2 - y^2$	$\sin^2\theta\cos 2\phi$	
d	2	$2z^2 - x^2 - y^2$ (abbreviated as z^2)	$(3\cos^2\theta - 1)$	

Table 2. The General Shapes of Both the General and Cubic Sets of f Orbitals

Major lobes	Shape	General set	Cubic set
2		z^3	x^3 y^3 z^3
4		xz^2 yz^2	none
6		$x(x^2 - 3y^2)$ $y(3x^2 - y^2)$	none
8		xyz $z(x^2 - y^2)$	xyz $x(z^2 - y^2)$ $y(z^2 - x^2)$ $z(x^2 - y^2)$

3. The p_x, p_y, and p_z orbitals all have the same shape with their nodes in the yz, xz, and xy planes, respectively. Note how the p orbitals are orthogonal to each other and to the s orbital.

4. There are 5 ($= 2l + 1$ for $l = 2$) distinct and mutually orthogonal d orbitals. In the conventionally used set of five d orbitals, four have the same general shape with four lobes and two nodes. The fifth d orbital, namely, the $d(2z^2 - x^2 - y^2) = d(z^2)$ orbital, has a unique shape. However, all possible pairs of the d orbitals are orthogonal. All possible shapes of d orbitals can be expressed as linear combinations of these two types of d orbitals by the following equation[17,18]:

(11) $\quad d = a\phi_{z^2} + (1 - a^2)^{1/2}\phi_{x^2 - y^2} \quad \sqrt{3}/2 = 0.866025 \le a \le 1$

In equation (11) ϕ_{z^2} refers to the function of the $d(z^2)$ atomic orbital and $\phi_{x^2-y^2}$ refers to the function of the $d(x^2 - y^2)$ atomic orbital, taken as a representative of one of the four d orbitals with four major lobes. Two different sets of five orthogonal *equivalent* d orbitals can be constructed by choosing five orthogonal linear combinations of the $d(z^2)$ and $d(x^2 - y^2)$ orbitals using equation (11). These are called the *oblate* and *prolate* sets of five equivalent *d* orbitals since they are oriented toward the vertices of an oblate and prolate pentagonal antiprism, respectively. The fivefold symmetry of the equivalent set of five *d* orbitals makes them inconvenient to use since relatively few molecules have the matching fivefold symmetry.

5. Table 2 depicts two different sets of the seven *f* orbitals, which are distinguished by their numbers of major lobes. The major lobes of the *f* orbitals with six and eight major lobes are directed toward the vertices of a regular hexagon and cube, respectively. The cubic set of *f* orbitals is used for certain highly symmetrical structures, e.g., those of cubic (O_h point group), octahedral (also O_h point group), or icosahedral (I_h point group) symmetry, since in the cubic set of *f* orbitals there are two triply degenerate subsets, namely, the x^3, y^3, z^3 subset and the $x(z^2 - y^2)$, $y(z^2 - x^2)$, $z(x^2 - y^2)$ subset. In less symmetrical structures the general set of *f* orbitals is used.

10.2.3. Polyhedra for Coordination Numbers from Four to Nine

The choice of favored coordination polyhedra, or even the assignment of a coordination polyhedron to a given chemical structure, would appear to be very complicated in view of the large number of topologically distinct polyhedra with as few as seven vertices. However, the properties of atomic orbitals coupled with an assumption of maximum symmetry for a given hybrid of atomic orbitals makes this problem both tractable and interesting.[19,20]

The coordination "polyhedra" for coordination numbers 2 and 3 are trivial, namely, linear or bent for coordination number 2 and trigonal planar or pyramidal (minus the apex) for coordination number 3. For coordination number 4 the two possibilities are tetrahedral using sp^3 hybrids and square planar using $sp^2d(x^2 - y^2)$ hybrids. Polyhedra for coordination numbers 5 through 8 can be generated by the following processes:

1. Addition of one or two *d* orbitals to a four-orbital spherical sp^3 manifold to give five- and six-coordinate polyhedra, respectively
2. Subtraction of one or two *d* orbitals from a nine-orbital spherical sp^3d^5 manifold to give eight- and seven-coordinate polyhedra, respectively

The irreducible representations, Γ_σ, for the hybrid orbitals corresponding to σ-bonds from a central atom directed toward the vertices of polyhedra for coordination numbers four to nine are shown in Table 3, in which polyhedra indicated in boldface cannot be formed using only *s*, *p*, and *d* orbitals. The polyhedra listed in Table 3 are depicted in Figure 1.

The following points are of interest concerning polyhedra for coordination numbers five through nine.

1. *Coordination number five.* The two possible polyhedra for coordination number five are the D_{3h} trigonal bipyramid for which the sp^3 manifold is supplemented by the $d(z^2)$ orbital with two opposite major lobes and the C_{4v} square pyramid for which

Table 3. The Irreducible Representations for the Hybrid Orbitals Corresponding to Polyhedra for Coordination Numbers Four to Nine

Polyhedron	G	v	e	f	Γ_σ
Tetrahedron	T_d	4	6	4	$A_1(s) + T_2(x,y,z)$
Square pyramid	C_{4v}	5	8	5	$2A_1(s,z,z^2) + B_1(x^2 - y^2) + E(x,y,xz,yz)$
Trigonal bipyramid	D_{3h}	5	9	6	$2A_1'(s,z^2) + E'(x,y,x^2 - y^2,xy) + A_2''\ (z)$
Trigonal prism	D_{3h}	6	9	5	$A_1'(s,z^2) + E'(x,y,x^2 - y^2,xy) + A_2''\ (z)$ $+ E''(xz,yz)$
Pentagonal pyramid	C_{5v}	6	10	6	$2A_1(s,z,z^2) + E_1(x,y,xz,yz) + E_2(x^2 - y^2,xy)$
Octahedron	O_h	6	12	8	$A_{1g}(s) + E_g(z^2,x^2 - y^2) + T_{1u}(x,y,z)$
Bicapped tetrahedron	C_{2v}	6	12	8	$3A_1(s,z,z^2,x^2 - y^2) + 2B_1(x,xz) + B_2(yz)$
Capped octahedron	C_{3v}	7	15	10	$3A_1(s,z,z^2) + 2E(x,y,x^2 - y^2,xy,xz,yz)$
Pentagonal bipyramid	D_{5h}	7	15	10	$2A_1'(s,z^2) + E_1'(x,y) + E_2'\ (x^2 - y^2,xy) + A_2''\ (z)$
4-Capped trigonal prism	C_{2v}	7	13	8	$3A_1(s,z,z^2,x^2 - y^2) + A_2(xy) + 2B_1(x,xz)$ $+ B_2(y,yz)$
Hexagonal pyramid	C_{6v}	7	12	7	$2A_1(s,z,z^2) + B_1\boxed{x(x^2 - 3y^2)} + E_1(x,y,xz,yz)$ $+ E_2(x^2 - y^2,xy)$
3-Capped trigonal prism	C_{3v}	7	12	7	$3A_1(s,z,z^2) + 2E(x,y,x^2 - y^2,xy,xz,yz)$
Cube	O_h	8	12	6	$A_{1g(s)} + T_{2g}(xy,xz,yz) + A_{2u}\boxed{xyz} + T_{1u}(x,y,z)$
3,3-Bicapped trigonal prism	D_{3h}	8	15	9	$2A_1'(s,z^2) + E'(x,y,x^2 - y^2,xy) + 2A_2''(z,\boxed{z^3}\ )$ $+ E''(xz,yz)$
Square antiprism	D_{4d}	8	16	10	$A_1(s,z^2) + B_2(z) + E_1(x,y) + E_2(x^2 - y^2,xy)$ $+ E_3(xz,yz)$
Bisdisphenoid ("D_{2d} dodecahedron")	D_{2d}	8	18	12	$2A_1(s,z^2) + 2B_2(z,xy) + 2E(x,y,xz,yz)$
Hexagonal bipyramid	D_{6h}	8	18	12	$2A_g(s,z^2) + E_{2g}(x^2 - y^2) + A_{2u}(z)$ $+ B_{2u}\boxed{x(x^2 - 3y^2)} + E_{1u}(x,y)$
Tricapped trigonal prism	D_{3h}	9	21	14	$2A_1'(s,z^2) + 2E'(x,y,x^2 - y^2,xy) + A_2''\ (z)$ $+ E''(xz,yz)$
Capped square antiprism	C_{4v}	9	20	13	$3A_1(s,z,z^2) + B_1(x^2 - y^2) + B_2(xy)$ $+ 2E(x,y,xz,yz)$

the sp^3 manifold is supplemented by the $d(x^2 - y^2)$ orbital with four coplanar major lobes. Both of these coordination polyhedra are found in five-coordinate ML_5 metal complexes.[21] The possibility of a continuous transformation of a trigonal bipyramid $sp^3d(z^2)$ hybrid to a square pyramid $sp^3d(x^2 - y^2)$ hybrid through linear combinations of z^2 and $x^2 - y^2$ orbitals by equation (4) can relate to the stereochemical nonrigidity of five-coordinate complexes by Berry pseudorotation processes as follows[22–25]:

Trigonal bipyramid $\longrightarrow$ Square pyramid $\longrightarrow$ Trigonal bipyramid

D_{3h} $\qquad\qquad\qquad$ C_{4v} $\qquad\qquad\qquad$ D_{3h}

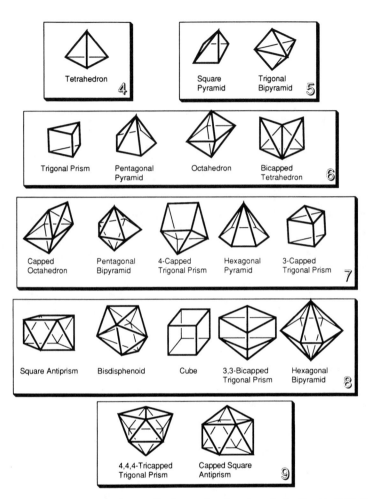

Figure 1. The polyhedra for coordination numbers four through nine listed in Table 3.

In addition, the locations of the major lobes in a z^2 or an $x^2 - y^2$ orbital is related to the geometry of the five-coordinate polyhedron arising when these d orbitals are added to the sp^3 hybrid of a tetrahedron. A tetrahedron has no more than three coplanar vertices and no pair of "opposite" vertices, i.e., a pair of vertices connected by a straight line through the center so that the corresponding X–M–X angle is 180°. However, adding a z^2 orbital, which has two major lobes opposite each other, to an sp^3 hybrid generates the trigonal bipyramid in which there is a pair of opposite vertices, namely, the two axial vertices. Similarly, adding an $x^2 - y^2$ orbital, which has four coplanar major lobes, to an sp^3 hybrid generates the square pyramid, which has four coplanar vertices, namely, the four basal vertices. Formation of a trigonal bipyramid or a square pyramid by addition of a z^2 or $x^2 - y^2$ orbital, respectively, to an sp^3 manifold is a consequence of maximizing the overlap of the hybrid orbitals with the ligand orbitals at the vertices of these polyhedra.

Pentagonal planar coordination is a conceivable alternative to the trigonal bipyramid and square pyramid for coordination number five and is actually found, albeit with some distortion, in a few tellurium complexes of sulfur ligands such as the xanthato complex $[Te(S_2COEt)_3]^-$ with one monodentate and two bidentate xanthato ligands.[26] Pentagonal planar coordination, like square planar coordination, can only use the two p orbitals that lie in the plane of the pentagon leading to $sp^2d^2(x^2 - y^2, xy)$ hybridization using the two d orbitals that have their major lobes in the plane of the pentagon.

2. *Coordination number six.* The very symmetrical (O_h point group) regular octahedron (Figure 1) is overwhelmingly favored for coordination number 6; it corresponds to the hybrid $sp^3d^2(x^2 - y^2, z^2)$ formed by adding the $(x^2 - y^2, z^2)$ pair of d orbitals to the sp^3 hybrid. Note that the major lobes of the d orbitals involved in the octahedral hybrids are directed toward the vertices of the octahedron. The D_{3h} trigonal prism (Figure 1) corresponding to $sp^3d^2(xz, yz)$ hybridization is found for tris(ethylenedithiolate) derivatives of early and middle transition metals.[27] The major lobes of the two d orbitals participating in trigonal prismatic hybridization are pointed toward the vertices although not as directly as in the case of the octahedron. A distorted pentagonal pyramid corresponding to $sp^3d^2(xy, x^2 - y^2)$ hybridization is found for certain complexes of post-transition elements such as Te(IV) and Sb(III). The pentagonal pyramid is the only six-vertex polyhedron in which five of the vertices are coplanar; it is therefore not surprising that it is formed using the only pair of d orbitals that have a total of eight major lobes, all of which are coplanar. Note also that the pair of d orbitals $(xy, x^2 - y^2)$ involved in pentagonal pyramidal hybridization is the same as those involved in the pentagonal planar $sp^2d^2(xy, x^2 - y^2)$ hybridization discussed above.

The final distinctive pair of d orbitals that can be added to an sp^3 manifold for a six-coordinate polyhedron is the (z^2, xz) pair corresponding to a bicapped tetrahedron (Figure 1) similar to that found in the metal cluster $Os_6(CO)_{18}$.[28] This polyhedron is not favorable for a coordination polyhedron since the two capping vertices are located further from the central metal atom than the remaining four vertices. However, the distortion of a regular $H_2M(CO)_2L_2$ octahedron toward an $M(CO)_2L_2$ tetrahedron in metal carbonyl dihydride derivatives of the type $H_2Fe(CO)_2L_2$ [L = CO[29] and PPh(OEt)$_2^{30}$] is related to bicapped tetrahedral stereochemistry.

3. *Coordination number seven.* The seven-coordinate polyhedron of maximum symmetry is the D_{5h} pentagonal bipyramid (Figure 1) with $sp^3d^3(xy, x^2 - y^2, z^2)$ hybridization. This polyhedron is commonly found in seven-coordinate complexes.[31] Other polyhedra found in seven-coordinate complexes include the capped octahedron with $sp^3d^3(z^2, xz, yz)$ or $sp^3d^3(z^2, xy, x^2 - y^2)$ hybridization and the 4-capped trigonal prism with $sp^3d^3(z^2, xy, xz)$ or $sp^3d^3(x^2 - y^2, xy, xz)$ hybridization. Note that the choice

of d orbitals for the sp^3d^3 hybrid again affects the choice of coordination polyhedron. No seven-vertex polyhedra with more than two symmetry elements have been found which are formed by $sp^3d^3(xy,xz,yz)$ hybrids.

Polyhedra with six coplanar vertices cannot be formed from hybrids using only s, p, and d orbitals. The smallest such polyhedron is the seven-vertex hexagonal pyramid, which requires the $f(x(x^2 - 3y^2))$ orbital with six major lobes pointed toward the vertex of a hexagon (Table 2).

4. *Coordination number eight*. The common coordination polyhedra for coordination number eight[32-34] are the bisdisphenoid or "D_{2d} dodecahedron" corresponding to $sp^3d^4(z^2,xy,xz,yz)$ hybridization and the square antiprism corresponding to sp^3d^4 $(x^2 - y^2,xy,xz,yz)$ hybridization (Figure 1). An $x^2 - y^2$ orbital with four major lobes is removed from the sp^3d^5 manifold to form the bisdisphenoid and a z^2 orbital with only two major lobes is removed from the sp^3d^5 manifold to form the square antiprism. Removal of the z^2 orbital from the sp^3d^5 manifold to form the square antiprism $sp^3d^4(x^2 - y^2,xy,xz,yz)$ hybrid corresponds to removal of electron density from the square faces of the square antiprism. As in the case of five-coordinate complexes, the possibility of a continuous transformation of a bisdisphenoid $sp^3d^4(z^2,xy,xz,yz)$ hybrid to a square antiprism $sp^3d^4(x^2 - y^2,xy,xz,yz)$ hybrid through linear combinations of "missing" z^2 and $x^2 - y^2$ orbitals can relate to the stereochemical nonrigidity of eight-coordinate complexes.[35]

Three eight-vertex polyhedra of relatively high symmetry, namely, the cube, the hexagonal bipyramid, and the D_{3h} 3,3-bicapped trigonal prism (Figure 1), cannot be formed using solely s, p, and d atomic orbitals (Table 2). Such polyhedra with nine or fewer vertices which cannot be formed from an sp^3d^5 nine-orbital manifold are called *forbidden coordination polyhedra*, or more specifically, *spd-forbidden coordination polyhedra*.[36] All of these eight-vertex polyhedra have symmetry point groups that can be expressed as direct products $R \times C_s'$. For the cube and hexagonal bipyramid the group $C_s' = C_i$ and $R = O$ and D_6, respectively, whereas for the 3,3-bicapped trigonal prism the group $C_s' = C_s$ and $R = D_3$. The nonidentity operation of C_s' can conveniently be called the *primary involution*. The character of the primary involution in the reducible representation of the direct product group $R \times C_s'$ corresponding to Γ_σ is equal to the number of vertices that remain fixed when the primary involution is applied. If the primary involution is an inversion, as is the case with $O_h = O \times C_i$ for the cube or $D_{6h} = D_6 \times C_i$ for the hexagonal bipyramid, its character is necessarily zero since *no* vertices of a polyhedron remain fixed by an inversion. Therefore, the reducible representation Γ_σ of an eight-vertex polyhedron with an inversion center contains equal numbers of even and odd irreducible representations. This corresponds to a hybridization using four symmetrical and four antisymmetrical atomic orbitals. Since only three orbitals of the sp^3d^5 manifold are antisymmetrical (namely, the three p orbitals), an eight-vertex polyhedron with an inversion center cannot be formed using only s, p, and d orbitals. Similarly in the case of the 3,3-bicapped trigonal prism, the primary involution is the horizontal reflection, σ_h, which also leaves no vertices fixed and thus has a character of zero so that the same arguments apply.

The f orbitals (Table 2) required in the hybridization of forbidden polyhedra relate to their shapes. Thus, the $f(xyz)$ orbital, with eight major lobes pointed toward the vertices of a cube, is used to form the sp^3d^3f hybrids of a cube. An $f(x(x^2 - 3y^2))$ orbital with six major lobes pointed toward the vertices of a regular hexagon is used to form the hybrids of a hexagonal pyramid or hexagonal bipyramid. An $f(z^3)$ orbital with its two major lobes along the z axis toward the two capping vertices is used to form the sp^3d^3f hybrids of a 3,3-bicapped trigonal prism. Actinides are the only elements in the periodic table for which the f orbitals can play a significant role in their covalent chemical bonding.[37] Cubic coordination is found in a few actinide complexes[38] such as $[Et_4N]_4[U(NCS)_8]$. In addition, hexagonal bipyramidal coordination is found in uranyl complexes with three small bite bidentate ligands such as $UO_2(NO_3)_3^-$ in which the uranyl oxygen atoms occupy axial positions. No examples of eight-coordinate complexes with 3,3-bicapped trigonal prismatic stereochemistry are known, even in actinide chemistry where an $f(z^3)$ orbital could be available for covalent chemical bonding. Thus, the 3,3-bicapped trigonal prism appears to be a very unfavorable coordination polyhedron.

5. *Coordination number nine.* The deltahedron of maximum symmetry is the D_{3h} tricapped trigonal prism (Figure 1). Either this nine-vertex polyhedron or the nine-vertex capped square antiprism can be formed using only a nine-orbital sp^3d^5 manifold and thus are feasible nine-vertex coordination polyhedra. The small number of nine-coordinate complexes including the hydrides[39,40] ReH_9^{2-} and TcH_9^{2-} generally use the tricapped trigonal prism.

10.2.4. Polyhedral Isomerizations

A polyhedral isomerization may be defined as a deformation of a specific polyhedron P_1 until its vertices define a new polyhedron P_2. Of particular interest are sequences of two polyhedral isomerization steps $P_1 \rightarrow P_2 \rightarrow P_3$ in which the polyhedron P_3 is combinatorially equivalent to the polyhedron P_1 although with some permutation of its vertices not necessarily the identity permutation. In this sense two polyhedra P_1 and P_2 may be considered to be *combinatorially equivalent* whenever there are three one-to-one mappings V, E, and F from the vertex, edge, and face sets of P_1 to the corresponding sets of P_2 such that incidence relations are preserved. Thus, if a vertex, edge, or face α of P_1 is incident to or touches on a vertex, edge, or face β of P_1, then the images of α and β under V, E, or F are incident in P_2.[41]

Consider a polyhedral isomerization sequence $P_1 \rightarrow P_2 \rightarrow P_3$ in which P_1 and P_3 are combinatorially equivalent. Such a polyhedral isomerization sequence may be called a *degenerate* polyhedral isomerization with P_2 as the *intermediate polyhedron*. Structures undergoing such degenerate isomerization processes are often called *fluxional*.[42] A degenerate polyhedral isomerization with a planar intermediate "polyhedron" (actually a polygon) may be called a *planar polyhedral isomerization*. The simplest example of a planar polyhedral isomerization is the interconversion of two enantiomeric tetrahedra ($\mathcal{P}_1$ and $\mathcal{P}_3$) through a square planar intermediate $\mathcal{P}_2$. Except for this simplest example, planar polyhedral isomerizations are unfavorable owing to excessive intervertex repulsion.

Polyhedral isomerizations may be treated from either the macroscopic or microscopic points of view. The earliest work in this area pioneered by Balaban *et al.*,[43] Muetterties,[44–46] Gielen,[47–54] Musher,[55,56] Klemperer,[57–59] and Brocas[60,61] focused on the *macroscopic* picture, namely, relationships between different permutational isomers. Such relationships may be depicted by reaction graphs called *topological representations* in which the vertices correspond to different permutational isomers and the edges correspond to single degenerate polyhedral isomerization steps. Subsequent work treats the *microscopic* picture in which the details of polyhedral topology are used to elucidate possible single polyhedral isomerization steps, namely, which types of isomerization steps are possible.

Consider an ML_n compound having n ligands. There are a total of $n!$ permutations of the ligand sites or the cluster vertices. These permutations form a group of order $n!$ called the *symmetric group*[62,63] and conventionally designated as S_n or less conventionally as P_n (to avoid confusion with improper rotations[64] also designated as S_n). The symmetric group S_n is the automorphism group corresponding to the symmetry of the *complete graph* K_n, which consists of n vertices with an edge between every pair of vertices for a total of $n(n-1)/2$ edges.

Now consider the symmetry point group G (or, more precisely, the framework group[65]) of the above ML_n coordination compound. This group has $|G|$ operations of which $|R|$ are proper rotations so that $|G|/|R| = 2$ if the compound is achiral and $|G|/|R| = 1$ if the compound is chiral (i.e., has no improper rotations). The $n!$ distinct permutations of the n sites in the coordination compound or cluster are divided into $n!/|R|$ right cosets[66] which represent the permutational isomers since the permutations corresponding to the $|R|$ proper rotations of a given isomer do not change the isomer but merely rotate it in space. This leads naturally to the concept of *isomer count*, I, namely,

$$(12) \qquad\qquad I = n!/|R|$$

if all vertices are distinguishable. Similarly, the quotient

$$(13) \qquad\qquad E = n!/|G| = I/2$$

for a given *chiral* polyhedron corresponds to the number of enantiomeric pairs. The isomer count I corresponds to the number of vertices in the corresponding topological representation. The degree of a vertex of a topological representation corresponds to the number of new permutational isomers generated in a single step from the isomer represented by the vertex in question; this is called the *connectivity*, δ, of the vertex. Topological representations can be classified by the number of vertices in the polyhedra participating in the rearrangements.

The concept of a topological representation is conveniently illustrated by the topological representations for rearrangements of coordination polyhedra having four and five vertices. More complicated topological representations for coordination polyhedra having six and eight vertices are discussed elsewhere.[67]

The only combinatorially distinct four-vertex polyhedron is the regular tetrahedron so that nonplanar isomerizations of tetrahedra are not possible. However, a tetrahedron can be converted to its mirror image (enantiomer) through a square planar intermediate. The isomer count for the tetrahedron, I_{tet}, is $4!/|T| = 24/12 = 2$ and the isomer count for the square, I_{sq}, is $4!/|D_4| = 24/8 = 3$. A topological representation of this process is a $K_{2,3}$ bipartite graph, which is derived from the trigonal bipyramid by deletion of the three equatorial–equatorial edges (Figure 2). In Figure 2 the two axial vertices (labeled T_d) correspond to the two tetrahedral isomers and the three equatorial vertices (labeled D_{4h}) correspond to the three square planar isomers. The connectivities of the tetrahedral (δ_{tet}) and square planar (δ_{sq}) isomers are 3 and 2, respectively, in accord with the degrees of the corresponding vertices of the $K_{2,3}$ graph. Thus,

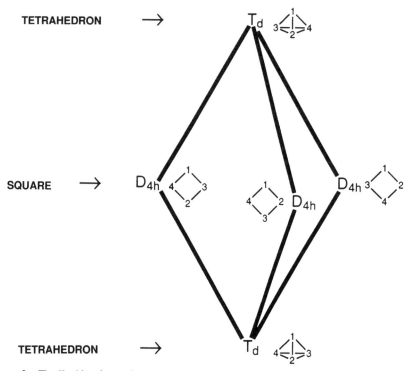

Figure 2. The $K_{2,3}$ bipartite graph as a topological representation of the degenerate planar isomerization of a tetrahedron (T_d) to its enantiomer through a square planar intermediate (D_{4h}). The isomers corresponding to the vertices of the $K_{2,3}$ bipartite graph are depicted next to the vertex labels.

$I_{tet}\delta_{tet} = I_{sq}\delta_{sq} = 6$; this is an example of the *closure* condition $I_a\delta_a = I_b\delta_b$ required for a topological representation with vertices representing more than one type of polyhedron.

The topological representation for rearrangements of five-coordinate polyhedra is somewhat more complicated than that for four-coordinate polyhedra. Thus, the two combinatorially distinct five-vertex polyhedra are the trigonal bipyramid and the square pyramid (Figure 1). The conversion of a trigonal bipyramid into an isomeric trigonal bipyramid through a Berry pseudorotation process involving a square pyramid intermediate has been discussed above. Some interesting graphs (Figure 3) are found in the topological representations for this process. The trigonal bipyramid has an isomer count $I = 5!/|D_3| = 120/6 = 20$ corresponding to 10 enantiomeric pairs. A given trigonal bipyramid isomer can be described by the labels of its two axial positions (i.e., the single pair of vertices not connected by an edge) with a bar used to distinguish enantiomers. In a single degenerate isomerization of a trigonal bipyramid through a square pyramid intermediate, both axial vertices of the original trigonal bipyramid become equatorial vertices in the new trigonal bipyramid leading to a connectivity of three for isomerizations of trigonal bipyramids. The corresponding topological representation thus is a 20-vertex graph in which each vertex has degree 3. However, additional properties of isomerizations of trigonal bipyramids exclude the regular (I_h) dodecahedron as a topological representation unless double group form is used to produce pseudohexagonal faces. A graph suitable for the topological representation of isomerizations of trigonal bipyramids is the Desargues–Levy graph, depicted in Figure 3 (left).[43]

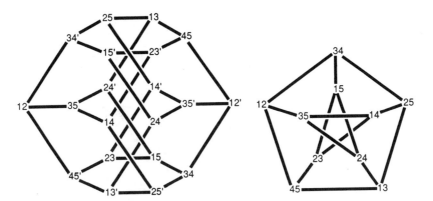

Figure 3. Topological representations for the isomerizations of trigonal bipyramids through dsd processes (Berry pseudorotations). The two digits represent labels for the axial positions, with primes used to indicate enantiomers. (Left) Desargues– Levy graph as a topological representation for dsd isomerizations of the 20 trigonal bipyramid isomers. (Right) Petersen's graph as a topological representation for dsd isomerizations of the 10 enantiomer pairs of the trigonal bipyramid isomers.

Less complicated but nevertheless useful topological representations can be obtained by using each vertex of the graph to represent a set of isomers provided that each vertex represents sets of the same size and interrelationship and each isomer is included in exactly one set. A simple example is the use of Petersen's graph (Figure 3, right) as a topological representation of isomerizations of the 10 trigonal bipyramid enantiomer pairs ($E = 5!/|D_{3h}| = 120/12 = 10$) by Berry pseudorotation processes. The use of Petersen's graph for this purpose relates to its being the odd graph O_3; an *odd graph O_k* is defined as follows[68]: its vertices correspond to subsets of cardinality $k - 1$ of a set S of cardinality $2k - 1$ and two vertices are adjacent if and only if the corresponding subsets are disjoint.

Now consider microscopic approaches to polyhedral rearrangements. Such approaches dissect such polyhedral rearrangements into elementary steps relating to the detailed topology of individual polyhedra. The most important elementary step is the *diamond–square–diamond* process which was first recognized in a chemical context by Lipscomb in 1966[69] as a generalization of the Berry pseudorotation process discussed above in connection with rearrangements of five-vertex trigonal pyramids. Such a diamond–square–diamond process or "dsd process" in a polyhedron occurs at two triangular faces sharing an edge and can be depicted as follows:

p_1 p_2 p_3

In this process a configuration such as p_1 can be called a *dsd situation* and the edge AB can be called a *switching edge*. If a, b, c, and d are taken to represent the degrees of the vertices A, B, C, and D, respectively, in p_1, then the *dsd type* of the switching edge AB can be represented as $ab(cd)$. In this designation the first two digits refer to the degrees of the vertices joined by AB but contained in the faces (triangles) having AB as the common edge (i.e., C and D in p_1). The quadrilateral face formed in structure p_2 may be called a *pivot face*.

In his pioneering paper Lipscomb[69] described some possible framework rearrangements of polyhedra having from 5 to 12 vertex atoms. Fifteen years later[70] I reexamined this question in light of advances in known experimental information on polyhedral chemical systems as well as improved understanding in polyhedral topology. Subsequently[71] I developed a mathematical approach for examining all *possible* nonplanar rearrangements of polyhedra having few (i.e., ≤ 6) vertices using a method developed by Gale[72] in 1956 for studying d-dimensional polytopes having only a few more than the minimum $d + 1$ vertices. This work[71] confirmed the crucial role of dsd processes conjectured so successfully by Lipscomb[69] and also provided insight for more detailed study of isomerizations of polyhedra having seven[73] and eight[74] vertices.

Consider a polyhedron having e edges. Such a polyhedron has e distinct dsd situations, one corresponding to each of the e edges acting as the switching edge. Application of the dsd process at each of the dsd situations in a given polyhedron leads in each case to a new polyhedron. In some cases the new polyhedron is identical to the original polyhedron. In such cases the switching edge can be said to be *degenerate* and the dsd type of a degenerate edge $ab(cd)$ can be seen by application of the process $p_1 \rightarrow p_2 \rightarrow p_3$ to satisfy the following conditions:

(14) $c = a - 1$ and $d = b - 1$ or $c = b - 1$ and $d = a - 1$

A dsd process involving a degenerate switching edge represents a pathway for a degenerate polyhedral isomerization of the polyhedron. A polyhedron having one or more degenerate edges is inherently fluxional whereas a polyhedron without degenerate edges is inherently rigid.

10.3. POLYHEDRAL BORANES: 3D AROMATICITY

10.3.1. Graph Theory and Hückel Theory

A classical concept in organic chemistry is the aromaticity of 2D planar delocalized hydrocarbons exhibiting unusual chemical stability of which benzene is the prototype.[75] More recently, inorganic chemists discovered 3D *closo*-borane anions, also of unusual chemical stability, of which $B_nH_n^{2-}$ ($6 \leq n \leq 12$) are the prototypes.[76] These anions have structures based on the deltahedra depicted in Figure 4, which are the deltahedra having only degree 4 and 5 vertices except for the one degree 6 vertex in the 11-vertex edge-coalesced icosahedron, since an 11-vertex deltahedron having only degree 4 and 5 vertices is topologically impossible.[77] The unusual stability of these so-called "electron-deficient" deltahedral boranes has led to the concept of 3D aromaticity.[78] A qualitative graph-theory-derived model for chemical bonding in such delocalized systems[79] demonstrates the analogy between the aromaticity in the planar polygonal hydrocarbons and that in deltahedral boranes. Since these properties are based on neighborhood relationships such as the presence or absence of chemical bonds between pairs of atoms or the connectivity with the molecular structure, such properties may be considered to be related to the topology of the molecule. Tensor surface harmonic theory[80–83] has also been used very effectively to describe the delocalized chemical bonding in deltahedral boranes.

In the graph-theory-derived model for the chemical bonding topology of deltahedral boranes and other delocalized systems, the topology of a chemical bonding network is represented by a graph in which the vertices correspond to atoms or orbitals participating in the bonding and the edges correspond to bonding relationships. Methods discussed in detail elsewhere[79,84–87] indicate that each eigenvalue x_k of the

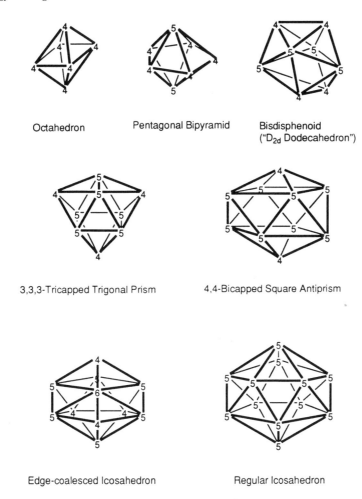

Octahedron   Pentagonal Bipyramid   Bisdisphenoid ("D$_{2d}$ Dodecahedron")

3,3,3-Tricapped Trigonal Prism   4,4-Bicapped Square Antiprism

Edge-coalesced Icosahedron   Regular Icosahedron

Figure 4. The deltahedra found in the deltahedral boranes $B_nH_n^{2-}$. The numbers in the vertices indicate their degrees.

adjacency matrix of such a graph corresponds to a Hückel molecular orbital with energy E_k given by

$$(15) \qquad E_k = \frac{\alpha + x_k\beta}{1 + x_kS}$$

In equation (15) α is the standard Coulomb integral, assumed to be the same for all atoms of a given type, β is the resonance integral, taken to be the same for all bonds, and S is the overlap integral between atomic orbitals on neighboring atoms. Positive and negative eigenvalues x_k from equation (15) thus correspond to bonding and antibonding orbitals, respectively.

The atoms at the vertices of a polygonal or polyhedral molecule can be classified as light atoms or heavy atoms. A *light atom* such as the boron and/or carbon atoms found at the vertices of planar hydrocarbons as well as deltahedral boranes and carboranes uses only its *s* and *p* orbitals for chemical bonding and therefore has only the four valence orbitals of its sp^3 manifold. A *heavy atom* such as a transition metal or posttransition element uses *s*, *p*, and *d* orbitals for chemical bonding and therefore has the nine valence orbitals of its sp^3d^5 manifold. A *normal vertex atom* in a polygonal or polyhedral molecule uses three of its k^2 valence orbitals ($k = 2$ for a light atom and $k = 3$ for a heavy atom) for the skeletal bonding within the polygon or polyhedron; these three orbitals are called *internal orbitals*.

The use of three internal orbitals for intrapolygonal or intrapolyhedral skeletal bonding by a normal vertex atom leaves one orbital in the case of a light vertex atom or six orbitals in the case of a heavy vertex atom for bonding to an atom or group external to the polygon or polyhedron. Such orbitals are called *external orbitals*. The single external orbital of a light vertex atom can bond to a single monovalent external group such as hydrogen, halogen, alkyl, aryl, alkoxy, dialkylamino, nitro, or cyano. This relates to the stoichiometries C_nH_n, $B_nH_n^{2-}$, and $C_2B_{n-2}H_n$ for the planar polygonal hydrocarbons, the cage borane dianions, and the cage carboranes, respectively.

In almost all polygonal and polyhedral molecules, each vertex atom has the electronic configuration of the next rare gas, which is neon in the case of the light vertex atoms boron and carbon. As a result of this rule, each external orbital of the vertex atom must be filled by an electron pair with the electrons coming from the vertex atom and/or an external group. This provides a method for calculating the number of electrons provided by various vertex groups to the polygonal or polyhedral skeleton; such electrons are called *skeletal electrons*. For example, a BH vertex in a polyhedral borane functions as a donor of two skeletal electrons and a CH vertex in a polygonal hydrocarbon or polyhedral carborane functions as a donor of three skeletal electrons. The external hydrogen atoms in BH and CH vertices can be replaced by other monovalent groups without affecting the number of skeletal electrons donated by the vertex in question.

The two extreme types of skeletal chemical bonding in molecules formed by polygonal or polyhedral clusters of atoms including planar aromatic hydrocarbons and polyhedral boranes (as well as various types of metal clusters discussed later in this chapter) can be called *edge-localized* and *globally delocalized*.[88–90] An edge-localized polygon or polyhedron has two-electron two-center bonds along each edge. A globally delocalized polygon or polyhedron has a multicenter bond involving all of the vertex atoms (hence the adjective "global"); such global delocalization is a feature of fully aromatic systems whether 2D such as benzene, C_6H_6, or 3D such as the deltahedral borane anions $B_nH_n^{2-}$ ($6 \leq n \leq 12$).

Consideration of the properties of vertex groups leads to the following very simple rule to determine whether polygonal or polyhedral molecules exhibit delocalized bonding or edge-localized bonding:

Table 4. Delocalized versus Localized Bonding and the "Matching Rule" (assumes three internal orbitals per vertex atom)

Structure type	Vertex degrees	Matching	Localization	Examples
Planar polygons	2	No	Delocalized	Benzene, $C_5H_5^-$, $C_7H_7^+$
"Simple polyhedra"	3	Yes	Localized	Polyhedranes: C_4H_4, C_8H_8, $C_{20}H_{20}$
Deltahedra	4, 5 (6)	No	Delocalized	Polyhedral boranes and carboranes

Delocalization occurs when there is a mismatch between the vertex degree of the polygon or polyhedron and the number of internal orbitals provided by the vertex atom.

This rule is illustrated in Table 4 for normal vertex atoms providing three internal orbitals.

The "matching rule" cited above implies that fully edge-localized bonding occurs in a polyhedral molecule in which all vertices have degree 3. Such is the case for the polyhedranes $C_{2n}H_{2n}$ such as tetrahedrane ($n = 2$), cubane ($n = 4$), and dodecahedrane ($n = 10$) in which the vertex degrees are all three which match the three available internal orbitals leading to edge-localized bonding represented by the $3n$ two-center carbon–carbon bonds of the skeleton. In the planar polygonal molecules $C_nH_n^{(n-6)+}$ ($n = 5, 6, 7$), the vertex degrees are all two and thus do not match the available three internal orbitals thereby leading to globally delocalized 2D aromatic systems. Furthermore, polyhedral molecules having all normal vertex atoms are globally delocalized if all vertices of the polyhedron have degrees 4 or larger; the simplest such polyhedron is the regular octahedron with six vertices, all of degree 4. Tetrahedral chambers in deltahedra, which lead to isolated degree 3 vertices, provide sites of localization in an otherwise delocalized molecule provided, of course, that all vertex atoms use the normal three internal orbitals.

10.3.2. Application to the Deltahedral Boranes

A major achievement of the graph-theory-derived approach to the chemical bonding topology of globally delocalized systems is the demonstration of the close analogy between the bonding in 2D planar aromatic systems such as benzene and that in 3D deltahedral boranes and carboranes. In such systems with n vertices the three internal orbitals on each vertex atom are partitioned into two twin internal orbitals (called *tangential* in some other methods[91]) and a unique internal orbital (called *radial* in some other methods[91]). Pairwise overlap between the $2n$ twin internal orbitals is responsible for the formation of the polygonal or deltahedral framework and leads to the splitting of these $2n$ orbitals into n bonding and n antibonding orbitals. The magnitude of this splitting can be designated as $2\beta_s$ where β_s refers to the parameter

β in equation (15). This portion of the chemical bonding topology can be described by a disconnected graph G_s having $2n$ vertices corresponding to the $2n$ twin internal orbitals and n isolated K_2 components; a K_2 component has only two vertices joined by a single edge. The dimensionality of this bonding of the twin internal orbitals is one less than the dimensionality of the globally delocalized system. Thus, in the case of the 2D planar polygonal systems, the pairwise overlap of the $2n$ twin internal orbitals leads to the σ-bonding network, which may be regarded as a set of 1D bonds along the perimeter of the polygon involving adjacent pairs of polygonal vertices. The n bonding and n antibonding orbitals thus correspond to the σ-bonding and σ^*-antibonding orbitals, respectively. In the case of the 3D deltahedral systems, the pairwise overlap of the $2n$ twin internal orbitals results in bonding over the 2D surface of the deltahedron, which may be regarded as topologically homeomorphic to the sphere.[92]

The equal numbers of bonding and antibonding orbitals formed by pairwise overlap of the twin internal orbitals are supplemented by additional bonding and antibonding orbitals formed by the global mutual overlap of the n unique internal orbitals. In the case of the 2D planar polygonal hydrocarbons such as benzene, the unique internal orbitals on the vertex carbon atoms are the single p_z orbitals, namely, the p orbital perpendicular to the plane of the polygon, designated as the xy plane. These p orbitals are *uninodal* orbitals, i.e., they have a single node, namely, the node in the xy plane. The 2D aromaticity in the planar polygonal hydrocarbons can also be called *uninodal orbital aromaticity* since the orbitals participating in the global mutual overlap have this single node. Similarly, in the case of the 3D deltahedral boranes of the type $B_nH_n^{2-}$ the unique internal orbitals on the vertex boron atoms are sp hybrids with the other sp hybrid on each boron atom directed toward the atomic orbital of the external hydrogen atom. The 3D aromaticity in the deltahedral boranes can also be called *anodal orbital aromaticity* since the sp hybrids functioning as unique internal orbitals have no nodes.

The bonding topology of the n unique internal orbitals, whether the uninodal p orbitals in the planar polygonal aromatic hydrocarbons or the anodal sp hybrids in the 3D deltahedral boranes, can be described by a graph G_c in which the vertices correspond to the vertex atoms of the polygon or deltahedron, or equivalently their unique internal orbitals, and the edges represent pairs of overlapping unique internal orbitals. The energy parameters of the additional molecular orbitals arising from such overlap of the unique internal orbitals are determined from the eigenvalues of the adjacency matrix $\mathbf{A_c}$ of the graph G_c using β, or more specifically β_c, as the energy unit [equation (15)]. In the case of the 2D aromatic system benzene, the graph G_c is the C_6 cyclic graph (the 1-skeleton of the hexagon) which has three positive $(+2, +1, +1)$ and three negative $(-2, -1, -1)$ eigenvalues corresponding to the three π-bonding and three π^*-antibonding orbitals, respectively. In this connection the spectra of the cyclic graphs C_n all have odd numbers of positive eigenvalues[93] leading to the familiar $4k + 2$ $(k = \text{integer})$ π-electrons[94] for planar aromatic hydrocarbons. The total benzene

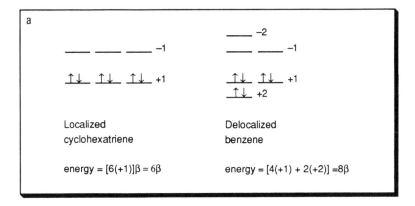

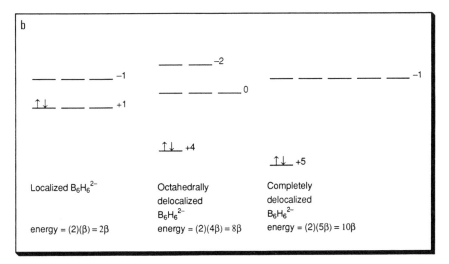

Figure 5. (a) The aromatic stabilization of benzene, C_6H_6. (b) The aromatic stabilization of $B_6H_6^{2-}$.

skeleton thus has 9 bonding orbitals (6σ and 3π) which are filled by the 18 skeletal electrons that arise when each of the CH vertices contributes 3 skeletal electrons. Twelve of these skeletal electrons are used for the σ-bonding and the remaining 6 electrons for the π-bonding.

Figure 5a illustrates how the delocalized bonding in benzene from the C_6 overlap of the unique internal orbitals, namely, the p orbitals, leads to aromatic stabilization. In a hypothetical localized "cyclohexatriene" structure in which the interactions between the p orbitals on each carbon atom are pairwise interactions, the corresponding graph G consists of three disconnected line segments (i.e., $3 \times K_2$). This graph has three

+1 eigenvalues and three −1 eigenvalues. Filling each of the corresponding three bonding orbitals with an electron pair leads to an energy of 6β from this π bonding. In a delocalized "benzene" structure in which the delocalized interactions between the p orbitals on each carbon atom are described by the cyclic C_6 graph, filling the three bonding orbitals with an electron pair each leads to an energy of 8β. This corresponds to a resonance stabilization of $8\beta - 6\beta = 2\beta$ arising from the delocalized bonding of the carbon p orbitals in benzene. We will see below how similar ideas can be used to describe the 3D aromaticity in deltahedral boranes.

An important question is the nature of the core bonding graph G_c for the deltahedral boranes $B_nH_n^{2-}$. The two limiting possibilities for G_c are the complete graph K_n and the deltahedral graph D_n and the corresponding core bonding topologies can be called the *complete* and *deltahedral* topologies, respectively. In the complete graph K_n each vertex has an edge going to every other vertex leading to a total of $n(n-1)/2$ edges. For any value of n the complete graph K_n has only one positive eigenvalue, namely, $n-1$, and $n-1$ negative eigenvalues, namely, -1 each. The deltahedral graph D_n is identical to the 1-skeleton of the deltahedron depicted in Figure 4 for the corresponding value of n ($6 \leq n \leq 12$). Thus, two vertices of D_n are connected by an edge if, and only if, the corresponding vertices of the deltahedron are connected by an edge.

The graphs D_n for the deltahedra of interest with six or more vertices (Figure 4) all have at least four zero or positive eigenvalues. However, in all cases there is a unique positive eigenvalue that is much more positive than any other of the positive eigenvalues. This unique positive eigenvalue can be called conveniently the *principal eigenvalue* and corresponds to the fully symmetric $A_{(1)(g)}$ irreducible representation of the symmetry group of G_c. The molecular orbital corresponding to the principal eigenvalue of G_c may be called the *principal* core orbital. Since deltahedral boranes of the stoichiometry $B_nH_n^{2-}$ have $2n + 2$ skeletal electrons of which $2n$ are used for the surface bonding, as noted above, there are only two skeletal electrons available for core bonding corresponding to a single core bonding molecular orbital and a single positive eigenvalue for G_c. Thus, deltahedral boranes are 3D aromatic systems having $4k + 2 = 2$ core bonding electrons where $k = 0$ analogous to the $4k + 2$ π electrons where $k = 0$ ($C_3H_3^+$), $k = 1$ ($C_5H_5^-$, C_6H_6, $C_7H_7^+$), or $k = 2$ ($C_8H_8^{2-}$) for planar 2D polygonal aromatic systems. Furthermore, only if G_c is taken to be the corresponding complete graph K_n will the simple model given above for globally delocalized deltahedra give the correct number of skeletal electrons in all cases, namely, $2n + 2$ skeletal electrons for $6 \leq n \leq 12$. Such a model with complete core bonding topology can be used as a working basis for the chemical bonding topology of deltahedral boranes as well as related metal clusters. However, deltahedral core bonding topology can also account for the observed $2n + 2$ skeletal electrons in the $B_nH_n^{2-}$ deltahedral boranes if there is a mechanism for raising the energies of all of the core molecular orbitals other than the principal core orbital to antibonding energy levels.

Figure 5b shows how the delocalized bonding in $B_6H_6^{2-}$ arising from overlap of the unique internal orbitals, namely, the radial sp hybrids on each boron atom, can lead to aromatic stabilization. In a hypothetical localized structure in which the interactions between the radial sp hybrids are pairwise interactions, the spectrum of the corresponding graph G_c is three disconnected line segments (i.e., $3 \times K_2$). The spectrum of this disconnected graph has three $+1$ eigenvalues and three -1 eigenvalues. Filling one of the bonding orbitals with the available two core bonding electrons leads to an energy of 2β from the core bonding. In a completely delocalized structure in which the core bonding is described by the complete graph K_6, this electron pair is in a bonding orbital with an eigenvalue of $+5$ corresponding to an energy of $(2)(5\beta) = 10\beta$ (Figure 5b). The aromatic stabilization of completely delocalized $B_6H_6^{2-}$ is thus $10\beta - 2\beta = 8\beta$ assuming the same β unit for both the localized and complete delocalized structures. In an octahedrally delocalized $B_6H_6^{2-}$ in which the core bonding is described by the deltahedral graph D_6 corresponding to the 1-skeleton of the octahedron, the core bonding electron pair is in a bonding orbital with an eigenvalue of $+4$ corresponding to an energy of $(2)(4\beta) = 8\beta$ (Figure 5b). The aromatic stabilization of octahedrally delocalized $B_6H_6^{2-}$ is thus $8\beta - 2\beta = 6\beta$. Thus, the aromatic stabilization of $B_6H_6^{2-}$ is considerable regardless of whether the delocalized core bonding is considered to have the complete topology represented by the complete graph K_6 or the octahedral topology represented by the deltahedral graph D_6.

There are several implications of the bonding model for delocalized deltahedral structures with n vertices using complete core bonding topology described by the corresponding K_n complete graph:

1. The overlap of the n unique internal orbitals to form an n-center core bond may be hard to visualize since its topology corresponds to that of the complete graph K_n which for $n \geq 5$ is nonplanar by Kuratowski's theorem[95] and thus cannot correspond to the 1-skeleton of a polyhedron realizable in 3D space. However, the overlap of these unique internal orbitals does not occur along the edges of the deltahedron or those of any other 3D polyhedron. For this reason, the topology of the overlap of the unique internal orbitals in the core bonding of a deltahedral cluster need not correspond to a graph representing a 1-skeleton of a 3D polyhedron. The only implication of the K_n graph description of the bonding topology of the unique internal orbitals is that the deltahedron is topologically homeomorphic to the sphere.

2. The equality of the interactions between all possible pairs of unique internal orbitals required by the K_n model for the core bonding is obviously a very crude assumption since in any deltahedron with five or more vertices all pairwise relationships of the vertices are not equivalent. The example of the nonequivalence of the cis and $trans$ vertex pairs in an octahedral structure such as $B_6H_6^{2-}$ has already been discussed. However, the single eigenvalue of the K_n graph is so strongly positive that severe inequalities in the different vertex

pair relationships are required before the spectrum of the graph representing precisely the unique internal overlap contains more than one positive eigenvalue.

3. Vertices of degree 4 or greater appear to be essential for the stability of deltahedral boranes of the type $B_nH_n^{2-}$. Thus, although the borane anions $B_nH_n^{2-}$ ($6 \leq n \leq 12$) are very stable, the five-boron deltahedral borane $B_5H_5^{2-}$ based on a trigonal pyramidal structure with two (apical) degree 3 vertices has never been prepared. Such degree 3 vertices lead to two-electron two-center bonds along each of the three edges meeting at the degree 3 vertex and leave no internal orbitals from degree 3 vertices for the multicenter core bond. However, the dicarbaborane $C_2B_3H_5$ isoelectronic with the carbon atoms in the degree 3 apical vertices of the trigonal bipyramid can be isolated.[96] The carbon–boron bonds along the 6 B–C edges of the C_2B_3 trigonal bipyramid in the isolable species 1,5-$C_2B_3H_5$ can be interpreted as edge-localized bonds leading to three-coordinate boron atoms similar to the B–C bonds and boron environment in trimethylboron $(CH_3)_3B$.

10.3.3. Fluxionality in Deltahedral Boranes

The microscopic approach to the study of polyhedral isomerizations can be used to study the isomerizations in the deltahedra found in boranes (Figure 4) thereby providing some insights regarding their fluxionality. Using the methods outlined in Section 10.2.4, the borane deltahedra depicted in Figure 4 can be very easily checked for the presence of one or more degenerate edges with the following results:

1. *Tetrahedron.* No dsd process of any kind is possible since the tetrahedron is the complete graph K_4. A tetrahedron is therefore inherently rigid.

2. *Trigonal bipyramid.* The three edges connecting pairs of equatorial vertices are degenerate edges of the type 44(33). A dsd process using one of these degenerate edges as the switching edge and involving a square pyramid intermediate corresponds to the Berry pseudorotation discussed above.

3. *Octahedron.* The highly symmetrical octahedron has no degenerate edges and is therefore inherently rigid.

4. *Pentagonal bipyramid.* The pentagonal bipyramid has no degenerate edges and thus by definition is inherently rigid. However, a dsd process using a 45(44) edge of the pentagonal bipyramid (namely, an edge connecting an equatorial vertex with an axial vertex) gives a capped octahedron. The capped octahedron is a low-energy polyhedron for ML_7 coordination complexes but a forbidden polyhedron for boranes and carboranes because of its tetrahedral chamber.[97]

5. *Bisdisphenoid.* The eight-vertex bisdisphenoid has four pairwise degenerate edges, which are those of the type 55(44) located in the subtetrahedron consisting of the degree 5 vertices of the bisdisphenoid (Figure 4). Thus, two

successive or more likely concerted (parallel) dsd process involving opposite 55(44) edges (i.e., a pair related by a C_2 symmetry operation) converts one bisdisphenoid into another bisdisphenoid through a square antiprismatic intermediate. Thus, a bisdisphenoid, like the trigonal bipyramid discussed above, is inherently fluxional.

6. *4,4,4-Tricapped trigonal prism.* The three edges of the type 55(44) corresponding to the "vertical" edges of the trigonal prism are degenerate. A dsd process using one of these degenerate edges as the switching edge involves a C_{4v} 4-capped square antiprism intermediate. Nine-vertex systems are therefore inherently fluxional.

7. *4,4-Bicapped square antiprism.* This polyhedron has no degenerate edges and therefore is inherently rigid.

8. *Edge-coalesced icosahedron.* The four edges of the type 56(45) are degenerate. This 11-vertex deltahedron is therefore inherently fluxional.

9. *Icosahedron.* This highly symmetrical polyhedron, like the octahedron, has no degenerate edges and is therefore inherently rigid.

This simple analysis indicates that in deltahedral structures the 4-, 6-, 10-, and 12-vertex structures are inherently rigid; the 5-, 8-, 9-, and 11-vertex structures are inherently fluxional; and the rigidity of the 7-vertex structure depends on the energy difference between the two most symmetrical 7-vertex deltahedra, namely, the pentagonal bipyramid and the capped octahedron. This can be compared with experimental fluxionality observations by boron-11 nuclear magnetic resonance on the deltahedral borane anions $B_nH_n^{2-}$ $(6 \leq n \leq 12)$[98] where the 6-, 7-, 9-, 10-, and 12-vertex structures are found to be rigid and the 8- and 11-vertex structures are found to be fluxional. The only discrepancy between experiment and these very simple topological criteria for fluxionality arises in the 9-vertex structure $B_9H_9^{2-}$.

The discrepancy between the predictions of this simple topological approach and experimental data for $B_9H_9^{2-}$ has led to the search for more detailed criteria for the rigidity of the deltahedral boranes. In this connection Gimarc and Ott have studied orbital symmetry methods particularly for the five-,[99] seven-,[100] and nine[101]-vertex borane and carborane structures, recognizing that a topologically feasible dsd process is orbitally forbidden if crossing of occupied and vacant molecular orbitals (i.e., a "HOMO–LUMO crossing") occurs during the dsd process as illustrated in Figure 6 for the single dsd process of the trigonal bipyramid. For such an orbitally forbidden process, which occurs in the five- and nine-vertex deltahedral boranes and carboranes, the activation barrier separating initial and final structures is likely to be large enough to prevent this polyhedral isomerization. Some selection rules have been proposed for distinguishing between symmetry-allowed and-forbidden processes in deltahedral boranes, carboranes, and related structures. Thus, Wales and Stone[102] distinguish between symmetry-allowed and-forbidden processes by observing that a HOMO–LUMO crossing occurs if the proposed transition state has a single atom lying on a

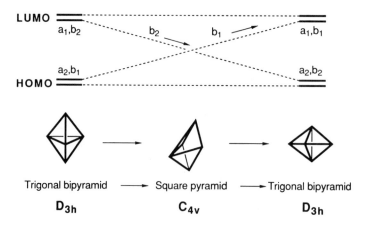

Figure 6. The HOMO–LUMO crossing occurring during the dsd rearrangement (Berry pseudorotation) of a trigonal bipyramid through a square pyramid intermediate.

principal C_n rotational axis where $n \geq 3$. A more detailed selection rule was observed by Mingos and Johnston.[103] If the four outer edges of the two fused triangular faces (i.e., the "diamond") are symmetry equivalent, then a single dsd process results in a *pseudorotation* of the initial polyhedron by 90° as follows:

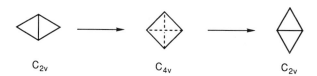

However, if the edges are not symmetry equivalent, then the rearrangement results in a *pseudoreflection* of the initial polyhedron which can be indicated as follows:

Pseudorotations are symmetry forbidden and have larger activation energies than pseudoreflections, which are symmetry allowed.

10.3.4. Electron-Rich (Hyperelectronic) Polyhedral Boranes

Electron-rich or hyperelectronic polyhedral systems are those containing more than the $2n + 2$ skeletal electrons required for globally delocalized n-vertex deltahedra without vertices of degree 3. In the case of boron hydride derivatives[104,105] there are well-known families of *nido* compounds having $2n + 4$ skeletal electrons and *arachno* compounds having $2n + 6$ skeletal electrons. In the *nido* polyhedra all but one of the faces are triangular; the unique nontriangular face may be regarded as a hole. Analogously the *arachno* polyhedra have either two nontriangular faces or one large nontriangular face (i.e., two holes or one large bent hole). Thus, successive additions of electron pairs to a closed $2n + 2$ skeletal electron deltahedron result in successive punctures of the deltahedral surface to give holes (faces) having more than three edges by a process conveniently called *polyhedral puncture*. The open polyhedral networks can also be considered to arise by excision of one or more vertices along with all of the edges leading to them from a closed deltahedron having $m > n$ vertices by a process conveniently called *polyhedral excision*. Figure 7 shows the *nido* and *arachno* polyhedra (and polyhedral fragments) derived from the octahedron, pentagonal bipyramid, and the regular icosahedron.

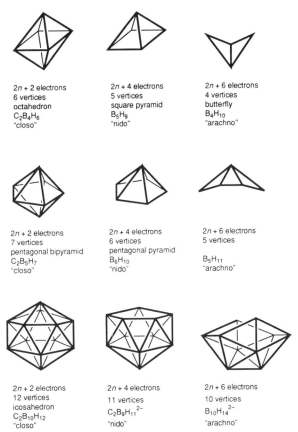

$2n + 2$ electrons	$2n + 4$ electrons	$2n + 6$ electrons
6 vertices	5 vertices	4 vertices
octahedron	square pyramid	butterfly
$C_2B_4H_6$	B_5H_9	B_4H_{10}
"closo"	"nido"	"arachno"

$2n + 2$ electrons	$2n + 4$ electrons	$2n + 6$ electrons
7 vertices	6 vertices	5 vertices
pentagonal bipyramid	pentagonal pyramid	
$C_2B_5H_7$	B_6H_{10}	B_5H_{11}
"closo"	"nido"	"arachno"

$2n + 2$ electrons	$2n + 4$ electrons	$2n + 6$ electrons
12 vertices	11 vertices	10 vertices
icosahedron		
$C_2B_{10}H_{12}$	$C_2B_9H_{11}{}^{2-}$	$B_{10}H_{14}{}^{2-}$
"closo"	"nido"	"arachno"

Figure 7. Examples of electron-rich (hyperelectronic) polyhedra with $2n + 4$ (*nido*) and $2n + 6$ (*arachno*) skeletal electrons obtained by removal of vertices from closed deltahedra.

Treatment of the skeletal bonding topology in electron-rich polyhedra, even the *nido* polyhedra with only one nontriangular face and $2n + 4$ skeletal electrons, is considerably more complicated than the treatment of the skeletal bonding topology discussed above. The vertex atoms in electron-rich polyhedra may be divided into the following two sets: border vertex atoms which are vertices of the one face containing more than three edges (i.e., they are at the border of the single hole) and interior vertex atoms which form vertices of only triangular faces. For example, in a square pyramid (Figure 7), which is the simplest example of a *nido* polyhedron, the four basal vertices are the border vertices since they all border the square "hole," i.e., the base of the square pyramid. However, the single apical vertex of the square pyramid is an interior vertex since it is a vertex where only triangular faces meet. The external and twin internal orbitals of the border vertex atoms are taken to be sp^2 hybrids. The unique internal orbitals of the border vertex atoms will thus be p orbitals. The external and unique internal orbitals of the interior vertex atoms are taken to be sp hybrids in accord with the treatment of closed deltahedra discussed in Section 10.3.2. The twin internal orbitals of the interior vertex atoms must therefore be p orbitals. Note that in the *nido* polyhedra the hybridization of the border vertex atoms is the same as that of the vertex atoms of polygonal systems whereas the hybridization of the interior vertex atoms is the same as that of the vertex atoms of deltahedral systems. A chemical consequence of the similar vertex atom hybridizations in polygons and the borders of *nido* polyhedra is the ability of both planar polygonal hydrocarbons (e.g., cyclopentadienyl and benzene) and the border atoms of *nido* carboranes[106] to form chemical bonds with transition metals of similar types involving interaction of the transition metal with *all* of the atoms of the planar polygon or the border atoms of the polygonal hole of the *nido* polyhedron.

Nido polyhedra can be classified into two fundamental types: the pyramids with only one interior vertex (the apex) and the nonpyramids with more than one interior vertex. If n is the total number of vertices and v is the number of interior vertices in a *nonpyramidal nido* polyhedron, the interactions between the internal orbitals which generate bonding orbitals are of the following three different types:

1. The $2(n - v)$ twin internal orbitals of the border atoms and the $2v$ twin internal orbitals of the interior atoms interact along the polyhedral surface to form n bonding orbitals and n antibonding orbitals.

2. The v unique internal orbitals of the interior vertex atoms all interact with each other at the core of the structure in a way that may be represented by the complete graph K_v to give a single bonding orbital and $v - 1$ antibonding orbitals.

3. The $n - v$ unique internal orbitals of the border atoms interact with each other across the surface of the hole in a way that may be represented by the complete graph K_{n-v} to give a single bonding orbital and $n - v - 1$ antibonding orbitals.

The above interactions in *nido* systems of the first two types correspond to the interactions found in the closed deltahedral systems discussed in the previous section whereas the interactions of the third type can only occur in polyhedra containing at least one hole such as the nonpyramidal *nido* systems. Furthermore, the values of v and $n - v$ in the second and third types of interactions are immaterial as long as they both are greater than one, since any complete graph K_i ($i > 1$) has exactly one positive eigenvalue, namely, $i - 1$. The total number of skeletal bonding orbitals in nonpyramidal *nido* systems with n vertices generated by interactions of the three types listed above are n, 1, and 1, respectively, leading to a total of $n + 2$ bonding orbitals holding $2n + 4$ skeletal electrons in accord with experimental observations.

Pyramidal *nido* polyhedra having only one interior vertex require a somewhat different treatment because the eigenvalue of the one-vertex no-edge complete graph K_1 is zero leading to ambiguous results for the second type of interaction listed above. This difficulty can be circumvented by realizing that the only types of pyramids relevant to delocalized borane and metal cluster chemistry are square pyramids, pentagonal pyramids, and hexagonal pyramids and bonding schemes for these types of pyramids can be constructed which are completely analogous to well-known[107] transition metal complexes of cyclobutadiene, cyclopentadienyl, and benzene, respectively. In applying this analogy the interior vertex atom plays the role of the transition metal and the planar polygon of the border vertex atoms plays the role of the planar polygonal ring in the metal complexes. Furthermore, the $n - 1$ unique internal orbitals of the border vertex atoms interact cyclically leading to three "submolecular" orbitals which may be used for bonding to the single interior vertex atom as represented by the three nonnegative eigenvalues of the corresponding C_{n-1} cyclic graph ($n = 5, 6, 7$). Of these three polygonal orbitals, one orbital, the A_1 orbital, has no nodes perpendicular to the polygonal plane whereas the other two remaining orbitals, the degenerate E orbitals, each have one node perpendicular to the polygonal plane with the two nodes from the pair of degenerate E orbitals being mutually perpendicular.

The following three interactions are used to generate the skeletal bonding orbitals in *nido* pyramids:

1. The $2(n - 1)$ twin internal orbitals of the border atoms interact along the edges of the base of the pyramid to form $n - 1$ bonding orbitals and $n - 1$ antibonding orbitals analogous to the σ-bonding and σ^*-antibonding orbitals, respectively, of planar polygonal hydrocarbons.

2. The unique internal orbital of the single interior vertex atom (the apex of the pyramid) interacts with the A_1 orbital to give one bonding orbital and one antibonding orbital.

3. The twin internal orbitals of the apex of the pyramid interact with the two orthogonal E orbitals in two separate pairwise interactions to give two bonding and two antibonding orbitals.

The total number of skeletal bonding orbitals in pyramidal *nido* systems generated by these three interactions are $n - 1$, 1, and 2, respectively, leading to a total of $n + 2$ bonding orbitals holding $2n + 4$ skeletal electrons. Thus, the graph-theoretical treatment of nonpyramidal and pyramidal *nido* polyhedra with n vertices leads to the prediction of the same numbers of skeletal bonding orbitals, namely, $n + 2$, in accord with experimental observations. However, the partitionings of these bonding orbitals are different for the two types of *nido* systems, namely, $(n, 1, 1)$ for the nonpyramidal systems and $(n - 1, 1, 2)$ for the pyramidal systems.

The process of polyhedral puncture, which forms *nido* polyhedra with one hole and $2n + 4$ skeletal electrons from closed deltahedra with $2n + 2$ skeletal electrons, can be continued further to give polyhedral fragments containing two or more holes or one larger hole. In boron chemistry polyhedra or polyhedral fragments having $2n + 6$ and $2n + 8$ skeletal electrons are called *arachno* and *hypho* structures, respectively. Some examples of *arachno* structures are included in Figure 7. Formation of a new hole by such polyhedral puncture splits the complete graph formed by interactions at the polyhedral core between the unique internal orbitals of the interior vertex atoms into two smaller complete graphs. One of these new complete graphs involves interaction at the polyhedral core between the unique internal orbitals of the vertex atoms which are still interior atoms after creation of the new hole or expansion of the existing hole. The second new complete graph involves interaction above the newly created hole between the unique internal orbitals of the vertex atoms which have become border atoms of the newly created hole. Since each new complete graph contributes exactly one new skeletal bonding orbital to the polyhedral system, each application of polyhedral puncture to give a stable system requires addition of two skeletal electrons.

10.3.5. Tensor Surface Harmonic Theory

The graph-theory-derived model for the skeletal bonding of a globally delocalized deltahedral borane with n vertices with complete core bonding topology outlined in Section 10.3.2 uses the corresponding complete graph K_n to describe the topology of the multicenter core bond. The precise topology of the cluster deltahedron does not enter directly into such models but only the absence of degree 3 vertices. In other words, graph-theory-derived models of the skeletal bonding of globally delocalized deltahedral clusters consider such deltahedra to be topologically homeomorphic to the sphere.

The topological homeomorphism of a deltahedron to a sphere used in the graph-theory-derived models is also the basis of the tensor surface harmonic theory developed by Stone.[108-111] The tensor surface harmonic theory defines the vertices of a deltahedral borane as lying on the surface of a single sphere with the atom positions described by the standard angular coordinates θ and ϕ related to latitude and longitude. The second-order differential equations for the angular dependence of the molecular orbitals from the core bonding become identical to the equations for the angular dependence of the atomic orbitals obtained by solution of the Schrödinger equation, with both sets of equations making use of spherical harmonics $Y_{LM}(\theta,\phi)$.

In tensor surface harmonic theory as applied to deltahedral boranes the internal orbitals of the vertex atoms are classified by the number of nodes with respect to the radial vector connecting the vertex atom with the center of the deltahedron.[112] The unique internal orbitals (namely, sp hybrids) are nodeless (i.e., σ-type) and lead to core bonding molecular orbitals described by the *scalar* spherical harmonics $\Theta(\theta)\cdot\Phi(\phi) = Y_{LM}(\theta,\phi)$, which for deltahedra having n vertices correspond successively to a single nodeless S^σ orbital (Y_{00}), the three uninodal P^σ orbitals (Y_{10}, Y_{11c}, Y_{11s}), the five binodal D^σ orbitals ($Y_{20}, Y_{21c}, Y_{21s}, Y_{22c}, Y_{22s}$), the seven trinodal F^σ orbitals ($Y_{30}, Y_{31c}, Y_{31s}, Y_{32c}, Y_{32s}, Y_{33c}, Y_{33s}$), etc., of increasing energy (compare Section 10.2.2). The S^σ, P^σ, D^σ, F^σ orbitals, etc., correspond to the molecular orbitals arising from the n-center core bond of the deltahedron.

The twin internal orbitals are uninodal (i.e., π-type) and lead to surface bonding described by the *vector* surface harmonics. Two vector surface harmonic functions can be generated from each Y_{LM} as follows:

(16a)
$$V_{LM} = \nabla\, Y_{LM}$$

(16b)
$$\overline{V}_{LM} = r \times \nabla\, Y_{LM}$$

In equation (16) ∇ is the vector operator

(17)
$$\nabla = \left(\frac{\partial}{\partial\theta}\,,\, \frac{1}{\sin\theta}\,\frac{\partial}{\partial\phi} \right)$$

$\times$ is the vector cross-product, and the $\overline{V}_{LM}$ of equation (16b) is the parity inverse of the V_{LM} of equation (16a), corresponding to a rotation of each atomic π-function by 90° about the radial vector **r**. The V_{LM} and $\overline{V}_{LM}$ correspond to the equal numbers of bonding and antibonding surface orbitals in a globally delocalized deltahedral cluster leading to three P^π, five D^π, seven F^π, etc., bonding/antibonding orbital pairs of increasing energy and nodality. Since Y_{00} is a constant, $\nabla Y_{00} = 0$ so that there are no S^π or $\overline{S}^\pi$ orbitals.

The core and surface orbitals defined above by tensor surface harmonic theory can be related to the following aspects of the graph-theory-derived model for the skeletal bonding in boranes with the deltahedral structures depicted in Figure 4:

1. The lowest-energy fully symmetric core orbital (A_{1g}, A_g, A_1, or A_1' depending on the point group of the deltahedron) corresponds to the S^σ orbital in tensor surface harmonic theory. Since there are no S^π or $\overline{S}^\pi$ surface orbitals, this lowest-energy core orbital cannot mix with any surface orbitals, so that it cannot become antibonding through core–surface mixing.

2. The three core orbitals of next lowest energy correspond to P^σ orbitals in tensor surface harmonic theory. These orbitals can mix with the P^π surface orbitals so that the P^σ core orbitals become antibonding with corresponding lowering of the bonding energies of the P^π surface orbitals below the energies of the other surface orbitals. This is why graph-theory-derived models of skeletal bonding in globally delocalized n-vertex deltahedra, which use the K_n graph to describe the multicenter core bond, give the correct numbers of skeletal bonding orbitals even for deltahedra whose corresponding deltahedral graph D_n has more than one positive eigenvalue. In this way tensor surface harmonic theory can be used to justify important assumptions in the graph-theory-derived models.

10.3.6. Kekulé-Type Structures in Polyhedral Boranes

A central idea in the aromaticity in planar benzenoid hydrocarbons, which are constructed by edge-sharing fusion of hexagons of sp^2 carbon atoms, is the contribution of two or more different structures of equivalent energy consisting of alternating carbon–carbon single and double bonds known as *Kekulé structures* to a lower-energy averaged structure known as a *resonance hybrid*. Thus, benzene itself has the following two equivalent Kekulé structures:

This concept of Kekulé structures can be extended to the 3D deltahedral borane anions $B_nH_n^{2-}$ ($6 \leq n \leq 12$).[113] Such Kekulé structures make use of three-center B–B–B bonds instead of the carbon–carbon double bonds in benzenoid Kekulé structures. Lipscomb's semitopological method[114–117] for studying the electron and orbital balance in boron networks containing mixtures of B–B two-center and B–B–B three-center bonds is essential for extending the concept of Kekulé structures from 2D benzenoid hydrocarbons to 3D deltahedral boranes.

Lipscomb's semitopological methods make the following assumptions:

1. Only the $1s$ orbital of hydrogen and the four sp^3 orbitals of boron are used.
2. Each external B–H bond is regarded as a typical two-center two-electron single bond requiring the hydrogen orbital, one hybridized boron orbital, and one electron each from the hydrogen and boron atoms. Because of the very small electronegativity difference between hydrogen and boron, these bonds are assumed to be nonpolar. In polynuclear boron hydrides, every boron atom may form zero or one but never more than two such external bonds.
3. Each B–H–B three-center "bridge" bond corresponds to a filled three-center localized bonding orbital requiring the hydrogen orbital and one hybrid orbital from each boron atom.

4. The orbitals and electrons of any particular boron atom are allocated to satisfy first the requirements of the external B–H single bonds and the bridge B–H–B bonds. The remaining orbitals and electrons are allocated to framework molecular orbitals.

The relative numbers of orbitals, electrons, hydrogen, and boron atoms as well as bonds of various types can be expressed in a systematic way. For a boron hydride B_pH_{p+q} containing s *bridging* hydrogen atoms, x *extra* B–H bonds in terminal BH_2 groups rather than BH groups, t three-center B–B–B bonds, and y two-center B–B bonds, balancing the hydrogen atoms leads to $s + x = q$ assuming that each boron atom is bonded to at least one hydrogen atom. Since each boron atom supplies four orbitals but only three electrons, the total number of three-center bonds in the molecule is the same as the number of boron atoms, namely, $s + t = p$. This leads to the following equations of balance:

(18a) $2s + 3t + 2y + x = 3p$ (orbital balance with 3 orbitals/BH vertex)

(18b) $s + 2t + 2y + x = 2p$ (electron balance with 2 *skeletal* electrons/BH vertex)

The two-center B–B bonds and three-center B–B–B bonds in polyhedral boranes can be components of Kekulé-type structures similar to the C–C single and C=C double bonds in planar hydrocarbons. First consider boranes of the stoichiometry $B_nH_n^{2-}$ ($6 \leq n \leq 12$) with one terminal hydrogen on each boron atom based on the deltahedra depicted in Figure 4. Such deltahedral boranes cannot have any terminal BH_2 groups or three-center B–H–B bonds and have two "extra" electrons for the -2 charge on the ion so that $s = x = 0$ in the equations of balance [equation (18)], which then reduce to the following equations in which n is the number of boron atoms in the deltahedron corresponding to p in equation (18):

(19a) $3t + 2y = 3n$ (orbital balance for $B_nH_n^{2-}$)

(19b) $2t + 2y = 2n + 2$ (electron balance for $B_nH_n^{2-}$)

Solving the simultaneous equation (19) leads to $y = 3$ and $t = n - 2$ implying the presence of three two-center B–B bonds and $n - 2$ three-center B–B–B bonds. Since a deltahedron with n vertices has $2n - 4$ faces, the $n - 2$ three-center B–B–B bonds will cover exactly half of the faces. In that sense a Kekulé-type structure for the deltahedral boranes $B_nH_n^{2-}$ has exactly half of the faces covered by three-center B–B–B bonds just like a Kekulé structure for a benzenoid hydrocarbon has half of the edges covered by C=C double bonds.

O'Neill and Wade[118] have discussed such localized bonding schemes using two-center B–B bonds and three-center B–B–B bonds with the following basic assumptions:

1. Each skeletal atom is assumed to participate in three skeletal bonds in addition to the external bond, typically to a hydrogen atom.
2. Each edge of the skeletal B_n polyhedron must correspond to a two-center B–B bond or a three-center B–B–B bond.
3. A pair of boron atoms cannot be simultaneously bonded to each other by both a two-center B–B bond and one or two three-center B–B–B bonds since these arrangements would require too close an alignment of the atomic orbitals involved.
4. Cross-polyhedral interactions, which are significantly longer than polyhedral edge interactions, are considered to be nonbonding.
5. When individual bond networks do not match the symmetry of the polyhedron in question, resonance between plausible canonical forms needs to be invoked.

These assumptions, particularly assumption 3, pose certain restrictions on the combinations of two-center B–B and three-center B–B–B bonds meeting at polyhedral vertices of various degrees (Figure 8)[105]:

1. Degree 3 vertices: Only three two-center B–B bonds along the polyhedral edges corresponding to edge-localized bonding or three three-center B–B–B bonds in the polyhedral faces are possible.

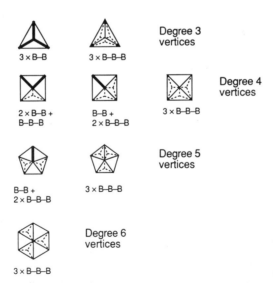

Figure 8. The bonding networks by which boron atoms at vertices of degrees 3 to 6 can bond to their skeletal neighbors. Two-center B–B bonds are indicated by bold edges whereas three-center B–B–B bonds are indicated by dashed lines meeting in the center of a face.

2. Degree 4 vertices: At least one three-center bond must meet at each degree 4 vertex since there are not enough internal orbitals to form exclusively two-center B–B bonds along each of the four edges of a degree 4 vertex.
3. Degree 5 vertices: A minimum of two three-center bonds must meet at each degree 5 vertex.
4. Degree 6 vertices: All three internal bonds at each degree 6 vertex must be three-center B–B–B bonds.

Now let us consider similar localized structures for *nido* boranes, which have $2n + 4$ skeletal electrons and are formally derived from the hypothetical $B_nH_n^{4-}$ tetraanions by protonation of one or more B–B single bonds. The structures of the *nido* boranes may also be derived from a $B_{n+1}H_{n+1}^{2-}$ deltahedron by removal of a single vertex.[91,119–122] For example, the *nido* boron hydrides B_5H_9 and B_6H_{10} are formally derived from pyramidal $B_5H_5^{4-}$ and $B_6H_6^{4-}$ by tetraprotonation. Generalization of equation (18) to $B_nH_n^{4-}$ with $s = x = 0$ leads to the equations

(20a)
$$3t + 2y = 3n \quad \text{(orbital balance for } B_nH_n^{4-}\text{)}$$

(20b)
$$2t + 2y = 2n + 4 \quad \text{(electron balance for } B_nH_n^{4-}\text{)}$$

Solving the simultaneous equation (20) leads to $y = 6$ and $t = n - 4$. Note that the square pyramid in the hypothetical $B_5H_5^{4-}$ and the known B_5H_9 is the smallest possible *nido* polyhedron as well as the smallest *nido* structure with a B–B–B three-center bond. Thus, $t = 1$ and $y = 6$ for the hypothetical $B_5H_5^{4-}$, which on tetraprotonation gives pentaborane-9, B_5H_9, with $t = 1$, $y = 2$, and $s = 4$, with the fewest boron atoms for a stable neutral borane without an insulating BH_2 vertex. Note also that for a neutral *nido* borane B_nH_{n+4} such as B_5H_9, B_6H_{10}, B_8H_{12}, and $B_{10}H_{14}$, $y = 2$ and $s = 4$ for all n assuming, of course, the absence of BH_2 vertices.

Addition of two more skeletal electrons to the *nido* structures gives the *arachno* structures formally derived from $B_nH_n^{6-}$ with structures based on a $B_{n+2}H_{n+2}^{2-}$ deltahedron by removal of a pair of vertices. Generalization of equation (18) to $B_nH_n^{6-}$ with $s = x = 0$ leads to the equations

(21a)
$$3t + 2y = 3n \quad \text{(orbital balance for } B_nH_n^{6-}\text{)}$$

(21b)
$$2t + 2y = 2n + 6 \quad \text{(electron balance for } B_nH_n^{6-}\text{)}$$

Solving the simultaneous equation (21) leads to $y = 9$ and $t = n - 6$. Note that an *arachno* structure *without BH_2 vertices* must contain at least seven boron atoms to have a B–B–B three-center bond. Hexaprotonation of hypothetical $B_nH_n^{6-}$ gives neutral *arachno* boron hydrides B_nH_{n+6}. However, the common *arachno* neutral boron hydrides including B_4H_{10}, B_5H_{11}, B_6H_{12}, B_8H_{14}, and B_9H_{15} but not *i*-B_9H_{15} have BH_2

Table 5. Feasibility of Deltahedral B_nH_n Species with 0,–2, and –4 Charges

| Deltahedron | Formula | Degeneracies | | Existence of Kekulé structure for $B_nH_n^z$ | | |
		HOMO	LUMO	$z = 0$	$z = -2$	$z = -4$
Octahedron	B_6H_6	3	3	0	+	0
Pentagonal bipyramid	B_7H_7	2	2	–	+	–
Bisdisphenoid	B_8H_8	1	1	+	+	+
Tricapped trigonal prism	B_9H_9	1	1	+	+	+
Bicapped square antiprism	$B_{10}H_{10}$	2	2	–	+	–
Edge-coalesced icosahedron	$B_{11}H_{11}$	1	1	+	+	+
Icosahedron	$B_{12}H_{12}$	4	4	0	+	0

vertices[123] whose boron atoms cannot participate fully in the delocalization similar to the CH_2 groups in hydrocarbons such as cyclohexane with only two internal orbitals.

O'Neill and Wade[118] consider the feasibility of *deltahedral* structures isoelectronic and isolobal with B_nH_n which are either neutral such as the B_nX_n halides,[124] have a –2 charge such as the stable deltahedral borane anions $B_nH_n^{2-}$, or have a –4 charge such as the 8-vertex species $(C_5H_5)_4Ni_4B_4H_4$[125] using the following criteria:

1. The feasibility of drawing a satisfactory Kekulé-type structure using two-center B–B and three-center B–B–B bonds.
2. The degeneracies of the highest occupied and lowest unoccupied molecular orbitals (HOMOs and LUMOs, respectively).

The latter criterion relates to the closed-shell configuration for the dinegative anions $B_nH_n^{2-}$ and thus the requirements of a nondegenerate HOMO for neutral B_nH_n also to have a closed-shell configuration and a nondegenerate LUMO for tetranegative $B_nH_n^{4-}$ also to have a closed-shell configuration. The conclusions from this study are summarized in Table 5. From these observations the deltahedral species B_nH_n ($n = 8$, 9, and 11) are seen to be potentially stable with 0, –2, and –4 charges whereas the deltahedral species B_nH_n ($n = 6, 7, 10,$ and 12) are seen to be stable only with a –2 charge. This is in approximate accord with the stability of the neutral halide species B_nX_n.[124]

10.4. METAL CARBONYL CLUSTERS

10.4.1. Metal Carbonyl Vertices and Electron Counting in Metal Carbonyl Clusters

Metal clusters[126] are compounds containing a metal skeleton in which three or more metal atoms are joined by metal–metal bonds. They may be characterized by the geometry of their metal skeletons. Some examples of neutral binary metal carbonyl cluster skeletons are given in Figure 9. The transition metal vertices in metal carbonyl

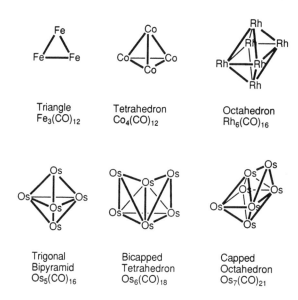

Figure 9. Some metal skeletons of neutral binary metal carbonyl clusters.

clusters use d orbitals as well as s and p orbitals leading to a nine-orbital sp^3d^5 manifold. In globally delocalized transition metal carbonyl clusters, three orbitals from each transition metal vertex are used for internal skeletal bonding just like in the polygonal hydrocarbons and the deltahedral boranes leaving $9 - 3 = 6$ valence orbitals on each transition metal vertex atom for bonding to one or generally more external ligands or groups with the following being the most common in transition metal cluster vertices:

1. A single external orbital bonding to a carbonyl group, an isocyanide ligand $(R–N^+\equiv C^-:)$, or a trivalent phosphorus ligand $[R_3P, (R_2N)_3P, (RO)_3P$, etc.]
2. Three external orbitals forming a bond to all of the carbon atoms in a planar pentagonal (cyclopentadienyl), hexagonal (benzene), or heptagonal (tropylium or cycloheptatrienyl) ring
3. A single external orbital containing a lone electron pair without bonding to an external group; such an external orbital is conveniently called a *nonbonding external orbital*

In all known cases at least some of the external orbitals of heavy vertex atoms are nonbonding; most frequently the number of such nonbonding external orbitals is three. The common types of vertex groups involving transition metals (M) as vertex atoms are metal tricarbonyl vertices $M(CO)_3$ and cyclopentadienylmetal vertices C_5H_5M; both of these types of vertex groups have three nonbonding external orbitals.

The transition metals in metal carbonyls usually have the electronic configuration of the next rare gas, which corresponds to 18 valence electrons.[127] Thus, well-known mononuclear metal carbonyls having this favored 18 (valence) electron configuration include $Cr(CO)_6$, $Fe(CO)_5$, and $Ni(CO)_4$. A similar principle applies to the transition metal vertices in metal carbonyl clusters. This provides a means for calculating the number of electrons provided by various transition metal vertex groups to the metal cluster skeleton; such electrons are called *skeletal electrons*. For example, consider a $Co(CO)_3$ vertex group using three internal cobalt orbitals for the skeletal bonding leaving six external cobalt orbitals requiring a total of 12 electrons to attain the favored 18-electron rare gas configuration. Of these 12 electrons 2 come from each of the three carbonyl groups leaving 6 electrons to be provided by the cobalt atom. Since a neutral cobalt atom has 9 valence electrons, this leaves 3 ($= 9 - 6$) electrons for the polygonal or polyhedral skeleton. Thus, a $Co(CO)_3$ group is a donor of 3 skeletal electrons and may be regarded as isoelectronic and isolobal to a C–H vertex, such as found in the carboranes $C_2B_{n-2}H_n$ or even benzene, C_6H_6. This isoelectronic analogy was first recognized by Wade[128] and subsequently extended by Hoffmann[129] to the concept of isolobality. Table 6 summarizes the skeletal electrons donated by important types of transition metal vertices and their isoelectronic and isolobal relationships to carbon and boron vertices. In Table 6 the hydrogen atoms in B–H and C–H vertices can be replaced by other monovalent groups such as halogen, alkyl, aryl, nitro, cyano, etc.; the carbonyl groups in $M(CO)_3$ vertices can be replaced by other 2-electron donor ligands such as tertiary phosphines and isocyanides; and the hydrogen atoms on the C_5H_5 rings in the C_5H_5M vertices can be replaced partially or completely by other monovalent groups, notably methyl or trimethylsilyl. Also note that the details of the distribution of carbonyl groups on the metal cluster skeleton are rarely significant for electron counting since both terminal and bridging carbonyls are 2-electron donors (except for some rather rare types of dihapto bridging carbonyls which are almost never found in carbonyl derivatives of the middle to late transition metals forming clusters).

Some metal carbonyl clusters have vertices of post-transition elements such as phosphorus, sulfur, and their heavier congeners.[130] Let us consider an alkyl-phosphinidene vertex R–P. The phosphorus atom of a *normal* alkylphosphinidene vertex uses only its s and p orbitals for chemical bonding leading to four sp^3 tetrahedrally hybridized orbitals, one of which is an external orbital for bonding to the alkyl group and the remaining *three* of which are internal orbitals for the cluster skeletal bonding. However, in addition, *hypervalent* alkylphosphinidene vertices are

Table 6.   Skeletal Electrons Donated by Transition Metal Vertices

Skeletal electrons	B or C vertex	$M(CO)_3$ vertex	C_5H_5M vertex
2	B–H	Fe, Ru, Os	Co, Rh, Ir
3	C–H	Co, Rh, Ir	Ni, Pd, Pt

also possible with five $d(x^2 - y^2)sp^3$ square pyramidally hybridized orbitals, one of which is an external orbital for bonding to the alkyl group and the remaining *four* of which are internal orbitals for the cluster skeletal bonding. Thus, the phosphorus atom in an alkylphosphinidene vertex can contribute either three or four internal orbitals to the cluster skeletal bonding in contrast to boron or carbon vertices with energetically inaccessible *d* orbitals which can contribute only three internal orbitals to the cluster skeletal bonding. Both normal and hypervalent alkylphosphinidene vertices contribute four skeletal electrons to the cluster bonding since only one of the five phosphorus valence electrons is required for external bonding to the alkyl groups in either case. Thus, the counting of skeletal electrons in alkylphosphinidene metal clusters does *not* require previous determination whether the alkylphosphinidene vertices are normal or hypervalent. Alkylarsinidene (RAs:) and bare sulfur (S:) vertices are isoelectronic and isolobal with alkylphosphinidene vertices. Bare phosphorus (P:) vertices are donors of three skeletal electrons since two of the five phosphorus valence electrons are required for the external lone pair.

A number of theoretical methods have been developed for describing metal carbonyl cluster skeletal bonding leading to a number of different electron-counting schemes. Some of these electron-counting schemes count only the apparent skeletal electrons whereas others count both skeletal and external electrons. Transition metal vertices are assumed to use three internal orbitals for skeletal bonding leaving six external orbitals requiring 12 electrons for each vertex. Therefore, a globally delocalized deltahedron with n vertices which has $2n + 2$ skeletal electrons will have $12n + 2n + 2 = 14n + 2$ total electrons. For example, a six-vertex globally delocalized octahedron with $(2)(6) + 2 = 14$ skeletal electrons has $(14)(6) + 2 = 86$ total electrons. The polyhedral skeletal electron pair theory of Wade[91] and Mingos[131,132] uses essentially equivalent topological ideas but frequently uses total electron counts rather than skeletal electron counts making the comparison between electron counts for polyhedra with different numbers of vertices but similar general topologies less obvious. However, the theories of Wade and Mingos[91,131,132] arrive at the $2n + 2$ skeletal electron count for deltahedra with no vertices of degree 3 similar to the graph-theory-derived method discussed in this chapter.

Another method of interest is the topological electron-counting theory of Teo.[133,134] This theory starts from Euler's theorem for polyhedra in the form

(22) $$e = v + f - 2$$

where e, v, and f are the numbers of edges, vertices, and faces in the polyhedron in question. Assuming that each atom on the surface of the polyhedron has the tendency to attain the 18-electron rare gas configuration as discussed above and that each edge can be considered as an edge-localized two-center two-electron metal–metal bond, the total electron count N for the cluster is

(23) $$N = 18v - 2e$$

The N electrons will fill the $N/2$ energetically low-lying metal cluster valence molecular orbitals (CVMO) so that the number of CVMOs can be obtained as

$$\text{(24)} \qquad\qquad \text{CVMO} = N/2 = 9v - e = 8v - f + 2$$

derived from equations (22) and (23). However, for a delocalized metal cluster, such as globally delocalized metal deltahedra, not all metal–metal interactions can be considered to be edge-localized. An adjustment factor X is introduced into equation (24) reflecting delocalization leading to the equation

$$\text{(25)} \qquad\qquad \text{CVMO} = N/2 = 9v - e + X = 8v - f + 2 + X$$

where X is the number of "extra" electron pairs "in excess" of the "18-electron rule." An alternative interpretation of X is that it is the number of "false" metal–metal bonds or, in molecular orbital terminology, the number of "missing" antibonding cluster orbitals if each polyhedral edge is considered as an edge-localized bond.

Using this terminology, Teo[133,134] has derived several rules for determining this adjustment factor X of which the following are easiest to understand:

1. $X = 0$ for all polyhedra in which all vertices have degree 3 such as the tetrahedron, cube, and dodecahedron. These are the polyhedra exhibiting edge-localized bonding.
2. Capping a face of a polyhedron having n edges leads to an increase in X by $n - 3$. Hence, capping a triangular, quadrilateral (including square), and pentagonal faces increases X by 0, 1, and 2, respectively.
3. $X = 0$ for all pyramids.
4. The values of X for bipyramids can be determined by capping the corresponding pyramids or by considering the molecular orbitals of the polygon forming the "belt" of the bipyramid leading to $X = 0$ for the trigonal bipyramid, $X = 1$ for a tetragonal bipyramid, and $X = 2$ for a pentagonal bipyramid. Elongation of a trigonal bipyramid can increase its X value from 0 to 2.

These rules have been justified by considering the molecular orbitals of the polyhedra in question.[133,134]

10.4.2. Electron-Poor (Hypoelectronic) Metal Carbonyl Clusters

The availability of d orbitals on transition metal vertices leads to the possibility of electron-poor or hypoelectronic transition metal clusters with n vertices having *less* than $2n + 2$ *apparent*, skeletal electrons. Such electron-poor clusters form deltahedra containing tetrahedral chambers, i.e., deltahedra with one or more vertices of degree 3 (Figure 9). The simplest examples of such deltahedra are the capped tetrahedra, of which the trigonal bipyramid (i.e., the monocapped tetrahedron) with five vertices is the smallest. The capped tetrahedra consist of a series of fused tetrahedral chambers with faces in common. An example of a cluster based on a bicapped tetrahedron is

$Os_6(CO)_{18}$ (Figure 9) which has 12 (i.e., $2n$) skeletal electrons. The simplest deltahe-dron in which tetrahedral chambers do not occupy the whole volume of the polyhedron is the capped octahedron with 7 vertices; this polyhedron is found in $Os_7(CO)_{21}$ which has 14 (i.e., $2n$) skeletal electrons (Figure 9).

The following properties of tetrahedral chambers in capped deltahedra are significant:

1. Tetrahedral chambers in capped deltahedra require localized bonding like isolated tetrahedra. Tetrahedral chambers formed by capping deltahedra hav-ing all vertices of degree > 3 may be regarded as islands of localization in an otherwise delocalized system.

2. Atoms at vertices of triangular faces being capped require more than three internal orbitals oriented inward toward the cluster polyhedron. Therefore, such atoms cannot be light atoms (e.g., boron or carbon) having only s and p orbitals in an sp^3 manifold, since this total of four valence orbitals cannot all be oriented on the same side of the plane bisecting the light atom (i.e., in the same "halfspace"). For this reason the analogies between the transition metal vertices in metal clusters and the boron vertices in polyhedral boranes as, for example, outlined in Table 6, break down.

3. The capping vertex contributes the usual number of skeletal electrons but no skeletal orbitals. Therefore, capping is a good remedy for "electron poverty."

There are two opposite or dual processes for converting closed deltahedra with n vertices which require $2n + 2$ skeletal electrons into polyhedra suitable for systems having a larger or smaller number of skeletal electrons relative to the number of vertices. Thus, electron-rich polyhedra having more than $2n + 2$ skeletal electrons are formed from deltahedra by either polyhedral excision or the equivalent polyhedral puncture. In polyhedral excision a vertex and all of its incident edges are removed so that more electrons than bonding orbitals are lost. Electron-poor polyhedra having less than $2n + 2$ skeletal electrons are formed by polyhedral capping in which a triangular face is capped with a new vertex to add electrons to the structure without adding bonding orbitals.

The bicapped tetrahedron (Figure 9) consisting of three fused tetrahedral cham-bers (i.e., an "analogue" of anthracene based on 3D tetrahedra rather than 2D hexa-gons) such as that found in $Os_6(CO)_{18}$ is a good example for illustrating the concept of apparent skeletal electron counting for electron-poor polyhedra. Since the bicapped tetrahedron has 12 edges, 24 skeletal electrons are required for edge-localized bonding as suggested by the three tetrahedral chambers. However, in a model assuming such edge-localized bonding there are six "extra" internal orbitals arising from the two vertices of degree 4 and the two vertices of degree 5 [i.e., $(2)(4 - 3) + (2)(5 - 3) = 6$] which provide 12 of these 24 skeletal electrons. Therefore, the apparent skeletal electron count for the bicapped tetrahedron is $24 - 12 = 12$. Five of the nine valence orbitals in the sp^3d^5 manifold of the degree 5 osmium vertices can be directed inward toward the edges of the bicapped tetrahedron to participate in skeletal rather than external bonding leading to the matching of the numbers of internal orbitals and vertex degrees required for edge-localized bonding.

10.4.3. Metal Carbonyl Clusters Having Interstitial Atoms

Many metal carbonyl clusters have interstitial atoms or groups located in the center of the polyhedron. Such interstitial atoms may be a light atom such as boron, carbon, or nitrogen; a post-transition element such as germanium, tin, or antimony; or a transition metal. Interstitial atoms most frequently provide all of their valence electrons as skeletal electrons since all of their valence orbitals are necessarily internal orbitals because of the location of the interstitial atom in the center of the polyhedron. Exceptions to this rule may occur when some of the valence electrons of the interstitial atom occupy orbitals of symmetries which cannot mix with any of the molecular orbitals arising from the polyhedral skeletal bonding.

Interstitial atoms create certain volume requirements for the surrounding polyhedron.[135] Thus, an interstitial carbon atom cannot fit into a tetrahedron but fits into an octahedron as exemplified by $Ru_6(CO)_{17}C$. An interstitial transition metal cannot fit into an octahedron but fits into a larger polyhedron such as a 12-vertex polyhedron. The volume of a polyhedron containing an interstitial atom can be increased by decreasing the number of edges. In the case of a deltahedron this can be done by converting pairs of triangular faces into single quadrilateral faces by rupture of the edge separating the two triangular faces, i.e.,

This process corresponds to the "diamond–square" portion of the diamond–square–diamond process involved in polyhedral rearrangements. For example, rupture of six edges in this manner from an icosahedron can give a cuboctahedron as follows:

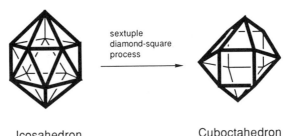

Icosahedron   Cuboctahedron

An n-vertex nondeltahedron derived from an n-vertex deltahedron by volume expansion through edge-rupture in this manner and containing an interstitial atom may function as a globally delocalized $2n + 2$ skeletal electron system like the original n-vertex deltahedron. Such nondeltahedra can conveniently be called *pseudodeltahedra*; they have only triangular and quadrilateral faces with a limited number of the latter. In an uncentered polyhedron having some faces with more than three edges, such faces may be regarded as holes in the otherwise closed polyhedral surface[136]; such polyhedra generate *nido* and *arachno* structures in the boron hydrides (Figure 7). In a centered pseudodeltahedron the interstitial atom in the center may be regarded as plugging up the surface holes arising from the nontriangular faces so that globally delocalized bonding is possible.

An example of electron counting in a metal carbonyl cluster with an interstitial atom is illustrated by $Ru_6C(CO)_{17}$,[137] which has the 14 skeletal electrons required for a globally delocalized octahedron by the following electron-counting scheme:

6 $Ru(CO)_3$ vertices: $(6)(2) =$	12 electrons
Interstitial carbon atom	4 electrons
"Deficiency" of one CO group	−2 electrons
Total skeletal electrons	14 electrons

10.4.4. Metal Carbonyl Clusters Consisting of Fused Octahedra

Numerous polycyclic benzenoid hydrocarbons, of which naphthalene and anthracene are familiar examples, can be constructed by edge-sharing fusion of benzene rings (carbon hexagons). Similarly, metal carbonyl cluster polyhedra can be fused by sharing of edges or faces to give 3D analogues of polycyclic aromatic hydrocarbons. The most extensive series of such metal carbonyl derivatives having multiple polyhedral cavities are obtained from metal octahedra; such structures can be viewed as pieces of an infinite 3D bulk metal lattice. Indeed, Teo[138] has shown that the Hume-Rothery rule[139] for electron counting in brasses can be extended to close-packed high-nuclearity metal clusters. Other aspects of the fusion of cluster polyhedra have been treated by Mingos[140,141] and by Slovokhotov and Struchkov.[142] Figure 10 depicts some fused octahedra in metal carbonyl clusters, mainly rhodium carbonyl derivatives, as analogues of benzenoid aromatic hydrocarbons.

Electron counting in metal carbonyl clusters having fused octahedra can be illustrated by the face-sharing naphthalene analogue $Rh_9(CO)_{19}^{3-}$. The structure of $Rh_9(CO)_{19}^{3-}$ (Ref. 143) consists of a pair of octahedra having a (triangular) face in common (Figure 10) analogous to naphthalene which consists of two carbon hexagons with an edge in common. The face-shared pair of octahedra has 9 vertices, 21 edges, and 14 faces like the tricapped trigonal prism (Figure 1 or 4), which is the nine-vertex deltahedron found in systems having $2n + 2 = 20$ skeletal electrons ($n = 9$) so that a bonding scheme having a K_9 complete graph for the core bonding is reasonable for a face-sharing fused pair of octahedra just as it is for the tricapped trigonal prism. However, in the fused pair of octahedra the three rhodium vertices common to both octahedra use four internal orbitals, whereas the six rhodium vertices belonging to only one of the octahedra use the normal three internal orbitals. This leads to the following electron-counting scheme for $Rh_9(CO)_{19}^{3-}$:

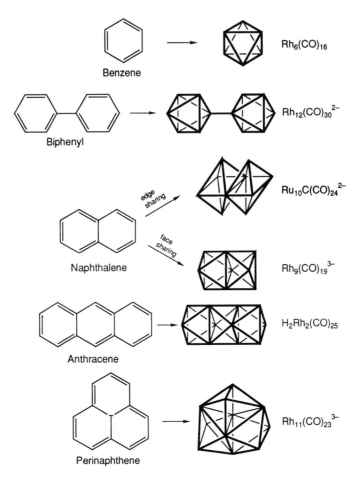

Figure 10. Analogies between the fusion of metal octahedra in metal carbonyl clusters and the fusion of benzene rings in planar polycyclic aromatic hydrocarbons.

a. Source of Skeletal Electrons

6 Rh(CO)$_2$ groups present in only one octahedron and therefore using only 3 internal orbitals: (6)(1) =	6 electrons
3 Rh(CO)$_2$ groups common to both octahedra and therefore using 4 internal orbitals: (3)(3) =	9 electrons
1 "extra" CO group	2 electrons
−3 charge on ion	3 electrons
Total available skeletal electrons	20 electrons

b. Use of Skeletal Electrons

9 Rh–Rh surface bonds: (9)(2) =	18 electrons
1 9-center core bond: (1)(2) =	2 electrons
Total skeletal electrons required	20 electrons

10.5. COINAGE METAL CLUSTERS

10.5.1. Anisotropic Valence Orbital Manifolds in Coinage Metals and Other Post-Transition Elements

A specific feature of the chemical bonding in some systems containing the late transition and early post-transition metals observed by Nyholm[144] as early as 1961 is the shifting of one or two of the outer p orbitals to such high energies that they no longer participate in the chemical bonding and the accessible spd valence orbital manifold is no longer isotropic. If one p orbital is so shifted to become antibonding, then the accessible anisotropic spd orbital manifold contains only eight orbitals (sp^2d^5) and has the geometry of a torus or doughnut (Figure 11a) in which the "missing" p orbital responsible for the hole in the doughnut. This toroidal sp^2d^5 manifold can bond only in the two dimensions of the plane of the ring of the torus. Filling this sp^2d^5 manifold of eight orbitals with electrons leads to the 16-electron configuration found in square planar complexes of the d^8 transition metals such as Rh(I), Ir(I), Ni(II), Pd(II), Pt(II), and Au(III). The locations of the four ligands in these square planar complexes can be considered to be points on the surface of the torus corresponding to the sp^2d^5 manifold. The toroidal sp^2d^5 manifold can also lead to trigonal planar and pentagonal planar coordination for three- and five-coordinate complexes, respectively (Figure 11a). The x, y, and z axes for a toroidal sp^2d^5 manifold are conventionally chosen so that the missing p orbital is the p_z orbital.

In some structures containing the late transition and post-transition metals, particularly the $5d$ metals Pt, Au, Hg, and Tl, *two* of the outer p orbitals are raised to antibonding energy levels. This leaves only one p orbital in the accessible anisotropic spd orbital manifold, which now contains seven orbitals (spd^5) and has cylindrical geometry extending in one axial dimension much farther than in the remaining two dimensions (Figure 11b). Filling this seven-orbital spd^5 manifold with electrons leads to the 14-electron configuration found in two-coordinate linear complexes of d^{10} metals such as Pt(0), Cu(I), Ag(I), Au(I), Hg(II), and Tl(III). The raising of one or particularly two outer p orbitals to antibonding levels has been attributed to relativistic effects.[145]

The p orbitals which are raised to antibonding levels as noted above can participate in $d\sigma \rightarrow p\sigma^*$ or $d\pi \rightarrow p\pi^*$ bonding in complexes of metals with toroidal sp^2d^5, cylindrical spd^5, and spherical sd^5 manifolds depending on the symmetry of the overlap (Figure 11c). Such bonding has been suggested by Dedieu and Hoffmann[146] for Pt(0)–Pt(0)

Toroidal Trigonal Planar

Toroidal Pentagonal Planar

Toroidal Square Planar

Cylindrical Linear

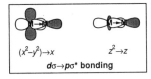

$(x^2-y^2)\rightarrow x$ $z^2\rightarrow z$

$d\sigma\rightarrow p\sigma^*$ bonding

$xz\rightarrow z$

$d\pi\rightarrow p\pi^*$ bonding

Figure 11. (a) Toroidal trigonal planar, square planar, and pentagonal planar coordination derived from sp^2d^5 manifolds; (b) cylindrical linear coordination derived from an spd^5 manifold; (c) idealized examples of $d\sigma\rightarrow p\sigma^*$ and $d\pi\rightarrow p\pi^*$ bonding found in coinage metal derivatives with toroidal and cylindrical valence orbital manifolds.

dimers on the basis of extended Hückel calculations. This type of surface bonding, like, for example, the $d\pi\rightarrow p\pi^*$ backbonding in metal carbonyls, does not affect the electron bookkeeping in the late transition and post-transition metal clusters but accounts for the bonding rather than nonbonding distances between adjacent metal vertices in such clusters.

10.5.2. Centered Gold Clusters

An important class of gold clusters includes the *centered* gold clusters (Figure 12).[147] Such clusters containing n gold atoms consist of a central gold atom surrounded by $n - 1$ peripheral gold atoms. The peripheral gold atoms all have a seven-orbital cylindrical spd^5 manifold of bonding orbitals and can be divided into the following two types:

1. Belt gold atoms which form a puckered hexagonal or octagonal belt around the center gold atom
2. Distal gold atoms which appear above or below the belt gold atoms

The topology of the centered gold clusters can be considered to be either spherical or toroidal depending on whether the center gold atom uses a nine-orbital spherical sp^3d^5

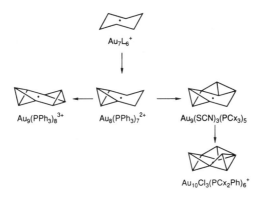

Figure 12. (a) The structures of some toroidal centered gold clusters of the general formula $Au_nL_yX_{n-1-y}^{(y-5)+}$ with the center gold atoms shown by a dot (•); (b) the structures of some spherical gold clusters of the general formula $Au_nL_yX_{n-1-y}^{(y-7)+}$ with the center gold atoms shown as Au.

manifold or an eight-orbital toroidal sp^2d^5 manifold of bonding orbitals. This distinction between spherical and toroidal centered gold clusters was first recognized by Mingos and co-workers.[148] Centered gold clusters have also been described as "porcupine compounds" since the central gold atom corresponds to the body of the porcupine and the peripheral gold atoms (with cylindrical geometry as noted above) correspond to the quills of the porcupine.[149]

The following features of centered gold clusters make their systematics very different from other metal cluster compounds such as metal carbonyl clusters:

1. The volume enclosed by the peripheral gold atoms must be large enough to contain the center gold atom. Thus, the volume of a cube of eight peripheral gold atoms is not large enough to contain a ninth center gold atom without some distortion. Therefore, centered cube gold clusters of the stoichiometry $Au_9L_8^+$ such as $Au_9(PPh_3)_8^+$ (Ref. 150) are distorted from the ideal O_h symmetry to lower symmetry such as D_3. However, the volume of an icosahedron of 12 peripheral gold atoms is large enough to contain a 13th central gold atom without any distortion as shown by the structure[151] of $Au_{13}Cl_2(PMe_2Ph)_{10}^{3+}$. The peripheral gold polyhedron of a spherical gold cluster containing fewer than

13 total gold atoms is generally based on an undistorted icosahedral fragment which has a large enough volume for the center gold atom.

2. The overlap topology at the core of a centered Au_n cluster from the $n-1$ unique internal orbitals of the peripheral gold atoms is not that of a K_{n-1} complete graph as in other globally delocalized metal clusters.[88-90,97,152] Instead the overlap topology of the unique internal orbitals of the peripheral gold atoms corresponds to the polyhedron formed by the peripheral gold atoms. This arises from the sharper "focus" of the cylindrical seven-orbital spd^5 manifold of the peripheral gold atoms relative to the spherical four-orbital sp^3 and nine-orbital sp^3d^5 manifolds. Thus, the number of positive eigenvalues of the graphs corresponding to the peripheral gold polyhedra relates to the number of bonding orbitals in the centered gold clusters.

3. The center gold atom has 11 valence electrons. All but one of these electrons are needed to fill its five d orbitals. The remaining electron is in the spherically symmetric s orbital, which is the orbital of the center gold atom overlapping with the unique internal orbitals of the cylindrical spd^5 manifold of the peripheral gold atoms. This overlap lowers the energy of the lowest (fully symmetric) cluster bonding orbital without adding any new bonding orbitals. The center gold atom is therefore a donor of one skeletal electron.

4. Mingos and co-workers[148] have observed a $12p + 16$ electron rule for toroidal centered gold clusters and a $12p + 18$ electron rule for spherical centered gold clusters where $p = n - 1$ is the number of peripheral gold atoms. These numbers count not only the skeletal electrons but also the 10 electrons needed to fill the five d orbitals of each peripheral gold atom and the 2 electrons needed for one bond from each peripheral gold atom to an external L or X group. The $12p$ terms in Mingos's total electron numbers thus correspond to nonskeletal electrons involving only the peripheral gold atoms leaving 16 or 18 electrons for a center gold atom with toroidal or spherical geometry, respectively. This corresponds exactly to the number of electrons required to fill the eight-orbital toroidal sp^2d^5 manifold or the nine-orbital spherical sp^3d^5 manifold, respectively. Subtracting 10 from the 16 or 18 electrons allocated to the center gold atom for its five d orbitals leaves 6 or 8 skeletal electrons for toroidal or spherical gold clusters, respectively.

5. Consider L to be a two-electron donor ligand (e.g., tertiary phosphines or isocyanides) and X to be a one-electron donor ligand [e.g., halogen, pseudo-halogen, $Co(CO)_4$]. Then the above considerations give centered toroidal clusters (Figure 12a) the general formula $Au_nL_yX_{n-1-y}^{(y-5)+}$ and centered spherical clusters (Figure 12b) the general formula $Au_nL_yX_{n-1-y}^{(y-7)+}$.

10.5.3. Coinage Metal Alkyls and Aryls

Coinage metal alkyls and aryls of the stoichiometry $(RM)_x$ (M = Cu, Ag; R = alkyl or aryl) clearly do not have simple monomeric structures with one-coordinate coinage metal atoms. Instead such alkyls have oligomeric structures (Figure 13) often with three-center M–C–M bonds analogous to the three-center Al–C–Al bonds in dimeric aluminum alkyls such as trimethylaluminum dimer. The bridging alkyl and aryl groups in these structures are donors of three apparent skeletal electrons. One of these three apparent skeletal electrons is a "real" electron arising from the carbon atom of the neutral alkyl or aryl group directly bonded to the two metal atoms through the three-center M–C–M bond. The other two apparent skeletal electrons are "virtual" electrons arising indirectly from the third atomic orbital in the three-center bond which otherwise would be an external orbital requiring an electron pair for the closed-shell electronic configuration.

Planar polygonal coinage metal rings M_n are found in many copper and silver alkyls. The alkyl group in the tetrameric copper alkyl[153] $[CuCH_2SiMe_3]_4$ (Figure 13) can only bond to the copper atom through the alkyl carbon atom so that the Me_3SiCH_2 ligand is a monodentate ligand, although it bridges two copper atoms so that the cluster

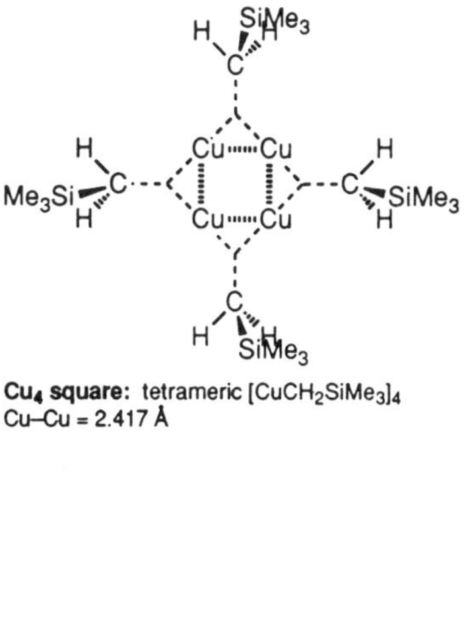

Cu₄ square: tetrameric $[CuCH_2SiMe_3]_4$
Cu–Cu = 2.417 Å

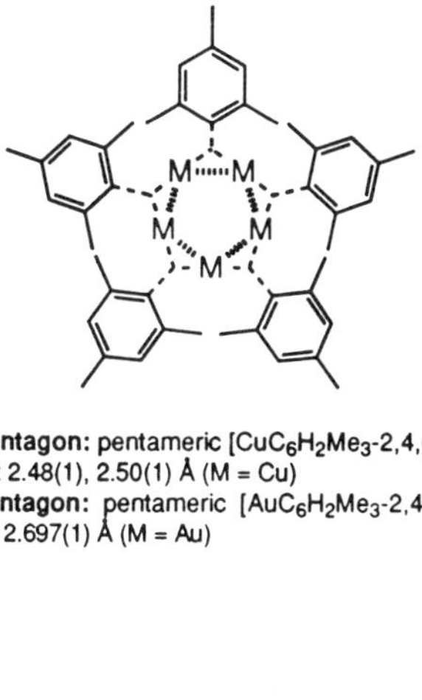

Cu₅ pentagon: pentameric $[CuC_6H_2Me_3\text{-}2,4,6]_5$
Cu–Cu: 2.48(1), 2.50(1) Å (M = Cu)
Au₅ pentagon: pentameric $[AuC_6H_2Me_3\text{-}2,4,6]_5\cdot2THF$
Au–Au: 2.697(1) Å (M = Au)

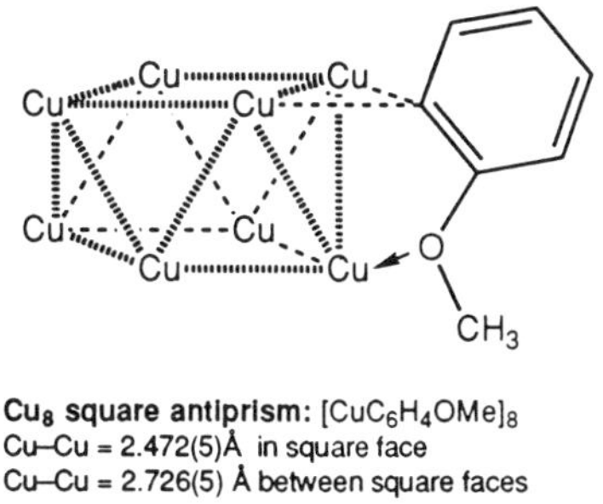

Cu₈ square antiprism: $[CuC_6H_4OMe]_8$
Cu–Cu = 2.472(5)Å in square face
Cu–Cu = 2.726(5) Å between square faces

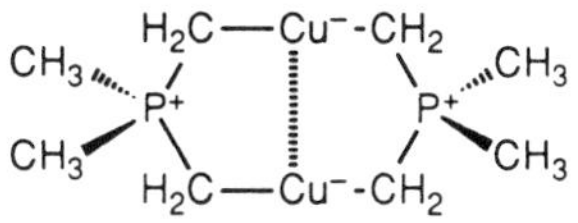

Cu₂ dimer:
$[Cu(CH_2)_2P(CH_3)_2]_2$
Cu–Cu = 2.843(3) Å

Figure 13. Some examples of coinage metal alkyl structures.

can be more precisely written $[Cu(CH_2SiMe_3)_{2/2}]_4$. The aryl group in the pentameric aryls[154] $[MC_6H_2Me_3\text{-}2,4,6]_5$ (Figure 13) is an analogous monodentate ligand so that these aryls can be written as $[M(C_6H_2Me_3\text{-}2,4,6)_{2/2}]_5$. Each coinage metal atom in the structures in Figure 13 can be considered to use a seven-orbital cylindrical spd^5 bonding orbital manifold with two linear sp hybrids. These linear sp hybrids form three-center two-electron Cu–C–Cu bonds with an alkyl or aryl carbon atom and an adjacent coinage metal atom similar to the three-center B–H–B bonds in boron hydrides (Section 10.3.6). The two neutral bridging alkyl or aryl groups associated with each coinage metal atom formally donate a total of three apparent electrons (see above) so that each coinage metal atom has $11 + 3 = 14$ valence electrons, which is exactly enough to fill its seven-orbital cylindrical spd^5 bonding manifold. The short metal–metal distances indicate appreciable metal–metal interactions in the coinage metal polygons in $[CuCH_2SiMe_3]_4$ and $[MC_6H_2Me_3\text{-}2,4,6]_5$ (Figure 13) in accord with $d\sigma{\to}p\sigma^*$ and $d\pi{\to}p\pi^*$ backbonding into the coinage metal antibonding p orbitals not used for the cylindrical spd^5 bonding orbital manifold (Figure 11c).

A related principle can be used to construct the octameric copper aryl $[CuC_6H_4OMe\text{-}2]_8$ (Figure 13).[155] In this case the copper atoms form a square antiprism. The 2-methoxyphenyl groups cap the eight triangular faces by forming three-center two-electron Cu–C–Cu bonds with two of the three copper vertices of the face being capped and a $CH_3O{\to}Cu$ dative bond with the third copper atom of the face so that each 2-methoxyphenyl group is shared with three copper atoms leading to the formulation $[Cu(C_6H_4OMe\text{-}2)_{3/3}]_8$. Each Cu_4 square face of the square antiprism in $[CuC_6H_4OMe\text{-}2]_8$ is closely related to the Cu_4 square in $[CuCH_2SiMe_3]_4$ discussed above. The ability of each of the eight 2-methoxyphenyl groups to function as a bidentate ligand bonding to metal atoms both through an aryl carbon and the methoxy oxygen acts as the "glue" to join two Cu_4 squares into the Cu_8 square antiprism in $[CuC_6H_4OMe\text{-}2]_8$. Bidentate aryl ligands are also found in the tetrameric $[CuC_6H_3(CH_2NMe_2)\text{-}2\text{-}Me\text{-}5]_4$ with a Cu_4 arrangement between square planar and tetrahedral.[156,157]

Another type of bidentate alkyl group is obtained by deprotonation of the phosphonium methylide $(CH_3)_3P{=}CH_2$ and occurs in the dimeric alkyl $[Cu(CH_2)_2P(CH_3)_2]_2$ (Figure 13).[158] In this structure each copper atom has a cylindrical spd^5 manifold with the linear sp hybrids forming two-center two-electron bonds with the alkyl carbons of two different $(CH_3)_2P(CH_2)_2$ groups.

10.5.4. Coinage Metal Derivatives of Metal Carbonyls

Some interesting clusters are known which combine the features of coinage metal clusters and metal carbonyls.[159] Such heterobimetallic clusters can often be obtained from various reactions of Cu(I) or Ag(I) compounds with the $Fe(CO)_4^{2-}$ dianion. Thus, a series of clusters is known (Figure 14) having structures constructed from macropolygonal or macropolyhedral networks with Fe–M'–Fe edges (M' = Cu, Ag) and typically $Fe(CO)_4$ vertices such as the triangular cluster[160] $Cu_3[Fe(CO)_4]_3^{3-}$, the but-

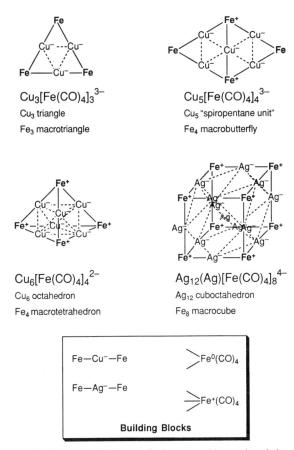

Figure 14.   Some metal skeletons of coinage metal iron carbonyl clusters.

terfly cluster[160] $Cu_5[Fe(CO)_4]_4^{3-}$, the tetrahedral cluster[161] $Cu_6[Fe(CO)_4]_4^{2-}$, and the paramagnetic centered cubic cluster[162] $Ag_{12}(Ag)[Fe(CO)_4]_8^{4-}$. In these structures the graph of the polygon or polyhedron formed by the coinage metal vertices is the line graph of the graph of the macropolygon or polyhedron formed by the transition metal vertices. In this context the *line graph L(G)* of a graph G is constructed by taking the edges of G as vertices of L(G) and joining two vertices in L(G) when the corresponding edges in G have a common vertex.[163]

The iron carbonyl vertices of these coinage metal iron carbonyl clusters (Figure 14) are always $Fe(CO)_4$ units, which are of two types, namely, divalent (degree 2) $Fe(CO)_4$ vertices forming two Fe–M′ bonds and trivalent (degree 3) $Fe(CO)_4$ vertices forming three Fe–M′ bonds. In order for the iron atoms in these $Fe(CO)_4$ vertices to have the favored 18-electron noble gas configuration, the divalent $Fe(CO)_4$ vertices must be formally neutral and the trivalent $Fe(CO)_4$ vertices must bear a formal $+1$ charge (Figure 14). In all cases the charge on the cluster anion is the algebraic sum of -1 for each coinage metal, M′, in the Fe–M′–Fe edges, zero for a bivalent $Fe(CO)_4$ group, and $+1$ for a trivalent $Fe(CO)_4$ group (Figure 14).

The large silver–iron carbonyl anion cluster unit $Ag_{12}[Fe(CO)_4]_8^{4-}$ is of particular interest. Its structure (Figure 14) consists of an Fe_8 macrocube with Fe–Ag–Fe edges found in the cluster anion $Ag_{12}(Ag)[Fe(CO)_4]_8^{4-}$, in which a 13th silver atom is located in the center of the cluster.[162] The 12 silver atoms at the edge midpoints of the Fe_8 macrocube form a cuboctahedron, which is the line graph of the cube. Since the outer $Ag_{12}[Fe(CO)_4]_8^{4-}$ unit has the expected closed-shell -4 charge arising from the algebraic sum of the formal -1 charges on the 12 Ag atoms and the formal $+1$ charges on the 8 trivalent $Fe(CO)_4$ vertices, the "central" silver atom in $Ag_{12}(Ag)[Fe(CO)_4]_8^{4-}$ is formally neutral, i.e., Ag^0. The observed ESR signal[162] in $Ag_{12}(Ag)[Fe(CO)_4]_8^{4-}$ indicates strong coupling of the unpaired electron with the central Ag atom but only loose coupling with the peripheral Ag atoms in accord with the dissection of $Ag_{12}(Ag)[Fe(CO)_4]_8^{4-}$ into a closed-shell outer $Ag_{12}[Fe(CO)_4]_8^{4-}$ unit with no unpaired electrons and a central Ag^0 atom on which the unpaired electron mainly resides. This interpretation of the structure and bonding in $Ag_{12}(Ag)[Fe(CO)_4]_8^{4-}$ resolves the problem of its "extra" 9 electrons relative to the predictions of most theories mentioned in the original paper on this cluster[162] and suggests the use of the $Ag_{12}[Fe(CO)_4]_8^{4-}$ unit to encapsulate other reactive neutral atoms besides Ag^0 in its cuboctahedral cavity.

10.6. POST-TRANSITION ELEMENT CLUSTERS

10.6.1. Bare Ionic Post-Transition Metal Clusters: Zintl Phases

The polyhedral boranes and carboranes (Section 10.3) may be regarded as boron clusters in which the single external orbital of each vertex atom is used for bonding to an external monovalent group. Analogously, the six external orbitals of each transition metal vertex atom in metal carbonyl clusters (Section 10.4) are used for bonding to ligands such as carbonyl groups, tertiary phosphines, or planar hydrocarbon rings such as cyclopentadienyl. The single external orbital on each gold atom in the centered gold clusters (Section 10.5) is used for bonding to a single tertiary phosphine, halide, or pseudohalide ligand. Electron-richer post-transition metals to the right of gold in the periodic table form clusters *without* external ligands bonded to the vertex atoms, conveniently called *bare* metal clusters. Anionic bare metal clusters were first observed by Zintl and co-workers in the 1930s,[164–167] who obtained the first evidence for anionic clusters of post-transition metals such as tin, lead, antimony, and bismuth through potentiometric titrations with alkali metals in liquid ammonia; for this reason such anionic post-transition metal clusters are often called *Zintl phases*. However, extensive structural information on these anionic post-transition metal clusters was obtained only in the 1970s by Corbett and co-workers,[168] who discovered that complexation with 2,2,2-crypt gave crystals of alkali metal derivatives of such clusters suitable for structure determination by X-ray diffraction. Corbett and co-workers[169] somewhat earlier also used X-ray crystallography to obtain definitive structural information on cationic post-transition metal clusters obtained as halometalate salts, particularly $AlCl_4^-$ salts, from highly acidic melts.

The rules for counting the number of skeletal electrons contributed by each vertex atom can be adapted to vertices consisting of post-transition metals lacking external groups through the following considerations:

1. The post-transition metals under consideration use a nine-orbital sp^3d^5 valence orbital manifold.

2. The post-transition metals (in their zero formal oxidation states) have a total of $10 + G$ valence electrons where G is the highest possible oxidation state of the post-transition metal. Thus, germanium, tin, and lead have $10 + 4 = 14$ valence electrons; arsenic, antimony, and bismuth have $10 + 5 = 15$ valence electrons; and selenium and tellurium have $10 + 6 = 16$ valence electrons.

3. If the clusters exhibit either 2D or 3D aromaticity as discussed in Section 10.3, three orbitals of each bare metal vertex atom will be required for the internal orbitals (two twin internal orbitals and one unique internal orbital). This leaves $9 - 3 = 6$ external orbitals. Each external orbital of the bare metal vertex atom must be filled with an electron pair thereby consuming $(2)(6) = 12$ electrons from each bare metal vertex atom.

4. As a result of these considerations, the number of skeletal electrons contributed by each bare metal vertex atom through its three internal orbitals must be $10 + G - 12 = G - 2$.

Application of this procedure to the post-transition metals forming clusters indicates that bare gallium, indium, and thallium vertices contribute one skeletal electron; bare germanium, tin, and lead vertices contribute two skeletal electrons; bare arsenic, antimony, and bismuth vertices contribute three skeletal electrons; and bare selenium and tellurium vertices contribute four skeletal electrons in 2D and 3D aromatic clusters. Thus, Ge, Sn, and Pb vertices are isoelectronic with BH, $Fe(CO)_3$, and C_5H_5Co vertices and As, Sb, and Bi vertices are isoelectronic with CH, $Co(CO)_3$, and C_5H_5Ni vertices in bare metal cluster compounds.

Some of the structures of the most important bare ionic post-transition metal clusters are depicted in Figure 15. Their chemical bonding topologies can be treated as follows:

1. *Square.* Bi_4^{2-}, Se_4^{2+}, and Te_4^{2+} isoelectronic and isolobal with the globally delocalized planar cyclobutadiene dianion with 14 skeletal electrons [e.g., for Bi_4^{2-}: $(4)(3) + 2 = 14$] corresponding to 8 electrons for the 4 σ-bonds and 6 electrons for the π-bonding.

2. *Butterfly.* $Tl_2Te_2^{2-}$ with $(2)(1) + (2)(4) + 2 = 12$ apparent skeletal electrons isoelectronic and isolobal with *neutral* cyclobutadiene but undergoing a Jahn-

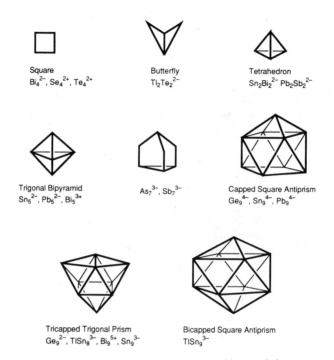

Figure 15.   The shapes of some ionic post-transition metal clusters.

Teller-like distortion to a butterfly structure as discussed by Burns and Corbett.[170]

3. *Tetrahedron.* $Sn_2Bi_2^{2-}$ and $Pb_2Sb_2^{2-}$ with $(2)(2) + (2)(3) + 2 = 12$ skeletal electrons for localized bonds along the edges of the tetrahedron analogous to organic tetrahedrane derivatives R_4C_4.

4. *Trigonal bipyramid.* Sn_5^{2-}, Pb_5^{2-}, and Bi_5^{3+} with 12 skeletal electrons [e.g., $(5)(2) + 2 = 12$ for Sn_5^{2-} and Pb_5^{2-}] analogous to the trigonal bipyramidal $C_2B_3H_5$ carborane.

5. *Seven-vertex systems.* As_7^{3-} and Sb_7^{3-} with the C_{3v} structure depicted in Figure 15. These anions have the correct $(4)(3) + (3)(1) + 3 = 18$ skeletal electrons for edge-localized bonds along the nine edges derived by considering the three vertices of degree 2 to use two internal orbitals each and the four vertices of degree 3 to use three internal orbitals each in accord with the requirement for matching the numbers of internal orbitals with the vertex degrees for edge-localized bonding.

6. *Capped square antiprism.* Ge_9^{4-}, Sn_9^{4-}, and Pb_9^{4-} with $(9)(2) + 4 = 22 = 2n + 4$ skeletal electrons required for an $n = 9$ vertex C_{4v} *nido* polyhedron having 12 triangular faces and one square face.

7. *Tricapped trigonal prism.* Ge_9^{2-} and $TlSn_8^{3-}$ with the $2n + 2 = 20$ skeletal electrons required for an $n = 9$ vertex globally delocalized D_{3h} deltahedron analogous to $B_9H_9^{2-}$ (Ref. 171); Bi_9^{5+} anomalously having $(9)(3) - 5 = 22$ rather than the expected 20 skeletal electrons suggesting[172] incomplete overlap of the unique internal orbitals directed toward the core of the deltahedron; E_9^{3-} (E = Ge, Sn, Pb) with $(9)(2) + 3 = 21$ skeletal electrons including one extra electron for a low-lying antibonding orbital analogous to radical anions formed by stable aromatic hydrocarbons such as naphthalene and anthracene.[173]

8. *Bicapped square antiprism.* $TlSn_9^{3-}$ with the $(1)(1) + (9)(2) + 3 = 22 = 2n + 2$ skeletal electrons required for $n = 10$ vertex globally delocalized D_{4d} deltahedron (the bicapped square antiprism in Figure 15) analogous to that found in the $B_{10}H_{10}^{2-}$ anion.[174]

10.6.2. Polyhedral Clusters Containing the Heavier Congeners of Boron: Novel Hypoelectronic Polyhedra in Indium and Thallium Intermetallics

The importance and stability of the deltahedral borane anions (Section 10.3) suggests that deltahedral clusters of the heavier congeners of boron, namely, aluminum, gallium, indium, and thallium, might likewise exhibit special stability. Known aluminum clusters containing isolated deltahedra are limited to the recently discovered[175] icosahedral $iBu_{12}Al_{12}^{2-}$, which appears to be less reactive than mononuclear organoaluminum compounds such as R_3Al (R = alkyl or aryl group). Structures containing gallium deltahedra are found in various intermetallic compounds containing gallium and alkali metals,[176–179] but such gallium deltahedra are always linked to the remainder of the structure through *exo* bonds from essentially all of the gallium vertices leading to infinite structures rather than discrete anions.

The chemical bonding topology of $iBu_{12}Al_{12}^{2-}$ is presumably analogous to that of the likewise icosahedral $B_{12}H_{12}^{2-}$ discussed above. Gallium deltahedra found in the alkali metal/gallium intermetallics include the Ga_8 bisdisphenoid, the Ga_{11} edge-coalesced icosahedron, and the Ga_{12} icosahedron, which appear to conform to the same electron-counting roles as the corresponding deltahedra (Figure 4) found in the deltahedral borane anions $B_nH_n^{2-}$ (Section 10.3).

The structures of clusters of indium and thallium exhibit some new features since these metals tend to form isolated naked clusters with no *exo* bonds from any polyhedral vertices. However, bare *homoatomic* clusters of the type M_n^{z-} (M = In, Tl) have too few skeletal electrons to form globally delocalized deltahedra having the required apparent $2n + 2$ skeletal electrons without excessively high negative charges or electron-donating interstitial atoms. For this reason intermetallic compounds of alkali metals with indium and thallium frequently are extremely hypoelectronic, i.e., they have much less than the apparent $2n + 2$ skeletal electrons for globally delocalized deltahedra. Such hypoelectronic indium and thallium polyhedra (Figure 16) are often very different from the capped deltahedra found in hypoelectronic metal carbonyl

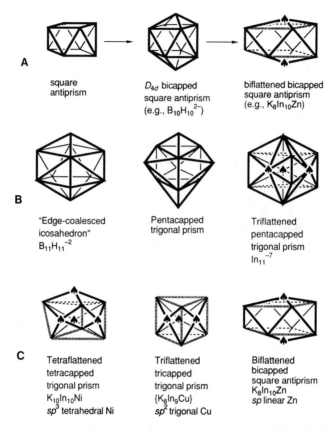

A

square antiprism — D_{4d} bicapped square antiprism (e.g., $B_{10}H_{10}^{2-}$) — biflattened bicapped square antiprism (e.g., $K_8In_{10}Zn$)

B

"Edge-coalesced icosahedron" $B_{11}H_{11}^{-2}$ — Pentacapped trigonal prism — Triflattened pentacapped trigonal prism In_{11}^{-7}

C

Tetraflattened tetracapped trigonal prism $K_{10}In_{10}Ni$ sp^3 tetrahedral Ni — Triflattened tricapped trigonal prism $\{K_8In_9Cu\}$ sp^2 trigonal Cu — Biflattened bicapped square antiprism $K_8In_{10}Zn$ sp linear Zn

Figure 16. (Top) Capping of a square antiprism to give a D_{4d} bicapped square antiprism followed by flattening the two caps to give the 10-vertex polyhedron found in $K_8In_{10}Zn$; (middle) chemically significant 11-vertex deltahedra; (bottom) the hypoelectronic polyhedra found in the centered indium clusters $K_{10}In_{10}M$ (M = Ni, Pd, Pt) and $K_8In_{10}Zn$ as well as that predicted for K_8In_9M (M = Cu, Ag, Au). Vertices of flattened pyramidal cavities are indicated by spades (♠) and the positions of edges broken during the flattening process are indicated by dashed and hashed lines.

clusters such as bicapped tetrahedral $Os_6(CO)_{18}$ and capped octahedral $Os_7(CO)_{21}$ (Figure 9).

The extreme hypoelectronicity of the indium and thallium clusters can be relieved by d-orbital participation from some of the vertex metal atoms. In the case of a *normal* bare post-transition metal vertex in a metal cluster such as those discussed in Section 10.6.1, the 12 external electrons may be divided into two types, namely, the 10 nonbonding d electrons and the 2 electrons of an external lone pair analogous to the B–H bonding pair in the polyhedral boranes $B_nH_n^{2-}$ ($6 \le n \le 12$). In this way a *normal* post-transition metal vertex such as indium may be considered to use a four-orbital sp^3 bonding manifold just like light vertex atoms such as boron or carbon. However, there is also the possibility of a *hypervalent* post-transition metal vertex in which one or more of the d orbitals of the vertex atom participate in the skeletal bonding. Since the valence d orbitals of post-transition metals are filled, their involvement in skeletal bonding provides a means for increasing the number of skeletal electrons by one electron pair for each d orbital. More specifically a *bare* hypervalent indium or thallium vertex using a five-orbital sp^3d manifold with only four rather than five nonbonding d orbitals would be a donor of $13 - 2$ (the external lone pair) $- 8$ (the 4 remaining nonbonding d orbitals) $= 3$ skeletal electrons. The participation of d orbitals of indium or thallium vertices in skeletal bonding in indium clusters is consistent with similar d orbital participation in known complexes of these metals[180] exhibiting coordination numbers greater than four such as MCl_5^{2-} (M = In, Tl). The use of hypervalent rather than normal indium vertices is a natural way of relieving the apparent electron poverty (hypoelectronicity) in indium cluster polyhedra.

The five valence orbitals $sp^3d(x^2 - y^2)$ on a hypervalent indium or thallium atom may be conveniently divided into one external orbital, *three* equivalent *triplet* internal orbitals, and one unique orbital. The external and unique internal orbitals form linear $sp(z)$ hybrid orbitals leaving two p orbitals and the $d(x^2 - y^2)$ orbital to form three equivalent hybrid orbitals for the surface bonding. Because of the larger number of surface orbitals generated by hypervalent vertex atoms relative to normal vertex atoms, some of the surface bonds will be three-center rather than two-center bonds without affecting the skeletal electron count.

The process of capping to form the capped deltahedra, such as those found in metal carbonyl derivatives with less than $2n + 2$ apparent skeletal electrons (Figure 9), can be supplemented in extremely electron-poor structures, such as those found in the indium clusters, by the following process conveniently called *flattening* since it consists of flattening the tetrahedral or pyramidal cavity formed by the capping vertex and the vertices of the face being capped:

Flattening opens at least some of the edges of the capped face and pushes the capping vertex closer to the center of the polyhedron so that it can bring the unique internal orbital of the capping vertex close enough to the center of the polyhedron to participate in the multicenter core bonding thereby increasing the delocalization of the system. The pyramidal cavity formed by capping square or rectangular faces can also be flattened thus converting, for example, the D_{2d} bicapped square antiprism found in $B_{10}H_{10}^{2-}$ to the 10-vertex polyhedron found in the centered indium cluster $K_8In_{10}Zn$ (Figure 16). In Figure 16 the dashed and hashed lines represent intervertex distances too long to be considered edges.

The flattening of the pyramidal cavities generated by selected capping vertices in hypoelectronic indium polyhedra with 10 and 11 vertices relates to the location of the hypervalent indium vertices. *Reasonable electron counts and bonding schemes are obtained if all of the indium vertices at apices of flattened pyramidal cavities are hypervalent and the remaining indium vertices are all normal.* The selective location of hypervalent indium atoms at apices of flattened pyramidal cavities may relate to the lower local curvature at such vertices which is better accommodated by the four internal orbitals (three triplet internal orbitals and one unique internal orbital) of a hypervalent sp^3d indium atom than the three internal orbitals (two twin internal orbitals and one unique internal orbital) of a normal sp^3 indium atom.

A number of the indium cluster anions also contain late transition metals (e.g., Ni, Pd, Pt) or post-transition elements (e.g., Zn) as interstitial atoms located in the center of the indium polyhedron. The five d orbitals of these interstitial atoms may be regarded as nonbonding orbitals and thus contain ten nonbonding electrons. For this reason an interstitial Ni, Pd, or Pt atom is a donor of zero skeletal electrons, an interstitial Cu, Ag, or Au atom is a donor of a single skeletal electron, and an interstitial Zn, Cd, or Hg atom is a donor of two skeletal electrons. This electron-counting scheme relates to the fact that the interstitial nickel atom in $Ni_{13}Sb_2(CO)_{24}^{4-}$ is also a donor of zero skeletal electrons.[181,182] Furthermore, the nonbonding role of the d electrons in such interstitial atoms means that only their s and p orbitals overlap with the unique internal orbitals (radial orbitals) of the vertices of the surrounding indium polyhedron. A result of the interstitial late transition or post-transition metal atoms in centered indium clusters is the splitting of the multicenter core bond into several multicenter bonds oriented around the interstitial atom in accord with a reasonable sp^n hybridization scheme, e.g., linear sp for Zn, tetrahedral sp^3 for Ni, Pd, Pt, etc. *In this sense the hypoelectronic anionic indium polyhedron can be regarded as a multidentate ligand encapsulating the interstitial atom.* Such an interpretation appears to be fully consistent with the observed electron counts in the centered hypoelectronic indium clusters which have so far been characterized.

10.7. POLYOXOMETALATES: BINODAL ORBITAL AROMATICITY

10.7.1. Structural Aspects of Polyoxometalates

The heteropoly- and isopolyoxometalates of early transition metals[183,184] have been known for more than a century and have been studied extensively. Their structures are characterized by networks of MO_6 octahedra in which the early transition metals M (typically M = V, Nb, Mo, W) appear in their highest oxidation states in which they have a a d^0 configuration. A characteristic of many, but not all, of such structures is their reducibility to highly colored mixed oxidation state derivatives often given the trivial names of molybdenum or tungsten "blues." The reducibility of early transition metal polyoxometalates requires the presence of MO_6 octahedra in which only one of the six oxygen atoms is a terminal oxygen atom.[185] Such an MO_6

octahedron can be related to mononuclear L_5MO species[186] in which there is an essentially nonbonding metal d orbital to receive one or two electrons. The reducibility of the originally d^0 polyoxometalates can be related to the delocalization of the added electrons(s) in molecular orbitals formed by interaction of these nonbonding d orbitals on each of the metal atoms in the MO_6 octahedra forming the polyoxometalate structure. Such an approach was used by Nomiya and Miya[187] in the form of a structural stability index based on interpenetrating loops of the type $-O-M-O-M-O-$ around the polyoxometalate cage. These authors suggested the analogy of closed loops of this type to macrocyclic π-bonding systems. The facile one-electron reducibility of a colorless or yellow polyoxomolybdate or polyoxotungstate to a highly colored mixed valence "blue" may thus be viewed as analogous to the one-electron reduction of benzenoid hydrocarbons such as naphthalene or anthracene to the highly colored corresponding radical anion. The graph-theory-derived methods outlined in Section 10.3 for the study of delocalization in hydrocarbons and boranes can also be used to study delocalization in polyoxometalates.[188]

The polyoxometalates of interest consist of closed networks of MO_6 octahedra in which M is a d^0 early transition metal such as V(V), Nb(V), Mo(VI), or W(VI). These networks may be described by the large polyhedron or *macropolyhedron* formed by the metal atoms as vertices. In general the edges of this macropolyhedron are M–O–M bridges and with rare exceptions there is no direct metal–metal (M–M) bonding.

The oxygen atoms in the polyoxometalates are of the following three types:

- O^t: terminal or external oxygen atoms, which are multiply bonded to the metal (one σ and up to two orthogonal π bonds) and directed away from the macropolyhedral surface
- O^b: bridging or surface oxygen atoms, which form some or all of the macropolyhedral edges
- O^i: internal oxygen atoms, which are directed toward the center of the macropolyhedron

The metal vertices of the macropolyhedron may be classified as $(\mu_n\text{-}O)_5MO$ or $cis\text{-}(\mu_n\text{-}O)_4MO_2$ vertices depending on the number and locations of the oxygen atoms. In the $cis\text{-}(\mu_n\text{-}O)_4MO_2$ vertices all nine orbitals of the sp^3d^5 manifold of M are used for the σ and π bonding to the two terminal oxygen atoms and σ bonding to the four bridging and internal oxygen atoms leaving no orbitals for direct or indirect overlap with other metal vertices of the metal macropolyhedron corresponding to a resonance hybrid depicted schematically in Figure 17a. The $cis\text{-}(\mu_n\text{-}O)_4MO_2$ vertices in polyoxometalates correspond to the saturated CH_2 vertices in cyclohexane and other cycloalkanes. In the $(\mu_n\text{-}O)_5MO$ vertices only eight of the nine orbitals of the sp^3d^5 manifold of M can be used for σ and π bonding to the single oxygen atom and σ bonding to the five bridging and internal oxygen atoms leaving one nonbonding d orbital [d_{xy} if the M≡O(terminal) axis is the z axis] as depicted in Figure 17a. Thus, a

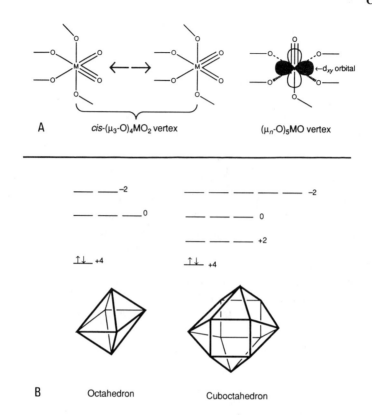

Figure 17. (a) The metal–oxygen bonds and nonbonding atomic orbitals in $cis\text{-}(\mu_n\text{-}O)_4MO_2^t$ and $(\mu_n\text{-}O)_5MO^t$ vertices in polyoxometalates; (b) the spectra of the octahedron and cuboctahedron showing the electron pairing on two-electron reduction of macrooctahedral and macrocuboctahedral polyoxometalates.

$(\mu_n\text{-}O)_5MO$ vertex with a nominally nonbonding d_{xy} orbital in a polyoxometalate is analogous to an unsaturated CH vertex with a nonbonding p orbital in a planar aromatic hydrocarbon such as benzene.

The nonbonding d_{xy} orbitals of the $(\mu_n\text{-}O)_5MO$ vertices in the reducible early transition metal polyoxometalates have two orthogonal nodes (see Figure 17a) and thus have improper fourfold symmetry. Matching this fourfold orbital symmetry with the overall macropolyhedral symmetry requires macropolyhedra in which a C_4 axis passes through each vertex. A true 3D polyhedron having C_4 axes passing through each vertex can have only O or O_h symmetry (the only point groups with multiple C_4 axes). The only two polyhedra having less than 15 vertices meeting these highly restrictive conditions are the regular octahedron and the cuboctahedron (Figure 17b). It is therefore not surprising that these two polyhedra form the basis of most specific early transition-metal polyoxometalate structures containing only $(\mu_n\text{-}O)_5MO$ vertices. Pope[183,185] calls such readily reducible structures containing only (μ_n-O)$_5$MO vertices *type I structures*.

The specific building blocks for type I polyoxometalate structures of interest are as follows:

A. *Octahedron*: $(MO^tO^b_{4/2}O^i_{1/6})_6 = M_6O^{n-}_{19}$ ($n = 8$, M = Nb, Ta; $n = 2$, M = Mo)

- O^t = one terminal oxygen atom per metal atom
- $O^b_{4/2}$ = one bridging oxygen atom along each of the 12 edges of the octahedron [i.e., (4/2)(6) = 12]
- $O^i_{1/6}$ = a single μ_6 oxygen in the center of the M_6 macrooctahedron shared equally among all six metal vertices

B. *Cuboctahedron (the so-called "Keggin structure")*: $(MO^tO^b_{4/2}O^i_{1/3})_{12}X^{n-} = XM_{12}O^{n-}_{40}$ ($n = 3$ to 7; M = Mo, W; X = B, Si, Ge, P, Fe^{III}, Co^{II}, Cu^{II}, etc.)

- O^t = one terminal oxygen per metal atom
- $O^b_{4/2}$ = one bridging oxygen along each of the 24 edges of the cuboctahedron [i.e., (4/2)(12) = 24]
- $O^i_{1/3}$ = an OM$_3$X oxygen bonded to three of the early transition metal atoms. The four oxygen atoms of this type surround the center of the cuboctahedron at the vertices of a tetrahedron. The heteroatom X is located in the center of the cuboctahedron with tetrahedral coordination to these oxygen atoms

The other type I structures considered by Pope[185] include $V_{10}O^{6-}_{28}$ formed by edge sharing of two V_6 macrooctahedra and $X_2M_{18}O^{6-}_{62}$ (the so-called "Dawson structure") formed by fusion of two M_{12} macrocuboctahedra.

Macropolyhedra that are less symmetrical than the octahedron and cuboctahedron are also found in polyoxometalate structures, particularly polyoxovanadate structures.[189] Of particular interest are macropolyhedra in which all vertices have degree 4 although not necessarily strict C_4 symmetry at each vertex. Boersma, Duijvestijn, and Göbel[190] have shown that three operations (conveniently called the BDG operations) are sufficient to convert the regular octahedron to all possible polyhedra in which all vertices have degree 4 (conveniently called *degree 4 polyhedra*). The BDG operations are depicted in Figure 18 with descriptive names describing their effects on readily recognizable polyhedra such as the octahedron and the square antiprism, the next smallest degree 4 polyhedron. The number of topologically distinct degree 4 polyhedra have been shown[189] to be 1, 0, 1, 1, 3, 3, 11, 18, 58, and 139 for polyhedra with 6, 7, 8, 9, 10, 11, 12, 13, 14, and 15 vertices, respectively. An important class of degree 4 polyhedra are the antiprisms with 2 staggered regular polygon faces with n vertices, $2n$ triangular faces, and D_{nd} symmetry. Other degree 4 polyhedra of actual or potential chemical significance are the cuboctahedron and the tricapped trigonal prism with the

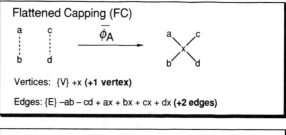

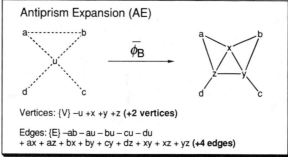

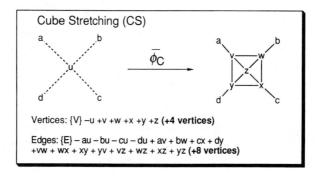

Figure 18. The three operations shown by Boersma, Duijvestijn, and Göbel (BDG) to be sufficient for conversion of a regular octahedron to all possible polyhedra having only degree 4 vertices.

3 "vertical" edges deleted. Figure 19 shows the smaller and more recognizable degree 4 polyhedra that can be generated from the regular octahedron by a small number of the BDG operations. Note that the "flattened capping" (FC) operation requires a quadrilateral face and thus cannot be applied to the regular octahedron so that there are no degree 4 polyhedra with 7 vertices in accord with the Federico[12] and Britton/Dunitz[13] listings of polyhedra.

These structures containing only $(\mu_n\text{-O})_5\text{MO}$ vertices can be contrasted with the nonreducible polyoxometalate structures containing only $cis\text{-}(\mu_n\text{-O})_4\text{MO}_2$ vertices (type II structures in the Pope nomenclature[185]). These structures are necessarily more open since only four of the six oxygens of the MO_6 octahedra can be bridging oxygens.

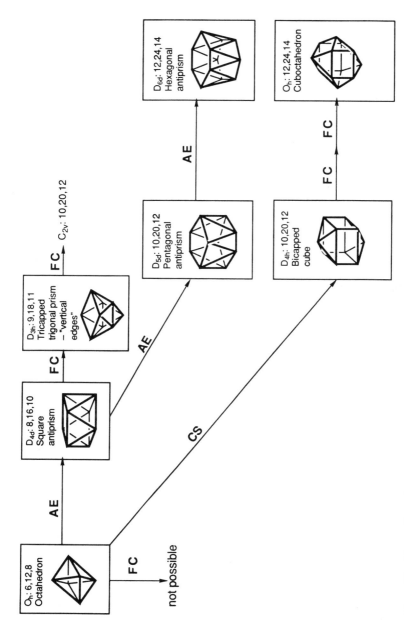

Figure 19. The smaller and more recognizable degree 4 polyhedra that can be generated from the regular octahedron by a small number of BDG operations depicted in Figure 18.

The most stable polyoxometalate structure is the icosahedral Silverton structure $M^{IV}(MoO_2^tO_{1/2}^bO_{3/3}^i)_{12}^{8-} = M^{IV}Mo_{12}O_{42}^{8-}$ (M = Ce, Th, U) in which the central metal atom forms an MO_{12} icosahedron with the interior oxygen atoms. The central metal atom is 12-coordinate and therefore is a large tetravalent lanthanide or actinide with accessible f orbitals. The oxygen atoms in the Silverton structure are of the following types:

- O_2^t = two terminal oxygens per metal atom
- $O_{1/2}^b$ = a total of six oxygen atoms [(1/2)(12) = 6] in $MMo_{12}O_{42}^{8-}$ which are located at the vertices of a large octahedron
- $O_{3/3}^i$ = an $OMMo_3$ μ_4-oxygen bonded to three of the molybdenum atoms and to the central metal atom M leading to icosahedral coordination of the latter

An icosahedron can be decomposed into five equivalent octahedra by partitioning the 30 edges of the icosahedron into five equivalent sets of six edges so that the midpoints of the edges in each set form a regular octahedron.[191] The vertices of each O_6^b large octahedron in $MMo_{12}O_{42}^{8-}$ are located above the midpoints of the corresponding octahedral set of six icosahedron edges.

10.7.2. Binodal Orbital Aromaticity

The Hückel approach (Section 10.3.1) can be used to treat the type I polyoxometalates as aromatic systems in which the d_{xy} orbitals on each of the transition metal vertices overlap to form molecular orbitals. Since these d_{xy} orbitals (Figure 17a) have two nodes, the resulting aromaticity can be called *binodal aromaticity*. Such binodal aromaticity is much weaker than anodal or uninodal orbital aromaticity since the metal atom vertices furnishing the orbitals participating in the delocalization are much farther apart being separated by M–O–M bridges rather than direct M–M bonds.

The topology of the overlap of the d_{xy} orbitals can be described by a graph G whose vertices and edges correspond to the vertices and edges, respectively, of the macropolyhedron (octahedron or cuboctahedron). Equation (15) can then be applied to these binodal aromatic systems to relate the spectrum of G to the energy parameters of the corresponding bonding and antibonding molecular orbitals. The weakness of the binodal orbital aromaticity in type I polyoxometalates translates into a low β parameter in equation (15).

Figure 17b illustrates the spectra of the octahedron and the cuboctahedron, which are the basic building blocks of the delocalized polyoxometalates $M_6O_{19}^{n-}$ and $XM_{12}O_{40}^{n-}$, respectively. The octahedron is thus seen to have the eigenvalues +4, 0, and −2 with degeneracies 1, 3, and 2, respectively, and the cuboctahedron is seen to have the eigenvalues +4, +2, 0, −2 with degeneracies 1, 3, 3, and 5, respectively. The most positive eigenvalue or *principal eigenvalue* of +4 for both polyhedra arises from the fact that each polyhedron corresponds to a regular graph of valence 4.[192] This highly positive principal eigenvalue corresponds to a highly bonding molecular orbital, which can accommodate the first two electrons on reduction of the initially d^0 polyoxometalates of the types $M_6O_{19}^{n-}$ and $XM_{12}O_{40}^{n-}$. The reported diamagnetism[193,194] of the two-electron reduction products of the $PW_{12}O_{40}^{3-}$, $SiW_{12}O_{40}^{4-}$, and $[(H_2)W_{12}O_{40}]^{6-}$ anions is in accord with the two electrons being paired in this lowest-lying molecular orbital. Thus, the overlap of the otherwise nonbonding d_{xy} orbitals in the $M_6O_{19}^{n-}$ and $XM_{12}O_{40}^{n-}$ d^0 early transition-metal polyoxometalates creates a low-lying bonding molecular orbital which can accommodate two electrons, thereby facilitating reduction of polyoxometalates of these types.

The spectrum of the cuboctahedron (Figure 17b) corresponding to the topology of the $XM_{12}O_{40}^{n-}$ derivatives not only has the single +4 eigenvalue but also the triply degenerate +2 eigenvalue corresponding to three additional bonding orbitals which can accommodate an additional six electrons. For this reason eight-electron reduction of the $XM_{12}O_{40}^{n-}$ d^0 early transition-metal derivatives might be expected to be favorable since eight electrons are required to fill the bonding orbitals of the cuboctahedron, i.e., the four bonding orbitals corresponding to the positive eigenvalues +4 and +2. However, experimental evidence indicates that when six electrons are added to a sufficiently stable $XW_{12}O_{40}$ derivative, rearrangement occurs to a more localized $XW_9^{VI}W_3^{IV}O_{40}^{n-}$ structure structure in which the three W^{IV} atoms form a bonded triangle[195] with $W-W = 2.50$ Å similar to the $W-W$ of 2.51 Å in the tungsten(IV) complex $[W_3O_4F_9]^{5-}$. This bonded W_3 triangle corresponds to one of the triangular faces of the W_{12} macrocuboctahedron in $XW_{12}O_{40}^{n-}$. This rearrangement of the $XW_{12}O_{40}^{n-}$ derivatives to a more localized structure on six-electron reduction is an indication of the weakness of the binodal orbital aromaticity in these polyoxometalates corresponding to a low value of β in equation (15). Thus, a configuration with three $W-W$ localized two-center two-electron σ bonds is more stable than a delocalized configuration with six electrons in the bonding molecular orbitals generated by binodal orbital overlap. Thus, if β_σ is defined by the equation

$$(26) \qquad \beta_\sigma = (\Delta E_{\text{bonding}} - \Delta E_{\text{antibonding}})/2$$

for a $W-W$ σ bond and β_d is the energy unit in equation (15) for overlap of the d_{xy} orbitals on the 12 tungsten atoms (Figure 17a), then $\beta_\sigma \gg \beta_d$.

The concept of binodal orbital aromaticity in reduced early transition-metal polyoxometalates may be related to their classification as mixed valence compounds. Robin and Day[196] divide mixed valence compounds into the following three classes:

- Class I: fully localized chemical bonding corresponding to an insulator in an infinite system
- Class II: partially delocalized chemical bonding corresponding to a semiconductor in an infinite system
- Class III: completely delocalized corresponding to a metal in an infinite system

ESR studies on the one-electron reduced polyoxometalates $M_6O_{19}^{n-}$ and $XM_{12}O_{40}^{n-}$ suggest class II mixed valence species.[197,198] Although such species are delocalized at accessible temperatures, they behave as localized systems at sufficiently low temperatures; this behavior is similar to that of semiconductors. This behavior is in accord with the much smaller overlap [i.e., lower β in equation (15)] of the metal d_{xy} orbitals associated with binodal orbital aromaticity as compared with the boron sp hybrid anodal internal orbitals in the deltahedral boranes $B_nH_n^{2-}$ or the carbon uninodal p orbitals in benzene.

REFERENCES

1. R. E. Merrifield and H. E. Simmons, *Topological Methods in Chemistry*, Wiley–Interscience, New York (1989).
2. A. T. Balaban, ed., *Chemical Applications of Graph Theory*, Academic Press, New York (1976).
3. R. B. King, ed., *Chemical Applications of Topology and Graph Theory*, Elsevier, Amsterdam (1983).
4. N. Trinajstić, *Chemical Graph Theory*, CRC Press, Boca Raton, Florida (1983).
5. R. B. King and D. Rouvray, eds., *Graph Theory and Topology in Chemistry*, Elsevier, Amsterdam (1987).
6. R. B. King, *Applications of Graph Theory and Topology in Inorganic Cluster and Coordination Chemistry*, CRC Press, Boca Raton, Florida (1993).
7. B. Grünbaum, *Convex Polytopes*, Interscience Publishers, New York (1967).
8. R. B. King, *J. Am. Chem. Soc. 91*, 7211 (1969).
9. F. Harary and E. M. Palmer, *Graphical Enumeration*, p. 224, Academic Press, New York (1973).
10. W. T. Tutte, *J. Combin. Theory Ser. B 28*, 105 (1980).
11. A. J. W. Duijvestijn and P. J. Federico, *Math. Comput. 37*, 523 (1981).
12. P. J. Federico, *Geom. Ded. 3*, 469 (1975).
13. D. Britton and J. D. Dunitz, *Acta Crystallogr. Sect. A 29*, 362 (1973).
14. H. G. Freedman, Jr., G. R. Choppin, and D. G. Feuerbacher, *J. Chem. Educ. 41*, 354 (1964).
15. C. Becker, *J. Chem. Educ. 41*, 358 (1964).
16. W. Smith and D. W. Clack, *Rev. Roum. Chim. 20*, 1243 (1975).
17. L. Pauling and V. McClure, *J. Chem. Educ. 47*, 15 (1970).
18. I. T. Keaveny and L. Pauling, *Isr. J. Chem. 10*, 211 (1972).
19. R. B. King, *Polyhedron 13*, 2005 (1994).
20. R. B. King, *J. Chem. Educ. 73*, 993 (1996).
21. E. L. Muetterties and R. A. Schunn, *Q. Rev. (London) 20*, 245 (1966).
22. R. S. Berry, *J. Chem. Phys. 32*, 933 (1960).
23. R. R. Holmes, *Acc. Chem. Res. 5*, 296 (1972).
24. P. C. Lauterbur and F. Ramirez, *J. Am. Chem. Soc. 90*, 6722 (1968).
25. B. W. Clare, M. C. Favas, D. L. Kepert and A. S. May, in: *Advances in Dynamic Stereochemistry* (M. Gielen ed.), Vol. 1, pp. 1–41, Freund, London (1985).
26. B. F. Hoskins and C. D. Pannan, *Aust. J. Chem. 29*, 2337 (1976).
27. R. Eisenberg, *Prog. Inorg. Chem. 12*, 295 (1970).
28. R. Mason, K. M. Thomas, and D. M. P. Mingos, *J. Am. Chem. Soc. 95*, 3802 (1973).
29. E. A. McNeill and F. R. Scholer, *J. Am. Chem. Soc. 99*, 6243 (1977).
30. L. G. Guggenberger, D. D. Titus, M. T. Flood, R. E. Marsh, A. A. Orio, and H. B. Gray, *J. Am. Chem. Soc. 94*, 1135 (1972).
31. D. L. Kepert, *Prog. Inorg. Chem. 25*, 41 (1979).
32. E. L. Muetterties and C. M. Wright, *Q. Rev. (London) 20*, 109 (1967).
33. S. J. Lippard, *Adv. Inorg. Chem. 8*, 109 (1967).
34. D. L. Kepert, *Prog. Inorg. Chem. 24*, 179 (1978).
35. E. L. Muetterties, *Inorg. Chem. 12*, 1963 (1973).

36. R. B. King, *Theor. Chim. Acta 64*, 453 (1984).
37. R. B. King, *Inorg. Chem. 31*, 1978 (1992).
38. G. Bombieri and G. De Paoli, in: *Handbook on the Physics and Chemistry of the Actinides* (A. J. Freeman and C. Keller, eds.), Vol. 3, pp. 75–141, Elsevier, Amsterdam (1985).
39. K. Knox and A. P. Ginsberg, *Inorg. Chem. 3*, 555 (1964).
40. S. C. Abrahams, A. P. Ginsberg, and K. Knox, *Inorg. Chem. 13*, 559 (1964).
41. X. Liu, D. J. Klein, T. G. Schmalz, and W. A. Seitz, *J. Comput. Chem. 12*, 1252 (1991).
42. F. A. Cotton, *Acc. Chem. Res. 1*, 257 (1968).
43. A. T. Balaban, D. Farcasiu, and R. Banica, *Rev. Roum. Chim. 11*, 1205 (1966).
44. E. L. Muetterties, *J. Am. Chem. Soc. 90*, 5097 (1968).
45. E. L. Muetterties, *J. Am. Chem. Soc. 91*, 1636, 4115 (1969).
46. E. L. Muetterties and A. T. Storr, *J. Am. Chem. Soc. 91*, 3098 (1969).
47. M. Gielen and J. Nasielski, *Bull. Soc. Chim. Belg. 78*, 339 (1969).
48. M. Gielen and J. Nasielski, *Bull. Soc. Chim. Belg. 78*, 351 (1969).
49. M. Gielen, C. Depasse-Delit, and J. Nasielski, *Bull. Soc. Chim. Belg. 78*, 357 (1969).
50. M. Gielen and C. Depasse-Delit, *Theor. Chim. Acta. 14*, 212 (1969).
51. M. Gielen, G. Mayence, and J. Topart, *J. Organometal. Chem. 18*, 1 (1969).
52. M. Gielen, M. de Clercq, and J. Nasielski, *J. Organometal. Chem. 18*, 217 (1969).
53. M. Gielen and N. Vanlautem, *Bull. Soc. Chim. Belg. 79*, 679 (1970).
54. M. Gielen, *Bull. Soc. Chim. Belg. 80*, 9 (1971).
55. J. I. Musher, *J. Am. Chem. Soc. 94*, 5662 (1972).
56. J. I. Musher, *Inorg. Chem. 11*, 2335 (1972).
57. W. G. Klemperer, *J. Chem. Phys. 56*, 5478 (1972).
58. W. G. Klemperer, *J. Am. Chem. Soc. 94*, 6940 (1972).
59. W. G. Klemperer, *J. Am. Chem. Soc. 94*, 8360 (1972).
60. J. Brocas, *Top. Curr. Chem. 32*, 43 (1972).
61. J. Brocas, in: *Advances in Dynamic Stereochemistry* (M. Gielen, ed.), Vol. 1, pp. 43–88, Freund Publishing Co., London (1985).
62. F. J. Budden, *The Fascination of Groups*, Cambridge University Press, London (1972).
63. C. D. H. Chisholm, *Group Theoretical Techniques in Quantum Chemistry*, Academic Press, New York (1976).
64. F. A. Cotton, *Chemical Applications of Group Theory*, John Wiley & Sons, New York (1971).
65. J. A. Pople, *J. Am. Chem. Soc. 102*, 4615 (1980).
66. S. Fujita, *Symmetry and Combinatorial Enumeration in Chemistry*, Springer-Verlag, Berlin (1991).
67. R. B. King, in: *Chemical Group Theory* (D. Bonchev and D. Rouvray, eds.), pp. 1–30, Gordon & Breach OPA, Amsterdam (1994).
68. N. L. Biggs, *Algebraic Graph Theory*, Cambridge University Press, London (1974).
69. W. N. Lipscomb, *Science 153*, 373 (1966).
70. R. B. King, *Inorg. Chim. Acta 49*, 237 (1981).
71. R. B. King, *Theor. Chim. Acta 64*, 439 (1984).
72. D. Gale, in: *Linear Inequalities and Related Systems* (H. W. Kuhn and A. W. Tucker, eds.), pp. 255–263, Princeton University Press, Princeton, NJ (1956).
73. R. B. King, *Inorg. Chem. 24*, 1716 (1985).
74. R. B. King, *Inorg. Chem. 25*, 506 (1986).
75. G. M. Badger, *Aromatic Character and Aromaticity*, Cambridge University Press, London (1969).
76. E. L. Muetterties and W. H. Knoth, *Polyhedral Boranes*, Marcel Dekker, New York (1968).
77. R. B. King and A. J. W. Duijvestijn, *Inorg. Chim. Acta 178*, 55 (1990).
78. J. Aihara, *J. Am. Chem. Soc. 100*, 3339 (1978).
79. R. B. King and D. H. Rouvray, *J. Am. Chem. Soc. 99*, 7834 (1977).
80. A. J. Stone, *Mol. Phys. 41*, 1339 (1980).
81. A. J. Stone, *Inorg. Chem. 20*, 563 (1981).

82. A. J. Stone and M. J. Alderton, *Inorg. Chem. 21*, 2297 (1982).

83. A. J. Stone, *Polyhedron, 3*, 1299 (1984).

84. K. Ruedenberg, *J. Chem. Phys. 22*, 1878 (1954).

85. H. H. Schmidtke, *J. Chem. Phys. 45*, 3920 (1966).

86. H. H. Schmidtke, *Coord. Chem. Rev. 2*, 3 (1967).

87. I. Gutman and N. Trinajstić, *Top. Curr. Chem. 42*, 49 (1973).

88. R. B. King, in: *Chemical Applications of Topology and Graph Theory* (R. B. King, ed.), pp. 99–123, Elsevier, Amsterdam (1983).

89. R. B. King, in: *Molecular Structure and Energetics* (J. F. Liebman and A. Greenberg, eds.), pp. 123–148, VCH Publishers, New York (1976).

90. R. B. King, *J. Math. Chem. 1*, 249 (1987).

91. K. Wade, *Adv. Inorg. Chem. Radiochem. 18*, 1 (1976).

92. M. J. Mansfield, *Introduction to Topology*, p. 40, Van Nostrand, Princeton, New Jersey (1963).

93. N. L. Biggs, *Algebraic Graph Theory*, p. 17, Cambridge University Press. London (1974).

94. E. Hückel, *Z. Phys. 76*, 628 (1932).

95. K. Kuratowski, *Fundam. Math. 15*, 271 (1930).

96. E. A. McNeill, K. L. Gallaher, F. R. Scholer, and S. H. Bauer, *Inorg. Chem. 12*, 2108 (1973).

97. R. B. King and D. H. Rouvray, *J. Am. Chem. Soc. 99*, 7834 (1977).

98. R. B. King, *Inorg. Chim. Acta 49*, 237 (1981).

99. B. M. Gimarc and J. J. Ott, *Inorg. Chem. 25*, 83 (1986).

100. J. J. Ott, C. A. Brown, and B. M. Gimarc, *Inorg. Chem. 28*, 4269 (1989).

101. B. M. Gimarc and J. J. Ott, *Inorg. Chem. 25*, 2708 (1986).

102. D. J. Wales and A. J. Stone, *Inorg. Chem. 26*, 3845 (1987).

103. D. M. P. Mingos and R. J. Johnston, *Polyhedron 7*, 2437 (1988).

104. E. L. Muetterties and W. H. Knoth, *Polyhedral Boranes*, Marcel Dekker, New York (1968).

105. E. L. Muetterties, ed., *Boron Hydride Chemistry*, Academic Press, New York (1975).

106. M. F. Hawthorne, *Acc. Chem. Res. 1*, 281 (1968).

107. R. B. King, *Transition Metal Organometallic Chemistry: An Introduction*, Academic Press, New York (1969).

108. A. J. Stone, *Mol. Phys. 41*, 1339 (1980).

109. A. J. Stone, *Inorg. Chem. 20*, 563 (1981).

110. A. J. Stone and M. J. Alderton, *Inorg. Chem. 21*, 2297 (1982).

111. A. J. Stone, *Polyhedron 3*, 1299 (1984).

112. R. L. Johnston and D. M. P. Mingos, *Theor. Chim. Acta 75*, 11 (1989).

113. R. B. King, *Croat. Chim. Acta 68*, 293 (1995).

114. R. E. Dickerson and W. N. Lipscomb, *J. Chem. Phys. 27*, 212 (1957).

115. W. N. Lipscomb, *Boron Hydrides*, Chapter 2, Benjamin, New York (1963).

116. I. R. Epstein and W. N. Lipscomb, *Inorg. Chem. 10*, 1921 (1971).

117. W. N. Lipscomb, in: *Boron Hydride Chemistry* (E. L. Muetterties, ed.), pp. 39–78, Academic Press, New York (1975).

118. N. O'Neill and K. Wade, *Polyhedron 3*, 199 (1984).

119. R. E. Williams, *Inorg. Chem. 10*, 210 (1970).

120. K. Wade, *Chem. Commun. 1971*, 792 (1971).

121. R. Rudolph and W. R. Pretzer, *Inorg. Chem. Commun. 11*, 1974 (1972).

122. R. E. Williams, *Adv. Inorg. Chem. Radiochem. 18*, 67 (1976).

123. W. W. Porterfield, M. E. Jones, and K. Wade, *Inorg. Chem. 29*, 2927 (1990).

124. J. A. Morrison, *Chem. Rev. 91*, 35 (1991).

125. J. R. Bowser, A. Bonny, J. R. Pipal, and R. N. Grimes, *J. Am. Chem. Soc. 101*, 6229 (1979).

126. B. F. G. Johnson, ed., *Transition Metal Clusters*, Wiley-Interscience, New York (1980).

127. J. E. Huheey, *Inorganic Chemistry: Principles of Structure and Reactivity*, 3rd ed., pp. 589–595, Harper & Row, New York (1983).

128. K. Wade, *Chem. Commun. 1971*, 792 (1971).
129. R. Hoffmann, *Angew. Chem. Int. Ed. Engl. 21*, 711 (1982), and references cited therein.
130. R. B. King, *New J. Chem. 13*, 293 (1989).
131. D. M. P. Mingos, *Nature Phys. Sci. 236*, 99 (1972).
132. D. M. P. Mingos, *Acc. Chem. Res. 17*, 311 (1984).
133. B. K. Teo, *Inorg. Chem. 23*, 1251 (1984).
134. B. K. Teo, G. Longoni, and F. R. K. Chung, *Inorg. Chem. 23*, 1257 (1984).
135. G. Ciani, L. Garlaschelli, A. Sironi, and S. Martinengo, *Chem. Commun. 1981*, 563 (1981).
136. R. B. King, *J. Am. Chem. Soc. 94*, 95 (1972).
137. A. Sirigu, M. Bianchi, and E. Benedetti, *Chem. Commun. 1969*, 596 (1969).
138. B. K. Teo, *Chem. Commun. 1983*, 1362 (1983).
139. W. Hume-Rothery, *The Metallic State*, p. 328, Oxford University Press, London (1931).
140. D. M. P. Mingos, *Chem. Commun. 1985*, 706 (1985).
141. D. M. P. Mingos, *Chem. Commun. 1985*, 1352 (1985).
142. Y. L. Slovokhotov and Y. T. Struchkov, *J. Organometal. Chem. 258*, 47 (1983).
143. S. Martinengo, A. Fumagalli, R. Bonfichi, G. Ciani, and A. Sironi, *Chem. Commun. 1982*, 825 (1982).
144. R. S. Nyholm, *Proc. Chem. Soc. 1961*, 273 (1961).
145. P. Pyykkö and J.-P. Desclaux, *Acc. Chem. Res. 12*, 276 (1979).
146. A. Dedieu and R. Hoffmann, *J. Am. Chem. Soc. 100*, 2074 (1978).
147. R. B. King, *Inorg. Chim. Acta 116*, 109 (1986).
148. C. E. Briant, K. P. Hall, A. C. Wheeler, and D. M. P. Mingos, *Chem. Commun. 1984*, 248 (1984).
149. R. B. King, *Prog. Inorg. Chem. 15*, 287 (1972).
150. J. G. M. van der Linden, M. L. H. Paulissen, and J. E. J. Schmitz, *J. Am. Chem. Soc. 105*, 1903 (1983).
151. C. E. Briant, B. R. C. Theobald, J. W. White, L. K. Bell, D. M. P. Mingos, and A. J. Welch, *Chem. Commun. 1981*, 201 (1981).
152. R. B. King, *Isr. J. Chem. 30*, 315 (1990).
153. J. A. J. Jarvis, R. Pearce, and M. F. Lappert, *J. Chem. Soc. Dalton Trans. 1977*, 999 (1977).
154. E. M. Meyer, S. Gambarotta, C. Floriani, A. Chiesi-Villa, and C. Guastini, *Organometallics 8*, 1067 (1989).
155. A. Camus, N. Marsich, G. Nardin, and L. Randaccio, *J. Organometal. Chem. 174*, 121 (1979).
156. G. van Koten and J. G. Noltes, *J. Organometal. Chem. 84*, 129 (1975).
157. J. M. Guss, R. Mason, A. I. Søtofte, G. van Koten, and J. G. Noltes, *Chem. Commun. 1972*, 446 (1972).
158. G. Nardin, L. Randaccio, and E. Zangrando, *J. Organometal. Chem. 74*, C23 (1974).
159. R. B. King, *Inorg. Chim. Acta 227*, 207 (1994).
160. G. Doyle, K. A. Eriksen, and D. Van Engen, *J. Am. Chem. Soc. 108*, 445 (1986).
161. G. Doyle, K. A. Eriksen, and D. Van Engen, *J. Am. Chem. Soc. 107*, 7914 (1985).
162. V. G. Albano, L. Grossi, G. Longoni, M. Monari, S. Mulley, and A. Sironi, *J. Am. Chem. Soc. 114*, 5708 (1992).
163. N. L. Biggs, *Algebraic Graph Theory*, p. 17, Cambridge University Press, London (1974).
164. E. Zintl, J. Goubeau, and W. Dullenkopf, *Z. Phys. Chem. Abt. A 154*, 1 (1931).
165. E. Zintl and A. Harder, *Z. Phys. Chem. Abt. A 154*, 47 (1931).
166. E. Zintl and W. Dullenkopf, *Z. Phys. Chem. Abt. B 16*, 183 (1932).
167. E. Zintl and H. Kaiser, *Z. Anorg. Allg. Chem. 211*, 113 (1933).
168. J. D. Corbett, *Chem. Rev. 85*, 383 (1985).
169. J. D. Corbett, *Prog. Inorg. Chem. 21*, 129 (1976).
170. R. C. Burns and J. D. Corbett, *J. Am. Chem. Soc. 103*, 2627 (1981).
171. L. J. Guggenberger, *Inorg. Chem. 7*, 2260 (1968).
172. R. B. King, *Inorg. Chim. Acta 57*, 79 (1982).
173. T. F. Fässler and M. Hunziker, *Inorg. Chem. 33*, 5380 (1994).
174. R. D. Dobrott and W. N. Lipscomb, *J. Chem. Phys. 37*, 1779 (1962).
175. W. Hiller, K.-W. Klinkhammer, W. Uhl, and J. Wagner, *Angew. Chem. Int. Ed. Engl. 30*, 179 (1991).

176. C. Belin and R. G. Ling, *J. Solid State Chem.* *48*, 40 (1983).
177. C. Belin and M. Tillard-Charbonnel, *Prog. Solid State Chem.* *22*, 59 (1993).
178. R. B. King, *Inorg. Chem.* *28*, 2796 (1989).
179. J. K. Burdett and E. Canadell, *J. Am. Chem. Soc.* *112*, 7207 (1990).
180. A. J. Carty and D. J. Tuck, *Prog. Inorg. Chem.* *19*, 243 (1975).
181. V. G. Albano, F. Demartin, M. C. Iapalucci, B. Longoni, A. Sironi, and V. Zanotti, *Chem. Commun.* *1990*, 547 (1990).
182. R. B. King, *Rev. Roum. Chim.* *36*, 379 (1991).
183. M. T. Pope, *Heteropoly and Isopoly Oxometalates*, Springer-Verlag, Berlin (1983).
184. V. W. Day and W. G. Klemperer, *Science* *228*, 533 (1985).
185. M. T. Pope, *Inorg. Chem.* *11*, 1973 (1972).
186. C. J. Ballhausen and H. B. Gray, *Inorg. Chem.* *1*, 111 (1962).
187. K. Nomiya and M. Miwa, *Polyhedron 3*, 341 (1984).
188. R. B. King, *Inorg. Chem.* *30*, 4437 (1991).
189. R. B. King, *J. Mol. Struct., THEOCHEM 336*, 165 (1995).
190. H. J. Boersma, A. J. W. Duijvestijn, and F. Göbel, *J. Graph Theory 17*, 613 (1993).
191. R. B. King and D. H. Rouvray, *Theor. Chim. Acta 69*, 1 (1986).
192. N. L. Biggs, *Algebraic Graph Theory*, p. 14, Cambridge University Press, London (1974).
193. R. A. Prados and M. T. Pope, *Inorg. Chem. 15*, 2547 (1976).
194. G. M. Varga, Jr., E. Papaconstatinou, and M. T. Pope, *Inorg. Chem. 9*, 662 (1970).
195. Y. Jeannin, J. P. Launay, and M. A. S. Sedjadi, *Inorg. Chem. 19*, 2933 (1980).
196. M. B. Robin and P. Day, *Adv. Inorg. Chem. Radiochem. 10*, 247 (1967).
197. J. P. Lennay, M. Fournier, C. Sanchez, J. Livage, and M. T. Pope, *Inorg. Nucl. Chem. Lett. 16*, 257 (1980).
198. J. N. Barrows and M. T. Pope, *Adv. Chem. Ser. 226*, 403 (1990).

Index